PO-3634

27602-1001

LIQUID INTERFACES IN CHEMISTRY AND BIOLOGY

LIQUID INTERFACES IN CHEMISTRY AND BIOLOGY

Alexander G. Volkov
David W. Deamer

Department of Chemistry and Biochemistry, University of California,
Santa Cruz, California

Darrell L. Tanelian
Vladislav S. Markin

Department of Anesthesiology and Pain Management, University of Texas,
Southwestern Medical Center, Dallas, Texas

A Wiley-Interscience Publication

JOHN WILEY & SONS, INC.

New York · Chichester · Weinheim · Brisbane · Singapore · Toronto

Library of Congress Cataloging-in-Publication Data:

Liquid interfaces in chemistry and biology / Alexander G. Volkov ...
   [et al.].
       p.   cm.
     "A Wiley-Interscience publication."
     Includes bibliographical references (p. –) and index.
     ISBN 0-471-14872-5 (cloth : acid-free paper)
     1. Surface chemistry.   2. Liquid–liquid interfaces.   I. Volkov,
   Alexander G. (Alexander George)
   QD509.L54L55   1998
   541.3'3—dc21

CONTENTS

D. CHEMISTRY AT LIQUID INTERFACES

8. Interfacial catalysis / 324

PREFACE

Phenomena involving interfaces have fascinated man since the beginning of recorded history. Over 2000 years ago ancient philosophers contemplated the interface between oil and water. Pliny the Elder (23–79 B.C.) documented his observations in the treatise "*Historie Naturalis*" and noted that waves could be dampened by pouring oil into water (Pliny, 1962). The mechanism underlying this effect is still not clear even today.

Titus Lucretius Carus (99–55 B.C.) discussed the ability of various liquids to mix with one another in his book "*De Rerum Natura*" and "crying teardrops of wine" were described in which droplets of alcohol-containing fluids appear to crawl up the side of a glass (Lucretius, 1953). In physical chemistry this is known as the Marangoni effect. Leonardo da Vinci (1452–1519) repeated Pliny the Elder's experiments with wave dampening and described the rise of a liquid in a small bore tube. This became known as capillarity because the tubes that were used were as fine as a hair. (The word *capillus* means hair in Latin.)

These and many other observations over time have culminated in the field known as surface science, which seeks to understand the physics and chemistry of interfaces. In most experimental situations only processes occurring in bulk phases are considered while those taking place at the interface between different phases go largely unnoticed. Yet the interfacial region can have totally unexpected properties, not intuitively apparent from a knowledge of bulk phase properties. Examples of interfacial phenomena include the ability to produce large (10×10^6 V m^{-1}) electrical fields at an interface due to asymmetric charge and dipole distribution; dramatic effects on solute distribution in a bulk phase due to differences in the energies required to partition a solute between the two phases; and development of catalytic sites at the interface of a dispersed two-phase system which allows reactions to take place that would not proceed in the bulk phases alone.

Processes occurring at the interface between two immiscible liquids are fundamental to life since cells and organelles are defined by membranes consisting of lipid bilayers that separate fluid compartments. Virtually all energy conversion processes in living organisms occur at these interfaces. The properties of liquid–liquid interfaces are also fundamental to a variety of industries including pharmaceuticals, cosmetics, paints, detergents, oil extraction processes, and mining.

Despite their obvious interest and significance, liquid–liquid interfaces are often poorly understood. This may stem from the fact that the field is relatively new, and until recently there has been little appreciation of the importance of interfaces to chemistry and biology. For these reasons we have dedicated this book to presenting the fundamental concepts and principal applications of surface science.

This book begins with a discussion of the thermodynamics of liquid interfaces. In Part A we introduce the principles of classical thermodynamics, discuss the measurement of interfacial tension, and present adsorption at liquid interfaces according Gibbs and Hansen's methods. Special attention is given to electrified interfaces in Part B, which includes interfacial potentials, electrocapillarity, and ion resolution. Part C addresses the structure of interfaces, focusing on specific models of interfacial adsorption and a detailed presentation of the electric double layer. In this section the reader will learn about laser photochemistry at liquid interfaces and recent methods for visualizing interface structure by molecular dynamics. The chemistry of interfacial catalysis is presented in Part D, which includes the theory of ion transfer across interfaces, electrolysis and electrocatalysis at ITIES, and biochemical reactions and photoelectrochemistry with environmental applications. Part E is devoted to biological membranes and presents the topics of membrane thermodynamics, transport, and mechanics. In it we first introduce the structure and properties of biological membranes, then focus on membrane electrostatics and transport of ions and non-electrolytes. The mechanical properties of membranes are presented together with practical applications such as membrane self-organization, electroporation, membrane fusion, and mechanosensitivity. Tables of fundamental constants, useful relations, and quantities can be found in the appendix.

In summary, this book presents both the theory of surface science and its applications in modern biology and chemistry. The authors intend this basic text to be used as an introduction to surface science for scientists and engineers, as a textbook for students, and as reference resource for experts in the field.

LIQUID INTERFACES IN CHEMISTRY AND BIOLOGY

A

THERMODYNAMICS OF INTERFACES

are called the thermal equations of state. If an internal parameter is the internal energy U, then the equation

$$U = U(a_1, \ldots, a_n; T) \tag{1.6}$$

is called the equation of energy or the caloric equation of state.

The total number of these equations is equal to the number of degrees of freedom of the system, or the number of independent variables that determine this state.

Simple systems play a special role in thermodynamics. These are single-phase systems with a constant number of particles. The state of such systems is determined by a single external parameter, for example by the volume V and by the temperature T. They are described by one thermal and one caloric equation of state

$$P = P(V, T), \tag{1.7}$$

$$U = U(V, T). \tag{1.8}$$

For example, in the elementary case of an ideal gas, the thermal equation of state is the familiar Clapeyron equation:

$$PV = nRT, \tag{1.9}$$

where n is the number of moles of the gas and R is the gas constant, equal to $8.31 \, \text{J} \, \text{mol}^{-1} \, \text{K}^{-1}$. It is impossible to derive the equation of state from general principles of thermodynamics. Instead, this equation must be determined either experimentally or by methods of statistical physics based on specific models.

Even if we cannot determine an explicit thermal equation of state, the mere fact of its existence leads to important conclusions. For instance, suppose we solve the equation of state (1.7) for the volume:

$$V = V(P, T). \tag{1.10}$$

Any change of the variables P and T produces a change in volume which can be presented as a differential:

$$dV = \left(\frac{\partial V}{\partial P} \right)_T dP + \left(\frac{\partial V}{\partial T} \right)_P dT. \tag{1.11}$$

The coefficients in front of differentials of pressure dP and temperature dT play a very important role in thermodynamics. These are partial derivatives which describe the change of volume in thermodynamic processes. The first coefficient $(\partial V/\partial P)_T$ gives the isothermal change of volume related to changes of pressure at constant temperature, and characterizes the isothermal compressibility of the material.

The second coefficient $(\partial V/\partial T)_P$ gives the change of volume with temperature at constant pressure and thereby characterizes volumetric expansion of the body. Obviously, each of these coefficients depends on temperature and pressure.

A similar relationship exists between pressure and temperature. The partial derivative $(\partial P/\partial T)_V$ characterizes the rate of change of pressure with temperature at constant volume. However, this value is not independent, but can be expressed by the previous two derivatives. If we assume volume in (1.11) to be constant ($dV = 0$) it follows that

$$\left(\frac{\partial P}{\partial T}\right)_V = -\frac{\left(\dfrac{\partial V}{\partial T}\right)_P}{\left(\dfrac{\partial V}{\partial P}\right)_T}. \tag{1.12}$$

By changing the order of differentiation, one can find

$$\left(\frac{\partial P}{\partial T}\right)_V = -\left(\frac{\partial P}{\partial V}\right)_T\left(\frac{\partial V}{\partial T}\right)_P \tag{1.13}$$

or

$$\left(\frac{\partial P}{\partial V}\right)_T\left(\frac{\partial V}{\partial T}\right)_P\left(\frac{\partial T}{\partial P}\right)_V = -1. \tag{1.14}$$

In this way we can define useful coefficients that describe relationships between the volume, temperature, and pressure in thermodynamic processes. For instance, the coefficient of thermal expansion is the ratio of the derivative $(\partial V/\partial T)_P$ to the volume V_0 at a given pressure at 0 °C (273.15 K):

$$\alpha = \frac{1}{V_0}\left(\frac{\partial V}{\partial T}\right)_P. \tag{1.15}$$

Specifying a volume V_0 that corresponds to a particular temperature 0 °C is required only for gases. For liquid and solid bodies, which expand much less as T increases, the index "0" can be omitted without loss of accuracy.

Other useful thermal coefficients include compressibility

$$\beta = -\frac{1}{V_0}\left(\frac{\partial V}{\partial P}\right)_T \tag{1.16}$$

and thermal coefficient of pressure

$$\gamma = \frac{1}{P_0}\left(\frac{\partial P}{\partial T}\right)_V. \tag{1.17}$$

Using the identity (1.14) one can find the relationship between the coefficients to be

$$\alpha = P_0 \beta \gamma. \tag{1.18}$$

If the volume and pressure are referred to the zero temperature, the same quantities are called the thermodynamic coefficients of expansion, compression, and pressure:

$$\alpha_t = \frac{1}{V}\left(\frac{\partial V}{\partial T}\right)_P, \qquad \beta_t = -\frac{1}{V}\left(\frac{\partial V}{\partial P}\right)_T, \qquad \gamma_t = \frac{1}{P}\left(\frac{\partial P}{\partial T}\right)_V. \tag{1.19}$$

These are related by an equation similar to (1.18).

Sometimes the isothermal modulus of compression is used instead of the coefficient of compression

$$K_T = -V\left(\frac{\partial P}{\partial V}\right)_T = -\left(\frac{\partial P}{\partial \ln V}\right)_T. \tag{1.20}$$

The isothermal modulus of compression K_T is always positive since any body is compressible to a certain degree. However, the coefficient of thermal expansion α can have any sign. It is known, for example, that for water in the range from 0 to 4 °C it is negative.

The First Principle of Thermodynamics. The first principle of thermodynamics is that the internal energy of a system is a function of its state, and can only be changed by external influences. The change of internal energy ΔU is equal to the sum of heat transferred to the system and the work performed *on this system*. If W is defined as the work performed *by the system*, then the first principle can be expressed mathematically as

$$Q = \Delta U + W. \tag{1.21}$$

Therefore the first principle reflects the law of energy conservation. The first principle also states that it is impossible to build an operating engine that could perform more work than the energy transferred to the system from outside. Such an engine is called a perpetual motion machine of the first kind.

For an elementary process, when an infinitesimal amount of heat δQ is transferred, an infinitesimal amount of work δW is performed and the internal energy undergoes an infinitesimal change dU, the equation of the first principle can be expressed as follows

$$\delta Q = dU + \delta W. \tag{1.22}$$

Thus, if the system has passed from the known state 1 to the known state 2, it is possible to find the total change of internal energy, but not the work performed or the heat transferred, because these quantities depend on the path of transition from one state to another. For example, if we first compressed the body and then heated it up, or if we first heated it up and then compressed it, the mechanical work of compression would be different, even though the changes of internal energy in the two cases are identical. Transferred heat and the work performed are not functions of state, and their differential quantities are designated by the symbols δQ and δW to distinguish them from differentials of quantities that are functions of state, such as the internal energy dU or volume dV. We also note that in thermodynamics the sign of the work term is subject to definition. Here we will consider work to be positive when it is performed *by the system* on the external environment.

The equation of the first principle in the form (1.22) is valid both for equilibrium and nonequilibrium processes. If we consider only equilibrium, or quasistatic, processes, and use the expression for elementary work (1.2) then the equation of the first principle will take the form

$$\delta Q = dU + \sum_i A_i \, da_i. \tag{1.23}$$

In the case of a simple system we obtain

$$\delta Q = dU + P \, dV. \tag{1.24}$$

Heat Capacity. When heat is transferred to a body, its temperature changes and the ratio of these two quantities determines an important characteristic of the body. The ratio of an infinitesimal amount of heat received by a body to a corresponding increase in its temperature is called its *heat capacity* (C)

$$C = \frac{\delta Q}{dT}. \tag{1.25}$$

As noted earlier, the amount of heat transferred to a body depends on the process, and therefore heat capacity is also process-dependent.

Using the concept of heat capacity the equation for internal energy (1.24) can be rewritten as follows:

$$C \, dT = dU + P \, dV. \tag{1.26}$$

Because the internal energy is a function of temperature and pressure, one can write

$$dU = \left(\frac{\partial U}{\partial T}\right)_V dT + \left(\frac{\partial U}{\partial V}\right)_T dV, \tag{1.27}$$

and therefore

$$C\,dT = \left(\frac{\partial U}{\partial T}\right)_V dT + \left[\left(\frac{\partial U}{\partial V}\right)_T + P\right] dV. \qquad (1.28)$$

We can now consider a few specific processes. If a body is heated up at constant volume $(dV = 0)$ the heat capacity corresponding to this process is equal to

$$C_V = \left(\frac{\partial U}{\partial T}\right)_V. \qquad (1.29)$$

If a constant pressure is maintained in the heating process, then from (1.28) we obtain heat capacity at constant pressure:

$$C_P = \left(\frac{\partial U}{\partial T}\right)_V + \left[\left(\frac{\partial U}{\partial V}\right)_T + P\right]\left(\frac{\partial V}{\partial T}\right)_P. \qquad (1.30)$$

It is obvious that these two heat capacities are different, and their difference is equal to

$$C_P - C_V = \left[\left(\frac{\partial U}{\partial V}\right)_T + P\right]\left(\frac{\partial V}{\partial T}\right)_P. \qquad (1.31)$$

Heat capacity at constant pressure cannot be less than heat capacity at a constant volume, hence the condition $C_P \geq C_V$ should always hold.

Heat capacity is only one example of caloric quantities. One can also define the heat of isothermal change of any external parameter of a system as the amount of heat necessary to increase this parameter by one unit with all other parameters held constant:

$$l_{a_i} = \left(\frac{\partial Q}{\partial a_i}\right)_{a_1,\ldots,a_n;\,T}. \qquad (1.32)$$

For example, the heat of isothermal expansion is equal to

$$l_V = \left(\frac{\partial Q}{\partial V}\right)_{a_1,\ldots,a_n;\,T}. \qquad (1.33)$$

The subscripts specify which parameters are kept constant during differentiation.

The Second Principle of Thermodynamics. This fundamental principle specifies the direction of possible thermal processes in nature. P. Clausius defined this principle by stating that "Heat cannot spontaneously pass from a cooler to a warmer body." This statement assumes that no other bodies participate in the process, and that work is not performed.

A number of equivalent formulations of the second principle can be derived from Clausius's postulate. Among them there is the Thomson–Planck postulate, which states that it is impossible to construct a machine that lifts a load at the expense of cooling a single thermal reservoir. This postulate is equivalent to Ostwald's rule that it is impossible to construct a perpetual motion device that transforms heat derived from a single thermal source into work. The only devices that are possible use the equilibration of temperature between a source and a sink, as conventional thermal machines do.

Based on the second principle of thermodynamics, one can prove that during an equilibrium process the ratio $\delta Q/T$ is a differential of a certain function of state S called entropy:

$$\frac{\delta Q}{T} = dS. \tag{1.34}$$

Sometimes this statement alone is called the second principle.

The entropy characterizes the degree of order in a system. If the number of states available for a system is Γ, then the entropy of this system is

$$S = k_B \ln \Gamma, \tag{1.35}$$

where k_B is the Boltzmann constant.

Combining (1.34) and (1.23) we obtain the basic thermodynamic equation for equilibrium processes

$$T\,dS = dU + \sum_i A_i\,da_i. \tag{1.36}$$

In a simple system

$$T\,dS = dU + P\,dV. \tag{1.37}$$

If nonequilibrium processes occur in a system, and the system receives a certain amount of heat δQ_{ne}, the equality (1.34) does not hold, and the system instead obeys the inequality

$$dS > \frac{\delta Q_{ne}}{T}. \tag{1.38}$$

Therefore, the entropy of a system increases during adiabatic nonequilibrium processes.

This rule can be illustrated by a system consisting of two bodies at different temperatures $T_1 < T_2$ that are completely isolated from the external world, but can exchange heat between themselves. In the process of heat exchange each body receives or donates a certain amount of heat, so that

$$dS_1 = \frac{\delta Q_1}{T_1} \quad \text{and} \quad dS_2 = \frac{\delta Q_2}{T_2}. \tag{1.39}$$

By virtue of the law of conservation of energy, $\delta Q_1 = -\delta Q_2$, and we find that the total change of entropy in the system is equal to

$$dS = dS_1 + dS_2 = \delta Q_1 \left(\frac{1}{T_1} - \frac{1}{T_2} \right). \tag{1.40}$$

Because the first body receives heat, δQ_1 is positive, and the inequality (1.38) is recovered. This particular example proves that entropy increases during nonequilibrium processes in adiabatically closed systems.

Combining the inequality (1.38) with the first principle, we obtain the basic inequality of thermodynamics of nonequilibrium processes:

$$T \, dS > dU + \sum_i A_i \, da_i \tag{1.41}$$

The remarkable feature of the basic equation of thermodynamics (1.36) is the fact that all the parameters represent functions of state. This allows us to carry out useful transformations of parameters of the system and to find relationships between them. For example, the internal energy of a simple system, being a function of state, can be defined by any pair of variables. However, it follows from equation (1.37) that the most natural parameters for the function U are the variables S and V:

$$U = U(S, V). \tag{1.42}$$

The derivatives with respect to these variables have the meaning of temperature and pressure

$$\left(\frac{\partial U}{\partial S} \right)_V = T \quad \text{and} \quad \left(\frac{\partial U}{\partial V} \right)_S = -P. \tag{1.43}$$

If we take the second derivatives and use the fact that mixed derivatives do not depend on the order of differentiation,

$$\frac{\partial^2 U}{\partial V \, \partial S} = \frac{\partial^2 U}{\partial S \, \partial V}, \tag{1.44}$$

then it is easy to show that

$$\left(\frac{\partial T}{\partial V}\right)_S = -\left(\frac{\partial P}{\partial S}\right)_V. \tag{1.45}$$

This equality is called the reciprocity relationship, or Maxwell's equation.

The Third Principle of Thermodynamics. This rule, also called the thermal theorem of Nernst, concerns processes at low temperature. It states that as the temperature approaches 0 K, the entropy of an equilibrium system ceases to depend on any thermodynamic parameters of state. It follows that the derivative of entropy with respect to any thermodynamic parameter at 0 K is equal to zero. Similarly, all thermal coefficients at 0 K also approach zero.

1.2. THERMODYNAMIC POTENTIALS

The method of thermodynamic potentials was developed by G. Gibbs and has become a very effective tool for analyzing thermodynamic systems. This method starts from the basic equation of thermodynamics, derives a number of functions of state which have the dimension of energy, and allows us to find relationships between parameters of a given system.

The internal energy U in variables S and V is called *a thermodynamic potential*, because the value of pressure,

$$P = -\left(\frac{\partial U}{\partial V}\right)_S, \tag{1.46}$$

is found from it in the same manner that a force is found from potential energy in mechanics. However, the function $U = U(S, V)$ has an important disadvantage because its variable S cannot be measured directly. For this reason other thermodynamic potentials are usually more convenient.

Enthalpy. The internal energy of an isolated system does not spontaneously vary, but if a system receives a certain amount of heat, its internal energy changes by the same amount as long as no mechanical work is performed on it. However, the process itself can vary. For instance, a system can be heated up and simultaneously expand to overcome the external pressure. In this case we cannot say that the total amount of heat received by the system is stored as internal energy because part of it is spent working against external forces. However, if the external pressure is maintained constant, it is possible to compose a function that exactly describes the amount of heat received.

Suppose that the pressure P is constant. Then in the equation of the first principle (1.24) the factor P can be brought in under the sign of the differential:

$$\delta Q = \mathrm{d}U + \mathrm{d}(PV) = \mathrm{d}(U + PV). \tag{1.47}$$

The sum in the parenthesis is called *enthalpy*

$$H = U + PV. \tag{1.48}$$

This function can be considered without imposing restrictions on the way the pressure P changes. However, it is especially useful when studying isobaric processes in which P is constant, because the enthalpy change represents the amount of heat received by the system. In this sense it replaces the internal energy in an isobaric process.

In a general case the differential of enthalpy is

$$dH = dU + P\,dV + V\,dP. \tag{1.49}$$

Substituting the expression for the differential of the internal energy (1.37), we obtain

$$dH = T\,dS + V\,dP. \tag{1.50}$$

This equation proves that at a constant pressure ($dP = 0$) the enthalpy change is equal to the amount of heat received by the system.

The enthalpy, as well as the internal energy, is a function of state and therefore it can be determined by any pair of parameters of the system: S and P, or T and P, or T and V, and so on. However, its natural variables are S and P. The derivatives of enthalpy with respect to these variables give other parameters of state:

$$\left(\frac{\partial H}{\partial S}\right)_P = T \quad \text{and} \quad \left(\frac{\partial H}{\partial P}\right)_S = V. \tag{1.51}$$

Therefore the enthalpy, expressed in variables S and P, is a thermodynamic potential.

Using the second derivatives, we obtain another Maxwell equation:

$$\left(\frac{\partial T}{\partial P}\right)_S = \left(\frac{\partial V}{\partial S}\right)_P. \tag{1.52}$$

We derived the enthalpy in (1.47) by the procedure called Legendre's transformation. Transformations of this type are widely used in thermodynamics to derive thermodynamic potentials and other useful relationships.

***Example 1: Comparison of Heat Capacities* C_P and C_V.** In the previous section we defined the heat capacity at constant volume C_V and at constant pressure C_P. The first quantity had a simple analytical formula (1.29), while the second was defined by a more complex expression (1.30). Now we are in a position to give a simpler definition to C_P. Because the amount of heat received by the system at constant pressure is equal to the change in enthalpy, we can write

$$C_P = \left(\frac{\partial H}{\partial T}\right)_P. \tag{1.53}$$

The difference between two heat capacitances is given by equation (1.31) which is not very useful for practical applications. A more convenient approach is to define the derivative $(\partial U/(\partial V)_T$ by considering entropy as a function of the variables T and V. Its differential is:

$$dS = \left(\frac{\partial S}{\partial T}\right)_V dT + \left(\frac{\partial S}{\partial V}\right)_T dV. \tag{1.54}$$

We shall substitute this differential in the expression for the internal energy (1.37):

$$dU = T\left(\frac{\partial S}{\partial T}\right)_V dT + \left[T\left(\frac{\partial S}{\partial V}\right)_T - P\right]dV. \tag{1.55}$$

One can conclude that

$$\left(\frac{\partial U}{\partial T}\right)_V = T\left(\frac{\partial S}{\partial V}\right)_T - P = T\left(\frac{\partial P}{\partial T}\right)_V - P, \tag{1.56}$$

and therefore

$$C_P - C_V = T\left(\frac{\partial P}{\partial T}\right)_V\left(\frac{\partial V}{\partial T}\right)_P. \tag{1.57}$$

Using the definitions of the coefficient of thermal expansion α and of the isothermal modulus of compression K_T, we obtain

$$C_P - C_V = \frac{\alpha^2 K_T V_0^2 T}{V}. \tag{1.58}$$

Recall that V_0 is the volume of the body at $0\,°C$ and V is the volume at a given temperature T.

Solid and liquid bodies expand relatively little with change of temperature, so that $V_0 \approx V$ and

$$C_P - C_V \approx \alpha^2 K_T VT. \tag{1.59}$$

The heat capacities C_P and C_V refer to bodies of arbitrary size. The same quantities per unit mass are specific heat capacities, designated c_P and c_V. If the body density is ρ_m, then substituting $1/\rho_m$ for V, we obtain the difference of the specific heat capacities:

$$c_P - c_V \approx \frac{\alpha^2 K_T T}{\rho_m}. \tag{1.60}$$

As an illustration we can consider water at 0 °C (Sivukhin, 1990). In this case $\alpha = -6.1 \times 10^{-5}\,\mathrm{K^{-1}}$, $K_T = 2 \times 10^9$ Pa and $\rho_m = 10^3\,\mathrm{kg\,m^{-3}}$. The negative sign of the coefficient of thermal expansion α reflects the fact that between 0 and 4 °C water volume does not increase with temperature but rather decreases. However, in equation (1.60) the quantity α is raised to the second power, and its sign does not influence the difference $c_P - c_V$ which remains positive. Using the numbers given above, we find that $c_P - c_V = 2\,\mathrm{J\,kg^{-1}\,K^{-1}}$. At the same time $c_P = 4.2 \times 10^3\,\mathrm{J\,kg^{-1}\,K^{-1}}$. Thus, the difference between heat capacities c_P and c_V is less than one thousandth of their values.

The reason for such a small difference is explained by the strong interaction of molecules in condensed bodies. If the coefficient of thermal expansion is small, virtually all heat delivered to the body goes to change the internal energy and only a very small portion is spent working against external pressure.

It should be noted that water represents a thermal anomaly in which the isobaric coefficient of thermal expansion changes sign at 0 °C. At temperatures below 4 °C the derivative $(\partial V/\partial T)_P$ is negative, and above 4 °C it is positive. At 4 °C the coefficient of thermal expansion becomes zero, and at this unique point $C_P = C_V$.

Helmholtz Free Energy. In previous sections we described the process of heat transfer to a system. Now we will consider how mechanical work can be performed on a system. If the system is adiabatically isolated, all work performed on it goes to change its internal energy. However, systems are rarely adiabatically isolated and more often are in thermal contact with their surroundings so that work performed on the system can escape to the surroundings in the form of heat. But if the temperature of the system is maintained constant, it is possible to construct a new function, or a new thermodynamic potential, the change of which would be equal to the work performed.

Suppose that the temperature of a system is maintained constant. Then in the expression for the internal energy (1.37) the temperature can be brought in under the sign of the differential:

$$dU = d(TS) - P\,dV. \tag{1.61}$$

Hence, the work performed on the system is equal to

$$-P\,dV = d(U - TS). \tag{1.62}$$

Performing a Legendre's transformation results in the Helmholtz free energy:

$$F = U - TS. \tag{1.63}$$

This is a function of state and we can consider its dependence on any pair of variables. However, its natural variables are T and V. By calculating the differential for both parts of equation (1.63) we obtain

$$dF = dU - T\,dS - S\,dT = -S\,dT - P\,dV. \tag{1.64}$$

From this it follows that in isothermal processes (when $dT = 0$), the work performed on the system is equal to a decrease of its Helmholtz free energy. In other words, this function plays a role in the potential energy for isothermal processes. Therefore the Helmholtz free energy, expressed in the variables T and V is a thermodynamic potential. In these variables, the derivatives have the physical meaning of entropy and pressure:

$$\left(\frac{\partial F}{\partial T}\right)_V = -S \quad \text{and} \quad \left(\frac{\partial F}{\partial V}\right)_T = -P. \tag{1.65}$$

Examining the second derivatives, we can find another Maxwell equation:

$$\left(\frac{\partial S}{\partial V}\right)_T = \left(\frac{\partial P}{\partial T}\right)_V. \tag{1.66}$$

In any process when the temperature and the volume of the system do not change ($dT = 0$ and $dV = 0$), the Helmholtz free energy remains constant.

There is a simple relationship between the internal energy and the Helmholtz free energy. Using (1.63) and (1.65), we obtain

$$U = F - T\left(\frac{\partial F}{\partial T}\right)_V. \tag{1.67}$$

This relationship is called the first Gibbs–Helmholtz equation. It is very useful when direct calculation of internal energy is difficult. Calculating free energy is usually much easier, and then with the help of (1.67) one can find internal energy. There is an interesting conclusion following from this equation: if the Helmholtz free energy does not depend on temperature, it is equivalent to the internal energy.

Gibbs Free Energy. We have considered thermodynamic potentials in variables S and V (internal energy), S and P (enthalpy), T and V (Helmholtz free energy). There is one more pair of variables T and P. Using the Legendre's transformation one can construct the thermodynamic potential for these variables which is called the Gibbs free energy:

$$G = U - TS + PV. \tag{1.68}$$

Calculating its differential, one finds that

$$dG = -S\,dT + V\,dP. \tag{1.69}$$

It follows that if $dT = 0$ and $dP = 0$, then $dG = 0$. In other words, this relationship replaces the law of internal energy conservation for systems under isothermic–isobaric conditions.

The Gibbs free energy is frequently used in various applications, because in the majority of thermodynamic processes the variables are temperature and pressure. The derivatives of Gibbs free energy give entropy and volume:

$$\left(\frac{\partial G}{\partial T}\right)_P = -S \quad \text{and} \quad \left(\frac{\partial G}{\partial P}\right)_T = V. \tag{1.70}$$

Examining the second derivatives, one can obtain another Maxwell equation:

$$\left(\frac{\partial S}{\partial P}\right)_T = -\left(\frac{\partial V}{\partial T}\right)_P. \tag{1.71}$$

Equation (1.68) can be rewritten in a form that shows the relationship between the Gibbs free energy and other thermodynamic potentials:

$$G = F + PV = H - TS. \tag{1.72}$$

The last expression in equation (1.72) provides a convenient way to calculate the enthalpy from known Gibbs free energy. Using the first relationship (1.70), one obtains the second Gibbs–Helmholtz equation:

$$H = G - T\left(\frac{\partial G}{\partial T}\right)_P. \tag{1.73}$$

The advantage of this equation stems from the fact that direct calculation of the Gibbs free energy is always easier than calculation of the enthalpy.

1.3. GENERALIZATION OF THERMODYNAMIC POTENTIALS

Natural Variables of Thermodynamic Potentials. Every thermodynamic potential has its natural variables: internal energy—entropy and volume; enthalpy—entropy and pressure; Helmholtz free energy—temperature and volume; Gibbs free energy—temperature and pressure. This does not mean that we cannot use other variables for these functions. For example, we can consider internal energy as a function of temperature and volume to derive the following relationships:

$$\left(\frac{\partial U}{\partial T}\right)_V = C_V \quad \text{and} \quad \left(\frac{\partial U}{\partial V}\right)_T = -P + T\left(\frac{\partial P}{\partial T}\right)_V. \tag{1.74}$$

However, when natural variables are not used, one obtains information in an inconvenient form. It is therefore more useful to express a thermodynamic function in terms of its natural variables, because it is easier to find all thermodynamic quantities of the system.

For example, if the Gibbs free energy $G = G(T, P)$ is expressed as a function of T and P, we can immediately find the entropy, $S = -(\partial G/\partial T)_P$, the heat capacitance, $C_P = T(\partial S/\partial T)_P$, and the volume, $V = -(\partial G/\partial P)_T$. The last formula defines the relationship between P, V, and T, which is the thermal equation of state. However, thermodynamics does not give explicit expressions for thermodynamic potentials, but only the relationships between them and their derivatives. To find a specific form of a thermodynamic potential one needs to know the equation of state and the dependence of the heat capacity on temperature.

Thermodynamic properties of bodies can also be calculated in statistical physics based on molecular kinetics, but the calculations proceed in the opposite direction. That is, thermodynamic functions are first determined from the laws of statistics. Then, knowing thermodynamic potentials in terms of their natural variables, the total thermodynamic information of the system can be calculated including the thermal equation of state, the heat capacitance, the modulus of compression, the coefficient of thermal expansion, and so on. However, if the thermodynamic potentials are not presented in terms of their natural variables, calculations become complicated and additional boundary conditions must be imposed. We shall illustrate this by the following example (Rumer and Ryvkin, 1980).

Example 2. The internal energy is given in the form $U(T, V)$. Let us switch to the Helmholtz free energy, for which T and V are natural variables. For this purpose we shall use the first Gibbs–Helmholtz equation (1.67), transformed as:

$$\frac{\partial}{\partial T}\left(\frac{F}{T}\right) = -\frac{U}{T^2}. \tag{1.75}$$

Now we integrate over temperature:

$$F(T, V) = -T\int_0^T \frac{U(\tau, V)}{\tau^2}\, d\tau + T\varphi(V). \tag{1.76}$$

Here $\varphi(V)$ is an arbitrary function of volume, arising as a result of integration of equation (1.75) using one variable. This function is unknown. To find it, we shall calculate the entropy:

$$S = -\left(\frac{\partial F}{\partial T}\right)_V = \int_0^T \frac{U(\tau, V)}{\tau^2}\, d\tau + \frac{U(\tau, V)}{T} - \varphi(V). \tag{1.77}$$

By designating

$$\psi(T, V) = \int_0^T \frac{U(\tau, V)}{\tau^2}\, d\tau + \frac{U(\tau, V)}{T}, \tag{1.78}$$

we obtain

$$S = \psi(T, V) - \varphi(V). \tag{1.79}$$

Now we need a boundary condition. According to the third principle of thermodynamics the entropy at absolute zero is zero. Therefore

$$\varphi(V) = \psi(0, V). \tag{1.80}$$

In this way we determined the required function and can find the Helmholtz free energy.

Similar calculations can be carried out if the Helmholtz free energy is given as a function of S and V, enthalpy as a function of P and T, or Gibbs free energy as a function of S and P. In other words, these transformations can be carried out in thermodynamic potentials if T is replaced by S or vice versa, with the third principle providing the missing boundary condition.

The situation is quite different if P is replaced by V or vice versa. Thermodynamic potentials presented in this way cannot provide useful information about the system (Rumer and Ryvkin, 1980).

In summary, the presentation of thermodynamic potentials using their natural variables is preferable because it allows all thermodynamic properties of the system to be easily determined.

Work of Arbitrary Character. When considering a simple system, we have defined work in terms of mechanical compression or expansion. The general expression for work can be presented as $\delta W = A \, da$. The physical meaning of the quantities A and a is determined by properties of the system. For example, if we stretch a spring, then a is its length and A is the force of stretching. If an electrical field is applied to a system, then $a = D$ and $A = -E$ and so on.

Example 3: Mechanical Stretching of a Bilayer Lipid Membrane. Assume that a membrane has an initial area A_0 and that external forces are applied to its edges. The origin of these forces can be the tension generated by the border where a planar lipid membrane contacts its solid support, or the tension can be created in a closed vesicular membrane stretched by osmotic pressure.

Suppose that the membrane is fluid and the tension is isotropic. The work, made *by the membrane*, is equal to

$$\delta W = -\gamma \, dA. \tag{1.81}$$

Here the letter A refers to the area of the membrane and γ is the tension.

Let us assume that the tension γ in the membrane is related to an increase of its area δA by a linear equation similar to Hook's law:

$$\gamma = \Gamma_H \frac{\Delta A}{A_0}. \tag{1.82}$$

These thermodynamic potentials are sometimes called the generalized enthalpy and generalized Gibbs free energy. The relationship between them is the same as in a simple system:

$$H = G + TS. \tag{1.95}$$

Thus the generalized enthalpy is the thermodynamic potential with variables S and A_i, while the generalized Gibbs free energy uses the variables T and A_i. The differentials of these functions are

$$dH = T\,dS + \sum_i a_i\,dA_i, \tag{1.96}$$

$$dG = -S\,dT + \sum_i a_i\,dA_i. \tag{1.97}$$

Hence the parameters of the system can be expressed as

$$T = \left(\frac{\partial H}{\partial S}\right)_{A_1,\dots,A_n}, \quad S = -\left(\frac{\partial G}{\partial T}\right)_{A_1,\dots,A_n},$$

$$a_i = \left(\frac{\partial H}{\partial A_i}\right)_{A_1,\dots,A_n;\,S} = \left(\frac{\partial G}{\partial A_i}\right)_{A_1,\dots,A_n;\,T}. \tag{1.98}$$

In different applications there is a need for intermediate forms of thermodynamic potentials. For example, to describe dielectrics in an electrical field, L. D. Landau used the modified Helmholtz free energy $\bar{F}$ obtained from Legendre's transformation of electrical variables:

$$\bar{F} = F - (\boldsymbol{E}\cdot\boldsymbol{D}). \tag{1.99}$$

where F is the Helmholtz free energy, $\boldsymbol{E}$ is the electrical field and $\boldsymbol{D}$ is the electrical induction. Employing thermodynamic potentials with several degrees of freedom allows us to analyze complex systems in which several forces act simultaneously.

Example 4: Stretching a Charged Membrane. Suppose that external forces stretch a membrane that is charged by an external source such as an applied voltage. The area of the membrane is equal to A, its tension γ is isotropic, the charge is q and the electrical potential difference is ϕ. During stretching and charging the external forces perform the work $\gamma\,dA + \phi\,dq$, and the membrane itself performs the work $\delta W = -\gamma\,dA - \phi\,dq$. In isothermal conditions this work is equal to a loss of the Helmholtz free energy:

$$dF = \gamma\,dA + \phi\,dq. \tag{1.100}$$

We can perform Legendre's transformation on electrical variables by introducing the modified Helmholtz free energy:

$$\bar{F} = F - \phi q. \tag{1.101}$$

Its natural variables are A and ϕ, and its differential is

$$d\bar{F} = \gamma \, dA - q \, d\phi. \tag{1.102}$$

This equation can be useful for the analysis of the dependence of membrane tension on membrane potential. Calculating second derivatives of the new Helmholtz free energy, we obtain:

$$\left(\frac{\partial \gamma}{\partial \phi} \right)_{T,A} = - \left(\frac{\partial q}{\partial A} \right)_{T,\phi}. \tag{1.103}$$

In order to include the specific electrical capacitance of the membrane C, its charge can be presented as $q = AC\phi$. Hence

$$\left(\frac{\partial q}{\partial A} \right)_{T,\phi} = C\phi + A\phi \left(\frac{\partial C}{\partial A} \right)_{T,\phi}. \tag{1.104}$$

The product $C\phi$ gives the charge density σ on the membrane. To calculate the derivative $(\partial C / \partial A)_{T,\phi}$ we need more detailed information on the structure of the membrane.

Let us consider a particular case. We will assume that the membrane is incompressible, so that the increase in area is accompanied by a reduction of its thickness h: $Ah = V = $ constant. Specific capacitance is then

$$C = \frac{\varepsilon_0 \varepsilon}{h} = \frac{\varepsilon_0 \varepsilon A}{V}, \tag{1.105}$$

and hence,

$$\left(\frac{\partial C}{\partial A} \right)_{T,\phi} = \frac{C}{A}. \tag{1.106}$$

Substituting this into equation (1.103) we obtain:

$$\left(\frac{\partial \gamma}{\partial \phi} \right)_{T,A} = -2\sigma. \tag{1.107}$$

Thus, if an electrical potential is imposed on an incompressible membrane, its tension decreases with proportionality coefficient equal to -2σ.

Minimum of Thermodynamic Functions and Conditions of Equilibrium.

If one knows the thermodynamic potentials of a system it is possible to find conditions of equilibrium and stability. However, we must now extend the definition of thermodynamic functions because they were earlier defined only for equilibrium conditions. At equilibrium all internal parameters are functions of external parameters and temperature, which is why equilibrium states can be described by a very small number of parameters. Simple systems can be described by two parameters only—temperature T and pressure P, which should be constant through the whole volume. In a nonequilibrium state the number of parameters can be very large so that description of such system becomes practically impossible.

To avoid this difficulty, thermodynamics usually considers nonequilibrium states that can be described by one or a very few additional parameters x_i. In that case the nonequilibrium state is reduced to equilibrium, but with a larger number of parameters and corresponding generalized forces X_i that maintain the system in the nonequilibrium state. Thermodynamic functions of a nonequilibrium system are assumed to be equivalent to those of the equilibrium system with additional "maintaining" forces:

$$F = F(T, V, x_i). \tag{1.108}$$

Such functions are called the Landau functions for thermodynamic potentials, named after the theoretical physicist L. D. Landau, who often used this definition.

At equilibrium a given thermodynamic potential reaches its extreme value. To illustrate this statement, consider a simple isolated system with a constant number of particles. We assume that in the process of approaching equilibrium the quantities U, V, and N are kept constant. The basic thermodynamic inequality for nonequilibrium processes in such a system has the form

$$T \, dS > dU + P \, dV. \tag{1.109}$$

Because $dU = 0$ and $dV = 0$ we find that $dS > 0$. This means that the entropy of an isolated system in a nonequilibrium process increases and tends to a maximum when the system reaches equilibrium. It also means that the first variation at the equilibrium is equal to zero, and the second is negative:

$$dS = 0, \qquad d^2 S < 0. \tag{1.110}$$

Therefore we have found the condition of equilibrium of an isolated system and the condition of its stability.

Now consider an isochoric and isothermal system (T, V, N are constant). The basic inequality of thermodynamics of nonequilibrium processes with independent variables V and T has the form

$$dF < -S \, dT - P \, dV. \tag{1.111}$$

For the system under consideration

$$dF < 0. \tag{1.112}$$

Thus, for a nonequilibrium process in an isochoric and isothermal system, the Helmholtz free energy decreases and at equilibrium reaches a minimum so that

$$dF = 0, \tag{1.113}$$

and the condition of stability for this state is

$$d^2 F > 0. \tag{1.114}$$

Consider now an isothermal and isobaric system (T, p, N are constant). The basic inequality of thermodynamics of nonequilibrium processes in variables P and T has the form

$$dG < -S \, dT + V \, dP. \tag{1.115}$$

For this system

$$dG < 0. \tag{1.116}$$

Hence at equilibrium the Gibbs free energy reaches a minimum, and the condition of equilibrium is

$$dG = 0, \tag{1.117}$$

while the condition of its stability is

$$d^2 G > 0. \tag{1.118}$$

The elucidation of conditions of equilibrium in more complex cases requires new concepts, and in particular the concept of chemical potential to be discussed in the next section.

1.4. CHEMICAL POTENTIAL

Systems with Variable Numbers of Particles. The internal energy of a closed system can be determined, for example, by its entropy and volume, but if the system is open, the amount of material in it can change. In a multicomponent system the mass of different components can change due to chemical reactions between them, and in this process the internal energy of the system can change.

In a simple one-component system this dependence can be accounted for by adding to the differential of internal energy a term proportional to the increase in the number of particles dN:

$$dU = T\,dS - P\,dV + \mu\,dN. \tag{1.119}$$

The factor μ that appeared in this equation has the dimension of energy (J) if N is expressed as the number of particles, or energy per mole (J mol^{-1}), if it is expressed in moles, and it is referred to as the chemical potential of a given substance. Its formal definition is given by the following equation:

$$\mu = \left(\frac{\partial U}{\partial N}\right)_{S,V}. \tag{1.120}$$

The concept of chemical potential was introduced into thermodynamics by G. Gibbs and has played an extremely important role. If we now derive the other thermodynamic potentials, starting with definition (1.120), the term $\mu\,dN$ appears in all expressions. Hence:

$$dH = T\,dS + V\,dP + \mu\,dN, \tag{1.121}$$

$$dF = -S\,dT - P\,dV + \mu\,dN, \tag{1.122}$$

$$dG = -S\,dT + V\,dP + \mu\,dN. \tag{1.123}$$

Proceeding from these equations, chemical potential can be expressed using any thermodynamic potential:

$$\mu = \left(\frac{\partial H}{\partial N}\right)_{S,P} = \left(\frac{\partial F}{\partial N}\right)_{T,V} = \left(\frac{\partial G}{\partial N}\right)_{T,P}. \tag{1.124}$$

The previous formulas for thermodynamic potentials are presented here as differentials, but can also be in an integrated form. The internal energy U is the most convenient for this purpose, because it is a function of extensive variables, and is a homogeneous function of the first order. According to Euler's theorem for homogeneous functions, equation (1.119) can be transformed to

$$U = TS - PV + \mu N. \tag{1.125}$$

Using the definitions of other thermodynamic potentials, we can obtain their integral forms:

$$H = TS + \mu N, \tag{1.126}$$

$$F = -PV + \mu N, \tag{1.127}$$

$$U = \mu N. \tag{1.128}$$

It is obvious from the last equality that the chemical potential represents the Gibbs free energy per single particle, or per mole of a reactant.

In the more general case of a complex multicomponent system the differential of internal energy in a quasistatic process can be presented as

$$dU = T\,dS - \sum_i A_i\,da_i + \sum_k \mu_k\,dN_k, \tag{1.129}$$

so that the chemical potentials of the components are given by the derivatives

$$\mu_k = \left(\frac{\partial U}{\partial N_k}\right)_{S,\,a_i,\,N_j}. \tag{1.130}$$

The differentials of other thermodynamic potentials also have an additional term $\sum_k \mu_k\,dN_k$. This sum $\sum_k \mu_k N_k$ is actually an integral form of the generalized Gibbs free energy of a multicomponent system:

$$G = \sum_k \mu_k N_k. \tag{1.131}$$

Therefore it is possible to write the integral forms of other thermodynamic potentials:

$$U = TS - \sum_i A_i a_i + \sum_k \mu_k N_k, \tag{1.132}$$

$$F = -\sum_i A_i a_i + \sum_k \mu_k N_k, \tag{1.133}$$

$$H = TS + \sum_k \mu_k N_k. \tag{1.134}$$

The Gibbs–Duhem Equation. In a multicomponent system the internal energy can be written as

$$U = TS - PV + \sum_k \mu_k N_k. \tag{1.135}$$

Differentiating this equality,

$$dU = T\,dS + S\,dT - P\,dV - V\,dP + \sum_k \mu_k\,dN_k + \sum_k N_k\,d\mu_k, \tag{1.136}$$

and comparing the result with (1.129), we obtain the Gibbs–Duhem equation which plays an important role in thermodynamics:

$$S\,dT - V\,dP + \sum_k N_k\,d\mu_k = 0. \tag{1.137}$$

In a single-component system it is convenient to introduce internal energy, $u = U/N$, volume, $v = V/N$, and entropy, $s = S/N$, per particle. Then from the Gibbs–Duhem equation one can find

$$d\mu = -s\,dT + v\,dP. \tag{1.138}$$

Chemical potential is a function of temperature and pressure:

$$\mu = \mu(T, P) = u - sT + vP. \tag{1.139}$$

From this function it is possible to find specific volume and specific entropy:

$$s = -\left(\frac{\partial \mu}{\partial T}\right)_P \quad \text{and} \quad v = \left(\frac{\partial \mu}{\partial P}\right)_T. \tag{1.140}$$

Grand Thermodynamic Potential Ω. By including the number of particles N_k, as a variable, we can perform Legendre's transformation in thermodynamic functions and switch from extensive variables N_k to intensive variables μ_k. This procedure can be carried out with any thermodynamic potential, but the Helmholtz free energy is the most convenient. The function obtained as a result of its transformation is called the grand potential and is designated

$$\Omega = F - \sum_k \mu_k N_k. \tag{1.141}$$

Its differential is

$$d\Omega = -S\,dT - P\,dV - \sum_k N_k\,d\mu_k. \tag{1.142}$$

Thus, the natural variables of the grand potential are temperature, volume, and chemical potential of the components. For the derivatives of the grand potential we find:

$$\left(\frac{\partial \Omega}{\partial T}\right)_{V,\mu_k} = -s, \qquad \left(\frac{\partial \Omega}{\partial V}\right)_{T,\mu_k} = -P, \qquad \left(\frac{\partial \Omega}{\partial \mu_k}\right)_{T,V,\mu_i \neq \mu_k} = -N_k. \tag{1.143}$$

The grand potential has a very simple expression. Using the definition of F given above, we find

$$\Omega = -PV. \tag{1.144}$$

The grand thermodynamic potential is convenient for analyzing a system where T, V, and μ_i are maintained constant. In these variables the basic inequality of thermodynamics takes the form

$$d\Omega < -S\,dT - P\,dV - \sum_k N_k\,d\mu_k. \tag{1.145}$$

If T, V, and μ are constant, then it follows immediately that

$$d\Omega \leq 0. \tag{1.146}$$

Therefore at equilibrium the grand thermodynamic potential reaches a minimum, so that

$$d\Omega = 0, \tag{1.147}$$

and the condition of stability of the equilibrium state is

$$d^2\Omega > 0. \tag{1.148}$$

Phase Equilibrium and Phase Transitions. In previous sections we dealt with homogeneous bodies. Now we will consider systems, consisting of several phases in equilibrium with one another. A phase is a homogeneous part of a system that differs in its properties from other parts and is separated from them by a distinct boundary. Examples of phases include gel and liquid crystalline regions in lipid membranes, ice and water in contact, and oil–water emulsions. A *qualitative* transformation of a substance from one state to another is called a phase transition.

For two phases α and β to be in equilibrium with each other, they should have equal temperature and pressure:

$$P^\alpha = P^\beta, \quad T^\alpha = T^\beta. \tag{1.149}$$

However, this is not enough. To derive one more condition we must consider the Gibbs free energy. If the chemical potentials of solute particles in the two phases are μ^α and μ^β, and their amounts are N^α and N^β so that the total number of particles is constant, then

$$G = G^\alpha + G^\beta = N^\alpha \mu^\alpha + N^\beta \mu^\beta = N^\alpha(\mu^\alpha - \mu^\beta) + N\mu^\beta. \tag{1.150}$$

The only free parameter of the system is the number of particles in one of the phases. Therefore the change in Gibbs free energy is

$$\delta G = (\mu^\alpha - \mu^\beta)\delta N^\alpha. \tag{1.151}$$

Because this change should be equal to zero, we conclude that

$$\mu^\alpha = \mu^\beta. \tag{1.152}$$

This equality of chemical potentials of solute particles in two contacting phases is the additional condition for equilibrium.

A basic goal in the study of phase transitions is finding the relationship between temperature and pressure at the point at which the phase transition occurs. In principle, equation (1.152) determines this relationship if the chemical potentials for each phase are known. However, the relationship can also be described in a more general form.

Equation (1.152) shows that the chemical potential of a substance does not change during a phase transition, but this is not true for the derivatives of chemical potential. These derivatives $(\partial\mu/\partial T) = -s$ and $(\partial\mu/\partial P) = v$ during a phase transition change abruptly because the volume and the entropy per particle in coexisting phases (for example in liquid and its vapor) are different. The transitions satisfying these conditions are called phase transitions of the first kind.

The change of entropy at phase transition means that the substance absorbs (or releases) a certain amount of heat, called the latent heat of transition or simply the heat of transition from state α to state β:

$$\lambda = T(s^\beta - s^\alpha). \tag{1.153}$$

This quantity can be referred to a single molecule or to one mole. Thus, during a phase transition of the first kind, there is an abrupt change of specific volume in parallel with absorption (or release) of heat of transition.

Equation (1.153) for the heat of transition can be presented in another form. If the phase transition occurs at a constant pressure, then the amount of heat introduced to the system is equal to a change in its enthalpy, and therefore the heat of transition is

$$\lambda = H^\beta - H^\alpha. \tag{1.154}$$

Consider now the relationship between the temperature and pressure at which the phase transition occurs. By substituting the expression for the differential of chemical potential (1.138) into equation (1.153), we obtain

$$-s^\alpha \, dT + v^\alpha \, dP = -s^\beta \, dT + v^\beta \, dP, \tag{1.155}$$

and

$$\frac{dP}{dT} = \frac{s^\beta - s^\alpha}{v^\beta - v^\alpha}. \tag{1.156}$$

Using expression (1.153), we finally obtain

$$\frac{dP}{dT} = \frac{\lambda}{T(v^\beta - v^\alpha)}. \tag{1.157}$$

This is the Clausius–Clapeyron equation which determines a curve of phase equilibrium on the plane P–T. To integrate this equation and find an explicit function $P(T)$ or $T(P)$ we need to know the thermal equation of state $V = V(P, T)$ for each phase and the dependence of the heat of transition on temperature. This is a difficult problem that can be solved analytically in only a few cases.

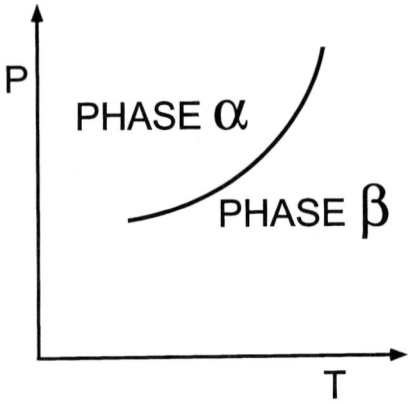

Figure 1.1. The curve representing the first-order phase transition from the phase α (liquid) to the phase β (vapor) on the $P-T$ plane. Note the positive slope of the curve.

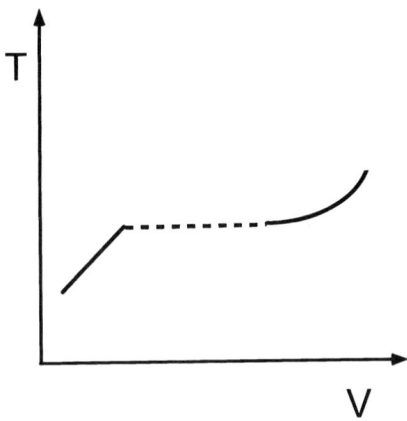

Figure 1.2. The curve of a first-order phase transition depicted on the $T-V$ plane. The volume undergoes a discontinuity jump when the transition temperature is attained at a given fixed pressure.

The curve of phase equilibrium on a plane $P-T$ has a monotonic shape with either a positive or negative slope. It is determined by the relationship of specific volumes before and after a phase transition. For example, during evaporation of water the specific volume of the vapor is larger than the specific volume of water and the transition involves absorption of heat. In this case $\partial P/\partial T > 0$ (Fig. 1.1). However, when ice melts, the opposite relationship holds: $v^\beta < v^\alpha$ and $\partial P/\partial T < 0$.

The curve of a phase transition of the first kind looks different on the plane $T-V$ (Fig. 1.2). If the pressure in the system is maintained constant, then at the phase transition the volume of the system experiences a finite change, which is reflected by a horizontal part of the graph (Fig. 1.2).

The phase transition graphed on the plane $P-V$ has a special interest. The van der

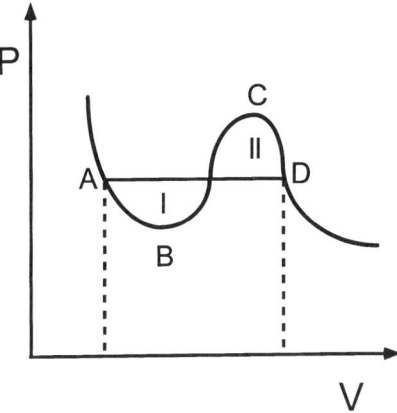

Figure 1.3. Van der Waals isotherm.

Waals gas law is a classic example. Its equation of state is

$$\left(P + \frac{a}{V^2}\right)(V - b) = RT, \tag{1.158}$$

where a and b are constants. The isotherm defined by this equation is shown in Fig. 1.3. The curve has a non-monotonous part ABCD where the phase transition occurs. To the left of this curve we have a liquid, to the right a gas.

The curve in Fig. 1.3 describes a continuous transition from a liquid state to a gas, and the pressure in this process should change non-monotonically. However, the real phase transition does not follow this isotherm, but rather finds a short cut and occurs on a horizontal line, or isobar. To explain this (Rumer and Ryvkin, 1980) we assume that the compression of a gas at constant temperature corresponds to movement along the isotherm from right to left. When the volume is decreased to a certain degree, the gas begins to condense and the system will simultaneously contain a gas and a liquid phase in equilibrium.

The isotherm in Fig. 1.3 does not describe this state because it corresponds to a homogeneous phase. However, it does define the points between which the phase transition actually occurs at constant temperature and pressure. At point D a gas exists, and at point A a liquid exists, both phases being in equilibrium. The chemical potentials of this substance at points A and D are identical, $\mu_A = \mu_D$, so that the Gibbs free energy does not change:

$$F_A + PV_A = F_D + PV_D. \tag{1.159}$$

From here it follows that

$$F_A - F_D = P(V_D - V_A). \tag{1.160}$$

On the other hand, if we slowly compress the system along the isotherm DCBA then the work performed in this process would be equal to the change of the Helmholtz free energy:

$$F_A - F_D = -\int_{DCBA} P \, dV = \int_{ABCD} P \, dV \tag{1.161}$$

Comparing expressions (1.160) and (1.161) we see that the integral along the curve ABCD is equal to the area of a rectangle, limited from above by the straight line AD, which means that I and II on Fig. 1.3 are equal.

The phase equilibrium between two states of a substance is therefore determined by the equality of areas I and II in the pressure and volume diagram at constant temperature. This statement is called Maxwell's rule.

1.5. MULTICOMPONENT SYSTEMS

Solutions. Homogeneous systems consisting of two or more mixed components having variable concentrations are called solutions. The component that constitutes the major part of a solution is called the solvent and all other components are dissolved substances or solutes. If a volume V contains two or more substances and the number of moles of component i is n_i, the total number of moles of all substances is $n = \Sigma_i n_i$. The components can also be given in numbers of molecules N_i so that the total number of molecules is $N = \Sigma_i N_i$.

The composition of a solution is described by the concentration of its components

$$c_i = \frac{n_i}{V}. \tag{1.162}$$

The SI dimension of this quantity is mol m^{-3} or mol dm^{-3} to replace the older dimension mol l^{-1}. The concentration can also be expressed in particles per unit volume:

$$c_i = \frac{N_i}{V}. \tag{1.163}$$

The dimension of this quantity is m^{-3} or dm^{-3}.

The mole fraction is the ratio of the number of moles of a given component to the total moles in the solution:

$$x_i = \frac{n_i}{\sum_k n_k}. \tag{1.164}$$

This quantity can be presented in a different way:

$$x_i = \frac{N_i}{\sum_k N_k} = \frac{c_i}{\sum_k c_k}. \tag{1.165}$$

Mole fractions are dimensionless quantities that obey the equation

$$\sum_{i=1}^{k} x_i = 1. \tag{1.166}$$

Therefore, if the number of components in the solution is equal to k, then only $k - 1$ of them are independent.

The composition of a solution can also be described by a mole ratio r_i which is the number of moles of solute i per mole of solvent which is usually designated by the number 1. Then

$$r_i = \frac{n_i}{n_1}. \tag{1.167}$$

It is obvious that $r_1 = 1$ and $r_i > x_i$.

Another characteristic of a solution is its molality m_i, which is the number of moles of the component i per kilogram of solvent:

$$m_i = \frac{1000 n_i}{n_1 M_1}. \tag{1.168}$$

Here M_1 is the molecular weight of the solvent.

Simple relationships exist between different characteristics of a solution:

$$x_i = \frac{r_i}{\sum_k r_k}, \qquad x_1 \sum_k r_k = 1. \tag{1.169}$$

For very dilute solutions all characteristics either become equal, $r_i \approx x_i$, or proportional to each other: $x_i \propto c_i \propto m_i \propto r_i$.

Earlier we mentioned that the state of a simple system is determined by two parameters, for example, by temperature T and pressure P. This is true for a one-component simple system. If the simple system contains k components, to define its state we need to know its composition in terms of molar fractions. Because among k mole fractions only $k - 1$ are independent, the total number of variables determining the state of a multicomponent system is equal to $k + 1$.

Partial Characteristics. Partial characteristics of components play an important role in the theory of solutions, as illustrated by the partial volume. The thermal equation of state of a solution can be presented as

$$V = V(P, T, n_1, n_2, n_3, \ldots). \tag{1.170}$$

The differential of volume can be written as

$$dV = \left(\frac{\partial V}{\partial P}\right)_{T, \{n_i\}} dP + \left(\frac{\partial V}{\partial P}\right)_{P, \{n_i\}} dT + \\ \left(\frac{\partial V}{\partial n_1}\right)_{P, T, \{n_i\}} dn_1 + \left(\frac{\partial V}{\partial n_2}\right)_{P, T, \{n_i\}} dn_2 + \cdots. \tag{1.171}$$

The subscript $\{n_i\}$ means that, when calculating the derivative of a given function, the amounts of all components except that in the derivative are kept constant. The derivative of volume is called a partial volume of this component:

$$v_k = \left(\frac{\partial V}{\partial n_k}\right)_{P, T, \{n_i\}}. \tag{1.172}$$

Because the volume is an extensive quantity and is also a homogeneous function of the first order with respect to the amounts of the different components, we can use the Euler theorem for homogeneous functions and write:

$$V = \sum_i v_i n_i. \tag{1.173}$$

This simply means that the total volume is composed of the partial volumes of individual components. However, partial volumes are not equal to specific volumes of pure components, and in some cases partial volumes can even be negative. For example, the partial volume of $MgSO_4$ in dilute aqueous solutions is negative. The negative sign of the partial volume means that addition of the given component to a solvent causes the volume of the solution to decrease, rather than increase.

Formula (1.173) can be generalized to any extensive property of a system and can be presented as a sum of partial quantities. We consider here a few particular examples.

The chemical potential of a given component represents its partial Gibbs free energy. Using the definition of Gibbs free energy, we obtain

$$\mu_i \equiv g_i = u_i - Ts_i + Pv_i. \tag{1.174}$$

Here u_i is the partial internal energy, and s_i is the partial entropy of the component i.

The partial entropy of the component i is equal to

$$s_i = -\left(\frac{\partial \mu_i}{\partial T}\right)_{P, \{n_k\}}, \tag{1.175}$$

and the partial volume is

$$v_i = \left(\frac{\partial \mu_i}{\partial P}\right)_{T, \{n_k\}}. \tag{1.176}$$

Partial enthalpy is

$$h_i = u_i + Pv_i = \mu_i - T\left(\frac{\partial \mu_i}{\partial T}\right)_{P, \{n_k\}}, \tag{1.177}$$

and finally, the partial heat capacity is

$$C_i = T\left(\frac{\partial s_i}{\partial T}\right)_{P, \{n_k\}} = \left(\frac{\partial h_i}{\partial T}\right)_{P, \{n_k\}}. \tag{1.178}$$

The total quantities are composed of partial quantities:

$$S = \sum_i n_i s_i, \qquad H = \sum_i n_i h_i, \qquad \text{and so on.} \tag{1.179}$$

Other characteristics of interest include average molar quantities, which are extensive quantities calculated per mole of mixture regardless of the type of component. For example, the average molar entropy is

$$\bar{s} = \frac{S}{n} = \frac{1}{n} \sum_i n_i s_i = \sum_i x_i s_i. \tag{1.180}$$

By analogy the average molar volume is

$$\bar{v} = \frac{V}{n} = \frac{1}{n} \sum_i n_i v_i = \sum_i x_i v_i. \tag{1.181}$$

An extensive quantity can be referred to a unit volume of the mixture. The result is the density of this quantity, such as the enthalpy density:

$$h = \frac{H}{V} = \frac{1}{V} \sum_i n_i h_i = \sum_i c_i h_i. \tag{1.182}$$

Chemical Potential and the Composition of a Solution. In the previous section we derived the Gibbs–Duhem equation, which can be rewritten using average molar quantities and molar fractions:

$$-\bar{s}\, dT + \bar{v}\, dP - \sum_i x_i\, d\mu_i = 0. \tag{1.183}$$

Chemical potential of any component is a function of temperature, pressure, and of the composition of the solution:

$$\mu_i = \mu_i(T, P, x_1, x_2, \ldots). \tag{1.184}$$

Therefore its differential is

$$d\mu_i = v_i\, dP - s_i\, dT + \hat{D}\mu_i. \tag{1.185}$$

Here the operator $\hat{D}$ describes the change of chemical potentials caused by the change of the composition of solution and is defined by the formula

$$\hat{D}\mu_i = \sum_k \frac{\partial \mu_i}{\partial x_k}\, dx_k. \tag{1.186}$$

Substituting the differential of chemical potential (1.185) in the Gibbs–Duhem equation and using the relationships between average molar and partial quantities, we obtain:

$$\sum_i x_i \hat{D}\mu_i = 0. \tag{1.187}$$

This important result defines the relationship between chemical potentials and molar fractions of components.

Thermodynamic Properties of Ideal Solutions. An important role in thermodynamics is played by the concept of an ideal solution. The chemical potential of a component in an ideal solution is

$$\mu_i = \mu_i^0 + RT \ln x_i, \tag{1.188}$$

where the standard value of chemical potential μ_i^0 depends only on temperature and pressure, but not on structure. The equality (1.188) defines an ideal solution, and solutions that obey this law are referred to as ideal.

Using equations (1.175) and (1.188) we find that partial molar entropy of an ideal solution is equal to

$$s_i = -\frac{\partial \mu_i^0}{\partial T} - R \ln x_i, \tag{1.189}$$

and the partial molar entropy is

$$h_i = \mu_i^0 - T\frac{\partial \mu_i^0}{\partial T} = h_i^0. \tag{1.190}$$

The last quantity happens to be equal to its standard value and therefore it does not depend on concentration. The partial volume of a component in an ideal solution also does not depend on concentration:

$$v_i = \frac{\partial \mu_i^0}{\partial P} = v_i^0. \tag{1.191}$$

It follows from these formulas that any two ideal solutions in identical solvents would mix at constant temperature and pressure without expansion or compression. Hence, the volume of the solution can be presented as

$$V = \sum_i n_i v_i = \sum_i n_i v_i^0, \tag{1.192}$$

where v_i^0 is the value of v_i in an infinitely dilute solution.

The dependence of partial volume on pressure can be presented as

$$v_i = v_i^*(1 - \beta_i P), \tag{1.193}$$

where v_i^* corresponds to zero pressure, and β_i is a thermodynamic factor of compression of a component i in an ideal solution, which can be considered independent of pressure and composition. However, both quantities v_i^* and β_i depend on temperature.

The enthalpy of the whole solution is

$$H = \sum_i n_i h_i = \sum_i n_i h_i^0, \tag{1.194}$$

where h_i^0 is the partial molar enthalpy of the component i at infinite dilution. It follows that mixing of two ideal solutions does not result in absorption or release of heat. In other words, the heat of (ideal) dissolution is equal to zero.

The Gibbs free energy is equal to

$$G = \sum_i n_i \mu_i = \sum_i n_i \mu_i^0 + RT \sum_i n_i \ln x_i = n \left(\sum_i x_i \mu_i^0 + RT \sum_i x_i \ln x_i \right)$$

$$= V \left(\sum_i c_i \mu_i^0 + RT \sum_i c_i \ln x_i \right). \tag{1.195}$$

The Helmholtz free energy is equal to

$$F = V \left(\sum_i c_i \mu_i^0 + RT \sum_i c_i \ln x_i - P \right). \tag{1.196}$$

We can now define average molar characteristics, which are the extensive quantities per mole of the mixture regardless of the type of component:

$$\bar{g} = \sum_i x_i \mu_i^0 + RT \sum_i x_i \ln x_i = \sum_i x_i \mu_i, \tag{1.197}$$

$$\bar{s} = -\sum_i x_i \frac{\partial \mu_i^0}{\partial T} - R \sum_i x_i \ln x_i = \sum_i x_i s_i, \tag{1.198}$$

$$\bar{h} = \sum_i x_i \left(\mu_i^0 - T \frac{\partial \mu_i^0}{\partial T} \right) = \sum_i x_i h_i^0, \tag{1.199}$$

$$\bar{v} = \sum_i x_i \frac{\partial \mu_i^0}{\partial P} = \sum_i x_i v_i^0. \tag{1.200}$$

Preparation of an ideal solution from individual substances increases the entropy of the system by the quantity called the entropy of mixing. It should be clear from the formulas above that in a binary ideal solution with the mole fractions x and $1 - x$ the average molar entropy of mixing is

$$\Delta \bar{s} = -R[x \ln x + (1 - x) \ln (1 - x)]. \tag{1.201}$$

Note that in an ideal solution there is no need to assign the role of the solvent to one of the components. However, if the component 1 is in excess, it can be conveniently considered a solvent, and taking into account that $1 - x_1 \ll 1$, its partial characteristics can be expanded into a series:

$$\mu_1 \approx \mu_1^0 - RT \sum_{i \neq 1} x_i, \qquad (1.202)$$

$$s_1 \approx -\frac{\partial \mu_i^0}{\partial T} - R \sum_{i \neq 1} x_i, \qquad (1.203)$$

$$v_1 \approx v_1^0. \qquad (1.204)$$

In very dilute solutions the chemical potential can be described by approximate formulas:

$$\mu_i \approx \mu_i^0 + RT \ln r_i \approx \mu_i^0 + RT \ln m_i \approx \mu_i^0 + RT \ln c_i. \qquad (1.205)$$

In this approximation m and c are dimensional quantities. Because they appear in logarithmic form it is important to use standard chemical potentials when solving the equation.

Non-ideal Solutions. The behavior of real solutions deviates from the basic equation (1.188) for ideal solutions and their properties should be described by more complex formulas. However, the laws of ideal solutions are mathematically convenient. One way to save these formal expressions is to combine the non-ideal properties in a new parameter a_i, the activity of the given component that can replace concentration in defining chemical potential:

$$\mu_i = \mu_i^0 + RT \ln a_i. \qquad (1.206)$$

The activity a_i and the concentration c_i are related via the so-called coefficient of activity γ:

$$a_i = \gamma c_i. \qquad (1.207)$$

The idea to introduce the concept of activity instead of concentration belongs to Lewis and Rendal (1961). At a strong dilution the coefficients of activity tend to 1 and activities to concentrations.

   If the component 1 is a solvent, its chemical potential given by the formula (1.206) can be transformed to

$$\mu_1 = \mu_1^0 - \zeta RT \sum_{i \neq 1} r_i, \qquad (1.208)$$

where ζ is an osmotic coefficient. One should note that after the summation sign we have put not a mole fraction x_i but rather a molar ratio r_i . In diluted solutions the difference between them is insignificant; however, for the calculation of osmotic effects, which will be considered in the following section, the latter parameter is more convenient.

If chemical potential is known, it is possible to calculate thermodynamic functions of the solution, for example its Gibbs and Helmholtz free energy. It is important to understand the relationship between these free energies. In simple systems they differ by the quantity PV, and in liquids this term is much smaller than the term TS. The difference of Gibbs free energy from Helmholtz free energy in simple liquid systems can therefore be neglected.

Osmosis. If a solution β is separated from a pure solvent α by a membrane that is permeable to the solvent but not the solute, the solvent molecules cross the membrane into the solution and create pressure. The extra pressure at equilibrium is called osmotic pressure, Π, and the chemical potential of the solvent is identical in both phases:

$$\mu_1^\alpha = \mu_1^\beta. \tag{1.209}$$

Using equation (1.207), we obtain

$$\mu_1^0(P^\alpha, T) = \mu_1^0(P^\beta, T) - \zeta RT \sum_{i \neq 1} r_i. \tag{1.210}$$

Chemical potential can be expanded into a series of powers of the difference $P^\beta - P^\alpha$:

$$\mu_1^0(P^\beta, T) = \mu_1^0(P^\alpha, T) + (P^\beta - P^\alpha)\left(\frac{\partial \mu_1^0}{\partial P}\right) = \mu_1^0(P^\alpha, T) + (P^\beta - P^\alpha)v_1. \tag{1.211}$$

Recall that v_1 is the partial volume of the solvent. Substituting this expression into the previous equation (1.210) we find the osmotic pressure:

$$\Pi = P^\beta - P^\alpha = \frac{\zeta RT}{v_1} \sum_{i \neq 1} r_i. \tag{1.212}$$

Using the definition of molar ratio this equation can be transformed into another one:

$$\Pi = \frac{\zeta RT}{V_1} \sum_{i \neq 1} n_i. \tag{1.213}$$

Here V_1 is the volume of the solvent, not the solution. However, normally V_1 differs little from the total volume of the solution V and the familiar formula for osmotic pressure can be derived:

$$\Pi = \zeta RT \sum_{i \neq 1} c_i. \tag{1.214}$$

In an ideal solution $\zeta = 1$ and this formula gives the van't Hoff law:

$$\Pi = RT \sum_{i \neq 1} c_i. \tag{1.215}$$

There is an interesting exception to this law when the dissolved substance consists of very large molecules. Even if the molar fraction of such a substance remains much less than unity, the volume occupied by the solute can be comparable to the volume of the whole solution. This is the case for hemoglobin inside red blood cells. The corrections to the van't Hoff equation in this case can be significant, and one should use equation (1.213) even for ideal solutions with $\zeta = 1$.

Chemical Equilibrium. When chemical reactions occur between the components of a solution, the products undergo dissociation, ionization, and aggregation processes that change the number of particles in a closed system. Any chemical reaction can be represented by the equation

$$\sum_i \nu_i B_i = 0, \qquad (1.216)$$

where B_i is the chemical symbol for reactants and ν_i is the stoichiometric coefficient indicating how many different molecules arise or disappear in a reaction. For example, the reaction of water formation from hydrogen and oxygen

$$2H_2 + O_2 = 2H_2O \qquad (1.217)$$

can be rewritten as

$$2H_2O - 2H_2 - O_2 = 0, \qquad (1.218)$$

and we can find that $\nu_{H_2O} = 2$, $\nu_{H_2} = -2$, $\nu_{O_2} = 2$.

Suppose that the number of moles of substances participating in the reaction changes by dn_1, dn_2, and so on. These numbers are related by the equations

$$\frac{dn_1}{\nu_1} = \frac{dn_1}{\nu_1} = d\xi. \qquad (1.219)$$

The quantity ξ is called the chemical variable or the number of runs of the chemical reaction.

Usually chemical equilibrium is studied at constant pressure and temperature. Under these conditions the Gibbs free energy is at a minimum and the variable on which the equilibrium establishes itself is the chemical variable ξ. The differential of the Gibbs free energy is

$$dG = \sum_i \mu_i \, dn_i = \left(\sum_i \nu_i \mu_i \right) d\xi. \qquad (1.220)$$

At minimum free energy this differential is equal to zero and therefore

$$\sum_i \nu_i \mu_i = 0. \qquad (1.221)$$

This defines the condition of chemical equilibrium. If more than one reaction occurs in the system then the equilibrium is determined by a set of equations similar to (1.221). This equation can be written in a more specific way if the expression for chemical potential is known. For example in dilute solutions

$$\mu_i = \mu_i(P, T) + RT \ln x_i. \tag{1.222}$$

Substituting this expression in (1.221), we obtain another form of the condition of chemical equilibrium:

$$\prod_i x_i^{\nu_i} = K(P, T), \tag{1.223}$$

where

$$K(P, T) = \exp\left(-\frac{1}{RT} \sum_i \nu_i \mu_i^0\right) \tag{1.224}$$

is the constant of chemical equilibrium. In this expression the constant K is dimensionless, but if the molar fractions are replaced by concentrations c_i, then this constant assumes a dimension and the condition of chemical equilibrium can be written

$$K(P, T) = \prod_i c_i^{\nu_i}, \tag{1.225}$$

which is known as the law of acting masses.

In non-ideal solutions this equation should include activities instead of concentrations.

Example 5: Dissociation of Electrolytes. When an electrolyte $B_{\nu^+} A_{\nu^-}$ dissolves it forms cations B and anions A with charge numbers z_B and z_A:

$$B_{\nu^+} A_{\nu^-} \leftrightarrow \nu^+ B + \nu^- A. \tag{1.226}$$

According to the law of mass action:

$$\frac{c_B^{\nu^+} c_A^{\nu^-}}{c_{BA}} = K, \tag{1.227}$$

where K is the constant of dissociation, c_B and c_A are the concentrations of ions, and c_{BA} is the concentration of undissociated molecules.

We shall introduce also the degree of dissociation:

$$\alpha = \frac{c - c_{BA}}{c} = 1 - \frac{c_{BA}}{c}, \tag{1.228}$$

where c is the total concentration of dissolved electrolyte. Because $c_A = v^-(c - c_{BA})$ and $c_B = v^+(c - c_{BA})$, the condition of equilibrium takes the form

$$\frac{\alpha^{v^+ + v^-}}{1 - \alpha} c^{v^+ + v^- - 1} = (v^+)^{-v^+}(v^-)^{-v^-} K. \tag{1.229}$$

For a 1:1 electrolyte this condition can be simplified:

$$\frac{\alpha^2 c}{1 - \alpha} = K. \tag{1.230}$$

In this form it is called the Ostwald law of dilution which shows that with decreasing electrolyte concentration the degree of dissociation α increases and approaches unity. In other words, in very dilute solutions virtually all molecules of an electrolyte are dissociated.

Gibbs Phase Rule. We will now consider phase (or heterogeneous) equilibrium in a system consisting of n phases with k components. The first question is how many independent variables define such systems, and how many different phases can be simultaneously present. This problem was resolved by Gibbs, and the equation he derived is called the Gibbs phase rule.

We mentioned earlier that the state of each phase is determined by $k + 1$ variables which include temperature T, pressure P, and mole fraction $k - 1$. Although one might think that the state of n phases is determined by $n(k - 1)$ variables, in fact not all of them are independent, because the parameters of the system are related by certain conditions. First, the temperature of all phases is identical, which imposes $n - 1$ restrictions because all temperatures can be equated to the temperature of the first phase.

Analogously the pressure in all phases should be identical giving another $n - 1$ restrictions, and the chemical potentials of all components should be identical in all phases to give $k(n - 1)$ additional restrictions.

Thus, the total number of restrictions is $(k + 2)(n - 1)$ and therefore the number of independent variables is $n(k - 1) - (k + 2)(n - 1)$. This quantity cannot be negative, and therefore

$$n \leq k + 2. \tag{1.231}$$

This is the Gibbs phase rule that establishes that in a system comprised of k components the number of phases in equilibrium with each other cannot exceed $k + 2$.

The number of independent variables that can be changed without infringement of the equilibrium of a heterogeneous system is called the number of thermodynamic degrees of freedom f, which is equal to

$$f = k + 2 - n. \tag{1.232}$$

1.6. EXTERNAL FIELD

If a system is placed in an external field, the condition of its equilibrium changes. The condition of constant and uniform temperature in all parts of the system is still valid, but the pressure within the system will vary, and the chemical potential will not be uniform throughout the system. The quantity μ_i that we have dealt with so far represents the "internal" chemical potential, which reflects the interaction of particles within the system. But the imposition of an external field increases the potential energy u_{pot} of the particles in this field. The sum of these two quantities is called the total chemical potential of a given component:

$$\bar{\mu}_i = \mu_i + u_{pot}. \tag{1.233}$$

The equation of equilibrium in external field takes the form

$$\bar{\mu}_i = \text{constant}. \tag{1.234}$$

If the imposed field is electrical, the total chemical potential is called an electrochemical potential $\bar{\mu}_i$, defined as

$$\bar{\mu}_i = \mu_i + z_i F \phi \tag{1.235}$$

with the usual designation F for the Faraday constant and ϕ for the electrical potential. The concept of an electrochemical potential plays a central role in electrochemistry and chemical thermodynamics.

If the system is placed in a gravitational field, then $u_{pot} = M_i gh$, where M_i is the mass of one mole of a given component, g is the acceleration due to gravity and h is the position in the field. The total chemical potential is

$$\bar{\mu}_i = \mu_i + M_i gh + z_i F \phi. \tag{1.236}$$

If a homogeneous substance is placed in an external field, this substance, strictly speaking, cannot be considered as being of one phase, because a phase is defined as being absolutely homogeneous in its state and properties. Instead even simple systems in external fields should be considered to be continuous sequences of phases $\alpha, \beta, \gamma, \ldots$, with infinitesimally small differences between them.

The thermodynamic functions for each phase will now contain, instead of chemical potentials μ_i, the total chemical potentials $\bar{\mu}_i$:

$$dU^\alpha = T^\alpha \, dS^\alpha - P^\alpha \, dV^\alpha + \sum_i \bar{\mu}_i^\alpha \, dn^\alpha, \tag{1.237}$$

$$dH^\alpha = T^\alpha \, dS^\alpha + V^\alpha \, dP^\alpha + \sum_i \bar{\mu}_i^\alpha \, dn^\alpha, \tag{1.238}$$

$$dF^\alpha = - S^\alpha \, dT^\alpha - P^\alpha \, dV^\alpha + \sum_i \bar{\mu}_i^\alpha \, dn^\alpha, \tag{1.239}$$

$$dG^\alpha = - S^\alpha \, dT^\alpha + V^\alpha \, dP^\alpha + \sum_i \bar{\mu}_i^\alpha \, dn^\alpha, \tag{1.240}$$

$$d\Omega^\alpha = -S^\alpha\,dT^\alpha - P^\alpha\,dV^\alpha - \sum_i \bar{\mu}_i^\alpha\,dn^\alpha. \tag{1.241}$$

In the integral form these thermodynamic functions will look like

$$G^\alpha = \sum_i \bar{\mu}_i^\alpha n^\alpha, \qquad F^\alpha = \sum_i \bar{\mu}_i^\alpha n^\alpha - P^\alpha V^\alpha, \qquad \text{and so on.} \tag{1.242}$$

Example 6: A Single Component System in a Gravitational Field. Consider a pure substance in a gravitational field. The condition of a uniform total chemical potential can be written in the following form:

$$d\mu + Mg\,dh = 0. \tag{1.243}$$

For constant temperature we can write $d\mu = v\,dP$, where v is the partial molar volume of the given substance. After introducing the mass density

$$\rho = \frac{M}{v} \tag{1.244}$$

we obtain the barometric formula for hydrostatic pressure:

$$dP = -\rho g\,dh. \tag{1.245}$$

Note that in this derivation the mass density was not assumed to be uniform. To use the condition of incompressibility of liquids, equation (1.245) can be integrated:

$$P = -\rho g h + \text{constant}. \tag{1.246}$$

Example 7: Solutions in the Gravitational Field. If a solution is in a gravitational field then the condition of equilibrium includes uniform temperature and chemical potential throughout the solution. The latter statement can be written

$$d(\mu_i + M_i g h) = 0. \tag{1.247}$$

At constant temperature

$$d\mu_i = v_i\,dP + \hat{D}\mu_i, \tag{1.248}$$

where $\hat{D}\mu_i$ is determined by changes in the solution composition and is given by equation (1.186). Multiplying (1.247) by x_i and taking the sum over all components, we obtain

$$\sum_i x_i v\,dP + \sum_i x_i M_i g\,dh = 0. \tag{1.249}$$

In this derivation we used the equation $\sum_i x_i \hat{D}\mu_i = 0$ derived in section 1.5 under the heading "Chemical potential and the composition of a solution."

After introducing the average molar volume $\bar{v}$ and average molar weight,

$$\bar{M} = \sum_i x_i M_i, \tag{1.250}$$

we arrive at the final equation

$$\bar{v}\, dP + \bar{M}g\, dh = 0. \tag{1.251}$$

This is the equation of hydrostatics, but note that we did not assume uniformity of $\bar{v}$ or $\bar{M}$.

Eliminating the variable dP, we obtain

$$\hat{D}\mu_i + \left(M_i - v_i \frac{\sum\limits_k x_k M_k}{\sum\limits_k x_k v_k} \right) g\, dh = 0. \tag{1.252}$$

This equation permits us to describe changes in the composition of the solution caused by height or, more importantly, by centrifugal force. Unfortunately, for a general case direct integration is impossible, but it can be conducted in a few special cases. For example, for an ideal solution $\hat{D}\mu_i = RT\, d \ln x_i$. In highly dilute solutions equation (1.252) is reduced to

$$RT\, d \ln x_i + \left(M_i - v_i \frac{M_1}{v_1} \right) g\, dh = 0. \tag{1.253}$$

Here, as usual, the index 1 designates the solvent which is in excess. If the solution is incompressible, then we can integrate the previous equation and find that the concentration of components in two points α and β separated by the distance $h = h^\beta - h^\alpha$ are related by the following equality:

$$\frac{c_i^\beta}{c_i^\alpha} \approx \frac{x_i^\beta}{x_i^\alpha} \approx \exp\left[-\frac{gh}{RT} \left(M_i - v_i \frac{M_1}{v_1} \right) \right]. \tag{1.254}$$

Nonequilibrium States. The problem of describing a nonequilibrium state and calculating its thermodynamic characteristics is very difficult. It cannot be solved within the framework of conventional thermodynamics because it demands additional assumptions and concepts. One approach to calculating thermodynamic functions for these systems was developed by M. A. Leontovich (1983).

The internal parameters of a system at equilibrium are functions of the external parameters and temperature. To define a nonequilibrium state it is necessary to introduce one or more additional internal parameters ξ, such as the distribution and concentration of its components in space. If the solution is not in thermal equilibrium, then we must include the spatial distribution of temperature gradients.

To find such functions one can draw a parallel between a nonequilibrium state and a specific equilibrium condition. Suppose that an external field X is imposed on a system, which then reaches an equilibrium state where the additional parameter ξ has its equilibrium value. We will designate the initial equilibrium state by the symbol 0, the nonequilibrium state that we are interested in by the symbol 1, and the state that becomes equilibrium in the field X by the symbol 1^*.

To determine the necessary thermodynamic functions, consider the transition of a system from state 0 to state 1 along the path $0 \Rightarrow 1 \Rightarrow 1^*$. The external field X is gradually increased in intensity, so that the system passes in a quasistatic manner from the equilibrium state 0 to another equilibrium state 1^*. In this process the additional parameter ξ will reach the desired value.

Suppose that in the initial state 0 the system had internal energy U_0, entropy S_0, Helmholtz free energy F_0, and so on. In the equilibrium state 1^* these functions will have the values U^*, S^*, and F^*. The functions U^* and F^* will include the potential energy of the additional field X. In the final nonequilibrium state the same parameters are designated U, S, and F.

When calculating the energy in state 1^* the parameter ξ is a generalized force, linked to the external parameter X. The work performed during the change dX of the external parameter is equal to

$$\delta W_X = \xi \, dX, \tag{1.255}$$

and therefore the change of energy U^* is equal to

$$dU^* = T \, dS^* - P \, dV - \xi \, dX. \tag{1.256}$$

Parameter ξ changes when the field X is turned on, and because the field increases in a quasistatic manner, ξ is a function of this field.

Having reached the state 1^*, the external field X can be quickly turned off so that the internal parameters of the system, including entropy, have no time to change. The system will then pass to the nonequilibrium state 1, in which

$$S = S^*. \tag{1.257}$$

However, the energy of the system does change. The decrease in the external field from X to 0 at constant ξ means that the energy has changed by ξX:

$$U = U^* + \xi X. \tag{1.258}$$

If the energy is a function of internal parameters, its differential in the nonequilibrium state is

$$dU = T \, dS - P \, dV + X \, d\xi. \tag{1.259}$$

A similar formula holds for the Helmholtz free energy:

$$dF = - S \, dT - P \, dV + X \, d\xi. \tag{1.260}$$

The final result can be interpreted in the following way. The term $X\,d\xi$ represents work done on the system by the field X. Therefore any change in the energy of the nonequilibrium state as the additional parameter ξ changes is determined by the work of the external field in which the system comes to equilibrium. This interpretation often allows nonequilibrium problems to be solved by the simplest method. An important example is a multiphase system in which each phase is at equilibrium, but without equilibrium between phases. Nonequilibrium between phases can occur with respect to any parameter, such as temperature, pressure, or chemical potential. To bring this system to equilibrium without changing its properties we can in principle apply an additional field at the borders of phases, for instance, additional pressure. The thermodynamic potentials of the nonequilibrium system can be found as the sum of thermodynamic potentials of equilibrium parts, either calculated directly from the thermodynamic characteristics of each phase, or from the work performed by the additional field.

This approach has the limitation that the assumed equilibrium inside the phases is an approximation. On the other hand, it provides a simple description of otherwise complex nonequilibrium systems.

1.7. SURFACES

Surface Tension. While considering multiphase systems we assumed that the energy, entropy, and other extensive characteristics of systems were characteristics of homogeneous phases. This implied that the boundaries between phases were considered to be mathematical surfaces, and the properties of phases were uniform throughout a given phase. This assumption neglected possible inhomogeneity of phases near their boundaries, but is reasonable if the phases are large and surface phenomena are not of interest.

However, at the contact between two phases there is a transition layer where the density of energy, solutes, and other thermodynamic parameters are different from the bulk values (Fig. 1.4). The thickness of this layer is small (usually in the range of 1–10 nm), and is referred to as a surface phase or a two-dimensional phase, which cannot exist separately from the bulk phases. For this reason it is also called a non-autonomous phase to distinguish it from the "autonomous" bulk phases.

The boundary between two phases has special properties, and changing its area A involves mechanical work. The elementary work performed by the system when the surface changes by dA is equal to

$$\delta W = -\gamma\,dA. \tag{1.261}$$

The parameter γ is called the interfacial or surface tension. The term "surface tension" is commonly used for the boundary between a liquid and a gas phase, and the term "interfacial tension" for the boundary between two phases of different chemical composition, such as oil and water.

The physical meaning of γ is the energy of a surface per unit area or the force per unit length. Therefore the dimension of γ is $J\,m^{-2}$ or $N\,m^{-1}$. Surface tension is a positive quantity because otherwise the surface would increase spontaneously, resulting in the mixing of two phases.

Mechanical Interpretation of Surface Tension. In the bulk liquid phase, the forces acting on a given plane are identical in all directions, but at the surface this equality does not exist. Therefore, in the interfacial region we should distinguish the normal component of the pressure P_N normal to the interface and the tangential component of the pressure P_T parallel to the interface. Because these components differ from each other, the work done by a two-phase system during a change of volume cannot be calculated with the usual formula $\delta W = P\,dV$. Instead we can select a cylindrical volume near the boundary of two phases α and β with its axis perpendicular to the boundary (Fig. 1.4). The length of this cylinder is such that its ends are in the bulk phases. The normal component of the pressure is constant and does not depend on the coordinate z. The tangential component of the pressure varies with coordinate z as shown in Fig. 1.4. The area of the base of the cylinder is equal to A and the height is τ. The volume of the cylinder is $V = A\tau$ and therefore $dV = A\,d\tau + \tau\,dA$.

If the height of the cylinder is changed by $d\tau$ the work performed by the system is $P_N A\,d\tau$. If the cylinder expands laterally the work is $(\int_0^\tau P_T dz)\,dA$. Thus the total work performed when the cylinder volume changes is equal to

$$\delta W = P_N A\,d\tau + \left(\int_0^\tau P_T dz\right) dA = P_n\,dV - P_N \tau\,dA + \left(\int_0^\tau P_T dz\right) dA$$

$$= P_N dV - \left[\int_0^\tau (P_N - P_T)\,dz\right] dA. \tag{1.262}$$

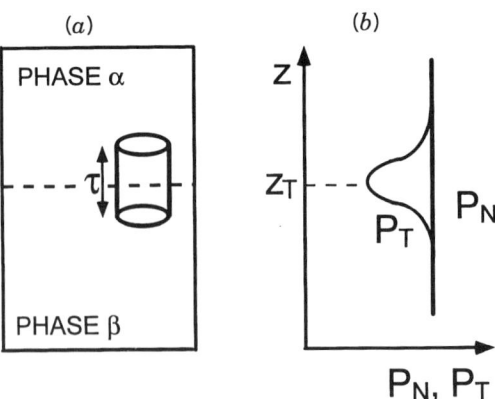

Figure 1.4. Mechanical properties of the border of two phases: (*a*) interface; (*b*) normal and tangential components of pressure at the interface.

Because $P_N = P_T$ in the bulk phases integration of this formula can be extended to infinity. Comparing this result with the expression for work in a two-phase system (1.261), we find the expression for surface tension:

$$\gamma = \int_{-\infty}^{\infty} (P_N - P_T)\, dz. \tag{1.263}$$

The transition from pressure to surface tension can be compared to the replacement of the physical system with a model in which pressure (to be more exact, the role of the difference $P_N - P_T$) is analogous to an infinitely thin elastic membrane stretched by the specific force γ. However, before this substitution is made we must decide where the membrane should be placed to create both the same force and the same moment. Suppose that it is located at a plane with coordinate z_t. Then its torque relative to the coordinate $z = 0$ is

$$\int_{-\infty}^{\infty} (P_N - P_T)z\, dz = z_t \gamma. \tag{1.264}$$

From here follows

$$z_t = \frac{\displaystyle\int_{-\infty}^{\infty} (P_N - P_T)z\, dz}{\displaystyle\int_{-\infty}^{\infty} (P_N - P_T)\, dz}. \tag{1.265}$$

The plane with coordinate z_t is called the surface of tension.

Thermodynamic Functions of a Two-Phase System. Now we can write down thermodynamic functions of a two-phase system that account for the special properties of its interface. Because the system is at equilibrium, the temperature and the chemical potentials of the two phases are identical. If the boundary between them is planar the pressure is also identical in both phases. Then the differential of internal energy can be presented as

$$dU = T\, dS - P\, dV + \gamma\, dA + \sum_i \mu_i\, dn_i. \tag{1.266}$$

It is clear from this expression that the surface tension can be defined by the derivative

$$\gamma = \left(\frac{\partial U}{\partial A}\right)_{S, V, \{n_i\}}. \tag{1.267}$$

The internal energy can be found by integration:

$$U = TS - PV + \gamma A + \sum_i \mu_i n_i. \tag{1.268}$$

We can also introduce other thermodynamic potentials. The Helmholtz free energy is

$$F = U - TS = -PV + \gamma A + \sum_i \mu_i n_i, \tag{1.269}$$

and its differential is

$$dF = -S\,dT - P\,dV + \gamma\,dA + \sum_i \mu_i\,dn_i. \tag{1.270}$$

The transition to enthalpy and to Gibbs free energy is not obvious because the expression for the internal energy contains two terms representing work: $-P\,dV$ and $\gamma\,dA$. By convention we will call the enthalpy (or the generalized enthalpy) the result of the Legendre transformation performed on the internal energy over both variables V and A:

$$H = U + PV - \gamma A = TS + \sum_i \mu_i n_i. \tag{1.271}$$

Then

$$dH = T\,dS + V\,dP - A\,d\gamma + \sum_i \mu_i\,dn_i. \tag{1.272}$$

Similarly, we will call the Gibbs free energy (or the generalized Gibbs free energy) the quantity

$$G = F + PV - \gamma A = \sum_i \mu_i n_i \tag{1.273}$$

with the differential

$$dG = -S\,dT + V\,dP - A\,d\gamma + \sum_i \mu_i\,dn_i. \tag{1.274}$$

Because two terms corresponding to mechanical work are present, we can introduce intermediate potentials such as the modified Gibbs free energy

$$\tilde{G} = F + PV = \gamma A + \sum_i \mu_i n_i \tag{1.275}$$

with the differential

$$d\tilde{G} = -S\,dT + V\,dP + \gamma\,dA + \sum_i \mu_i\,dn_i. \tag{1.276}$$

Similarly, we can introduce the modified enthalpy

$$\tilde{H} = U + PV = TS + \gamma A + \sum_i \mu_i n_i \qquad (1.277)$$

with the differential

$$d\tilde{H} = T\,dS + V\,dP + \gamma\,dA + \sum_i \mu_i\,dn_i. \qquad (1.278)$$

Last, we introduce the grand potential

$$\Omega = F - \sum_i \mu_i n_i = - PV + \gamma A \qquad (1.279)$$

with the differential

$$d\Omega = - S\,dT - P\,dV + \gamma\,dA - \sum_i n_i\,d\mu_i. \qquad (1.280)$$

The surface tension can be expressed through these potentials:

$$\gamma = \left(\frac{\partial F}{\partial A}\right)_{T,V,\{n_i\}} = \left(\frac{\partial \tilde{H}}{\partial A}\right)_{S,P,\{n_i\}} = \left(\frac{\partial \tilde{G}}{\partial A}\right)_{T,P,\{n_i\}} = \left(\frac{\partial \Omega}{\partial A}\right)_{T,V,\{\mu_i\}} . \quad (1.281)$$

Surface Functions. The thermodynamic potentials of a two-phase system, found above, contain contributions from bulk phases and from the interfacial boundary. It is important to differentiate between these contributions and find the role of the interface. This problem was solved by Gibbs (1928).

As before, we will consider a system with two phases α and β and total volume V, temperature T, interfacial area A and amounts of components n_i (Fig. 1.4). The transition layer at the boundary of two phases is highlighted by shading. Somewhere within the limits of this layer we can place the surface which will divide the volumes of phases V^α and V^β. The position of this surface is arbitrary, and for a planar boundary the position does not have thermodynamic significance. However, the normal to this surface is unambiguous for a flat or curved surface. The reason is that gradients of solutes or other thermodynamic properties of the system exist in the transition layer along the normal. Any mathematical surface with such a normal can be chosen to divide the two volumes, and it is reasonable to place it within the limits of the transition layer. After the dividing surface is chosen, the volumes V^α and V^β are fixed and satisfy the condition

$$V^\alpha + V^\beta = V. \qquad (1.282)$$

This condition was used by Gibbs. However, as will be shown in subsequent chapters, this condition is not necessary, and the thermodynamics of surfaces can be constructed without it. Nevertheless the method developed by Gibbs is the most straightforward and we will follow it in this chapter.

The bulk phases α and β contain a certain density of components and energy. We will define the extensive properties of the phase α by the following equations:

$$n_i^\alpha = c_i^\alpha V^\alpha, \qquad U^\alpha = u^\alpha V^\alpha, \qquad \text{and so on.} \qquad (1.283)$$

The sums of the quantities $n_i^\alpha + n_i^\beta$ and $U^\alpha + U^\beta$ are not equal to the total quantities n_i and U for the whole system. Their differences

$$n_i^\sigma = n - n_i^\alpha - n_i^\beta, \qquad (1.284)$$

$$U^\sigma = U - U^\alpha - U^\beta, \qquad (1.285)$$

are attributed to the dividing surface σ and are called surface excesses.

Because the choice of the dividing surface σ is arbitrary, the surface excesses are also arbitrary. To define the position of the dividing surface, one can impose the condition that the surface excess of one of the components (or energy, or entropy) is zero, so that all other surface excesses are fixed and can have either positive or negative values.

In a one-component system the dividing surface can be defined by the condition $n^\sigma = 0$, and this is referred to as an equimolar system. If the number of components is more than one, then in the Gibbs method there is no dividing surface for which all excesses n_i^σ equal zero. However, in Hansen's formal approach, which will be considered later, can be found a dividing surface for which both surface excesses are equal to zero. It is possible also to impose a condition in which a certain combination of surface excesses equals zero, often leading to simplification.

Surface excesses of thermodynamic potentials are particularly interesting. By removing bulk terms from equation (1.268) we can find the internal energy of the interface:

$$U^\sigma = TS^\sigma + \gamma A + \sum_i \mu_i n_i^\sigma. \qquad (1.286)$$

Note the disappearance of the volume term from this equation, which is a consequence of the Gibbs condition (1.283). The surface Helmholtz free energy can be written as

$$F^\sigma = \gamma A + \sum_i \mu_i n_i^\sigma. \qquad (1.287)$$

Only the second term of this equation depends on the choice of the dividing surface. Therefore, if the dividing surface is defined by the condition $\sum_i \mu_i n_i^\sigma = 0$, the surface tension will be equal to the Helmholtz free energy of the surface per unit area:

$$\gamma = \frac{F^\sigma}{A}. \qquad (1.288)$$

In this connection, the grand thermodynamic potential is especially convenient because at any dividing surface its surface value is

$$\Omega^\sigma = \gamma A. \tag{1.289}$$

Therefore, the surface tension is equal to the surface grand potential per unit area regardless of the choice of dividing surface.

The Gibbs free energy of the interface is equal to

$$G^\sigma = \sum_i \mu_i n_i^\sigma. \tag{1.290}$$

It is clear that the dividing surface can be chosen in such a way that the Gibbs free energy of the interface is zero. The modified Gibbs free energy of the surface is equivalent to the surface Helmholtz free energy:

$$\tilde{G}^\sigma = F^\sigma. \tag{1.291}$$

The differentials of these thermodynamic potentials are

$$dU^\sigma = T\,dS^\sigma + \gamma\,dA + \sum_i \mu_i\,dn_i^\sigma. \tag{1.292}$$

$$dF^\sigma = -S^\sigma\,dT + \gamma\,dA + \sum_i \mu_i\,dn_i^\sigma. \tag{1.293}$$

$$d\Omega^\sigma = -S^\sigma\,dT + \gamma\,dA - \sum_i n_i\,d\mu_i^\sigma. \tag{1.294}$$

$$dG^\sigma = -S^\sigma\,dT - A\,d\gamma + \sum_i \mu_i\,dn_i^\sigma. \tag{1.295}$$

$$d\tilde{G}^\sigma = dF^\sigma. \tag{1.296}$$

Curved Interfaces. The thermodynamics of curved surfaces is more complex than for planar surfaces. An important difference is that the area and tension depend on the dividing surface of a curved interface, instead of a planar interface. Because of this the term γA appearing in thermodynamic potentials is more complex.

The expression for the internal energy of a two-phase system can be written as

$$U = TS - P^\alpha V^\alpha - P^\beta V^\beta + \gamma A + \sum_i \mu_i n_i. \tag{1.297}$$

Here we took into account that the pressure in phases α and β is different. In calculating the differential of (1.297) we encounter another term which depends on the surface tension and the curvature of the dividing surface. However, there is a way to eliminate this term. If we place the dividing surface at the surface of tension, the additional term equals zero (Rowlinson and Widom, 1982; Rusanov, 1967). We will assume that the dividing surface is located just at this position, so that

$$dU = T\,dS - P^{\alpha}\,dV^{\alpha} - P^{\beta}\,dV^{\beta} + \gamma\,dA + \sum_{i} \mu_i\,dn_i. \tag{1.298}$$

Similarly for the Helmholtz free energy we have

$$F = U - TS = - P^{\alpha}V^{\alpha} - P^{\beta}V^{\beta} + \gamma A + \sum_{i} \mu_i n_i, \tag{1.299}$$

and

$$dF = - S\,dT - P^{\alpha}\,dV^{\alpha} - P^{\beta}\,dV^{\beta} + \gamma\,dA + \sum_{i} \mu_i\,dn_i, \tag{1.300}$$

and finally for the grand potential we obtain

$$\Omega = F - \sum_{i} \mu_i n_i = - P^{\alpha}V^{\alpha} - P^{\beta}V^{\beta} + \gamma A, \tag{1.301}$$

and

$$d\Omega = - S\,dT - P^{\alpha}\,dV^{\alpha} - P^{\beta}\,dV^{\beta} + \gamma\,dA - \sum_{i} n_i\,d\mu_i. \tag{1.302}$$

The derivation of the Gibbs free energy requires that Legendre's transformations are used for the variables describing the work of the system. The number of terms has increased because instead of one parameter P we have now two: P^{α} and P^{β}. For this reason the number of intermediate thermodynamic potentials analogous to the Gibbs free energy is increased. The Gibbs free energy is defined as before:

$$G = F + P^{\alpha}V^{\alpha} + P^{\beta}V^{\beta} - \gamma A = \sum_{i} \mu_i n_i. \tag{1.303}$$

Its differential is

$$dG = - S\,dT + V^{\alpha}\,dP^{\alpha} + V^{\beta}\,dP^{\beta} - A\,d\gamma - \sum_{i} \mu_i\,dn_i. \tag{1.304}$$

Among the intermediate potentials is a modified Gibbs free energy defined by the formula

$$\tilde{G} = F + P^{\alpha}V = (P^{\alpha} - P^{\beta})V^{\beta} + \gamma A + \sum_{i} \mu_i n_i. \tag{1.305}$$

Its differential is

$$d\tilde{G} = - S\,dT + V^{\alpha}\,dP^{\alpha} + (P^{\alpha} - P^{\beta})\,dV^{\beta} + \gamma\,dA + \sum_{i} \mu_i\,dn_i. \tag{1.306}$$

The surface versions of these thermodynamic potentials are given by the same formulas that were derived in the previous section.

The Laplace Equation. In the presence of a curved interface the pressure in phase α is different from the pressure in phase β. To find the relationship between them, suppose that the system is closed ($dn_i = 0$) and has a constant temperature and total volume. Then the equilibrium of the system is determined by the minimum Helmholtz free energy, the differential of which is equal to

$$dF = -P^\alpha \, dV^\beta - P^\beta \, dV^\beta + \gamma \, dA. \tag{1.307}$$

The right side of this equation contains three differentials. However, if phase β is a sphere with radius r, all three differentials can be expressed through this radius. The free energy becomes a function of a single parameter, and by varying this parameter we can find the equilibrium condition. For instance, in the equilibrium $dF = 0$, we obtain

$$P^\beta = P^\alpha + \gamma \frac{dA}{dV^\beta}. \tag{1.308}$$

The quantity dA/dV^β is the curvature of the interface. For a sphere $dA = 8\pi \, r \, dr$ and $dV = 4\pi \, r^2 \, dr$. Substituting dA and dV into (1.308), we obtain the Laplace formula

$$P^\beta - P^\alpha = \frac{2\gamma}{r}. \tag{1.309}$$

For an arbitrary surface with main radii of curvature r_1 and r_2 we have:

$$P^\beta - P^\alpha = \gamma \left(\frac{1}{r_1} + \frac{1}{r_2} \right). \tag{1.310}$$

Equation (1.310) represents the condition of mechanical equilibrium of two phases with a curved dividing surface. In the case of a flat interface the Laplace pressure is equal to zero.

Nucleus of a New Phase. When a homogeneous phase reaches a transition temperature, it can, in principle, separate into two phases, but the transition to a new phase state is usually delayed. For example, pure water can be overheated without boiling, and a steam can be overcooled without condensation. In this case the homogeneous system exists with a free energy exceeding the free energy of a non-uniform two-phase system. This state is called metastable.

The transition to a stable state is hindered by the fact that portions of the new phase have additional surface energy, which creates an energy barrier for the phase transition. For example, initiation of steam condensation is complicated because during the formation of small droplets their surface free energy grows faster (as r^2) than the volumetric free energy decreases (as r^3). Large drops, once they have reached a certain critical radius, find themselves in the opposite situation: the volumetric energy term decreases faster than the surface term grows. As a result, drops exceeding the critical size can increase in volume. Small aggregates of a new phase are called nuclei, and the aggregates that reach the critical size are called critical or equilibrium nuclei.

In thermodynamics of phase transitions, the work of the nucleus formation W plays an important role. This is the change of a pertinent thermodynamic potential during the formation of a nucleus of a given size. Within the framework of thermodynamics this quantity can be defined unambiguously for an equilibrium nucleus only when the total system is at equilibrium, even though this equilibrium is metastable. Any nucleus of arbitrary size makes the system nonequilibrium, and calculation of its thermodynamic potentials should be carried out according to the special procedure described above.

Following this procedure, we introduce the Gibbs dividing surface around the spherical nucleus, and consider the external phase α to be homogeneous up to the dividing surface. This determines the surface excesses of extensive properties which are usually attributed to the nucleus β. Now we have two homogeneous phases α and β which are *not in equilibrium* with each other. For this reason the chemical potentials of substances in phases μ_i^α and μ_i^β are *different*, as are the pressures in the two phases P^α and P^β. However, the mechanical equilibrium at the boundary is not altered and the pressure differential still obeys the Laplace formula (1.310). The surface tension of the nucleus is considered constant.

Now we assume that a new phase forms at constant pressure P^α, at constant temperature and constant chemical potentials. Then the state of the system can be conveniently described by the thermodynamic potential $\tilde{G}$ introduced by equation (1.306). In addition we must account for different chemical potentials in the different phases:

$$\tilde{G} = (P^\alpha - P^\beta)V^\beta + \gamma A + \sum_i \mu_i^\alpha n_i^\alpha + \sum_i \mu_i^\beta n_i^\beta. \qquad (1.311)$$

In the absence of a nucleus this potential is equal to the standard Gibbs free energy:

$$\tilde{G}_0 = \sum_i \mu_i^\alpha n_i. \qquad (1.312)$$

Thus the work of nucleus formation is equal to

$$W = \Delta\tilde{G} = \tilde{G} - \tilde{G}_0 = (P^\alpha - P^\beta)V^\beta + \gamma A + \sum_i (\mu_i^\beta - \mu_i^\alpha)n_i^\beta. \qquad (1.313)$$

It is clear that the work of nucleus formation at first increases with increasing radius, passes through a maximum and then decreases (Fig. 1.5). When the radius reaches

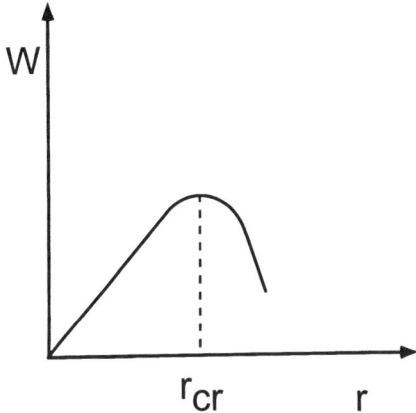

Figure 1.5. Work of nucleus formation reaches a maximum at a critical radius.

a critical value r_{cr}, the differential dW and therefore $d\tilde{G}$ are equal to zero. At this point, the system is in a metastable equilibrium where the chemical potentials in two phases are equal, and the last term in (1.313) equals zero. Using the Laplace formula, Gibbs found the work of formation of a critical nucleus:

$$W_{cr} = \tfrac{1}{3}\gamma A. \tag{1.314}$$

Now we can find the radius of a critical nucleus. For this purpose, note that the chemical potentials in equation (1.313) correspond not only to different phases, but also to different pressures. We can write them as functions $\mu_i^\alpha(P^\alpha)$ and $\mu_i^\beta(P^\beta)$, then bring them to the same pressure as the external phase P^α. Expanding the function $\mu_i^\beta(P)$ into a series of powers of $P^\beta - P^\alpha$ at the point $P = P^\alpha$ we obtain

$$\mu_i^\beta(P^\beta) = \mu_i^\beta(P^\alpha) + v_i^\beta(P^\beta - P^\alpha). \tag{1.315}$$

Chemical potentials at the same pressure are convenient because they characterize the degree of saturation of the phase α. Substitution of this formula in (1.313) simplifies the expression for the work of nucleus formation:

$$W = \gamma A + \sum_i [\mu_i^\beta(P^\alpha) - \mu_i^\alpha(P^\alpha)]n_i^\beta. \tag{1.316}$$

By equating this expression to the work of formation of a critical nucleus (1.314), we find the critical radius:

$$r_{cr} = \frac{2\gamma\bar{v}^\beta}{\sum_i [\mu_i^\alpha(P^\alpha) - \mu_i^\beta(P^\alpha)]x_i^\beta}. \tag{1.317}$$

Here $\bar{v}^{\beta}$ is the average molar volume. If the system contains only one component, it is simply the molar volume of the given substance in the phase β.

The ability to calculate the work of nucleus formation expands our understanding of new phase formation. In particular we can conclude that subcritical nuclei spontaneously appear in the new phase and disappear in the old phase until they reach the critical radius and the phase transition eventually occurs. It is also interesting to find the distribution of the nucleus radii, which is given by the Boltzmann law:

$$n = n_0 \exp\left(-\frac{W}{k_B T}\right), \tag{1.318}$$

where n_0 is a constant or, to be more exact, a quantity weakly dependent on radius or on W.

Linear Tension. In some cases a two-dimensional phase can be considered to be quasi-independent with the bulk phases contacting it acting as a thermostat. An obvious example is a lipid bilayer membrane which can be described by "two-dimensional thermodynamics" in complete conformity with the three-dimensional version. The role of pressure in this case is played by surface tension. The differential of internal energy is

$$dU = T\,dS + \gamma\,dA + \sum_i \mu_i\,dn_i, \tag{1.319}$$

and its integral form is

$$U = TS + \gamma A + \sum_i \mu_i n_i. \tag{1.320}$$

The other thermodynamic potentials can be defined in a similar way. For example, for the Gibbs free energy we have

$$G = \sum_i \mu_i n_i, \tag{1.321}$$

$$dG = -S\,dT - A\,d\gamma + \sum_i \mu_i\,dn_i. \tag{1.322}$$

Chemical potential μ_i depends on temperature and tension of the two-dimensional phase:

$$\mu_i = \mu_i(T, \gamma). \tag{1.323}$$

Its differential can be written

$$d\mu_i = -s_i\,dT - a\,d\gamma, \tag{1.324}$$

where s_i and a_i are specific entropy and specific area for a given component in the two-dimensional phase:

$$s_i = -\left(\frac{\partial \mu_i}{\partial T}\right)_\gamma, \qquad a_i = -\left(\frac{\partial \mu_i}{\partial \gamma}\right)_T. \qquad (1.325)$$

If the two-dimensional system consists of two phases, the border between them is a line on the plane or curved surface. As with other interfaces in three dimensions, this line has an excess energy, so that changing the length L of this line requires mechanical work:

$$\delta W = -f\,dL. \qquad (1.326)$$

The coefficient f is called the linear tension and its dimension is the ratio of energy to length, that is $J\,m^{-1} = N$. In this sense the border of two-dimensional phases is similar to a rubber band stretched by the force f. The linear tension is always positive.

The differential of internal energy of a two-dimensional system with a rectilinear border is

$$dU = T\,dS + \gamma\,dA + f\,dL + \sum_i \mu_i\,dn_i. \qquad (1.327)$$

In integral form it is

$$U = TS + \gamma A + fL + \sum_i \mu_i n_i. \qquad (1.328)$$

The other thermodynamic potentials can be found in the usual manner.

Linear tension can be presented as a derivative of appropriate thermodynamic potentials, for example,

$$f = \left(\frac{\partial U}{\partial L}\right)_{S, A, \{n_i\}}. \qquad (1.329)$$

Now let us consider two planar two-dimensional phases ν and ω with a curved border between them (Fig. 1.6) in which the surface tensions within the limits of each phase γ^ν and γ^ω will be different. The differential of internal energy in this case is

$$dU = T\,dS + \gamma^\nu\,dA^\nu + \gamma^\omega\,dA^\omega + f\,dL + \sum_i \mu_i\,dn_i, \qquad (1.330)$$

and the internal energy is

$$U = TS + \gamma^\nu A^\nu + \gamma^\omega A^\omega + fL + \sum_i \mu_i n_i. \qquad (1.331)$$

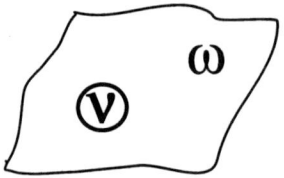

Figure 1.6. A two-dimensional two-phase system.

As in the three-dimensional bulk phase interface, there is a transition zone between the two phases where the properties are not uniform. We can introduce a dividing line as well as a line of tension, but we encounter the familiar problem of finding the position of the dividing line in the transition zone. To write the differential of energy in a simple form (1.331), we must place the line at the line of tension (Rusanov, 1967).

Now we can introduce the Helmholtz free energy of a two-dimensional two-phase system:

$$F = U - TS = \gamma^v A^v + \gamma^\omega A^\omega + fL + \sum_i \mu_i n_i. \tag{1.332}$$

Its differential is equal to

$$dF = -S\,dT + \gamma^v\,dA^v + \gamma^\omega\,dA^\omega + f\,dL + \sum_i \mu_i\,dn_i. \tag{1.333}$$

This thermodynamic potential makes it possible to find the relationship between surface tensions γ^v and γ^ω. Suppose that the two-dimensional system is closed, so that temperature T and total area of the system $A = A^v + A^\omega$ are constant ($dT = 0$, $dA = 0$). The equilibrium in a two-dimensional system is determined by the minimum Helmholtz free energy:

$$dF = \gamma^v\,dA^v + \gamma^\omega\,dA^\omega + f\,dL. \tag{1.334}$$

It follows that

$$\gamma^\omega - \gamma^v = f\frac{dL}{dA}. \tag{1.335}$$

The term dL/dA^v is the curvature of the dividing line. If the phase v is a circle with radius r then $dL = 2\Pi\,dr$ and $dA = 2\Pi r\,dr$, and we find the two-dimensional analog of the Laplace formula:

$$\gamma^\omega - \gamma^v = \frac{f}{r}. \tag{1.336}$$

Note that the surface tension in the external phase is higher than in the internal phase.

The Helmholtz free energy is useful for calculating surface tensions. For other purposes the modified Gibbs free energy, defined by equation (1.305), is appropriate:

$$\tilde{G} = F - \gamma^{\omega} A = (\gamma^{\nu} - \gamma^{\omega})A^{\nu} + fL + \sum_i \mu_i n_i. \tag{1.337}$$

Its differential is

$$d\tilde{G} = -S\,dT + (\gamma^{\nu} - \gamma^{\omega})\,dA^{\nu} - A\,d\gamma^{\omega} + f\,dL + \sum_i \mu_i\,dn_i. \tag{1.338}$$

This version of the Gibbs free energy describes nucleus formation in two-dimensional phases, which usually occurs at constant surface tension of the external phase γ^{ω}. The calculations are analogous to the three-dimensional case.

The work of formation of a two-dimensional nucleus is equal to

$$W = \Delta\tilde{G} = (\gamma^{\nu} - \gamma^{\omega})A^{\nu} + fL + \sum_i [\mu_i^{\nu}(\gamma^{\nu}) - \mu_i^{\omega}(\gamma^{\omega})]n_i^{\nu}. \tag{1.339}$$

The chemical potentials of components depend on surface tension in a given phase. In this derivation we assumed that the tension of the external phase γ^{ω} was constant, but no restrictions were imposed on tension inside the nucleus γ^{ν}. If mechanical equilibrium is maintained at the border of two phases, then γ^{ν} changes with nucleus radius.

When the nucleus reaches the critical size, the chemical potentials inside and outside become equal and the third term in the formula (1.339) disappears. Then it follows from geometrical considerations that

$$W_{cr} = \tfrac{1}{2}fL. \tag{1.340}$$

This is the analog of the Gibbs equation for the work of formation of a critical nucleus in a two-dimensional phase. Following the calculations for the three-dimensional case we find the radius of a critical nucleus:

$$r_{cr} = \frac{f\bar{a}^{\nu}}{\sum_i [\mu_i^{\omega}(\gamma^{\omega}) - \mu_i^{\nu}(\gamma^{\omega})]x_i^{\nu}}. \tag{1.341}$$

The symbol $\bar{a}^{\nu}$ designates the average molar area occupied by the substances in phase ν. If chemical potentials in the formula (1.341) are calculated per particle rather than per mole, then $\bar{a}^{\nu}$ is an average area per one particle. To illustrate this result, we will consider an actual example of nucleus formation in a two-dimensional phase.

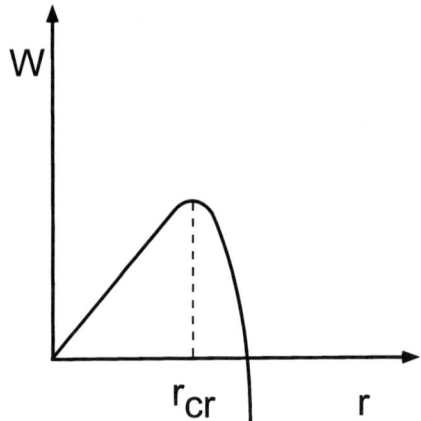

Figure 1.7. Work of pore formation in a membrane.

Example 8. Breakdown of Bilayer Lipid Membrane. A planar bilayer lipid membrane forming a barrier between two solutions is in a metastable state. Energetically it is more favorable for the lipid in the membrane to go to the torus at the edge of the aperture, but for this to occur a hole must first appear in the membrane. Expansion of this hole results in rupture of the membrane. The bilayer is metastable because hole formation is hindered by the energy barrier of nucleus formation.

Suppose a pore with radius r appears in a defined portion of the membrane with total area A which consists of the area of the pore A^p and the area of the lipid bilayer A^l. Because there is no lipid in the pore, its tension is zero: $\gamma^p = 0$. The tension and amount of lipid are constant, and the process occurs at constant temperature and pressure.

According to equation (1.339), the work of pore formation is

$$W = \gamma^l A^p + fL. \tag{1.342}$$

The dependence of this work on pore radius is presented in Fig. 1.7.

The critical radius cannot be calculated from (1.341) but using the condition of mechanical equilibrium (1.336) we find

$$r_{cr} = \frac{f}{\gamma^l}. \tag{1.343}$$

Substituting this formula into (1.342) gives the final expression for the work of formation of critical pore:

$$W_{cr} = \frac{\pi f^2}{\gamma^l}. \tag{1.344}$$

An important feature of pore formation distinguishes it from nucleus formation in saturated vapor. At the boundary of a subcritical pore there is no condition of mechanical equilibrium so that linear tension at the edge of the pore is counteracted by membrane tension. At small radii the linear tension prevails and the system is not at equilibrium. To produce equilibrium an additional field is required to balance the forces, such as a two-dimensional pressure p_x. The work of this external field would provide the work of pore formation.

The equilibrium of these three forces is expressed as

$$p_x = \frac{f}{r} - \gamma^l, \tag{1.345}$$

and hence the work of pore formation is

$$W = \int_0^{A^p} p_x \, dA^p = \int_0^r \left(\frac{f}{r} - \gamma^l\right) 2\pi r \, dr = 2\pi r f - \pi r^2 \gamma^l = fL - \gamma^l A^p. \tag{1.346}$$

As expected, this is identical to the result obtained directly from thermodynamic potential $\tilde{G}$.

1.8. IRREVERSIBLE PROCESSES

The thermodynamics of irreversible processes involves changes of state in systems that are not at equilibrium. Under these conditions, gradients of T, P, and μ lead to irreversible transport of mass, heat, and electrical charge within the system. The factors causing the irreversible processes are called forces X_i and the rates of these processes are called fluxes J_i. The relationship between fluxes and forces at small displacements from equilibrium are defined by phenomenological laws which have linear characteristics. This relationship is referred to as the thermodynamics of linear irreversible processes.

This aspect of thermodynamics is based on the concept of local equilibrium. It is assumed that small localized volumes of a system come to internal equilibrium much faster than overall equilibrium is established. The local equilibrium within the limited volumes is characterized by thermodynamic parameters that continuously change along with the system. The state of local equilibrium at any moment is determined by the basic equation of thermodynamics for equilibrium processes:

$$T \, dS = dU + P \, dV - \sum_i \mu_i \, dn_i. \tag{1.347}$$

The extensive variables S, U, V, and n are often defined per unit volume or unit mass.

The concept of local equilibrium is an approximation that reflects reality only if the gradients of thermodynamic parameters are not too large. The main assumption of the thermodynamics of linear irreversible processes is that the flux J_i can be caused by any force and the relationship between them is linear:

$$J_i = \sum_k L_{ik} X_k. \tag{1.348}$$

This relationship is determined by phenomenological (or kinetic) coefficients L_{ik}. Note that the diagonal term L_{ii} which defines the relationship between similar forces and fluxes such as diffusion, heat conduction, and electrical conduction, differs from non-diagonal (cross) terms describing superimposed phenomena like thermodiffusion, electrophoresis and others.

The phenomenological coefficients satisfy Onsager's relationship of reciprocity: with appropriate choice of fluxes J_i and forces X_i the matrix $\mathbf{L}_{ik}$ is symmetric:

$$\mathbf{L}_{ik} = \mathbf{L}_{ki}. \tag{1.349}$$

As fluxes occur the system approaches equilibrium and entropy increases at every point in the system. The rate of generation of entropy per unit volume is given by the equation

$$\sigma = \frac{1}{T} \sum_i J_i X_i. \tag{1.350}$$

The dimension of σ is $J\,K^{-1}\,m^{-3}\,s^{-1} = N K^{-1}\,m^{-2}\,s^{-1} = Pa\,K^{-1}\,s^{-1}$. The quantity σT characterizes local dissipation of energy.

Equation (1.350) defines the choice of forces and fluxes. This can be illustrated with heat flux according to Fourier's law:

$$J_1 = -\kappa\,\mathrm{grad}\,T. \tag{1.351}$$

If heat is transferred from a region with temperature T to a region with temperature $T + \Delta T$, the change of entropy is

$$\frac{J_1}{T + \Delta T} - \frac{J_1}{T} = -\frac{J_1 \Delta T}{T^2}. \tag{1.352}$$

If the distance between these parts is Δl, the entropy increase per unit volume is equal to

$$\sigma = -\frac{J_1}{T^2}\left(\frac{dT}{dl}\right) \tag{1.353}$$

or in the vector form:

$$\sigma = -\kappa\left(J_1 \cdot \frac{\mathrm{grad}\,T}{T}\right). \tag{1.354}$$

Thus the force linked to the flux J_1 is

$$J_1 = -\frac{1}{T}\,\mathrm{grad}\ T. \qquad (1.355)$$

Comparison with equation (1.348) shows that the phenomenological coefficient L_{11} is

$$L_{11} = \kappa T. \qquad (1.356)$$

The special case of nonequilibrium is a steady state. As in true equilibrium, the thermodynamic functions of the steady state do not change. But while the equilibrium state can be defined by a minimum of one thermodynamic potential, the steady state is determined by the principle of minimum entropy production (Defay *et al.*, 1966). This was formulated by I. Prigogine in the following way: if the system is in a steady state and an irreversible process occurs in this system, then the rate of entropy production in the whole system has a minimum value when a given external condition prevents the system from reaching equilibrium (Bazarov, 1964). The same law was formulated by Onsager as a principle of minimum dissipation of energy.

The formulas presented above constitute the mathematical foundation of the thermodynamics of irreversible processes. Their practical application depends on particular circumstances, and we will give a few examples.

Examples of Forces and Fluxes. Thermodynamic driving forces are proportional to gradients of temperature and pressure, and the electrochemical potential is the sum of concentration gradients and electrical potential. Therefore we have four main driving forces, which are coupled to fluxes of heat, mass, concentrations, and electrical charge. These cross-effects are listed in Table 1.1.

The first row of the table lists the effects which can be caused by a temperature gradient. In the diagonal box 11 we find heat conduction—the flux of heat caused by a temperature difference. Box 12 represents the thermomechanical effect of a pressure differential between two connected liquid volumes at different temperatures. If the vessels are separated by a porous partition, this effect is called thermo-osmosis. Box 13 represents thermodiffusion, or Soret's effect, which is the generation of diffusional flux caused by a temperature difference. Box 14 represents Seebeck's effect, the generation of an electromotive force in a circuit containing two different conductors with their junctions at different temperatures.

The second row presents fluxes caused by a pressure gradient. Box 21 represents the mechanocaloric effect which is the generation of a temperature difference between two connected vessels maintained at different pressures. Box 22 represents the familiar flow of a substance caused by a pressure difference, and box 23 is reverse osmosis which is the generation of a concentration difference by pressure when solutions are filtered through a microporous membrane.

Box 24 represents electrical charge transfer caused by a pressure difference. For

Because $\sigma = dS/dt$, the primary thermodynamic equation describing irreversible processes (1.350) can be presented as

$$\sigma = \frac{1}{T} JA. \tag{1.360}$$

Therefore the affinity A can be a driving force while the quantity J is the flux of matter through the chemical reaction. Note that at equilibrium the affinity of any chemical reaction is zero.

In conclusion, Table 1.1 shows that any force can cause any flux. However, this statement is valid only for quantities having identical tensor dimensions. The cross-term for quantities with different tensor dimensions is zero. The space fluxes are vectors, while the rate of a reaction is a scalar. Therefore space gradients of temperature, pressure, concentration. and potential have no *direct* influence on the rates of chemical reactions.

2

THE MEASUREMENT OF INTERFACIAL TENSION

In this chapter we will discuss interfacial phenomena between two immiscible liquids, and the air–liquid interface in the presence of surface-active compounds. We use the term *interface* for the boundary region which has a characteristic thickness and chemical, physical, and biological properties differing from those of bulk phases. We are interested in interfacial phenomena between two immiscible liquids or the air–liquid interface in the presence of surface-active compounds.

Interfacial tension is an important parameter in surface science, electrochemistry, and colloid chemistry. A liquid droplet has a tendency to contract as if it is surrounded by a hypothetical stretched "skin." If the drop of liquid is uninfluenced by external forces, such as gravity, it will adopt a truly spherical shape. The interfacial tension, γ, has units expressed in $N\,m^{-1}$ or $J\,m^{-2}$ (SI). Young first explained surface tension as resulting from the attraction between neighboring molecules (Young, 1805). These forces are isotropic in the interior of the bulk phase, but molecules at the interface are attracted less completely than in the bulk. The tendency of the surface area to decrease spontaneously results from the vectorial force acting on the molecules at the interface (Fig. 2.1). Because the free energy of a system tends to a minimum the surface of a pure phase or the interface between two immiscible liquids will always tend to contract spontaneously.

The interfacial and surface tensions for a variety of liquid interfaces are shown in Table 2.1. The interfacial tension is always positive (Landau and Lifshitz, 1958). If $\gamma < 0$ the interface between two immiscible liquids would tend to increase without limit, and the phases would mix and cease to exist separately. If $\gamma > 0$ the interface tends to become as small as possible for a given volume of two phases.

The interfacial tension can be altered by adding a surface-active solute. The molecules of a surfactant tend to concentrate at the interface rather than in the bulk phase and typically form a monomolecular layer.

Gibbs published his classic treatise about adsorption and capillarity in 1875, but his pioneering work was only appreciated years later, and experimental applications

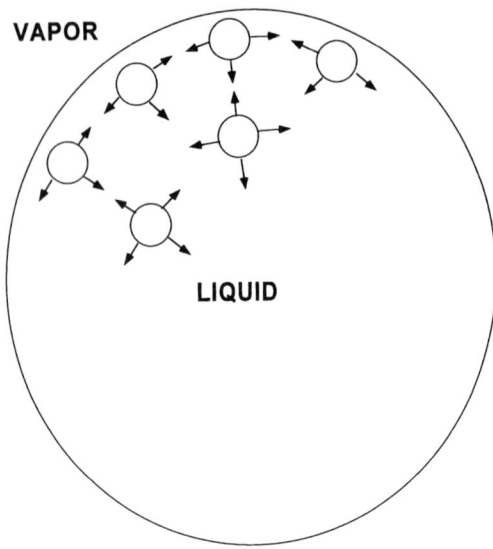

Figure 2.1. Attractive forces balance at the surface and in a liquid bulk.

of the Gibbs equation of adsorption became important only in the beginning of this century (Gibbs, 1928). The Gibbs equation is expressed by

$$\Gamma = -(c/RT)\, d\gamma/dc \qquad (2.1)$$

where Γ represents the excess number of adsorbed particles in a portion of solution containing a unit area in comparison with some homogeneous solution having exactly the same volume. Γ can be understood as the surface excess concentration of the surface-active solute (Abramson, 1981). It is assumed that the heterogeneous region at the interface is limited by a monolayer of the adsorbate, while the Gibbs plane separates this monolayer from the homogeneous solution.

The term $d\gamma/dc$ has a physical meaning of surface activity. If the interfacial tension γ does not change with changing concentration, $d\gamma/dc = 0$, $\Gamma = 0$ and such compounds are not surface active. But if interfacial tension decreases with increasing concentration of solute, $d\gamma/dc < 0$, $\Gamma > 0$ and this compound will be surface active. This is positive adsorption. If γ increases with increasing concentration, $d\gamma/dc > 0$, $\Gamma < 0$, this is negative adsorption, which is rare in nature.

The principal methods for studying adsorption of surface-active compounds at the liquid–air or the liquid–liquid interfaces are based on measurement of the interfacial tension, and a variety of methods can be used. Some examples are listed below:

- drop weight (Gugeshashvili, 1974; Gugeshashvili *et al.*, 1988; Harkins and Brown, 1919; Popov, 1977; Young and Harkins, 1928)
- the Wilhelmy plate (Deamer and Volkov, 1996; Wilhelmy, 1863)
- sessile drop profile (Kingery and Humerick, 1953)

Table 2.1. Interfacial, $\gamma_{o/w}$, and surface, γ, tension for liquid interfaces

Compound	Formula	$\gamma_{o/w}$, mN·m^{-1}	γ, mN·m^{-1}	t, °C
Carbon tetrachloride	CCl$_4$	45.0	26.66	20
Chloroform	CHCl$_3$	32.8	27.13	20
Nitromethane	CH$_3$NO$_2$	9.66	36.82	20
Tetrachloroethylene	C$_2$Cl$_4$	47.48	31.74	20
Ethyl ether	C$_4$H$_{10}$O	10.7	17.10	20
Trimethylethylene	C$_5$H$_{10}$	36.69	17.26	20
Methyl propyl ketone	C$_5$H$_{10}$O	6.28	24.15	20
Isopentane	C$_5$H$_{12}$	49.64	13.72	20
Bromobenzene	C$_6$H$_5$Br	39.82	36.26	20
Chlorobenzene	C$_6$H$_5$Cl	37.41	33.08	20
Nitrobenzene	C$_6$H$_5$NO$_2$	25.66	43.38	20
		24.8	43.1	25
Benzene	C$_6$H$_6$	35.00	28.86	20
		34.71	28.4	25
Aniline	C$_6$H$_7$N	5.77	42.58	20
Cyclohexanol	C$_6$H$_{12}$O	3.92	34.23	20
Ethyl propyl ketone	C$_6$H$_{12}$O	13.58	25.39	20
Methyl butyl ketone	C$_6$H$_{12}$O	9.73	25.49	20
n-Hexane	C$_6$H$_{14}$	51.10	18.43	20
n-Heptyl alcohol	C$_7$H$_{16}$O	8.0	26.5	20
		7.7	26.9	25
Styrene	C$_8$H$_8$	35.48	32.14	20
o-Xylene	C$_8$H$_{10}$	36.06	29.89	20
p-Xylene	C$_8$H$_{10}$	37.77	28.33	20
n-Octane	C$_8$H$_{18}$	50.81	21.77	20
n-Octyl alcohol	C$_8$H$_{18}$O	8.52	27.53	20
p-Cymene	C$_{10}$H$_{14}$	34.61	28.09	20
Water	H$_2$O		74.23	10
			72.8	20
			71.99	25
			67.94	50
			63.57	75
			58.91	100

- pendant drop profile (Kakiuchi *et al.*, 1988a,b; Popov *et al.*, 1983; Smith, 1944)
- capillary rise method (Sugden, 1921)
- Du Noüy Ring method (Davies and Rideal, 1963; Du Noüy, 1919)
- maximum bubble pressure (Adamson, 1990; Padday, 1969)
- the rotating drop method (Princen *et al.*, 1967)
- the touching drops method (Krotov *et al.*, 1995; Rusanov and Prokhorov, 1994)
- the pendant drop and a meniscus in a capillary tube (Evans, 1974)
- maximum pull on a cylinder (Jaycock and Parfitt, 1981)
- maximum pull on a cone (Rusanov and Prokhorov, 1994).

Methods for measuring the dynamic surface tension at a freshly formed or partially aged surface can be found in Joos and Vanuffelen (1995) and Adamson (1990).

Table 2.2. Suitability of methods of measuring of surface tension of pure liquids and solutions

Method	Pure liquids	Solutions
Drop weight or drop volume	Very satisfactory	Poor when ageing effects or slow adsorption suspected
Wilhelmy plate	Very quick, very accurate and convenient	Provides accurate results when ageing occurs
Sessile drop profile	Very satisfactory	Suitable for studying surface ageing
Pendant drop profile	Very satisfactory	Suitable for studying surface ageing
Capillary height	Very satisfactory	Difficult if contact angle is not 0°
Maximum pull on a cylinder	Very satisfactory	Satisfactory
Maximum pull on a cone	Very satisfactory	Satisfactory
Du Noüy ring	Satisfactory	Non suitable
Drops touching	Very satisfactory	Satisfactory
Maximum bubble pressure	Useful if other methods are difficult to use	Poor when ageing effects suspected

Harkins first compared different methods of surface tension measurements (Harkins, 1952), and some comments are summarized in Table 2.2. We will discuss two of the most precise methods of interfacial or surface tension measurements at liquid interfaces.

Drop Weight Method. The interfacial tension at the oil–water interface is usually measured by determining the weight of a drop which falls from the end of a capillary tube. This method for measuring interfacial tension between two immiscible liquids is based on the assumption that the drop weight of liquids, corrected for the weight of another displaced solution, is a good approximation of the force F retaining the *ideal* drop:

$$F \approx V(d_1 - d_2)g \qquad (2.2)$$

where V is the volume of the drop, d_1 and d_2 are the densities of the immiscible liquids, and g is the acceleration due to gravity, 9.807 m s^{-2}.

According to the Tate's law (Tate, 1864) this force acting on the drop is also equal to

$$F = 2\pi r\gamma \qquad (2.3)$$

where γ is interfacial tension and r is the radius of the tip, which is taken as the radius of the outside wall when the drop covers the bottom of the tip, or as the radius of the inside wall when the liquid exudes without wetting the bottom of the orifice. Solving the two equations gives

$$\gamma \approx V(d_1 - d_2)g/2\pi r. \qquad (2.4)$$

As a matter of fact, the real drop which falls is not an ideal sphere and its weight is slightly smaller than predicted by equation (2.4) since the liquid drop breaks apart some distance below the edge and the drop does not totally leave the tip. Two other problems of equation (2.4) arise from the fact that the boundary tension forces are not generally vertical and there is a pressure difference across the curved interface. Lohnstein, Harkins and Brown estimated the degree of approximation of the equation (2.4) (the deviation of the absolute value of γ may reach 35%) and proposed a method for applying the appropriate corrections (Harkins and Brown, 1919; Lohnstein, 1906, 1908, 1913). The relationship between γ and drop weight can be expressed by the equation:

$$\gamma = V(d_1 - d_2)gf/r \qquad (2.5)$$

where f is a correction factor from the Harkins–Brown table that is a function of $r/V^{1/3}$ and takes into account the deviation of the drop shape from an ideal sphere. Lando and Oakley made a regression analysis of Harkins and Brown drop-weight data and developed an analytical expression for correction factors employed in determining boundary tensions by the drop weight method (Lando and Oakley, 1967). Lando and Oakley provided tabulated correction factors at intervals small enough to minimize the need for interpolation (see Table 2.3). Recently, error analysis of the drop weight method was carried out (Earnshaw *et al.*, 1996) and recommended values of the corrected function, which are very close to the Lando and Oakley data, are plotted in Fig. 2.2.

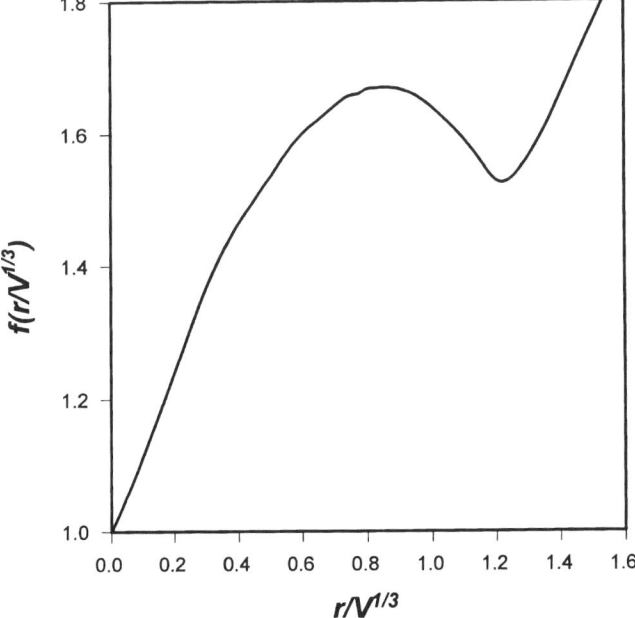

Figure 2.2. Dependence of a correction factor for the drop-weight method. Data are taken from Earnshaw *et al.*, (1996).

Table 2.3. Tabulated correction factors for the drop-weight-volume determination of surface and interfacial tensions

$r/V^{1/3}$	0	1	2	3	4	5	6	7	8	9
0.30	0.2166	2168	2169	2171	2173	2175	2176	2178	2180	2182
0.31	2183	2185	2187	2189	2191	2192	2194	2196	2197	2199
0.32	2201	2203	2204	2206	2208	2210	2211	2213	2215	2216
0.33	2218	2220	2221	2223	2225	2226	2228	2230	2231	2231
0.34	2235	2236	2238	2240	2241	2243	2245	2246	2248	2250
0.35	2251	2253	2255	2256	2258	2259	2261	2263	2264	2266
0.36	2267	2269	2271	2272	2274	2275	2277	2278	2280	2282
0.37	2283	2285	2286	2288	2289	2291	2292	2294	2296	2297
0.38	2299	2300	2302	2303	2305	2306	2308	2309	2311	2312
0.39	2314	2315	2317	2318	2320	2321	2323	2324	2326	2327
0.40	2328	2330	2331	2333	2334	2336	2337	2339	2340	2342
0.41	2343	2344	2346	2347	2349	2350	2351	2353	2354	2356
0.42	2357	2358	2360	2361	2363	2364	2365	2367	2368	2369
0.43	2371	2372	2374	2375	2376	2378	2379	2380	2382	2383
0.44	2384	2386	2387	2388	2390	2391	2392	2394	2395	2396
0.45	2397	2399	2400	2401	2403	2404	2405	2406	2408	2409
0.46	2410	2411	2413	2414	2415	2416	2418	2419	2420	2421
0.47	2423	2424	2425	2426	2428	2429	2430	2431	2432	2434
0.48	2435	2436	2437	2438	2440	2441	2442	2443	2444	2445
0.49	2447	2448	2449	2450	2451	2452	2454	2455	2456	2457
0.50	2458	2459	2460	2461	2463	2464	2465	2466	2467	2468
0.51	2469	2470	2471	2472	2474	2475	2476	2477	2478	2479
0.52	2480	2481	2482	2483	2484	2485	2486	2487	2488	2489
0.53	2490	2491	2492	2494	2495	2496	2497	2498	2499	2500
0.54	2501	2501	2502	2503	2504	2505	2506	2507	2508	2509
0.55	2510	2511	2512	2513	2514	2515	2516	2517	2518	2519
0.56	2520	2521	2522	2523	2524	2524	2525	2526	2527	2528
0.57	2529	2530	2531	2532	2533	2533	2534	2535	2536	2537
0.58	2538	2539	2540	2540	2541	2542	2543	2544	2545	2545
0.59	2546	2547	2548	2549	2550	2550	2551	2552	2553	2554
0.60	2554	2555	2556	2557	2558	2558	2559	2560	2561	2561
0.61	2562	2563	2564	2564	2565	2566	2567	2567	2568	2569
0.62	2570	2670	2571	2572	2573	2573	2574	2575	2575	2576
0.63	2577	2578	2578	2579	2580	2580	2581	2582	2582	2583
0.64	2584	2584	2585	2586	2586	2587	2588	2588	2589	2590
0.65	2590	2591	2591	2592	2593	2493	2594	2594	2595	2596
0.66	2596	2597	2597	2598	2599	2599	2600	2600	2601	2602
0.67	2602	2603	2603	2604	2604	2605	2605	2606	2607	2607
0.68	2608	2608	2609	2609	2610	2610	2611	2611	2612	2612
0.69	2613	2613	2614	2514	2615	2615	2616	2616	2617	2617
0.70	2618	2618	2618	2619	2619	2620	2620	2621	2621	2622
0.71	2622	2622	2623	2623	2624	2624	2625	2625	2625	2626
0.72	2626	2627	2627	2627	2628	2628	2630	2630	2630	2631
0.73	2630	2530	2631	2631	2631	2632	2632	2633	2633	2633
0.74	2634	2534	2634	2635	2635	2635	2635	2636	2636	2636
0.75	2637	2637	2637	2638	2638	2638	2638	2639	2639	2639
0.76	2640	2640	2640	2640	2641	2641	2641	2641	2642	2642
0.77	2642	2642	2642	2643	2643	2643	2643	2644	2644	2644
0.78	2644	2644	2645	2645	2645	2645	2645	2646	2646	2646
0.79	2646	2646	2646	2647	2647	2647	2647	2647	2647	2647
0.80	2648	2648	2648	2648	2648	2648	2648	2648	2648	2649
0.81	2649	2649	2649	2649	2649	2649	2649	2649	2649	2649
0.82	2650	2650	2650	2650	2650	2650	2650	2650	2650	2650

$r/V^{1/3}$	0	1	2	3	4	5	6	7	8	9
0.83	2650	2650	2650	2650	2650	2650	2650	2650	2650	2650
0.84	2650	2650	2650	2650	2650	2650	2650	2650	2650	2650
0.85	2650	2650	2650	2650	2650	2650	2650	2650	2650	2650
0.86	2650	2650	2649	2649	2649	2649	2649	2649	2649	2649
0.87	2649	2649	2649	2648	2648	2648	2648	2648	2648	2648
0.88	2648	2647	2647	2647	2647	2647	2647	2647	2647	2546
0.89	2646	2646	2646	2646	2645	2645	2645	2645	2645	2644
0.90	2644	2644	2644	2644	2643	2643	2643	2643	2643	2642
0.91	2642	2642	2642	2641	2641	2641	2641	2640	2640	2640
0.92	2640	2639	2639	2639	2639	2638	2638	2638	2637	2637
0.93	2637	2637	2636	2636	2636	2635	2635	2635	2634	2634
0.94	2634	2633	2633	2633	2632	2632	2632	2631	2631	2631
0.95	2630	2630	2629	2629	2629	2628	2628	2628	2627	2627
0.96	2626	2626	2626	2625	2625	2624	2624	2624	2623	2623
0.97	2622	2622	2621	2621	2621	2620	2620	2619	2619	2618
0.98	2618	2617	2617	2616	2616	2615	2615	2614	2614	2613
0.99	2613	2612	2612	2611	2611	2610	2610	2609	2609	2608
1.00	2608	2607	2607	2606	2606	2605	2605	2604	2604	2603
1.01	2602	2602	2601	2601	2600	2600	2599	2598	2598	2597
1.02	2597	2596	2595	2595	2594	2594	2593	2592	2592	2591
1.03	2590	2590	2589	2589	2588	2587	2587	2586	2585	2585
1.04	2584	2583	2583	2582	2581	2581	2580	2579	2579	2578
1.05	2577	2576	2576	2575	2574	2574	2573	2572	2572	2571
1.06	2570	2569	2569	2568	2567	2566	2566	2565	2564	2563
1.07	2563	2562	2561	2560	2560	2559	2558	2557	2556	2556
1.08	2555	2554	2553	2552	2552	2551	2550	2549	2548	2547
1.09	2547	2546	2545	2544	2543	2542	2542	2541	2540	2539
1.10	2538	2537	2536	2536	2535	2534	2533	2532	2531	2530
1.11	2529	2529	2528	2527	2526	2525	2524	2523	2522	2521
1.12	2520	2519	2518	2518	2517	2516	2515	2514	2513	2512
1.13	2511	2510	2509	2508	2507	2506	2505	2504	2503	2502
1.14	2501	2500	2499	2498	2497	2496	2495	2494	2493	2492
1.15	2491	2490	2489	2488	2487	2486	2485	2484	2483	2482
1.16	2480	2479	2478	2477	2476	2475	2474	2473	2472	2471
1.17	2470	2469	2468	2466	2465	2464	2463	2462	2461	2460
1.18	2459	2457	2456	2455	2454	2453	2452	2451	2449	2448
1.19	2447	2446	2445	2444	2442	2441	2440	2439	2438	2437
1.20	2435	2434								

The precision of this method is 0.01 mN m^{-1} if measurements are made in a thermostat. The remarkable accuracy of this method arises from the fact that the surface excess Γ is proportional to the surface activity $d\gamma/dc$ and the relatively small changes of γ due to changes of the concentration of the surface-active compound are more important than the absolute values of the decreasing interfacial tension with increasing concentration.

Two important details of the drop-weight method should be noted. First, it is essential to purify all of the solvents and solutes, including water, to remove all possible impurities which could reduce interfacial tension. The apparatus must also be assembled so that the influence of vibration is minimized. Otherwise the drops will detach too soon and erroneous values of interfacial tension will be obtained.

The equipment for interfacial tension measurements consists of a micrometer, syringe, capillary tube, and container. The drops are delivered to the end of the

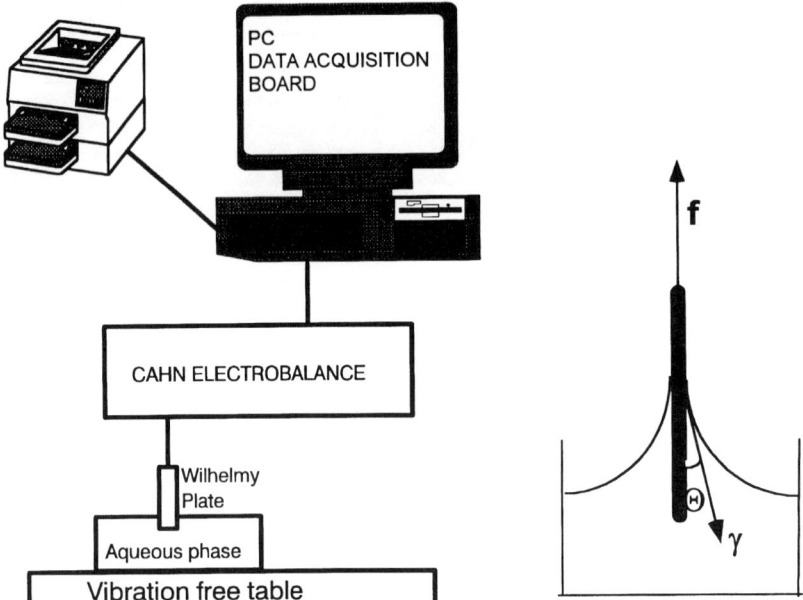

Fig. 2.3. Apparatus for measuring the time dependence of surface or interfacial tension by the Wilhelmy plate method. An IBM-compatible microcomputer (PC) with a multi I/O plug-in data acquisition board DAS-8pga (*Keithley MetraByte*) was interfaced with the electrobalance and used to record the digital data. The DAS-8pga features an 8-channel 12-bit successive approximation analog-to-digital converter with sample and hold.

capillary tube from a syringe, whose plunger is operated by the micrometer. The vertical capillary tube must be joined tightly to the syringe and the tip of the tube should be made on a watchmaker's lathe so that the end is sharp, regular, free from any nicks, and perpendicular to the tube, which can have vertical or U-shape. The choice between these shapes depends on wetting. Experimentally the tip from which the drops are formed must be either completely wetted or completely non-wetted by the liquid, otherwise the liquid drop will form with an uncertain radius. On the other hand, the interior of the tube must not be too hydrophobic or the drops will break off inside the tube instead of at the end. For this reason, Teflon cannot be used.

To a calibrate the syringe connected to the micrometer, drops of distilled water are permitted to fall into a container which is weighed before and after a certain number of drops has fallen. The volume of liquid per unit scale of a micrometer can then be estimated.

The drops should form no faster than one every two minutes for pure liquids and one every four minutes for three-component systems. If the drops do not form slowly, they will detach too soon and the measured interfacial tension will be in error. Slow formation of the drop is especially important just before the detachment.

The Wilhelmy Plate Method. Wilhelmy showed that the surface tension of a liquid can be found by measuring the maximum force required to pull a very thin plate vertically through the surface of liquid (Fig. 2.3). If the contact angle Θ is zero, the weight of the Wilhelmy plate will be equal to the sum:

$$W_{\text{total}} = W_{\text{plate}} + 2\gamma l \tag{2.6}$$

where $2l$ is the perimeter of the plate.

If the Wilhelmy plate is completely wetted by one liquid it is possible to use this method for measuring interfacial tension between two immiscible liquids.

This method is very simple to use in the study of chemical or photochemical reactions at gas–water or oil–water interfaces, or of surface adsorption of amphiphilic compounds or monolayers.

3

ADSORPTION AT LIQUID INTERFACES

Components of any solution behave differently at an interface than they do within the bulk solution. One example is that their concentration at the interface is different than within the bulk solution. This difference, called adsorption at liquid interfaces, manifests itself by a change in the capillary properties of the interface. A general theory of capillarity was developed more than a century ago by Josiah Willard Gibbs and presented in his classic treatise "*Equilibrium of Heterogeneous Substances*" (Gibbs, 1928). The thermodynamic theory of capillarity and electrocapillarity is designed to establish a general relationship between surface tension and other parameters of the system. Although considerable progress has been achieved in understanding the thermodynamics of surfaces, research in this field is still largely based on the Gibbs method. In his description of surface phenomena Gibbs used two fundamental ideas: the concept of a dividing surface and the concept of surface excesses.

In 1962 R. S. Hansen expanded upon the most fundamental aspects of Gibbs surface thermodynamics (Hansen, 1962). Like Gibbs, Hansen based his method on the concept of surface excesses, but used more sophisticated concepts of a dividing surface and a different reference system for the definition of these excesses. Hansen's concept of a dividing surface initially caused some confusion in the literature (Good, 1976, 1982, 1986; Markin and Volkov, 1988a; Motomura, 1977, 1980, 1986; Motomura and Aratono, 1987) but later proved to be very valuable for the theory of electrocapillarity (Markin and Volkov, 1989a, 1989b). To follow, we will present both Gibbs and Hansen approaches to the field of surface thermodynamics.

3.1. GIBBS METHOD FOR PLANAR INTERFACES

Let us consider two homogeneous liquid phases that are in contact, and contain $n_1, n_2,$ ..., n_m moles of 1, 2, ..., m components. Let us assume that the interface is a plane and its area is A. We shall designate the energy, entropy and volume of the whole

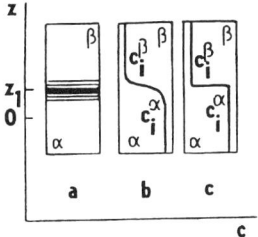

Figure 3.1. Diagram showing the distribution of substance i between the α and β phases (*a, b*) and Gibbs reference system (*c*).

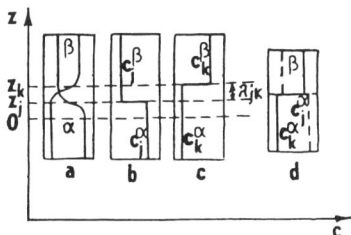

Figure 3.2. Distribution of components j and k between two immiscible liquids α and β phases in: (*a*) the real system; (*b, c*) Gibbs reference systems; (*d*) Hansen's reference system.

system as U, S, and V, respectively; the concentration of the i-th component in the bulk of the α and β phases by c_i^α and c_i^β, respectively; entropy per unit volume of the α and β phases by s^α and s^β; and surface tension by γ. With the planar boundary at equilibrium, the temperature T, pressure p, and chemical potentials μ_i are constant throughout the whole system. In the case of charged particles an electrochemical potential, $\bar{\mu}_i$, is maintained constant instead of a chemical potential.

Macroscopically, any interface appears clearly defined and precisely localized. However, at the molecular level, the interface is rather diffuse, such that the concentrations of the components and other properties of the system do not change abruptly at the interface boundary, but rather undergo gradual change (Fig. 3.1(*a*)). Therefore, at the molecular level the boundary between the two phases cannot be described as an infinitely thin mathematical surface. Instead, it is a finite transition layer with varying concentration which results in the development of surface tension.

In describing the properties of the interface, Gibbs rigorously defined the boundary between the two phases. He introduced the so-called *dividing surface*, which is a mathematical surface having no thickness. In Fig. 3.2 this surface is a plane shown by the dotted line. It lies parallel to the interfacial layer and can pass through any point, for example, through the point with coordinate z_j.

Next, this real system is replaced with an imaginary reference system in which both phases are presumed to be homogeneous everywhere, up to the dividing surface (Fig. 3.1(*c*)). The geometrical outline of the reference system coincides with the outline of the real system. If the surface divides the volume of the real system (and the reference system) into V^α and V^β parts, the overall entropy of the reference system will be $s^\alpha V^\alpha + s^\beta V^\beta$, and the amount of the substances $c_i^\alpha V^\alpha + c_i^\beta V^\beta$. These extensive characteristics are not necessarily equal to the entropy or to the amount of substances in the real system. The difference between them is

$$S^s = S - s^\alpha V^\alpha - s^\beta V^\beta, \tag{3.1}$$

$$n_i^s = n_i - c_i^\alpha V^\alpha - c_i^\beta V^\beta. \tag{3.2}$$

In order to make the extensive characteristics of the two systems identical, Gibbs attributed these excesses to the interface in the reference system, thereby converting them into extensive characteristics of the interface. They are called the surface excesses of the entropy and the i-th component. Dividing them by the area of the interface, A, we obtain the surface density of the excess of i-th component,

$$\Gamma_i = n_i^s/A, \tag{3.3}$$

known as the absolute adsorption of the i-th component; and the surface density of the excess in entropy

$$s^s = S^s/A. \tag{3.4}$$

These excesses depend on the position of the dividing surface. With proper choice of this surface, the adsorption of any component j can be made to be zero, $\Gamma_j = 0$. In this case, the values of other surface excesses at this specific position of the dividing surface will be referred to as the relative excesses or relative adsorptions and denoted by $\Gamma_{i(j)}$. In an analogous way we will refer to the relative surface density of the entropy excess $s_{(j)}^s$. Sometimes this particular choice of dividing surface is referred to by the convention $\Gamma_j = 0$. In the modern literature the relative surface excesses are sometimes designated $\Gamma_i^{(j)}$. However, Gibbs himself used the designation $\Gamma_{i(j)}$, which will be adopted in this book.

The Gibbs dividing surface can be chosen in such a way that adsorption of any given component, but only one component at a time, is zero. In Fig. 3.2, two Gibbs dividing surfaces are presented, one which corresponds to the convention $\Gamma_j = 0$, and the other to the convention $\Gamma_k = 0$. These surfaces do not coincide and one can find the distance λ_{jk} between these two surfaces by starting with an arbitrary dividing surface and measuring the coordinate z from this surface in the direction from α to β. Suppose that the concentration of component j in the α phase is greater, $c_j^\alpha > c_j^\beta$, and the surface excesses of components j and k at this surface are Γ_j and Γ_k.

If the dividing surface is displaced by a distance z, the surface excess of component j becomes $\Gamma_j - z(c_j^\alpha - c_j^\beta)$. It is easy to find that this excess is zero at the point $z_j = \Gamma_j/(c_j^\alpha - c_j^\beta)$. Accordingly, the dividing surface with zero adsorption of component k has the coordinate $z_k = \Gamma_k/(c_k^\alpha - c_k^\beta)$. The distance between these dividing surfaces, λ_{jk}, is given by the equation:

$$\lambda_{jk} = z_k - z_j = \Gamma_k/(c_k^\alpha - c_k^\beta) - \Gamma_j/(c_j^\alpha - c_j^\beta). \tag{3.5}$$

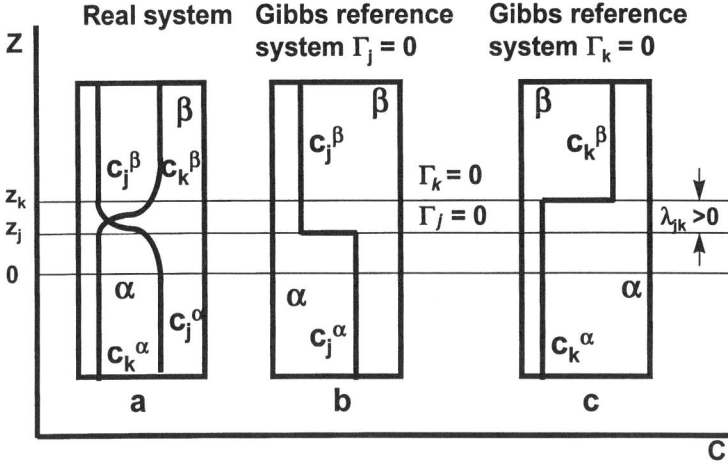

Figure 3.3. Distribution of components j and k between two immiscible liquid phases α and β when: (a) there is no overlapping of the masses of the components; (b, c) in Gibbs model.

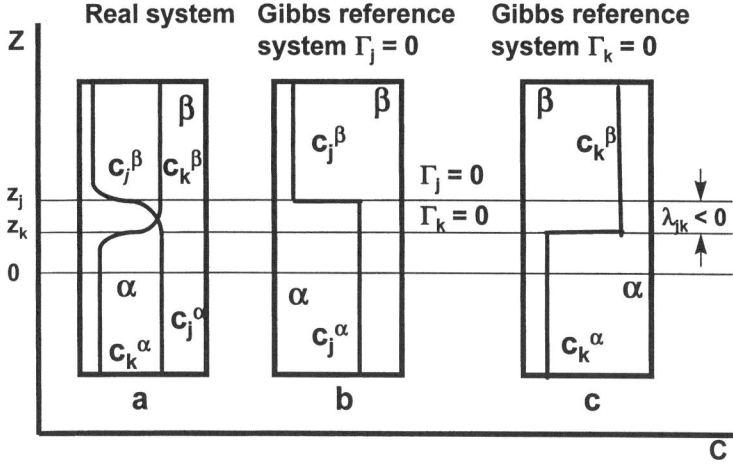

Figure 3.4. Distribution of components j and k between two immiscible liquid phases α and β in a real system with: (a) partial overlapping of the masses of the components; (b, c) Gibbs reference system.

In the case shown in Fig. 3.3 component j is concentrated mainly in the α phase, and component k in the β phase. At the position specified for the dividing surfaces with zero adsorption, the main bodies of substances j and k are separated from one to another, and the distance λ_{jk} is positive. If, as shown in Fig. 3.4, the bodies of substances j and k partially overlap, the distance between these planes is negative in accordance with the accepted definition.

If the initial dividing surface can be chosen such that the surface excess of one of the considered components becomes zero, the expression for the distance between these two surfaces simplifies:

$$\lambda_{jk} = z_k - z_j = \Gamma_{k(j)}/(c_k^\alpha - c_k^\beta) = -\Gamma_{j(k)}/(c_j^\alpha - c_j^\beta). \tag{3.6}$$

Very often either the α or β solvent is chosen as a reference substance. In this case the distance between the dividing surfaces with zero adsorption of the solvent is:

$$\lambda_{\alpha,\beta} = \frac{\Gamma_\beta}{c_\beta^\alpha - c_\beta^\beta} - \frac{\Gamma_\alpha}{c_\alpha^\alpha - c_\alpha^\beta} = \frac{\Gamma_{\beta(\alpha)}}{c_\beta^\alpha - c_\beta^\beta} = -\frac{\Gamma_{\alpha(\beta)}}{c_\alpha^\alpha - c_\alpha^\beta} \tag{3.7}$$

where c_α^α is the concentration of the α solvent in the α phase, c_α^β is the concentration of the α solvent in the β phase, etc.

If the solvents α and β are absolutely immiscible, i.e., $c_\beta^\alpha = c_\alpha^\beta = 0$, the equation for the distance between the planes with zero adsorption simplifies to:

$$\lambda_{\alpha,\beta} = -\frac{\Gamma_\beta}{c_\beta^\beta} - \frac{\Gamma_\alpha}{c_\alpha^\alpha} = -\frac{\Gamma_{\beta(\alpha)}}{c_\beta^\beta} = -\frac{\Gamma_{\alpha(\beta)}}{c_\alpha^\alpha}. \tag{3.8}$$

Gibbs demonstrated that surface excesses determine the surface tension between two phases. The energy of a two-phase system with a planar interface is usually represented as follows:

$$dU = T\,dS - P\,dV + \gamma\,dA + \sum_{i=1}^{m} \mu_i\,dn_i \tag{3.9}$$

Integrating this equation with the constant intensive parameters T, P, γ, and μ, we obtain

$$U = TS - PV + \gamma A + \sum_{i=1}^{m} \mu_i n_i. \tag{3.10}$$

Differentiating it and comparing the result with (3.9) we arrive at the following basic equation:

$$-S\,dT + V\,dP - A\,d\gamma - \sum_{i=1}^{m} n_i\,d\mu_i = 0 \tag{3.11}$$

which relates the surface tension γ to the intensive variables T, p, and μ_i. It is not very convenient that this equation contains the extensive parameters S, V, A, and n that characterize the total system including the bulk phases. It would be desirable to have extensive parameters that pertain to the surface only. To achieve this goal, one must use the Gibbs–Duhem equations for separate bulk phases:

$$-s^\alpha\,dT + dP - \sum_{i=1}^{m} c_i^\alpha\,d\mu_i = 0 \tag{3.12}$$

$$-s^\beta\,dT + dP - \sum_{i=1}^{m} c_i^\beta\,d\mu_i = 0. \tag{3.13}$$

Next we have to multiply equation (3.12) by V^α and equation (3.13) by V^β and then subtract them from equation (3.11). The sum of the volumes of the α and β phases in the reference system equals the volume of the real system:

$$V^\alpha + V^\beta = V. \tag{3.14}$$

Therefore, we can exclude the characteristics of the bulk phases:

$$A\,d\gamma = -S^s\,dT - \sum_{i=1}^{m} n_i^s\,d\mu_i. \tag{3.15}$$

Dividing by the area of the interface A, we obtain the Gibbs adsorption equation:

$$d\gamma = -s^s\,dT - \sum_{i=1}^{m} \Gamma_i\,d\mu_i. \tag{3.16}$$

If the dividing surface is specified by zero adsorption of component j, then the Gibbs adsorption equation takes the form:

$$d\gamma = -s_{(j)}^s\,dT - \sum_{i\neq 1} \Gamma_{i(j)}\,d\mu_i. \tag{3.17}$$

The j component is called a reference substance. Summation is performed over all substances except the reference substance and traditionally the reference substance j is underlined in the Gibbs adsorption equation. Furthermore, there is no need to exclude the term $\Gamma_{i(j)}\,d\mu_i$ when $i = j$ because this term automatically becomes zero, since $\Gamma_{j(j)} = 0$.

Accounting for Surface Torques. In the previous section the energy of a two-phase system with a planar interface was represented by equation (3.9). However, if one follows Gibbs approach consistently and considers any changes in the system's properties, including changes in shape of the interface, then additional terms should be introduced to the right side of equation (3.9), to describe the change in the curvature of the boundary. By designating the principal curvatures of the boundary as c_x and c_y, and the corresponding radii as R_x and R_y, so that $c_x = 1/R_x$ and $c_y = 1/R_y$ we can write the additional terms as $AC_x\,dc_x$ and $AC_y\,dc_y$. The coefficients C_x and C_y designate mechanical torques at the interface. Unfortunately, it is customary to designate the principal curvature by the symbol "c," which is also used to designate the concentration. To prevent confusion, we shall supplement the principal curvatures by the subscripts x and y.

In the case of a planar interface, parameters c_x and c_y are equal to zero, but it does not mean that their differentials are also equal to zero. Therefore, in the general case, the energy of a two-phase system with a planar interface should be expressed as:

$$dU = T\,dS - P\,dV + \gamma\,dA + AC_x\,dc_x + AC_y\,dc_y + \sum_{i=1}^{m} \mu_i\,dn_i. \tag{3.18}$$

If account is taken of the additional terms related to a change in the curvature and if equation (3.18) is considered a starting-point, the derivation of the Gibbs adsorption equation follows along the same lines as before. Curvatures c_x and c_y are intensive parameters, and therefore the integration of equation (3.18) with fixed intensive parameters yields the usual result:

$$U = TS - PV + \gamma A + \sum_{i=1}^{m} \mu_i n_i. \tag{3.19}$$

By differentiating this equation and comparing it with equation (3.18) we obtain

$$-S\,dT + V\,dP - A\,d\gamma + AC_x\,dc_x + AC_y\,dc_y - \sum_{i=1}^{m} n_i\,d\mu_i = 0. \tag{3.20}$$

If we exclude the bulk characteristics as before, then the Gibbs adsorption equation can be written as:

$$d\gamma = -s^s\,dT + C_x\,dc_x + C_y\,dc_y - \sum_{i=1}^{m} \Gamma_i\,d\mu_i, \tag{3.21}$$

or with relative excesses:

$$d\gamma = -s^s_{(j)}\,dT + C_x\,dc_x + C_y\,dc_y - \sum_{i \neq j} \Gamma_{i(j)}\,d\mu_i. \tag{3.22}$$

It should be noted that the torque moments C_x and C_y depend on the position of the dividing surface and they change if one shifts this surface. In the case of a spherical interface they simultaneously become zero only at one surface, which is known as *the surface of tension*. At an arbitrarily shaped interface the situation is more complicated because at the surface of tension the terms $C_x\,dc_x$ and $C_y\,dc_y$ do not individually become zero, but rather their sum equals zero. For details see Markin *et al.* (1988).

Equation (3.22) with total differentials makes it possible to determine all coefficients as the derivatives of surface tension with respect to corresponding variables. For example,

$$\Gamma_{i(j)} = -\left(\frac{\partial \gamma}{\partial \mu_i}\right)_{T, c_x, c_y, \mu_i \neq \mu_j}. \tag{3.23}$$

The subscripts at the derivative indicate which variables, besides those that are part of the derivative, are kept constant. The presence in the subscript of the principal curvatures, c_x and c_y indicates that the dividing surface is maintained planar. At the same time coefficients C_x and C_y are defined as derivatives of the surface tension with respect to curvatures:

$$C_x = -\left(\frac{\partial \gamma}{\partial c_x}\right)_{T, c_y, \mu_i \neq \mu_j} \quad ; \quad C_y = -\left(\frac{\partial \gamma}{\partial c_y}\right)_{T, c_x, \mu_i \neq \mu_j}. \tag{3.24}$$

Here the derivatives are taken for $c_x = c_y = 0$. Thus, all coefficients in equation (3.22) are thermodynamically significant values.

Later we shall return to the discussion of the role of these additional terms with curvature in the theory of capillarity.

An Alternative Geometric Interpretation of the Gibbs Method. The position of the Gibbs dividing surface was defined by the convention of zero-adsorption of a specified component of the system. The displacement of this surface from one position to another resulted in a change of surface excesses so that adsorption of another component can become zero. However, the adsorption convention is not the only way in which one can define the position of a dividing surface. Because of the presence of the mechanical force γ and the torques C_x and C_y at the interface, which depend on the position of the dividing surface, the convention can be based on the force criteria. For example, the convention $C_x = C_y = 0$ determines the position of *the surface of tension* in the case of a planar interface. This is a unique surface, which displays the surface tension without any mechanical moments.

These intricacies may not be very important at a planar interface but they become essential considerations at curved interfaces. We will expand upon these mechanical factors later.

Whatever criterion is chosen—adsorption or force criterion—the position of the dividing surface can be attributed to a specific place *in the real system* and the reference system can be almost a replica of its prototype. However, there is no need to have this one-to-one correspondence between the real and reference systems as will be presented in the following example.

Let us use a force criterion and select a surface of tension as a dividing surface (Fig. 3.1(*a*)). In building our standard Gibbs reference system (panel (*b*)) let us say that the absolute surface excesses will be Γ_i. In this case the complete adsorption equation has the form of equation (3.16). If we now exclude dP from equations (3.12) and (3.13), find the differential dμ_j, and substitute it into equation (3.16), we obtain the following expression for γ:

$$d\gamma = -\left[s^s - \Gamma_j \frac{s^\beta - s^\alpha}{c_j^\beta - c_j^\alpha}\right]dT - \sum_{i \neq j}\left[\Gamma_i - \Gamma_j \frac{c_i^\beta - c_i^\alpha}{c_j^\beta - c_j^\alpha}\right]d\mu_i. \qquad (3.25)$$

It is easy to see that the terms inside the parenthesis represent the density of the relative excess of entropy, $s_{(j)}^s$, and relative adsorption, $\Gamma_{i(j)}$:

$$s_{(j)}^s = s^s - \Gamma_j \frac{s^\beta - s^\alpha}{c_j^\beta - c_j^\alpha}; \qquad \Gamma_{i(j)} = \Gamma_i - \Gamma_j \frac{c_i^\beta - c_i^\alpha}{c_j^\beta - c_j^\alpha}. \qquad (3.26)$$

Thus, we obtain the complete adsorption equation with relative excesses $s_{(j)}^s$ and $\Gamma_{i(j)}$. Since we have not shifted the dividing surface and have only carried out algebraic transformations, the physical sense of the term "relative excess" becomes clearer.

Next we shall consider the use of this equation for some specific cases. Let us assume that the reference substance j is present in one phase only, for example, in the α phase. This means that $c_j^{\beta} = 0$. We shall assume that the i-th component is also present only in one phase, the α phase. In that case from equation (3.26) we obtain:

$$\Gamma_{i(j)} = \Gamma_i - \Gamma_j c_i^{\alpha}/c_j^{\alpha}. \tag{3.27}$$

If the i- and j-th components are in different phases (j in the α phase and i in the β phase):

$$\Gamma_{i(j)} = \Gamma_i + \Gamma_j c_i^{\beta}/c_j^{\alpha}, \tag{3.28}$$

and the equations for relative excesses of substances simplify. This is not true for the excess of entropy, because s^{α} and s^{β} cannot be equal to zero.

Now consider the geometrical interpretation of these transformations. Since the position of the dividing surface in a real system is fixed, the total volume V_0 is divided into parts V_0^{α} and V_0^{β}, so that $V_0^{\alpha} + V_0^{\beta} = V_0$. This dividing surface cannot be displaced and will be used as the basis for the reference system (panel (c) in Fig. 3.5). Further derivation employs the adsorption criterion $\Gamma_j = 0$, and thereby we define the volumes $V_{(j)}^{\alpha}$ and $V_{(j)}^{\beta}$ which are not equal to V_0^{α} and V_0^{β}, but still must meet the requirements of preservation of substance,

$$c_j^{\alpha} V_{(j)}^{\alpha} + c_j^{\beta} V_{(j)}^{\beta} = n_j, \tag{3.29}$$

and volume

$$V_{(j)}^{\alpha} + V_{(j)}^{\beta} = V_0. \tag{3.30}$$

Now if we try to superimpose the reference system (c) with $\Gamma_j = 0$ as in Fig. 3.5 with the real system (a), so the dividing surfaces coincide, we find that these systems are displaced relative to each another along the normal to this surface. Another reference system with adsorption criterion $\Gamma_k = 0$ will be displaced a different distance (panel (d)). Their displacement relative to each other equals the previously derived parameter λ_{jk}. In the case shown in Fig. 3.5, this parameter is positive, since the bodies of the j and k substances are separated from each other. However, as presented in Fig. 3.6, when the bodies of the j and k substances overlap somewhat, the displacements of the reference systems with $\Gamma_j = 0$ and $\Gamma_k = 0$ occur in opposite directions, and the λ_{jk} parameter is negative.

Thus, we have shown that when constructing a Gibbs reference system, two conventions can be applied: a force convention and an adsorption convention. The first determines the position of the dividing surface in a real system, and the second determines the relationship between the volumes of phases in a reference system or the displacement of the reference system relative to the real system. In the last case the two systems cannot be superimposed.

We believe that the geometrical interpretation described above corresponds more precisely to the essence of the Gibbs method. Unfortunately, modern papers and textbooks on the thermodynamics of surface phenomena do not discuss the necessity

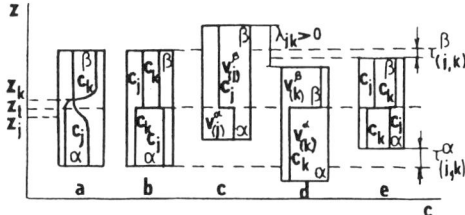

Figure 3.5. Distribution of components j and k between phases α and β in a real system without overlapping of the masses of the components and a geometrical interpretation of Gibbs and Hansen's methods: (a) real system; (b) primary Gibbs reference system with absolute excesses Γ_i; (c) Gibbs reference system with $\Gamma_j = 0$; (d) Gibbs reference system with $\Gamma_k = 0$; (e) Hansen's reference system with $\Gamma_j = \Gamma_k = 0$.

of simultaneous application of force and adsorption criteria in constructing a reference system. A rare exception are the publications by Good (1976, 1982, 1986) who stressed the importance of distinguishing a real system from a Gibbs reference system by defining the surface excesses.

3.2. HANSEN'S METHOD

In Gibbs method the volume of the system plays a special role and together with the pressure differential it disappears from the adsorption equation. It is assumed from the very beginning that the sum of volumes of homogeneous phases in the reference system equals to the volume of the real system (equation (3.14)). In textbooks of surface thermodynamics this assumption is considered to be so obvious that it is not even discussed. If follows from this assumption that the possibility of superimposing the reference system with the real system arises (if no use is made of an additional force criterion for choosing the dividing surface), since their outer contours are identical.

However, in the initial equation (3.11) all the differentials in it are equivalent. By using the Gibbs–Duhem equations (3.12) and (3.13) we can exclude any pair of differentials, not necessarily including the term with pressure. This is the approach that was used by Hansen (1962). According to his method, volume is not singled out from other extensive variables. From the geometric point of view this means that the volume of the reference system is not assumed to be equal to the volume of the real system. Therefore, together with the surface excesses of the substances and the entropy we can consider the surface excess of volume.

In his work, Hansen (1962) has limited himself to formal-algebraic transformations of equations and has not used a geometrical interpretation in his method. This limitation has hindered acceptance of his studies and even led to confusion in their interpretation (Good, 1986; Motomura, 1986). To follow, we shall give a detailed description of Hansen's reference system (Good, 1976, 1986).

Let us assume that an arbitrarily chosen dividing surface (for example, the surface of tension with coordinate (z_t) divides the volume of a real system, V_0, into V_0^{α} and

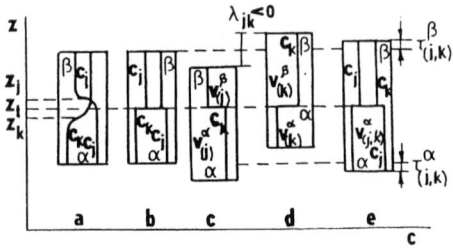

Figure 3.6. Distribution of components j and k between phases α and β in a real system with overlapping of the masses of the components and a geometrical interpretation of Gibbs and Hansen's methods: (a) real system; (b) primary Gibbs reference system with absolute excesses Γ_i; (c) Gibbs reference system with $\Gamma_j = 0$; (d) Gibbs reference system with $\Gamma_k = 0$; (e) Hansen's reference system with $\Gamma_j = \Gamma_k = 0$.

V_0^β (Fig. 3.5(a)). Based on this dividing surface, we construct a reference system with volumes V^α and V^β that are not supposed to sum to V_0 (Fig. 3.5(e)). This reference system defines the density of surface excesses of the entropy and of the substance:

$$s^s = (S - s^\alpha V^\alpha - s^\beta V^\beta)/A, \qquad \Gamma_i = (n_i - c_i^\alpha V^\alpha - c_i^\beta V^\beta)/A. \qquad (3.31)$$

At the same time it defines the surface excess of volume,

$$V^s = V_0 - V^\alpha - V^\beta, \qquad (3.32)$$

that can be translated into the surface density of the volume excess, v^s, which has a dimension of length τ^s:

$$\tau^s = v^s = V^s/A = (V_0 - V^\alpha - V^\beta)/A. \qquad (3.33)$$

Therefore the general adsorption equation can be presented in the form:

$$d\gamma = -s^s\,dT + \tau^s\,dP - \sum_{i=1}^{m} \Gamma_i\,d\mu_i. \qquad (3.34)$$

If we take into account possible changes in the curvature of the dividing surface, we should, in analogy with previous sections, include two additional terms which reflect the mechanical effects of torque. Due to the Gibbs–Duhem equations, any two differentials on the right-hand side of equation (3.34) linearly depend on the other variables in this equation. Any two differentials can be readily excluded from equation (3.34) by reducing the corresponding coefficients to zero. If we wish to exclude terms with the differentials $d\mu_j$ and $d\mu_k$, we should apply two conventions to the V^α and V^β phase volumes in the reference system:

$$n_j - c_j^\alpha V^\alpha - c_j^\beta V^\beta = 0,$$
$$n_k - c_k^\alpha V^\alpha - c_k^\beta V^\beta = 0. \qquad (3.35)$$

Since V^α and V^β are independent, the set of equations (3.35) is not overdefined and it can be solved. If these equations are satisfied, then

$$\Gamma_j = 0 \quad \text{and} \quad \Gamma_k = 0. \tag{3.36}$$

The conventions (3.35) determine a concrete Hansen's reference system, which we shall designate as (j, k). It is sometimes said that the symbol (j, k) determines the criterion for selecting a concrete Hansen's system. In this case, the j and k components are referred to as reference substances. The j and k indices can obviously be transposed; that is, the criterion (j, k) is equivalent to the criterion (k, j). In Figs. 3.5(e) and 3.6(e) Hansen's (j, k)-reference system is shown. It can be clearly seen that its volume differs from that of the real system.

The j or the k index can also stand for the entropy or the volume. The criterion which determines Hansen's reference system can then be expressed as (s, j), (v, k) or even (s, v). In Gibbs method, the reference system is defined by only a single index, (j); however, it is assumed that the surface volume excess equals zero. Therefore, Gibbs criterion (j) is equivalent to Hansen's criterion (v, j) or (j, v) and the Gibbs reference system represents only a special case in Hansen's reference system.

By analogy with Gibbs conventions, we shall designate the surface excesses in Hansen's (j, k) reference system, as $\Gamma_{i(j,k)}$. The adsorption equation then acquires the following form:

$$d\gamma = -s^s_{(j,k)} \, dT + \tau^s_{(j,k)} \, dP - \sum_{i \neq j,k} \Gamma_{i(j,k)} \, d\mu_i. \tag{3.37}$$

In this expansion of Gibbs relative excesses, the term $\Gamma_{i(j,k)}$ can be called Hansen's relative excess or the excess of substance i relative to j and k.

It should be noted that for the simple case of a two-component system the transformations that are related to the exclusion from the adsorption equation of two chemical potentials can be found in Gibbs work (1928, equations (578)–(584)). However, Gibbs did not develop this idea further and offered no geometrical interpretation for such transformations.

Transition to the (j, k) reference system formally corresponds to exclusion of the $d\mu_j$ and $d\mu_k$ differentials from the initial equation (3.16). By expressing them in the Gibbs–Duhem equations for homogeneous phases, equations (3.12) and (3.13), and by substituting them into equation (3.16), we obtain:

$$d\gamma = -(s^s + s^\alpha \tau^\alpha_{(j,k)} + s^\beta \tau^\beta_{(j,k)}) \, dT + (\tau^\alpha_{(j,k)} + \tau^\beta_{(j,k)}) \, dP$$

$$- \sum_{i \neq j,k} (\Gamma_i + c^\alpha_i \tau^\alpha_{(j,k)} + c^\beta_i \tau^\beta_{(j,k)}) \, d\mu_i \tag{3.38}$$

where

$$\tau^\alpha_{(j,k)} = (c^\beta_j \Gamma_k - c^\beta_k \Gamma_j)/(c^\alpha_j c^\beta_k - c^\alpha_k c^\beta_j),$$
$$\tau^\beta_{(j,k)} = (c^\alpha_k \Gamma_j - c^\alpha_j \Gamma_k)/(c^\alpha_j c^\beta_k - c^\alpha_k c^\beta_j). \tag{3.39}$$

The terms $\tau_{(j,k)}^\alpha$ and $\tau_{(j,k)}^\beta$ are symmetrical with respect to the j and k indices:

$$\tau_{(j,k)}^\alpha = \tau_{(k,j)}^\alpha, \qquad \tau_{(j,k)}^\beta = \tau_{(k,j)}^\beta. \tag{3.40}$$

These terms represent the coefficients for the transition from an arbitrary starting reference system to a concrete reference system, or from absolute excesses to relative ones.

Equation (3.38) shows that Hansen's surface excesses for all extensive properties in the reference system (j, k) are equal to:

$$s_{(j,k)}^s = s^s + s^\alpha \tau_{(j,k)}^\alpha + s^\beta \tau_{(j,k)}^\beta, \tag{3.41}$$

$$\tau_{(j,k)}^s = \tau_{(j,k)}^\alpha + \tau_{(j,k)}^\beta \tag{3.42}$$

$$\Gamma_{i(j,k)} = \Gamma_i + c_i^\alpha \tau_{(j,k)}^\alpha + c_i^\beta \tau_{(j,k)}^\beta. \tag{3.43}$$

The transition coefficients, $\tau_{(j,k)}^\alpha$ and $\tau_{(j,k)}^\beta$, play a significant role in describing surface (j, k) excesses, and therefore it is important to clarify their geometrical meaning. From the two relations presented in (3.35) we can determine the phase volumes in Hansen's reference system:

$$V_{(j,k)}^\alpha = (n_j c_k^\beta - n_k c_j^\beta)/(c_j^\alpha c_k^\beta - c_k^\alpha c_j^\beta),$$
$$V_{(j,k)}^\beta = (n_k c_j^\alpha - n_j c_k^\alpha)/(c_j^\alpha c_k^\beta - c_k^\alpha c_j^\beta). \tag{3.44}$$

Since

$$n_j = c_j^\alpha V_0^\alpha + c_j^\beta V_0^\beta + A\Gamma_j, \qquad n_k = c_k^\alpha V_0^\alpha + c_k^\beta V_0^\beta - A\Gamma_k, \tag{3.45}$$

we can substitute these expressions into the two preceding equations, and by taking equation (3.39) into account we obtain:

$$V_0^\alpha - V_{(j,k)}^\alpha = A\tau_{(j,k)}^\alpha,$$
$$V_0^\beta - V_{(j,k)}^\beta = A\tau_{(j,k)}^\beta. \tag{3.46}$$

Thus, the term $\tau_{(j,k)}^\alpha$ stands for the surface density of the volume excess in the α phase of the j, k-system and the term $\tau_{(j,k)}^\beta$ for the volume excess in the β phase. It is evident that the sum of these terms in equation (3.42) represents the density of the volume excess of the whole system $\tau_{(j,k)}^s$.

Hansen's relative excesses, like Gibbs excesses, are the derivatives of the surface tension with respect to the corresponding variables; the difference lies in the fact that when taking the derivative different parameters are kept constant:

$$s_{(j,k)}^s = -\left(\frac{\partial \gamma}{\partial T}\right)_{P,\mu_i \neq \mu_j, \mu_k} \tag{3.47}$$

$$\tau_{(j,k)}^s = +\left(\frac{\partial \gamma}{\partial p}\right)_{T,\mu_i \neq \mu_j, \mu_k} \tag{3.48}$$

$$\Gamma_{i(j,k)} = -\left(\frac{\partial \gamma}{\partial \mu_i}\right)_{T,P,\mu_i \neq \mu_j, \mu_k}. \tag{3.49}$$

Relationship Between Gibbs and Hansen's Parameters. Gibbs and Hansen's methods describe the same properties of liquid interfaces, yet they arrive at their respective solutions by using different reference systems. Comparison of Gibbs and Hansen's relative excesses shows that in the general case they represent quite different values. However, in certain special cases, Hansen's relative excesses become equal to Gibbs surface excesses. Let us assume that the reference substances, j and k, are only present in different phases: substance j is only in the α phase, and k only in the β phase. The transformation coefficients from an arbitrary Gibbs reference system to a specific Hansen's system will then be $\tau^{\alpha}_{(j,k)} = -\Gamma_j/c^{\alpha}_j$ and $\tau^{\beta}_{(j,k)} = -\Gamma_k/c^{\beta}_k$. Suppose that the i-th component is present only in the α phase,

$$\Gamma_{i(j,k)} = +\Gamma_i - \Gamma_j c^{\alpha}_i/c^{\alpha}_j. \tag{3.50}$$

It is clear that in the present case $\Gamma_{i(j,k)} = \Gamma_{i(j)}$, i.e., Hansen's surface excess is equivalent to the Gibbs excess.

This result is of importance for classical electrochemistry which deals with the interface of metal–aqueous electrolyte solutions. For example, in the Hg–H_2O system, each phase is virtually insoluble in the other. Thus, Hansen's equations become equivalent to Gibbs adsorption equations. Remember, this simplification holds true only for the excesses of substances and does not apply to $s^s_{(j,k)}$ and $\tau^s_{(j,k)}$ since neither the specific entropy nor the specific volume of the phases can be equal to zero.

Example. To illustrate the relationships described above we shall consider the following simple example. Let us assume that a two-phase system is composed of two solvents, α and β, which do not intermix with one another and two additional components, A and B, such that A is only present in the α phase and B only in the β phase (Fig. 3.7). This means that

$$c^{\alpha}_{\beta} = c^{\beta}_{\alpha} = c^{\alpha}_{B} = c^{\beta}_{A} = 0. \tag{3.51}$$

Now we will define Hansen's system by the (α, β) criterion. In the general case, the adsorption equation for a four-component system, where the planar dividing surface coincides with the surface of tension, can be expressed as follows:

$$d\gamma = -s^s_{(\alpha,\beta)} dT + \tau^s_{(\alpha,\beta)} dP - \Gamma_{A(\alpha,\beta)} d\mu_A - \Gamma_{B(\alpha,\beta)} d\mu_B. \tag{3.52}$$

If we use the conditions in (3.51) and consider only an isothermic–isobaric process ($dT = 0, dP = 0$), equation (3.52) simplifies to:

$$d\gamma = -\Gamma_{A(\alpha)} d\mu_A - \Gamma_{B(\beta)} d\mu_B. \tag{3.53}$$

At first glance equation (3.53) seems a little strange since the surface excesses $\Gamma_{A(\alpha)}$ and $\Gamma_{B(\beta)}$ turn out to be defined in different Gibbs systems; this is incompatible with the basic principles of Gibbs approach. Nevertheless, one frequently comes across special equations of this type in the literature (Parsons, 1954; Reid *et al.*, 1983). As can be seen from Hansen's analysis, this equation is approximate, since it is based

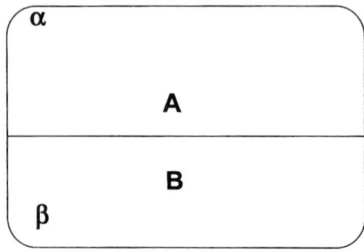

Figure 3.7. An example of the distribution of components A and B between α and β phases.

on an approximate condition (3.51); the equation in fact describes the limiting transition from Hansen's system to two Gibbs systems.

Physical Meaning of the Term τ^s. The choice of a concrete Hansen's reference system (j, k) is usually determined by those parameters which are not supposed to be controlled experimentally. They remain unfixed, free and automatically adjust to the independent variables. Elimination of the differentials of these dependent variables makes equation (3.39) especially useful for practical purposes. However, elimination of the variables $d\mu_j$ and $d\mu_k$, from equation (3.39) means that these variables should be considered dependent, while the rest are independent, and therefore equation (3.37) is most convenient to use. Correspondingly, Gibbs adsorption equation (3.16), in which the pressure differential, dP, is absent, does not prevent us from considering pressure as an independent variable that can be controlled in the experiment. In practice, however, when analyzing the dependence of surface tension on pressure it is more convenient to operate with Hansen's equation (3.39) rather than Gibbs adsorption equation. Equation (3.48), which defines $\tau^s_{(j,k)}$ as a derivative of the surface tension with respect to the pressure under isothermic conditions with fixed chemical potentials of all components except the reference substances j and k, directly follows from equation (3.39).

We initially introduced τ^s as a characteristic of the excess volume at the interface. Now we shall establish this relationship directly by using an intermediate thermodynamic potential for the whole system $\bar{G}(T, P, A, n_k)$. Its differential is:

$$d\bar{G} = -S\,dT + V\,dP + \gamma\,dA + \sum_i \mu_i\,dn_i. \qquad (3.54)$$

Cross-differentiation of this relation readily yields (Bridgman, 1952):

$$(\partial V/\partial A)_{T,p,n_i} = (\partial\gamma/\partial P)_{T,p,n_i}. \qquad (3.55)$$

The left-hand side of this equation represents a change in the volume of a closed system when the area of the interface changes, or according to Motomura's (1997) definition, "the volume of interface formation." As will be shown in the next section, in a two-component system the volume of interface formation is $\tau^s_{(\alpha,\beta)}$. Incidentally, this concept, although in a somewhat different form, can also be found in Gibbs work.

Note that in a system consisting of a large number of components, $\tau^s_{(j,k)}$ does not equal the volume of interface formation, although a definite relationship can be established between them.

Correlation Between $\tau^s_{(j,k)}$ and $\lambda_{j,k}$. Hansen's method employs a new parameter $\tau^s_{(j,k)}$, which has a dimension of length. Gibbs method also has a parameter of length, $\lambda_{j,k}$. The latter is equal to the distance between two dividing surfaces determined by the conventions $\Gamma_j = 0$ and $\Gamma_k = 0$. In a more precise interpretation $\lambda_{j,k}$ represents the relative displacement of two Gibbs reference systems. Naturally, the question arises: are these two parameters identical? This question was the center of discussion between Good (1976, 1982, 1986) and Motomura (1977, 1980, 1986). To compare equations (3.6) and (3.39) we shall take the same starting Gibbs reference system, relative to which we measure the initial absolute surface excesses Γ_i. Thus the volume excess becomes

$$\tau_{(j,k)} = \Gamma_j(c_k^\alpha - c_k^\beta)/(c_j^\alpha c_k^\beta - c_k^\alpha c_j^\beta) + \Gamma_k(c_j^\beta - c_j^\alpha)/(c_j^\alpha c_k^\beta - c_k^\alpha c_j^\beta) \qquad (3.56)$$

A comparison of equation (3.56) with equation (3.6) shows that these quantities are different. They can coincide only if the j and k components are selectively soluble in different phases; for example, the j component in the α phase, and the k component in the β phase. If $c_j^\beta = c_k^\alpha = 0$ the terms $\lambda_{j,k}$ and $\tau^s_{(j,k)}$ are described by the same equation

$$\lambda_{j,k} = \tau^s_{(j,k)} = -\Gamma_j/c_j^\alpha - \Gamma_k/c_k^\beta. \qquad (3.57)$$

Thus, the density of the volume excess τ^s defined by the thermodynamic equation (3.49) does not equal the distance between two Gibbs dividing surfaces. And although the term $\lambda_{j,k}$ has a simple and obvious meaning, it cannot be identified with the coefficient in front of the pressure differential in the adsorption equation. The density of the volume excess and the distance between two Gibbs dividing surfaces coincide only in specific cases. The signs of both $\lambda_{j,k}$ and $\tau^s_{(j,k)}$ are determined in the same way. The sign of $\tau^s_{(j,k)}$ characterizes the properties of the substance near the interface. This can be illustrated in a two-phase system, i.e., a system consisting of only two components, α and β, which are free of any admixtures. If the density of the substance near the interface is smaller than in the bulk of the phases, then with the formation of a new surface the volume of the system increases, i.e., $\tau^s_{(j,k)} > 0$. This situation is frequently encountered, as can be seen from the data summarized in Table 3.1. These data represent the surface tension and specific volume of interface formation for an oil–water system.

Table 3.1 shows that $\tau^s_{(\alpha,\beta)}$ is small and changes symbiotically with a change in surface tension, which correlates with the mutual solubility of water and oil.

The Symmetrized Form of the Adsorption Equation. In both Gibbs and Hansen's adsorption equations the differentials of the chemical potentials are formally equivalent to the differentials of temperature and pressure and can be treated similarly. This makes it possible to symmetrize the adsorption equation with regard

Table 3.1. Surface tension γ, specific volume of oil–water interface formation $\tau^s_{(\alpha,\beta)}$, molar volume of oil (v_{org}), Gibbs adsorption of oil $\Gamma_{org(w)}$, surface entropy s^s, and internal energy u^s at 298.15 K and 0.1 Mpa (Motomura et al., 1983).

Organic phase	$\tau^s_{(\alpha,\beta)}$ nm	V_{org} mm^{3}m^{-2}	$\Gamma_{org(w)}$ mole/m^2	γ mN/m	s^s mJ/Km^2	u^s mJ/m^2
Hexane	0.0227	131.6	-0.172	51.10	0.0916	77.7
Octane	0.0278	163.4	-0.170	50.81	0.0889	77.5
Decane	0.0296	195.7	-0.151	51.41	0.0868	77.3
Dodecane	0.0311	228.1	-0.136	51.85	0.0858	77.4
Cyclohexane	0.0304	108.7	-0.280	49.86	0.0905	76.8
Benzene	0.0106	89.4	-0.119	34.71	0.0604	52.0
Butyl bromide	0.0173	108.0	-0.160	36.76	0.0572	53.8
Hexyl bromide	0.0209	141.5	-0.148	39.85	0.0552	56.3

to all of its variables and thereby give it a simple and compact form. First, we shall introduce the formal definitions:

$$c^\alpha_s = s^\alpha, \qquad c^\beta_s = s^\beta, \qquad d\mu_s = dT; \tag{3.58}$$

$$c^\alpha_v = 1, \qquad c^\beta_v = 1, \qquad d\mu_v = -dP. \tag{3.59}$$

Besides the usual sum of all components, Σ_i we shall use an expanded sum Σ^{ex}_i in which the summation index i can acquire additional v and s values. In this case the Gibbs–Duhem equations can be represented in a symmetrized form:

$$\Sigma^{ex}_i c^\alpha_i \, d\mu_i = 0, \tag{3.60}$$

$$\Sigma^{ex}_i c^\beta_i \, d\mu_i = 0, \tag{3.61}$$

and the adsorption equation for an arbitrary Hansen's reference system acquires the following form:

$$d\gamma = -\Sigma^{ex}_i \Gamma_i \, d\mu_i \tag{3.62}$$

where Γ_s and Γ_v are surface excesses of the entropy and the volume, respectively. A similar attempt to symmetrize the Gibbs adsorption equation has been made by Rowlinson and Widom (1982). However, these authors only introduced a term containing entropy and temperature and omitted a term containing volume and pressure since they used Gibbs reference system instead of Hansen's reference system.

If the reference system uses Hansen's criterion (j, k), then the adsorption equation becomes:

$$d\gamma = -\Sigma^{ex}_{i \neq j,k} \Gamma_{i(j,k)} \, d\mu_i. \tag{3.63}$$

In this case, summation is carried out for all the i indexes except j and k. The excluded j and k indexes can refer both to the system's components and to its volume or entropy.

If we consider all variables in equation (3.62) as being independent of one another, then:

$$\Gamma_i = -(\partial\gamma/\partial\mu_i)_l. \tag{3.64}$$

Subscript l means that all variables, except the variable μ_i in the equation, remain constant. For a concrete Hansen's system we obtain from equation (3.57):

$$\Gamma_{i(j,k)} = -(\partial\gamma/\partial\mu_i)_{l\neq j,k}. \tag{3.65}$$

The subscript $l \neq j, k$ means that during differentiation all variables except j and k and those indicated in the equation remain constant, and l, j, and k can acquire any values in the expanded sense.

To switch from absolute excesses, Γ_i, to relative excesses, $\Gamma_{i(j,k)}$, one can use another, more formal approach. If we consider the variables j and k as functions of the remaining variables, then:

$$(\partial\gamma/\partial\mu_i)_{i\neq j,k} = (\partial\gamma/\partial\mu_i)_l + (\partial\gamma/\partial\mu_j)_l(\partial\mu_j/\partial\mu_i)_{l\neq k} + (\partial\gamma/\partial\mu_k)_l(\partial\mu/\partial\mu_i)_{l\neq j} \tag{3.66}$$

and therefore

$$\Gamma_{i(j,k)} = \Gamma_i + (\partial\mu_j/\partial\mu_i)_{l\neq k}\Gamma_j + (\partial\mu_k/\partial\mu_i)_{l\neq j}\Gamma_k. \tag{3.67}$$

We shall note some obvious properties of surface excesses $\Gamma_{i(j,k)}$. As mentioned above, these excesses are symmetrical with respect to the j and k indexes; that is their permutation does not change the magnitude or the sign of this parameter. If the i index coincides with either the j or k index, then

$$\Gamma_{j(j,k)} = \Gamma_{k(j,k)} = 0. \tag{3.68}$$

Now we shall determine the relationship between different surface excesses. We shall consider the excess of the same i component in both the (j, k) and (r, q) systems. Equation (3.67) contains absolute excesses $\Gamma_i, \Gamma_j, \Gamma_k$ in some arbitrary initial reference system. An example of such a system is Hansen's system (r, q). Then we obtain:

$$\Gamma_{i(j,k)} = \Gamma_{i(r,q)} + (\partial\mu_j/\partial\mu_i)_{l\neq k}\Gamma_{j(r,q)} + (\partial\mu_k/\partial\mu_i)_{l\neq j}\Gamma_{k(r,q)}. \tag{3.69}$$

Thus, the excess of the i component in the (j, k) system is expressed through the excess of the same component in the (r, q) system, and the excesses of the j and k reference substances are expressed in the same (r, q) system.

The general equation (3.69) makes it possible to answer a series of specific questions and to find special correlations. Let us assume, for example, that the new system differs from the old one only by one index, i.e., $q = k$. Then, according to equation (3.68), we have:

$$\Gamma_{i(j,k)} = \Gamma_{i(r,k)} + (\partial\mu_j/\partial\mu_i)_{l\neq k}\Gamma_{j(r,k)}. \tag{3.70}$$

One of the special cases of equation (3.70) has been described by Eriksson (1966, 1969, 1983). If we are interested in the changes in the $\Gamma_{i(j,k)}$ excess due to the permutation of the i and j indexes, then by assuming in equation (3.69) $r = i$ and $q = k$ we find

$$\Gamma_{i(j,k)} = +(\partial\mu_j/\partial\mu_i)_{l\neq k}\Gamma_{j(i,k)}. \tag{3.71}$$

In equations (3.70) and (3.71) one finds the derivatives $(\partial\mu_j/\partial\mu_i)_{l\neq k}$. Their calculation requires use of a symmetrized form of the Gibbs–Duhem equations (3.60) and (3.61). After obtaining the required derivative with the help of these equations, we have:

$$(\partial\mu_j/\partial\mu_i)_{l\neq k} = -(c_i^\alpha c_k^\beta - c_i^\beta c_k^\alpha)/(c_j^\alpha c_k^\beta - c_j^\beta c_k^\alpha). \tag{3.72}$$

Now we have a full set of quantities which are sufficient for converting surface excesses from one system into another. In a previous section we discussed such a conversion by transforming absolute excesses into relative excess. It was done in terms of volume concentration of the corresponding components and $\tau_{(j,k)}^\alpha$ and $\tau_{(j,k)}^\beta$. These two approaches are complementary, and either one or the other may be more convenient, depending on the problem.

There are several more special cases which are of interest. We previously analyzed the relationship between Gibbs and Hansen's excesses, $\Gamma_{i(j)}$ and $\Gamma_{i(j,k)}$, and in particular, the possibility of their being equal. Gibbs excess can be transformed into Hansen's by the simple relationship: $\Gamma_{i(j)} \equiv \Gamma_{i(j,v)}$. Analysis has shown that for $\Gamma_{i(j)}$ to coincide with $\Gamma_{i(j,k)}$ it is necessary that the i and j components be soluble in the same phase and only in this phase, and that the k component be soluble only in the opposite phase. Even if only one of the three components is present in both phases, this equality does not hold true.

Lastly, let us note that if one of the indexes in derivative (3.66) stands for the volume, the equation becomes considerably simplified. For example, if $k = v$, we have:

$$(\partial\mu_j/\partial\mu_i)_{l\neq v} = -(c_i^\alpha - c_i^\beta)/(c_j^\alpha - c_j^\beta) \tag{3.73}$$

Surface Excesses in Terms of Mole Fractions of the Components. The

equations derived above are expressed in terms of volume concentrations of components c_i. However, it is often more convenient to use mole fractions of components rather than concentrations, which are defined in the following way:

$$x_i^\alpha = n_i^\alpha/n^\alpha, \qquad x_i^\beta = n_i^\beta/n^\beta, \tag{3.74}$$

where n^α and n^β are the total number of moles for all the components present in the α and β phases, respectively:

$$n^\alpha = \sum_l n_l^\alpha, \qquad n^\beta = \sum_l n_l^\beta. \tag{3.75}$$

Mole fractions can also be expressed in terms of the concentrations of their components:

$$x_i^\alpha = c_i^\alpha / \sum_l c_l^\alpha, \qquad x_i^\beta = c_i^\beta / \sum_l c_l^\beta. \tag{3.76}$$

Mole fractions obey a simple relationship:

$$\sum_l x_i^\alpha = 1 \qquad \text{and} \qquad \sum_l x_i^\beta = 1. \tag{3.77}$$

If we express the Gibbs–Duhem equations through mole fractions they become:

$$\bar{s}^\alpha \, dT - \bar{v}^\alpha \, dp + \sum_i x_i^\alpha \, d\mu_i = 0, \tag{3.78}$$

$$\bar{s}^\beta \, dT - \bar{v}^\beta \, dp + \sum_i x_i^\beta \, d\mu_i = 0, \tag{3.79}$$

where the terms for the average molar entropy,

$$\bar{s}^\alpha = s^\alpha / \sum_l c_l^\alpha, \qquad \bar{s}^\beta = s^\beta / \sum_l c_l^\beta, \tag{3.80}$$

and for the average molar volume,

$$\bar{v}^\alpha = 1 / \sum_l c_l^\alpha, \qquad \bar{v}^\beta = 1 / \sum_l c_l^\beta, \tag{3.81}$$

are introduced.

If we carry out the same transformations one can see that the adsorption equations in both Gibbs and Hansen's presentations remain unchanged and relative excesses can be presented in the following form:

$$s_{(j,k)}^s = s^s + \bar{s}^\alpha \, \Theta_{(j,k)}^\alpha + \bar{s}^\beta \, \Theta_{(j,k)}^\beta, \tag{3.82}$$

$$\tau_{(j,k)} = \bar{v}^\alpha \, \Theta_{(j,k)}^\alpha + \bar{v}^\beta \, \Theta_{(j,k)}^\beta, \tag{3.83}$$

$$\Gamma_{i(j,k)} = \Gamma_i + x_i^\alpha \, \Theta_{(j,k)}^\alpha + x_i^\beta \, \Theta_{(j,k)}^\beta, \tag{3.84}$$

where

$$\Theta_{(j,k)}^\alpha = (x_j^\beta \, \Gamma_k - x_k^\beta \, \Gamma_k)/(x_j^\alpha x_k^\beta - x_k^\alpha x_j^\beta)$$

$$\Theta_{(j,k)}^\beta = (x_k^\alpha \, \Gamma_j - x_j^\alpha \, \Gamma_k)/(x_j^\alpha x_k^\beta - x_k^\alpha x_j^\beta). \tag{3.85}$$

The relation between $\tau_{(j,k)}^\alpha$ and Θ takes the form

$$\Theta_{(j,k)}^\alpha = \tau_{(j,k)}^\alpha \sum_i c_i^\alpha. \tag{3.86}$$

A similar equation holds true for the β phase.

These equations can be symmetrized if we introduce additional designations:

$$x_s^\alpha = \overline{s}^\alpha, \qquad x_s^\beta = \overline{s}^\beta \qquad\qquad (3.87)$$

$$x_v^\alpha = \overline{v}^\alpha, \qquad x_v^\beta = \overline{v}^\beta \qquad\qquad (3.88)$$

Then, the Gibbs–Duhem equations can be written in the following form:

$$\Sigma_i^{ex} x_i^\alpha \, d\mu_i = 0 \qquad\qquad (3.89)$$

$$\Sigma_i^{ex} x_i^\beta \, d\mu_i = 0. \qquad\qquad (3.90)$$

Correspondingly, equation (3.72) becomes:

$$(\partial\mu_j/\partial\mu_i)_{l \neq k} = -(x_i^\alpha x_k^\beta - x_i^\beta x_k^\alpha)/(x_j^\alpha x_k^\beta - x_k^\alpha x_j^\beta) \qquad (3.91)$$

Difficulties in Interpreting Hansen's Method. Hansen only formally described his method at the algebraic level. However, he noted that the coefficients which multiply the Gibbs–Duhem equations have the meaning of phase volumes. This set of circumstances, as well as the prevailing influence of Gibbs classical ideas, hindered acceptance of Hansen's contribution to surface thermodynamics and even led to some erroneous interpretations (Motomura, 1977, 1980, 1986). For example, it was widely thought that Hansen's criterion $\Gamma_j = 0$, $\Gamma_k = 0$ determines not a single, but rather two dividing surfaces, and $\tau_{(j,k)}$ is the distance between them, which is clearly wrong.

The appearance of erroneous interpretations of Hansen's method is hardly surprising since it is not always easy to master even the classical concepts of Gibbs dividing surface as has been pointed by Good (1986). Psychologically, the source of this type of error apparently lies in a subconscious wish to interpret the concept of surface excesses in structural terms. The concept of the Gibbs dividing surface often tempts scholars to ascribe some of the physical properties of the boundary of two phases to the dividing surface. As noted by Good, the conceptual problems of Gibbs thermodynamics for surfaces center on two questions: first, where is the substance that forms the non-zero excesses Γ_i spatially located. Second, what is the relationship between any set of $\Gamma_{i(j)}$ and the physical gradients of the composition in the interphase layer. The thermodynamic methods, however, are unable to provide an answer to these questions, which are related to the structure of the interphase layer.

Of course, one can attempt to compare the concrete values of Γ_i with different physical distributions of the i component in the interphase layer; for example, with a tightly or loosely packed layer of a substance at the interphase boundary. However, this is not always possible. Gibbs had pointed out that if the i-th component is a surface-active substance, the dividing surface $\Gamma_i = 0$ can lie so far away from the actual phase boundary that it will have no physical meaning.

Attempts to identify these abstract concepts with real physical entities can easily lead to erroneous conclusions. This is especially true of the more abstract Hansen's method. Since in this method the volume of the reference system is less than the volume of the real system by $V_{(j,k)}^s = A\tau_{(j,k)}^s$, one is tempted to "correct" this system by separating the two phases in the Hansen reference system along the dividing surface and pushing them a distance $\tau_{(i,k)}$ apart. This gap would be just big enough to hold the missing volume $V_{(j,k)}^s$, which in turn makes it tempting to "place" into this gap the surface excesses of all the remaining components. There is hardly any need to convince the reader that such an artificial picture has nothing in common with the real physical distribution of substances at the interface. Such a distortion can only lead to confusion in our attempts to understand interface phenomena.

3.3. OTHER DEFINITIONS OF SURFACE EXCESSES

Surface Excesses According to Frumkin. Gibbs surface excesses have also been interpreted by Frumkin (1919) for the practical purpose of explaining experimental data. His interpretation was repeatedly discussed by Seelich (1948), Overbeek (1952), Everett (1987) and others (Frumkin, 1979; Frumkin, et al., 1980; Verschaffeit, 1936a,b). While giving due credit to these scientists who elaborated on his interpretation we shall refer to it as the surface excesses according to Frumkin.

When the area of an interface increases, certain components of the system adsorb at this new interface. This causes a finite change, if only a small one, in the chemical composition of the phases in contact. In order for the chemical composition of the solutions to remain unchanged, we need to add Γ_i amounts of the corresponding components to the system for each unit increase in the area of the interface.

This concept can be expressed mathematically. Let us consider a two-phase system which consists of r components. Its Helmholtz free energy can be expressed as

$$dF = -S\,dT - P\,dV + \gamma\,dA + \sum_i \mu_i\,dn_i. \tag{3.92}$$

This system has r degrees of freedom corresponding to r independent differentials for the intensive variables in the adsorption equation. However, to define the system completely, we must assign three more extensive variables which determine the volume (or mass) of the two phases and the area of the interface. This can be done in different ways; the most convenient way is to choose the total volume, V, the area, A, and the total mass of one of the components, n_j. A new thermodynamic potential, $F - \sum_{i \neq j} \mu_i n_i$, can be obtained based on these variables. Its differential is

$$d\left(F - \sum_{i \neq j} \mu_i n_i\right) = -S\,dT - P\,dV + \gamma\,dA + \mu_j\,dn_j - \sum_{i \neq j} n_i\,d\mu_i. \tag{3.93}$$

By carrying out cross-differentiation we obtain:

$$(\partial\gamma/\partial\mu_i)_{T,V,A,n_j,[\mu_k \neq \mu_j]} = -(\partial n_i/\partial A)_{T,V,n_j,[\mu_k \neq \mu_j]}. \tag{3.94}$$

The left-hand side of this equation contains the derivative of the surface tension with respect to the chemical potential of the i-th component. The subscript $[\mu_k \neq \mu_j]$ signifies that when calculating this derivative the chemical potentials of all components should be kept constant except for the j-th component (and of course the i-th component that is explicitly present in the derivative). From Gibbs adsorption equation it follows that the surface tension is determined by the temperature and $r - 1$

chemical potentials of the system's components. Therefore, the subscripts V, A, and n_j have no effect on the derivative and can be omitted. Then we can obtain the Gibbs relative excess:

$$(\partial\gamma/\partial\mu_i)_{T,V,A,n_j[\mu_k\neq\mu_j]} = (\partial\gamma/\partial\mu_i)_{T,[\mu_k\neq\mu_j]} = -\Gamma_{i(j)}. \tag{3.95}$$

Such a set of variables fully determines the state of the system, and therefore all remaining intensive variables, P and μ_j, also do not change. Hence, equation (3.94) can be rewritten as

$$\Gamma_{i(j)} = (\partial n_i/\partial A)_{T,V,n_j,[\mu_k]}. \tag{3.96}$$

We have obtained the Gibbs excess $\Gamma_{i(j)}$ of component i relative to component j because we fixed the total mass of the j-th component. If we had fixed the volumes of the phases instead of the total mass, we would obtain the absolute excesses Γ_i. However, they are the relative excesses that are thermodynamically meaningful.

Equation (3.95) provides a better understanding of the definition of surface excesses as used by Frumkin. The relative excesses represent the amounts of the i-th components which should be added to the system for each unit increase in the surface area, in order that all chemical potentials remain unchanged, while the volume of the system remains constant. It follows that the temperature and pressure will also remain constant. If we want to keep the chemical composition of the phases constant rather than the chemical potentials of every component, we must superimpose the condition of constancy for one of the two intensive parameters: temperature or pressure. The remaining parameter in this case will automatically remain constant.

There is a curious nuance in connection with equation (3.94): the calculation of the derivative $\partial\gamma/\partial\mu_i$ implies that the pressure in the system changes. In principle this presents no difficulty. However, Overbeek (1952) considered this situation to be inconvenient and proposed another definition of surface excesses. He used three extensive variables to characterize the system, the area of the interface A and the masses of two arbitrary components, n_j and n_k. Having introduced a new thermodynamic potential $F + PV - \Sigma_{i\neq j,k}\mu_i n_i$ with differential:

$$d\left(F + PV - \sum_{i\neq j,k}\mu_i n_i\right) = -S\,dT + V\,dP + \gamma\,dA + \mu_j\,dn_j + \mu_k\,dn_k$$
$$- \sum_{i\neq j,k} n_i\,d\mu_i, \tag{3.97}$$

Overbeek arrived at the following equation:

$$(\partial\gamma/\partial\mu_i)_{T,P,[\mu_l\neq\mu_j]} = -(\partial n_i/\partial A)_{T,P,n_j,n_k,[\mu_l]}. \tag{3.98}$$

Overbeek proposed that these derivatives were surface excesses. They represent the amounts of these components which should be added to the system per unit interface area increase, such that temperature, pressure, masses of the selected j-th and k-th components, and the chemical potentials of all components remain unchanged. It should be noted that no conditions are imposed on the volume and therefore it can vary.

A comparison of equation (3.98) with equation (3.49) shows that the left-hand side of this equation contains Hansen's surface excess relative to the substances j and k:

$$\Gamma_{i(j,k)} = (\partial n_i/\partial A)_{T,p,n_j,n_k,[\mu_l]}. \tag{3.99}$$

Hence, Overbeek's definition provides an additional interpretation of Hansen's excesses. It is interesting to note that Overbeek arrived at his definition ten years before Hansen formulated his method.

This definition has an interesting implication for a closed two-component two-phase system. If the temperature and pressure are kept constant, then a change in the area of the interface has no effect on the chemical composition of the phases. However, the volume of the phases and the volume of the whole system in this case change.

At first glance this assertion seems to contradict the interpretation of Gibbs surface excesses according to Frumkin's method. If, at an arbitrary dividing surface in Gibbs reference system, the absolute excesses of two components equal Γ_1 and Γ_2, then it requires that additional amounts of the components be added to the system in order for the chemical composition of all phases to remain constant when there is a unit increase in the interface area. As we know, this does not change the volume of the system. However, a binary two-phase system possesses a remarkable property: any amount of these two substances can be added to it independently of the area of the interface and the compositions of the phases will remain unchanged when temperature and pressure are held constant. Therefore, when the interface increases there is no need for additional substances to be added to the system. However, if we do not add these substances the volume of the system changes with an increase in the area of the interface.

Frumkin's interpretation can be applied not only to surface excesses of the system's components, but also to the excesses of a system's volume and entropy. By using Overbeek's approach it can be readily shown that the values of $\tau_{(j,k)}^s$ and $s_{(j,k)}^s$ indicate how much the volume and entropy of the system should be increased per unit area increase in the interface such that at constant mass of the j-th and k-th components all the intensive parameters will remain constant. This interpretation applies to Hansen's excesses. Gibbs entropy excess can be analyzed in an analogous way, and one has to keep in mind that the Gibbs volume excess is always equal to zero.

Hansen's and Guggenheim's Methods. Another approach to the excesses of volume and entropy was developed by the Dutch physicist Guggenheim. His method of using a finite interfacial layer, *the Guggenheim method* (Boruvka *et al.*, 1985; Buff, 1956; Melrose, 1968; Verschaffeit, 1936a,b), is widely used in modern studies of surfaces thermodynamics. For example, in the Recommendations of IUPAC, compiled by Trasatti and Parsons (1988), this method is considered to be the only basic method for describing interphase boundaries. According to this method the entire interface region is regarded as a heterogeneous layer σ having thickness τ (Fig.

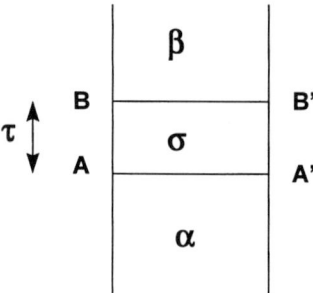

Figure 3.8. Model of an interphase layer of finite thickness.

3.8). The layer contains n_i moles of the i-th component, and its entropy is S^σ. By taking these quantities per unit of surface area,

$$s^\sigma = S^\sigma/A, \qquad \Gamma_i = n_i^\sigma/A, \qquad (3.100)$$

we can write the Gibbs adsorption equation as:

$$d\gamma = -s^\sigma\,dT + \tau\,dp - \sum_i \Gamma_i\,d\mu_i. \qquad (3.101)$$

This form of the Gibbs adsorption equation resembles Hansen's equation, but differs in that we have attained the characteristics of a real physical volume—a specific layer enclosed between two homogeneous phases. Incidentally, it is because of this circumstance that many researchers think highly of the method of the finite layer (Rusanov, 1967). However, other scientists, for example, Rowlinson and Widom (1982) believe that the advantages of Guggenheim's method are illusory since we cannot unambiguously measure and independently determine the thermodynamic properties of the surface layer, which is a non-autonomous phase. If we are to determine the properties of the surface layer as being the difference between the properties of the whole system and those of the α and β phases, such an approach differs only slightly from Gibbs equation and leads to the same experimental consequences. It is difficult not to agree with this assessment.

Hansen's equations, which are based on the elimination of two arbitrary differentials from the original adsorption equation, can also be obtained by starting with the equation for the finite layer as was done by Kakiuchi and Senda (1983b) in deriving the equation of electrocapillarity for a multicomponent system. They have carried out all the necessary transformations beginning with basic equation (3.11) instead of using Hansen's clear-cut definitions and readily available formulas.

3.4. VOLUME AND ENTROPY OF INTERFACE FORMATION

Definitions. The problem of determining the specific volume and entropy of formation of an interface between two immiscible liquids was first addressed by

Gibbs who defined the quantity $(\partial\gamma/\partial P)$ for a two-component system as a change in the volume when the interface area increases by unity. Later, Bridgman (1952) derived the relation $(\partial V/\partial A)_{T,p} = (\partial\gamma/\partial P)_{T,A}$ for two immiscible liquids, and it was first noted by Defay *et al.* (1966) that the term $(\partial\gamma/\partial P)$ has a dimension of length.

We have introduced Hansen's surface excesses of volume and entropy $\tau^s_{(j,k)}$ and $s^s_{(j,k)}$ with respect to the reference substances j and k. However, it would be interesting to define analogous terms which are not related to concrete reference substances j and k and which could be the absolute characteristics of the system. In 1980 Motomura introduced the concept of thermodynamic characteristics of interface formation which include volume, entropy, energy, and other parameters. He referred to the specific volume and entropy of interface formation as a change in the volume and entropy of a two-phase system when a unit area of the interface is formed from two given phases.

Let the specific volume of surface formation be denoted by $\tau^s_{(form)}$ and the specific entropy of surface formation by $s^s_{(form)}$. In accordance with the definition of specific volume of surface formation, this quantity should be found when the temperature, pressure, and total mass of all components in the entire system are kept constant:

$$\tau^s_{(form)} = (\partial V/\partial A)_{T,p,[n_i]}. \tag{3.102}$$

The symbol $[n_i]$ is defined as follows:

$$[n_i] = n_1, n_2, \ldots, n_r, \tag{3.103}$$

where r is the number of components. The term $\tau^s_{(form)}$ can be expressed through the dependence of surface tension on pressure, if we use the thermodynamic potential of the system which has the differential

$$d\bar{G} = -S\,dT + V\,dP + \gamma\,dA + \sum \mu_i\,dn_i. \tag{3.104}$$

Using cross-differentiation we obtain

$$\tau^s_{(form)} = (\partial V/\partial A)_{T,P,[n_i]} = (\partial\gamma/\partial P)_{T,A,[n_i]}, \tag{3.105}$$

and

$$s^s_{(form)} = (\partial S/\partial A)_{T,P,[n_i]} = -(\partial\gamma/\partial T)_{P,A,[n_i]}. \tag{3.106}$$

The subscripts of these derivatives give the most convenient conditions for experimentally studying the dependence of surface tension on temperature and pressure. Therefore, the volume and entropy of surface formation can be directly determined experimentally. Now, we can find out how these quantities are related to Gibbs and Hansen's surface excesses.

Gibbs Description. Consider an arbitrary Gibbs reference system having a total volume V, phase volumes V^α and V^β, and surface excesses Γ_i. If the masses of the i-th component in the α and β phases are n_i^α and n_i^β and their molar volumes are v_i^α

and v_i^β, then the total mass of the i-th component is

$$n_i = n_i^\alpha + n_i^\beta + \Gamma_i A, \tag{3.107}$$

and the volume of the system is

$$V = \sum_i (v_i^\alpha n_i^\alpha + v_i^\beta n_i^\beta). \tag{3.108}$$

The surface excess $\Gamma_i A$ can be assumed to be small in comparison with the mass of the phase; then for constant total mass of the i-th component

$$dn_i^\alpha + dn_i^\beta + \Gamma_i dA = 0. \tag{3.109}$$

If the properties of the interface remain unchanged, i.e., $\Gamma_i = $ constant, we have

$$(\partial V/\partial A)_{T,P,[n_i]} = \sum_k [v_k^\alpha (\partial n_k^\alpha/\partial A)_{T,P,[n_i]} + v_k^\beta (\partial n_k^\beta/\partial A)_{T,P,[n_i]}]. \tag{3.110}$$

To find the derivatives $(\partial n_k^\alpha/\partial A)_{T,P,[n_i]}$ and $(\partial n_k^\beta/\partial A)_{T,P,[n_i]}$ we use the condition of equality of chemical potential of uncharged component in both phases:

$$(d\mu_k^\alpha)_{T,P} = (d\mu_k^\beta)_{T,P}. \tag{3.111}$$

Now we express the chemical potentials μ_k^α in terms of the molar fractions x_j^α of components and take into account that

$$dx_j^\alpha = \sum_m (\delta_{jm} - x_j^\alpha)(1/n^\alpha)\, dn_m^\alpha, \tag{3.112}$$

where $n^\alpha = \Sigma n_m^\alpha$ and δ_{jm} is Kronecker's symbol which is 1 for $j = m$ and 0 for $j \neq m$. With these definitions we obtain:

$$\sum_m (1/n^\alpha)(\partial n_m^\alpha/\partial A)_{T,P,[n_i]} \sum_j (\delta_{jm} - x_j^\alpha)(\partial \mu_k^\alpha/\partial x_j^\alpha)_{T,P,[x_l \neq x_j]}$$

$$= \sum_m (1/n^\beta)(\partial n_m^\beta/\partial A)_{T,P,[n_i]} \sum_j (\delta_{jm} - x_j^\beta)(\partial \mu_k^\beta/\partial x_j^\beta)_{T,P,[x_l \neq x_j]}. \tag{3.113}$$

As follows from (3.109),

$$(\partial n_m^\alpha/\partial A)_{T,P,[n_i]} + (\partial n_m^\beta/\partial A)_{T,P,[n_i]} + \Gamma_m = 0. \tag{3.114}$$

Using the set of linear equations (3.113) and (3.114) one can readily find the derivatives in question, and by substituting them into equation (3.110) find the volume of interface formation.

The general solutions of equations (3.113) and (3.114) yield rather cumbersome expressions. The following examples will be used to illustrate their solutions.

Binary System. Let one assume that component 1 is the main component in the α phase, and component 2 the main one in the β phase, and that these liquids are partially soluble in each other. Under these conditions, the volume of interface formation becomes

$$\tau^s_{form} = -[\Gamma_1(c^\beta_2 - c^\alpha_2) + \Gamma_2(c^\alpha_1 - c^\beta_1)]/(c^\alpha_1 c^\beta_2 - c^\alpha_2 c^\beta_1) \equiv \tau^s_{(1,2)}. \qquad (3.115)$$

Thus, the specific volume of surface formation in a binary system equals Hansen's volume excess with respect to components 1 and 2. This is an obvious result which could have been obtained directly from equation (3.105). When considering the pressure derivative of surface tension one can note that according to the phase rule the state of a two-phase binary system is determined by two intensive parameters. If temperature and pressure are taken as such parameters, then:

$$(\partial\gamma/\partial P)_{T,A,[n_i]} = (\partial\gamma/\partial P) = \tau^s_{(1,2)}. \qquad (3.116)$$

This result was also obtained by a different method in equation (3.115).

Next, one can derive equations for the masses of components 1 and 2 which are adsorbed from the α and β phases per unit area on the dividing surface:

$$(\partial n^\alpha_1/\partial A)_{T,P,n_1 \cdot n_2} = x^\alpha_1(x^\beta_1 \Gamma_2 - x^\beta_2 \Gamma_1)/(x^\alpha_1 x^\beta_2 - x^\beta_1 x^\alpha_2) \qquad (3.117)$$

$$(\partial n^\beta_1/\partial A)_{T,P,n_1 \cdot n_2} = x^\beta_1(x^\alpha_2 \Gamma_1 - x^\alpha_1 \Gamma_2)/(x^\alpha_1 x^\beta_2 - x^\beta_1 x^\alpha_2) \qquad (3.118)$$

$$(\partial n^\alpha_2/\partial A)_{T,P,n_1 \cdot n_2} = x^\alpha_2(x^\beta_1 \Gamma_2 - x^\beta_2 \Gamma_1)/(x^\alpha_1 x^\beta_2 - x^\beta_1 x^\alpha_2) \qquad (3.119)$$

$$(\partial n^\beta_2/\partial A)_{T,P,n_1 \cdot n_2} = x^\beta_2(x^\alpha_2 \Gamma_1 - x^\alpha_1 \Gamma_2)/(x^\alpha_1 x^\beta_2 - x^\beta_1 x^\alpha_2). \qquad (3.120)$$

Once these quantities have been determined one can find the volume of surface formation, τ^s_{form}, and also the contribution of each of the phases, τ^α_{form} and τ^β_{form} to τ^s_{form}:

$$\tau^\alpha_{form} = v^\alpha_1(\partial n_1{}^\alpha/\partial A)_{T,P,n_1,n_2} + v^\alpha_2(\partial n^\alpha_2/\partial A)_{T,P,n_1,n_2}$$
$$= (c^\beta_1 \Gamma_1 - c^\beta_2 \Gamma_2)/(c^\alpha_1 c^\beta_2 - c^\alpha_2 c^\beta_1), \qquad (3.121)$$

and in analogous way

$$\tau^\beta_{form} = (c^\alpha_2 \Gamma_1 - c^\alpha_1 \Gamma_2)/(c^\alpha_1 c^\beta_2 - c^\alpha_2 c^\beta_1). \qquad (3.122)$$

It can be readily shown that

$$\tau^s_{form} = \tau^\alpha_{form} + \tau^\beta_{form}. \qquad (3.123)$$

Furthermore

$$\tau^\alpha_{form} = \tau^\alpha_{(1,2)} \qquad \text{and} \qquad \tau^\beta_{form} = \tau^\beta_{(1,2)}. \qquad (3.124)$$

It should be noted that the expression for the volume of surface formation contains no derivatives of the $\partial \mu_i / \partial x_k$ type, which are present in equations (3.113) and (3.114). This is a remarkable feature of the binary system and means that the molar fractions of the components and the chemical potentials of the binary system do not undergo any change during interface formation. They remain constant:

$$d\mu_i^\alpha = d\mu_i^\beta = 0. \tag{3.125}$$

This holds true only for a binary system. The addition of one more component alters this situation. Nevertheless, the results obtained for the binary system can be used as a reference for describing more complicated systems. For this purpose Hansen's method is very useful.

Hansen's Description. If the phase volumes in Hansen's reference system, defined by the convention $(1, 2)$, equal $V_{(1,2)}^\alpha$ and $V_{(1,2)}^\beta$, the volume of the real system is

$$V = V_{(1,2)}^\alpha + V_{(1,2)}^\beta + \tau_{(1,2)}^s A = \sum_{i=1,2} (v_i^\alpha n_i^\alpha + v_i^\beta n_i^\beta) + \tau_{(1,2)}^s A, \tag{3.126}$$

and the total amount of the i-th component is

$$n_i = n_i^\alpha + n_i^\beta + \Gamma_{i(1,2)} A. \tag{3.127}$$

If i equals 1 or 2, the last term in this sum disappears. In this case the volume of surface formation is

$$\tau_{\text{form}}^s = \tau_{(1,2)}^s + \sum_k [v_k^\alpha (\partial n_k^\alpha / \partial A)_{T,P,[n_i]} + v_k^\beta (\partial n_k^\beta / \partial A)_{T,p,[n_i]}]. \tag{3.128}$$

It should be noted that summation is carried over all the components, including 1 and 2, although they are not adsorbed in the Hansen's system under consideration. However, as the area of the interface changes, all the components, including 1 and 2, are redistributed between the phases. This leads to a change in the volume of the system. The main set of equations (3.113) and (3.114), for determining the derivatives $(\partial n_k^\alpha / \partial A)$ and $(\partial n_k^\beta / \partial A)$ varies only slightly. Equation (3.113) remains unchanged, while equation (3.114) now becomes:

$$(\partial n_m^\alpha / \partial A)_{T,p,[n_i]} + (\partial n_m^\beta / \partial A)_{T,p,[n_i]} + \Gamma_{m(1,2)} = 0. \tag{3.129}$$

By solving this set of equations, (3.113) and (3.129), and substituting the results into equation (3.110), one obtains the volume of surface formation. This can be illustrated by a ternary two-phase system.

Ternary Two-Phase System. Let us choose a system that consists of three components (1, 2, and 3) which may each be present in either the α and β phases. Components 1 and 2 are solvents which comprise the majority the α and β phases.

The solutions to Hansen's equations can be found without approximation; however, the final equations are very cumbersome. To simplify them, we shall consider dilute solutions: $x_2^\alpha \ll 1$, $x_3^\alpha \ll 1$, $x_1^\beta \ll 1$, $x_3^\beta \ll 1$. Furthermore, we will assume that the solutions are ideal, i.e.,

$$\mu_i = \mu_i^0 + RT \ln x_i. \tag{3.130}$$

With these assumptions, one can obtain the volume of surface formation in the form of a series expansion

$$\tau^s = \tau_{(1,2)}^s - [v_3^\alpha c_3^\alpha V_{(1,2)}^\alpha + v_3^\beta c_3^\beta V_{(1,2)}^\beta + (v_1^\beta - v_1^\alpha) c_1^\beta V_{(1,2)}^\beta + (v_2^\beta - v_2^\alpha) c_2^\alpha V_{(1,2)}^\alpha]$$
$$\times (x_3^\beta - x_3^\alpha) \Gamma_{3(1,2)} / (c_3^\alpha V_{(1,2)}^\alpha + c_3^\beta V_{(1,2)}^\beta). \tag{3.131}$$

In deriving this equation it is assumed that the volumes of phases α and β are comparable.

It is important to note that the volume of surface formation depends not only on intensive variables, but also on the extensive quantities $V_{(1,2)}^\alpha$ and $V_{(1,2)}^\beta$. However, if we neglect the mutual solubility of components 1 and 2 and assume that component 3 is present only in the α phase, the volume of surface formation will not depend on $V_{(1,2)}^\alpha$ and $V_{(1,2)}^\beta$. In this case,

$$\tau_{form}^s = \tau_{(1,2)}^s - v_3^\alpha \Gamma_{3(1,2)}. \tag{3.132}$$

This particular equation was derived by Motomura (1980). Now let us assume that the volume of one phase considerably exceeds that of another phase: $V^\alpha \ll V^\beta$. Then,

$$\tau_{form}^s = \tau_{(1,2)}^s - v_3^\beta \Gamma_{3(1,2)}. \tag{3.133}$$

In this case, the volume τ_{form}^s does not depend on the extensive variables since during the formation of new surface area adsorption takes place only from the large β phase.

Other Characteristics of Surface Formation. Until now we have derived equations only for the specific volume of surface formation. However, using these equations and substituting entropy for volumes, one can easily derive analogous expressions for the specific entropy of surface formation:

$$s_{form}^s = s_{(1,2)}^s + \sum_k [s_k^\alpha (\partial n_k^\alpha / \partial A)_{T,P,[n_i]} + s_k^\beta (\partial n_k^\beta / \partial A)_{T,P,[n_i]}]. \tag{3.134}$$

Equations (3.113), (3.133), and (3.134) make it possible to calculate the specific entropy of interface formation. Since the energy of the system can be represented as

$$U_{form} = \gamma + T s_{form}^s - P \tau_{form}^s, \tag{3.135}$$

one can extend this method to the specific energy of surface formation. It is assumed that the temperature and pressure in the system remain unchanged, and that the total masses of the components remain constant. For the specific Helmholtz free energy we obtain:

$$f_{\text{form}} = \gamma - P\tau^s_{\text{form}} \tag{3.136}$$

Elasticity of Surface Excesses. All previous calculations have been carried out under the assumption that the amount of substance adsorbed at the interface is small in comparison with the total amount of substance in the overall system. That is why, when differentiating the material balance equation, the surface excesses remain unchanged during surface extension, i.e.,

$$(\partial \Gamma_{k(1,2)}/\partial A)_{T,P,[n_i]} = 0. \tag{3.137}$$

This assumption is quite reasonable, provided the components have finite solubility, at least in one of the phases. However, sometimes the system can contain an impurity of such high surface activity that virtually all of the impurity is located at the interface. In this case expansion of the surface leads to a considerable change in at least one of the surface excesses describing this surfactant. Hence, the redistribution of the substances will now be described by equation (3.36) instead of equation (3.114),

$$(\partial n_k^\alpha/\partial A)_{T,P,[n_i]} + (\partial n_k^\beta/\partial A)_{T,P,[n_i]} + \Gamma_{k(1,2)} + A(\partial \Gamma_{k(1,2)}/\partial A)_{T,P,[n_i]} = 0. \tag{3.138}$$

The last term in this equation, $A(\partial \Gamma_{k(1,2)}/\partial A)_{T,P,[n_i]}$ may be called, according to Gibbs terminology, the elasticity of surface excess for the k-th component. Consequently, the analysis performed above refers to zero elasticity of surface excesses. The elasticity is maximal if all of the k-th component is located at the interface. Then, $\Gamma_{k(1,2)}A = $ constant and

$$A(\partial \Gamma_{k(1,2)}/\partial A)_{T,P,[n_i]} = -\Gamma_{k(1,2)}. \tag{3.139}$$

Therefore, the elasticity of any surface excess ranges from 0 to $-\Gamma_{k(1,2)}$.

Equation (3.113), for determining the derivatives $\partial n_k^\alpha/\partial A$ and $\partial n_k^\beta/\partial A$, also remains unchanged. The solution of these equations should be substituted into the expression for the specific volume of surface formation which now instead of being the same as equation (3.128) takes the form

$$\tau^s_{\text{form}} = \tau^s_{(1,2)} + A \left(\frac{\partial \tau^s_{(1,2)}}{\partial A} \right)_{T,P,[n_i]} + \sum_k \left[v_k^\alpha \left(\frac{\partial n_k^\alpha}{\partial A} \right)_{T,P,[n_i]} + v_k^\beta \left(\frac{\partial n_k^\beta}{\partial A} \right)_{T,P,[n_i]} \right]. \tag{3.140}$$

Thus, this expression also reveals the elasticity effect: the term $A(\partial \tau^s_{(1,2)}/\partial A)_{T,p,[n_i]}$ is the elasticity of the excess volume with respect to components 1 and 2.

Comparing these equations with those in the preceding section, we see that the sums $\Gamma_{k(1,2)} + A(\partial \Gamma_{k(1,2)}/\partial A)_{T,P,[n_i]}$ and, $\tau^s_{(1,2)} + A(\partial \tau^s_{(1,2)}/\partial A)_{T,P,[n_i]}$ take the place of the quantities $\Gamma_{k(1,2)}$ and $\tau^s_{(1,2)}$, respectively. Hence, the solutions obtained above will be valid if we make the corresponding substitutions.

Let us consider a ternary system in which component 3 is entirely located at the interface. This means that its elasticity is maximal and that

$$\Gamma_{k(1,2)} + A(\partial\Gamma_{k(1,2)}/\partial A)_{T,P,[n_i]} = 0. \tag{3.141}$$

It then follows from the relations of the preceding section that

$$\tau_{form}^s = \tau_{(1,2)}^s + A(\partial\tau_{(1,2)}^s/\partial A)_{T,P,[n_i]}. \tag{3.142}$$

This is an important result because if a monolayer of a poorly soluble substance is placed at the interface, the excess volume $\tau_{(1,2)}^s$, with respect to the two solvents, can be quite large, about the monolayer thickness. Nevertheless, the volume of surface formation, τ_{form}^s, is still small because the elasticity of the excess volume $\tau_{(1,2)}^s$ strongly reduces this sum.

In conclusion, note that the concept of the elasticity of surface excesses, which was described above, is a direct consequence of the Gibbs elasticity of films (Gibbs, 1928), which is defined by $A(\partial\gamma/\partial A)$. It can be demonstrated that these quantities are closely interrelated: films are mechanically elastic only if surface excesses possess elasticity. As an illustration, consider a ternary two-phase system in which one of the components possesses high surface activity. Hansen's adsorption equation for this system is of the form

$$d\gamma = -s_{(1,2)}^s \, dT + \tau_{(1,2)}^s \, dp - \Gamma_{3(1,2)} \, d\mu_3. \tag{3.143}$$

The chemical potential of the third component, μ_3 depends on temperature, pressure, and its surface excesses

$$\mu_3 = \mu_3(T, P, \Gamma_{3(1,2)}). \tag{3. 144}$$

From the preceding equation one finds the Gibbs elasticity for a film to be:

$$A(\partial\gamma/\partial A)_{T,P,[n_i]} = -A\Gamma_{3(1,2)}(\partial\mu_3/\partial A)_{T,P,[n_i]}$$
$$= -A(\partial\Gamma_{3(1,2)}/\partial A)_{T,P,[n_i]}(\partial\mu_3/\partial \ln \Gamma_{3(1,2)})_{T,P}. \tag{3.145}$$

Thus, the Gibbs elasticity of a film is proportional to the elasticity of the surface excess of a poorly soluble component. This relation is very useful for many model calculations.

Experimental Data for the Volume of Surface Formation.

Experimentally, the volume of surface formation is determined with the help of equation (3.105) from the relationship between surface tension and pressure. As mentioned above, experiments are usually carried out in such a way that the τ_{form}^s is determined directly. It should be noted that this quantity is very small (cf., Table 3.1). For the binary system shown in Table 3.1 this quantity is positive and is less than 0.04 nm. For binary systems of decanol–water and dodecanol–water the τ_{form}^s is negative and amounts to -0.0087 nm and -0.0680 nm, respectively (Lin *et al.*, 1979). More detailed data on

Table 3.2. Specific volume of hexane–water interface formation in the presence of surface active substances at 303.15 K and 0.1 Mpa (Motomura, 1986).

Surface-active substance	Molar fraction $\times 10^{-3}$	Concentration m mole/kg	τ^s_{form} (nm)
Tetradecanol	1	-	0.0128
Octadecanol	0.5	-	0.0162
Dioctadecyl ether	0.5	-	0.0121
Sodium dodecylsulfate	-	3	0.0256
Dodecylammonium chloride	-	3	0.0251

the volume of surface formation for various systems can be found in the papers by Motomura (1977, 1980) and Lin et al. (1979).

For ternary and multi-component systems containing highly surface-active substances adsorption can be considerable and therefore the magnitude of $v_3^\alpha \Gamma_{3(1,2)}$ in equation (3.132) becomes very large. According to Motomura, it can reach up to 1 nm. However, Hansen's excess of volume, $\tau^s_{(1,2)}$, should also increase sharply such that the experimental value of τ^s_{form} may not undergo any significant changes. In fact, as may be seen from the data in Table 3.2, the value of τ^s_{form} for a number of surface-active substances at the hexane–water interface is low and does not exceed 0.03 nm.

In conclusion, it should be noted that only equations for the specific volume of the surface formation have been derived. However, by direct substitution of volumes with entropy, one can readily obtain analogous equations for the specific entropy of surface formation.

Figure 3.9 shows the dependence of the surface tension between water and a homologous series of alkanes (hexane, octane, decane, dodecane) on pressure (from 0.1 to 150 MPa), and on temperature over the range of 270 K to 310 K. As clearly seen in this figure, surface tension linearly increases with pressure and decreases with temperature, although in a complicated manner. Figure 3.10 presents the analogous

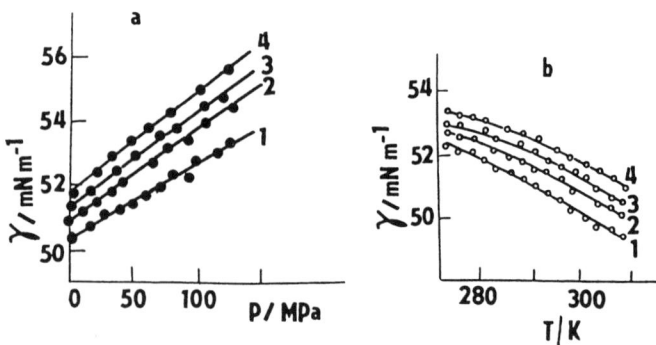

Figure 3.9. Variation of the interfacial tension of the oil–water interface with: (a) pressure at 298.15 K; (b) temperature under atmospheric pressure: (1) hexane; (2) octane; (3) decane; (4) dodecane (Motomura et al., 1983). With permission from Academic Press.

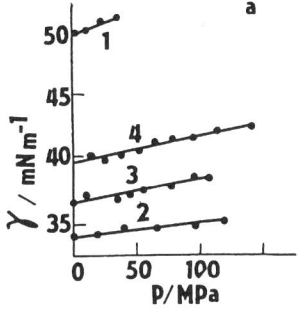
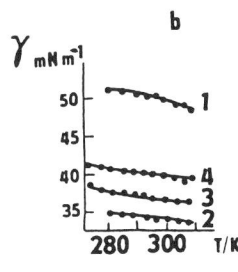

Figure 3.10. Variation of the interfacial tension of the oil–water interface with: (*a*) pressure at 298.15 K; (*b*) temperature under atmospheric pressure: (1) cyclohexane; (2) benzene; (3) butylbromide; (4) hexyl bromide (Motomura *et al.*, 1983). With permission from Academic Press.

dependence of surface tension between water and non-aqueous solvents with different structures. The curves are similar to those obtained for alkanes, differing only in their slope. The difference between the volume of surface formation of cyclohexane (0.0304 nm) and benzene (0.0106 nm) is considerable, although the difference between their molar volumes is small. The former is caused by the strong intermolecular interaction between benzene and water.

The data shown in Figs. 3.8 and 3.9 permit one to calculate the thermodynamic parameters and functions, which are summarized in Table 3.1.

3.5. CURVED INTERFACES

Hansen's method was initially developed only for flat interfaces. Later it was extended to curved boundaries (Markin and Volkov, 1988a). In most papers on capillary properties of curved surfaces, the authors consider only the simplest case—the spherical surface. This is because a description of a non-spherical interface involves a number of fundamental difficulties (Bridgman, 1952; Buff, 1956; Melrose, 1968) which can result in different definitions of surface tension (Boruvka *et al.*, 1985; Markin *et al.*, 1988; Trasatti and Parson, 1988). In this section we will use the traditional Gibbs definition of surface tension (Markin *et al.*, 1988).

Let us consider a two-phase system with a curved interface (Fig. 3.11) and assume that the sides of this system are oriented perpendicularly to the interface. The z-axis is also perpendicular to the interface and its positive direction is from the α phase to the β phase. Now draw an arbitrary dividing surface with coordinate z and area A. The geometrical properties of this surface are described by the main curvatures, c_x and c_y, which are positive if the α phase is convex; for instance, if it is a sphere. Instead of the principle curvatures, two other curvatures are often used—the average curvature

$$J = c_x + c_y, \tag{3.146}$$

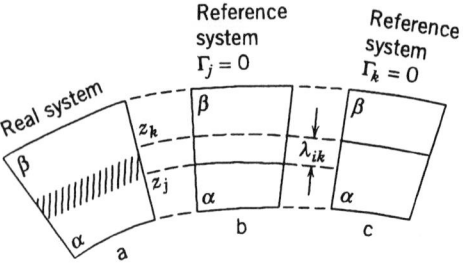

Figure 3.11. A two-phase system with a curved interface.

and the Gaussian curvature,

$$K = c_x c_y. \tag{3.147}$$

The value of parameters A, c_x, c_y, J, K depends on the choice of the interface, i.e., on its z-coordinate. This relationship is defined by the following equations (Markin *et al.*, 1988):

$$dA(z)/dz = J(z)A(z);$$
$$dJ(z)/dz = 2K(z) - J^2(z); \tag{3.148}$$
$$dK(z)/dz = -J(z)K(z).$$

Let us choose a certain dividing surface as a reference surface. For example, let it be the dividing surface with coordinate z_j on which the surface excess of component j is zero, $\Gamma_j = 0$. For the sake of brevity:

$$A(z_j) = A_j, \qquad J(z_j) = J_j, \qquad K(z_j) = K_j \tag{3.149}$$

From equation (3.148) we obtain

$$A(z) = [1 + (z - z_j)J_j + (z_j - z_j^2)K_j]A_j. \tag{3.150}$$

For a planar interface it was previously found that λ_{jk} stands for the distance between two dividing surfaces where the excesses of components j and k correspondingly become zero. Now one can find the same quantity for a curved interface. Let the coordinate of the second dividing surface be z_k and the volume enclosed between them be ΔV_{jk}. Then

$$\lambda_{jk} = z_k - z_j \tag{3.151}$$

and

$$\Delta V_{jk} = \int_{z_j}^{z_k} A(z)\,dz = [\lambda_{jk} + \tfrac{1}{2}J_j\lambda_{jk}^2 + \tfrac{1}{3}K_j\lambda_{jk}^3]A_j. \tag{3.152}$$

At the first surface the excess of component k is $\Gamma_k = \Gamma_{k(j)}$ and at the second surface s$\Gamma_k = 0$, therefore one can arrive at the following relation:

$$\Gamma_{k(j)} A_j + (c_k^\beta - c_k^\alpha) \Delta V_{jk} = 0. \tag{3.153}$$

By substituting ΔV_{jk} into this relation we obtain

$$\Gamma_{k(j)} + (c_k^\beta - c_k^\alpha)[\lambda_{jk} + \tfrac{1}{2} J_j \lambda_{jk}^2 + \tfrac{1}{3} K_j \lambda_{jk}^3] = 0. \tag{3.154}$$

If we assume that the terms $J_j \lambda_{jk}$ and $K_j \lambda_{jk}^2$ are small, a series solution can be used:

$$\lambda_{jk} \equiv \lambda_{jk}^0 [1 - \tfrac{1}{2} J_j \lambda_{jk}^0 + (J_j^2/2 - K_j/3)(\lambda_{jk}^0)^2] \tag{3.155}$$

where λ_{jk}^0 stands for the limiting value of λ_{jk} for a plane interface

$$\lambda_{jk}^0 = \Gamma_{k(j)}/(c_k^\alpha - c_k^\beta). \tag{3.156}$$

The terms in square brackets in equation (3.155) are the corrections for λ_{jk}^0 due to the curvature.

The sign of λ_{jk} and λ_{jk}^0 is determined in the same way as for the planar case. It is assumed that the j-th component is concentrated in the α phase (this can be the α solvent itself) and the k-th component in the β phase (for example, it can be the β solvent). If the main masses of the j and k components are separated from each other, as shown in Fig. 3.4, then $\lambda_{jk} > 0$. If these masses partially overlap (as shown in Fig. 3.5), $\lambda_{jk} < 0$. Finally, it should be noted that the term λ_{jk} is symmetrical with respect to the transposition of indices, $\lambda_{jk} = \lambda_{kj}$.

The transition from one dividing surface to another causes not only a change in the amount of the surface excess, but also changes the torques acting on the dividing surface. This, in principle, can be detected if the boundary is curved. However, in the planar case this effect usually remains unnoticed because the bending deformation is simply not considered. Nevertheless, the displacement of the dividing surface even in that case is not entirely without consequence. The complications arising from this displacement become more evident at curved interfaces where the displacement of the dividing surface results not only in a change in torques and curvatures, but also in a change in the area of the dividing surface and the magnitude of surface tension. The latter circumstance causes considerable inconvenience, and therefore, as is often noted in the literature (Rusanov, 1967), it is advisable to use the surface of tension as the dividing surface for a curved interface.

When one follows this rule, the geometrical interpretation of Gibbs method shown in Fig. 3.12 is modified. In the case of a real two-phase system of volume V_0 one draws the surface of tension t, which has area A_t, such that it divides the total volume into parts V_0^α and V_0^β (Fig. 3.12(a)). As a reference system, we use a system whose contour coincides with that of the initial real system (Fig. 3.12(b)). This reference system will be called the primary reference system and is related to the absolute surface excesses of component Γ_i. For this system, we can write the equation for the energy as follows:

$$dU = -P^\alpha \, dV_0^\alpha - P^\beta \, dV_0^\beta + T \, dS + \gamma \, dA_t + \sum_i \mu_i \, dn_i, \tag{3.157}$$

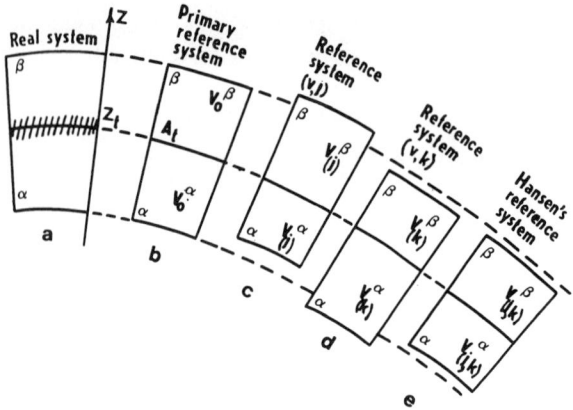

Figure 3.12. A novel geometrical interpretation of Gibbs and Hansen's methods for a curved surface.

hence

$$V_0^\alpha \, dP^\alpha + V_0^\beta \, dP^\beta - S \, dT - A_t \, d\gamma - \sum_i n_i \, d\mu_i = 0. \tag{3.158}$$

The Gibbs–Duhem equations hold in the α and β phases:

$$dP^\alpha - s^\alpha \, dT - \sum_i c_i^\alpha \, d\mu_i = 0 \tag{3.159}$$

$$dP^\beta - s^\beta \, dT - \sum_i c_i^\beta \, d\mu_i = 0. \tag{3.160}$$

Using the standard procedure described above, one can readily obtain the adsorption equation for the surface of tension. To do this, one can multiply equations (3.159) and (3.160) by V_0^α and V_0^β, and then subtract them from equation (3.158):

$$d\gamma = -s^s \, dT - \sum_i \Gamma_i \, d\mu_i. \tag{3.161}$$

At this dividing surface

$$s^s = (S - s^\alpha V_0^\alpha - s^\beta V_0^\beta)/A_t, \tag{3.162}$$

$$\Gamma_i = (n_i - c_i^\alpha V_0^\alpha - c_i^\beta V_0^\beta)/A_t. \tag{3.163}$$

During these algebraic transformations the pressure differentials disappear from the adsorption equation. However, the differentials of the chemical potentials of all components remain in this equation (if no additional restrictions are imposed on them). This is not surprising, since we have already eliminated two differentials, dP^α and dP^β, from the Gibbs–Duhem equation.

If this problem is approached by a more general method, equations (3.159) and (3.160) should be additionally multiplied by arbitrary volumes V^α and V^β and the equations thus obtained should be subtracted from equation (3.158). In this case the area of the dividing surface, A_t, remains unchanged. The terms V^α and V^β are the phase volumes in the reference system. Their magnitude is determined by a convention applied to the reference system. We shall designate these conventions in the general case by a subscript (conv) which can be replaced by a specific condition:

$$d\gamma = \tau^\alpha_{(conv)} \, dp^\alpha + \tau^\beta_{(conv)} \, dp^\beta - s^s_{(conv)} \, dt - \sum_i \Gamma_{i(conv)} \, d\mu_i. \qquad (3.164)$$

Here we use the following designations:

$$\tau^\alpha_{(conv)} = (V^\alpha_0 - V^\alpha_{(conv)})/A_t, \qquad (3.165)$$

$$\tau^\beta_{(conv)} = (V^\beta_0 - V^\beta_{(conv)})/A_t. \qquad (3.166)$$

The terms $\tau^\alpha_{(conv)}$ and $\tau^\beta_{(conv)}$ represent the surface density of the volume excess of the α and β phases in the reference system (conv), and they play a key role in determining other surface excesses. The surface volume excess of the entire system is:

$$\tau^s_{(conv)} = \tau^\alpha_{(conv)} + \tau^\beta_{(conv)}. \qquad (3.167)$$

All these quantities have the dimension of length. The other surface excesses are:

$$s^s_{(conv)} = (S - s^\alpha V^\alpha_{(conv)} - s^\beta V^\beta_{(conv)})/A_t = s^s + s^\alpha \tau^\alpha_{(conv)} + s^\beta \tau^\beta_{(conv)} \qquad (3.168)$$

$$\Gamma_{i(conv)} = (n_i - c^\alpha_i V^\alpha_{(conv)} - c^\beta_i V^\beta_{(conv)})/A_t = \Gamma_i + c^\alpha_i \tau^\alpha_{(conv)} + c^\beta_i \tau^\beta_{(conv)}. \qquad (3.169)$$

Now we shall demonstrate how these general equations can be applied to actual cases.

Gibbs Reference System (v, j). In this case, we select a reference system having the same volume as the real system, but require that the adsorption of the j-th component vanish (Fig. 3.12(c)). In mathematical language this means:

$$V^\alpha_{(v,j)} + V^\beta_{(v,j)} = V^\alpha_0 + V^\beta_0, \qquad (3.170)$$

$$\Gamma_{j(v,j)} = 0. \qquad (3.171)$$

At the same time

$$\tau^\alpha_{(v,j)} + \tau^\beta_{(v,j)} = 0,$$

$$\Gamma_j + c^\alpha_j \tau^\alpha_{(v,j)} + c^\beta_j \tau^\beta_{(v,j)} = 0. \qquad (3.172)$$

The solution of this set of equations yields the surface volume excesses:

$$\tau^\alpha_{(v,j)} = \Gamma_j/(c^\beta_j - c^\alpha_j); \qquad \tau^\beta_{(v,j)} = \Gamma_j/(c^\alpha_j - c^\beta_j); \qquad \tau^s_{(v,j)} = 0. \quad (3.173)$$

In this case the adsorption equation can be written in two ways:

$$d\gamma = \tau^\alpha_{(v,j)}\, d(P^\alpha - P^\beta) - s^s_{(v,j)}\, dT - \sum_{i \neq j} \Gamma_{i(v,j)}\, d\mu_i \qquad (3.174)$$

or

$$d\gamma = \tau^\beta_{(v,j)}\, d(P^\beta - P^\alpha) - s^s_{(v,j)}\, dT - \sum_{i \neq j} \Gamma_{i(v,j)}\, d\mu_i \qquad (3.175)$$

where

$$s^s_{(v,j)} = s^s + (s^\alpha - s^\beta)\tau^\alpha_{(v,j)} \qquad (3.176)$$

$$\Gamma_{i(v,j)} = \Gamma_i + (c^\alpha_i - c^\beta_i)\tau^\beta_{(v,j)}. \qquad (3.177)$$

It can be readily seen that these surface excesses do not differ from the planar case except that the adsorption equation includes an additional term with the differential of the pressure difference.

Hansen's Reference System (j, k). One can start with the dividing surface and select the volumes $V^\alpha_{(j,k)}$ and $V^\beta_{(j,k)}$ such that the surface excesses of the j-th and k-th components would become zero simultaneously:

$$\Gamma_j + c^\alpha_j \tau^\alpha_{(j,k)} + c^\beta_j \tau^\beta_{(j,k)} = 0,$$
$$\Gamma_k + c^\alpha_k \tau^\alpha_{(j,k)} + c^\beta_k \tau^\beta_{(j,k)} = 0. \qquad (3.178)$$

It is clear that these conventions coincide with the conventions applied in the planar case. Therefore, the solution is the same:

$$\tau^\alpha_{(j,k)} = (c^\beta_j \Gamma_k - c^\beta_k \Gamma_j)/(c^\alpha_j c^\beta_k - c^\alpha_k c^\beta_j);$$
$$\tau^\beta_{(j,k)} = (c^\alpha_k \Gamma_j - c^\alpha_j \Gamma_k)/(c^\alpha_j c^\beta_k - c^\alpha_k c^\beta_j). \qquad (3.179)$$

The adsorption equation takes the following form:

$$d\gamma = \tau^\alpha_{(j,k)}\, d(P^\alpha - P^\beta) + \tau^\beta_{(j,k)}\, dP^\beta - (s^s + s^\alpha \tau^\alpha_{(j,k)} + s^\beta \tau^\beta_{(j,k)})\, dT$$
$$- \sum_{i \neq j,k} (\Gamma_i + c^\alpha_i \tau^\alpha_{(j,k)} + c^\alpha_i \tau^\beta_{(j,k)})\, d\mu_i \qquad (3.180)$$

Thus, for a curved interface the terms $\tau^\alpha_{(j,k)}$ and $\tau^\beta_{(j,k)}$ also represent the transition coefficients from the absolute to the relative Hansen's excesses. If we wish to write the adsorption equation not in terms of volume concentrations of the components c_i, but in terms of molar fractions x_i, we should use the transition coefficient Θ_i.

The Reference System Defined by the (v^α, j) *Convention.* This convention means that

$$\tau^\alpha_{(v^\alpha, j)} = 0,$$
$$\Gamma_j + c_j^\beta \tau^\beta_{(v^\alpha, j)} = 0,$$
$$\text{(3.181)}$$

hence

$$\tau^\beta_{(v^\alpha, j)} = -\Gamma_j / c_j^\beta, \qquad (3.182)$$

and the adsorption equation takes the form

$$d\gamma = -\frac{\Gamma_j}{c_j^\beta} dP^\beta - \left(s^s - \frac{\Gamma_j}{c_j^\beta} s^\beta \right) dT - \sum_{i \neq j} \left(\Gamma_i - \frac{\Gamma_j c_i^\beta}{c_j^\beta} \right) d\mu_i. \qquad (3.183)$$

Earlier, we noted that for a planar interface Hansen's adsorption equation is especially convenient for analyzing the relation between surface tension and pressure. The same holds true for the curved interface although, in this case, the problem becomes more complicated.

Example: A Binary System. Let us consider the simplest case of a two-phase binary system consisting of α and β solvents. Then, in accordance with equation (3.180), we have:

$$\begin{aligned}
d\gamma &= \tau^\alpha_{(\alpha, \beta)} dP^\alpha + \tau^\beta_{(\alpha, \beta)} dP^\beta - s^s_{(\alpha, \beta)} dT \\
&= \tau^\alpha_{(\alpha, \beta)} d(P^\alpha - P^\beta) + \tau_{(\alpha, \beta)} dP^\beta - s^s_{(\alpha, \beta)} dT.
\end{aligned} \qquad (3.184)$$

It is clear from this relation that the thermodynamic definition of the surface excess of volume for a binary system is:

$$\tau_{(\alpha, \beta)} = \left(\frac{\partial \gamma}{\partial P^\beta} \right)_{T, (P^\alpha - P^\beta)}. \qquad (3.185)$$

In calculating these derivatives, the pressure and temperature differences between the phases must be kept constant. Furthermore

$$\tau^\alpha_{(\alpha, \beta)} = \left(\frac{\partial \gamma}{\partial (P^\alpha - P^\beta)} \right)_{T, P^\beta} = \left(\frac{\partial \gamma}{\partial P^\alpha} \right)_{T, P^\beta} \qquad (3.186)$$

$$\tau^\beta_{(\alpha, \beta)} = \left(\frac{\partial \gamma}{\partial (P^\beta - P^\alpha)} \right)_{T, P^\alpha} = \left(\frac{\partial \gamma}{\partial P^\beta} \right)_{T, P^\alpha}. \qquad (3.187)$$

If in equation (3.184) we ignore the term with a pressure difference, we obtain the equation for a planar interface. Therefore, the specifics of a curved interface are entirely determined by the term which contains the pressure drop $P^\alpha - P^\beta$. This

find that $\gamma/RK^{\alpha}_{comp} = (0.02 \text{ nm})/R$. Since the radius of an interface is always much greater than 0.02 nm, this ratio is small and we can ignore the contribution of compressibility. The term $1/K^{\alpha}_{comp}$ in braces in equation (3.200) can be neglected. For reasonable curvatures the coefficient $(1 - J\tau^{\alpha}_{(\alpha,\beta)})$ does not differ significantly from 1 and therefore this term plays a significant role only for very small droplets.

These examples show that Hansen's method not only extends and generalizes Gibbs method but gives us deeper insight into the nature of interfacial phenomena and provides a convenient apparatus for their description.

3.6. VARIABLES OF THE SYSTEM

The Phase Rule. Any thermodynamic system can be described by numerous parameters and it is very important to know how many of them are independent. The answer to this question is called the phase rule and it is one of the best known results of Gibbs thermodynamics for heterogeneous systems. For the case of planar interfaces it can be formulated as follows: if a system consists of r independent components and of n phases, then the number of its degrees of freedom (or the variance) is

$$f = r - n + 2. \tag{3.201}$$

This rule has been generalized to more complicated cases, in which it was demonstrated that in the absence of two-dimensional phase transitions, the presence of plane interfaces does not affect equation (3.201). This is because the new variable—surface tension—can be compensated by the Gibbs adsorption equation.

Electrocapillarity theory additionally takes into account the interfacial potential. However, a thermodynamic description of an electrolyte system with charged interfaces involves a number of specific features related to perfectly polarizable and non-polarizable interfaces. Some authors consider these to be qualitatively different, while others regard one system as a particular case of another.

Direct application of the phase rule (3.201) to electrolyte systems requires a correct estimate of the number of independent components of the system. For example, a solvent α may contain a strong electrolyte BA that dissociates completely into components A^- and B^+ such that α, A^-, B^+ are present simultaneously in the α phase and the number of independent components in this phase is two. If a solvent β contains a strong electrolyte DC dissociating into ions C^- and D^+ the number of independent components in this phase is also two. Let us bring these phases into contact and wait until equilibrium is established (Fig. 3.13). Suppose that unlike the solvents, α and β, the electrolytes BA and DC cannot pass from one phase to another. Then assuming in equation (3.201) $r = 4$, $n = 2$, we could conclude that the variance of the system is equal to four. If we have preset temperature, pressure, and initial electrolyte concentrations in two phases, it would seem that the list of independent variables were complete. However, this is not the case: it is well known that in such a system the interfacial potential can be easily varied by means of an external electric circuit (Markin and Volkov, 1989a).

Now let us assume that the electrolytes BA and DC can cross the interface (Fig.

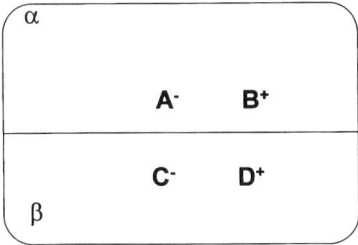

Figure 3.13. Partitioning of completely dissociated electrolytes BA and DC between two immiscible phases α and β in the presence of a perfectly polarizable interface.

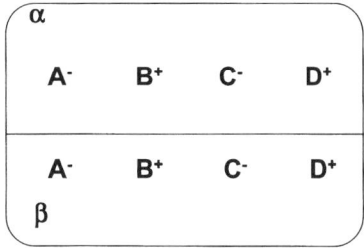

Figure 3.14. Partitioning of completely dissociated electrolytes BA and DC between two immiscible phases α and β in the presence of a reversible interface.

3.14). Assuming as before $r = 4$ and $n = 2$, we would have the same number of independent variables in the system, namely four. Therefore, at preset temperature, pressure, and initial concentrations of AB and CD, we have no reason to expect that the interfacial potential difference should undergo an independent change.

Estimating the number of independent variables for a two-phase electrolytic system was carried out by Parsons (1980) and later by Kakiuchi and Senda (1983b) for the particular case of a planar interface and absence of phase transition. Because of the importance of this problem for electrocapillarity theory, we shall consider its general case without imposing the above-mentioned restrictions. Furthermore, we will consider the methods of choosing independent variables of the system that are most suitable for the electrocapillarity theory.

Let the system consist of n electrolytic and non-electrolytic phases with plane interfaces. We shall designate the total number of different components in the system by r; not all components can be independent since homogeneous and heterogeneous reactions may occur in the system. The number of independent reactions will be designated by w. This means that the chemical equilibrium imposes w constraints on the variables of the system.

Let an arbitrary α phase contain r_t^α different components, and an arbitrary component i be present in n phases. Each α phase may, generally speaking, be described by temperature, pressure, potential, and r_t^α concentration variables. Before imposing constraints, one may formally consider that there are $\sum_\alpha r_t^\alpha$ concentration variables in the system. The chemical equilibrium imposes w constraints. The phase equilibrium, defined by the equality of the electrochemical potentials of particles in different phases, imposes on each i-th component $(n_i - 1)$ constraints, so that for all components there are $\Sigma_i(n_i - 1)$ constraints. Finally, in electrolytic systems it is necessary to take into account the electroneutrality condition. This condition is imposed on a system as a whole rather than on each phase separately. The reason for this is that ions can pass from one phase to another or into the interfacial layer, where their charges are compensated for by ions from the other phase. Therefore, the electroneutrality condition gives only one constraint. Thus the total number of constraints on the components of the α phase is $1 - r_t + \Sigma_i n_i + w$. Since $\Sigma_i n_i = \Sigma_\alpha r_t^\epsilon$, the number of independent components of the system is

$$r = r_t - 1 - w. \tag{3.202}$$

We also have to take into consideration that the Gibbs–Duhem equation is valid in each phase. This imposes n more constraints, bringing the number of independent concentration variables, for example in the Gibbs adsorption equation, to $r_t - n - 1 - w$. Temperature and pressure can form additional variables, so that the number of independent variables of the system defining its state becomes

$$f = r_t - n + 1 - w. \tag{3.203}$$

We have obtained the phase rule for a system with planar interfaces. If the interfaces are curved, then instead of being described by one pressure, the system will be described by n unequal pressures in each phase. In that case, the phase rule assumes the form

$$f = r_t - w. \tag{3.204}$$

It is not difficult to generalize this formula to the case when only some interfaces are curved and others are flat. Thus equation (3.203) gives a natural, convenient, and general representation of the phase rule in electrolytic systems.

Equation (3.203) can serve as a basis for description of electrocapillary phenomena. It easily resolves the difficulty mentioned at the beginning of this section. In the first example considered ($r_t = 6, n = 2, w = 0$), equation (3.203) gives the correct result ($f = 5$). This means that besides temperature, pressure, and electrolyte concentrations in two phases, it is possible to independently vary the interfacial potential difference $\Delta_\beta^\alpha \phi$. The error in the original reasoning lies in the fact that the number of independent components in this system is not four, but in accordance with equation (3.202), is five. This is understood to mean that besides the total amount of the electrolytes AB and CD introduced into the system, we may, without perturbing the equilibrium, vary arbitrarily the content of another component, say the ion A^-. In fact, if the electrode located in the α phase is reversible with respect to the ion A^-, then application of a potential difference to the interface means that a certain number of these ions are introduced into (or removed from) the α phase. Of course, additional ions do not remain in the bulk phase, but accumulate at the interface, forming the corresponding surface excesses. This, incidentally, indicates that either the number of ions A^- in the system, or the interfacial potential, may be used equally well as an additional variable. Another detail: if the electrode in the β phase is reversible with respect to D^+ ions, it can also introduce into the system (or remove from it) a certain number of ions. However, this quantity is not independent since additional D^+ ions must compensate for the charge of the A^- ions.

In the second example the number of independent degrees of freedom is also not four but five. To determine the ionic composition of one of the solutions (the composition of the second solution is determined automatically) it is necessary to preset the volume concentrations of three ions. Together with temperature and pressure, this uses up all five degrees of freedom. Hence, without changing volume concentrations of substances (or T and p) it is impossible to change the interfacial potential difference.

Consider another example: the α phase consists of r_t^α neutral and charged components, and the β phase of r_t^β components. There are no common components in the system, but a redox reaction between two redox couples from different phases occurs at the interface. In this case $w = 1$ and $f = r_t^\alpha + r_t^\beta - 2$. To determine the composition of the bulk phase one must preset the concentration of all solutes. Under the condition of electroneutrality this means that for the α phase one must preset $r_t^\alpha - 2$ variables. In the β phase one has to preset $r_t^\beta - 2$ variables. Together with temperature and pressure this means that $r_t^\alpha + r_t^\beta - 2$ variables must be preset. In other words, determining the composition of the bulk phases, T and p, we use up all the degrees of freedom. Consequently, the interfacial potential difference cannot now be imposed on the system from outside; to change this potential one must change the composition of the bulk phases.

Finally, suppose that in the system considered above not one but two redox reactions between the corresponding couples from different phases take place. Then $w = 2$ and the variance of the system is $f = r_t^\alpha + r_t^\beta - 3$. If, as before, we determine the composition of the phases, temperature, and pressure, we have to preset $r_t^\alpha + r_t^\beta - 2$ variables, whose number exceeds the variance of the system. Therefore such a system cannot be in equilibrium. Heterogeneous reactions will proceed in it until they change volume concentrations.

Choice of Independent Variables. The phase rule defines the number of independent variables that can be selected from a large set of variables describing the system under consideration. It is of utmost importance to have a convenient choice of concentration and other variables especially when operating with the electrocapillary equation. This problem was considered by Mohilner (1966), Parsons (1980), Kakiuchi and Senda (1983b). We shall describe how to choose concentration variables both for polarizable and non-polarizable interfaces.

In the most general representation, Gibbs adsorption includes the differentials of temperature, pressure, and electrochemical potentials of all the r_t components, i.e., it contains $r_t + 2$ differentials (aside from the surface tension differential):

$$d\gamma = -s^s\,dT + \tau^s\,dP - \sum_i \Gamma_i\,d\bar\mu_i \qquad (3.205)$$

However, in practice it is more convenient to operate with chemical potentials of neutral components, rather than with electrochemical potentials of ions. Let a two-phase system contain r_h neutral components, r_c types of cations, r_a types of anions, so that

$$r_t = r_h + r_c + r_a. \tag{3.206}$$

The number of independent degrees of freedom of the system is given by equation (3.202) and the number of independent differentials on the right side of equation (3.205) should be the same.

Individual components of the system, both charged and neutral, may be present in both phases, or in only one. Let us conditionally divide all charged particles into two groups α and β. The particles that can be present only in the α phase shall belong to the α group, and the particles that can be present only in the β phase shall belong to the β group. The rest of the particles will be distributed arbitrarily between two groups. Usually the components are divided into three groups (Kakiuchi and Senda, 1983b; Koenig, 1934; Mohilner, 1966; Parsons, 1980) which makes the formulas more bulky. We shall divide them into two groups. Let group α contain r_c^α types of cations, which we shall designate by j, and r_a^α types of anions, which we shall designate by k. Indexes l and m will be used to designate cations and anions in group β, for which the number of types of cations and anions are r_c^β and r_a^α respectively. The sum of these numbers give the total amount of different types of cations and anions in the entire system:

$$r_c^\alpha + r_c^\beta = r_c, \qquad r_a^\alpha + r_a^\beta = r_a. \tag{3.207}$$

In equation (3.205) all the ions are formally equivalent, but when electrical measurements are performed two of them become different from the others. Suppose that the potentials in the α and β solutions are measured with the use of reference electrodes R_α and R_β, with electrode R_α being reversible with respect to the cation j' (belonging to the α group), and the electrode R_β to anion m' (belonging to the β group). These ions j' and m' are called indicator ions (Mohilner, 1966).

Any system containing several types of anions and cations can be prepared in many different ways, but for thermodynamic analysis the solutions may be assumed to be formed of neutral binary salts. As usual, for the sake of brevity, we call "salts" any electrolyte, including acids and bases. Since group α contains r_c^α cations and r_a^α anions, from these ions it is formally possible to prepare $r_c^\alpha r_a^\alpha$ different binary salts. From these we shall choose the minimum combination sufficient for an unambiguous determination of the system. This set can be made up of $(r_c^\alpha + r_a^\alpha - 1)$ neutral salts and one indicator ion, so that the number of differentials corresponding to the ion groups remains equal to $r_c^\alpha + r_c^\alpha$. The salts are chosen as follows. Select an arbitrary anion k' from the α group, which together with the indicator cation j forms the indicator salt $j' k'$. In addition, include in the α groups $r_c^\alpha + r_a^\alpha - 2$ binary salts containing the cation j' or the anion k'. We shall designate these salts by $j' k$ and jk'.

Now construct the chemical formula of the salt jk. Let this salt be made up of cation C_j, and anion A_k with the stoichiometric coefficients v_{jk}^+ and v_{jk}^-. Then the formula for the salt is $(C_j)_{v_{jk}^+}(A_k)_{v_{jk}^-}$ and its dissociation equation is

$$(C_j)_{v_{jk}^+}(A_k)_{v_{jk}^-} \Leftrightarrow v_{jk}^+(C_j) + v_{jk}^-(A_k). \tag{3.208}$$

If z_j and z_k are the charge numbers of ions, then the condition of the electroneutrality of the salt is

$$v_{jk}^+ z_j + v_{jk}^- z_k = 0. \tag{3.209}$$

The chemical potentials of the salts consist of the electrochemical potentials of the ions contained in them:

$$\mu_{j'k'} = v_{j'k'}^+ \bar\mu_{j'} + v_{j'k'}^- \bar\mu_{k'} \tag{3.210}$$

$$\mu_{jk'} = v_{jk'}^+ \bar\mu_j + v_{jk'}^- \bar\mu_{k'} \qquad (j \neq j') \tag{3.211}$$

$$\mu_{jk'} = v_{j'k}^+ \bar\mu_{j'} + v_{j'k}^- \bar\mu_k \qquad (k \neq k'). \tag{3.212}$$

We have obtained $r_c^a + r_a^a - 1$ relations from which all the electrochemical potentials can be found except one, which shall be the electrochemical potential of the indicator ion j'. Similar transformations can be performed for β group ions. This will permit substitution of the differentials of the electrochemical potentials of ions by $r_c + r_a - 2$ differentials of neutral salts and 2 differentials of indicator ions, i.e., the total amount of differentials in the sum with respect to ions will as before be equal to $r_c + r_a$. As will be shown below, another variable may be excluded by virtue of the electroneutrality condition, so that the total number of ionic variables reduces to $r_c + r_a - 1$.

Further reduction of the number of variables in equation (3.205) can be achieved by using the Gibbs–Duhem equation in two bulk phases. Usually this is accomplished by means of a number of algebraic transformations (Girault and Schiffrin, 1984c). The final result, however, can be obtained at once without any transformations if Hansen's description of the interface is used and the α and β solvents are chosen as reference substances (Good, 1982, 1986; Hansen, 1962; Markin and Volkov, 1988a, 1988b). Then by definition the surface excesses Γ_α and Γ_β vanish and the rest assume the form of relative Hansen's surface excesses $\Gamma_{i(\alpha,\beta)}$. The number of differentials in (3.205) is reduced by two and becomes equal to $r_h + r_c + r_a - 1$. If chemical reactions do not occur in the system, this number is equal to the variance of the system and all the differentials in the Gibbs adsorption equation thus transformed are independent.

Hansen's relative excesses $\Gamma_{i(\alpha,\beta)}$ are expressed in terms of the absolute excesses of the same i-th components and their concentrations c_i^α and c_i^β (or mole fractions x_i^α and x_i^β) in the α and β phases according to the equation

$$\Gamma_{i(j,k)} = \Gamma_i + \Theta_{(j,k)}^\alpha x_i^\alpha + \Theta_{(j,k)}^\beta x_i^\beta, \tag{3.213}$$

where

$$\Theta_{jk}^\alpha = \frac{x_j^\beta \Gamma_k - x_k^\beta \Gamma_j}{x_j^\alpha x_k^\beta - x_k^\alpha x_j^\beta}; \qquad \Theta_{jk}^\beta = \frac{x_j^\alpha \Gamma_k - x_k^\alpha \Gamma_j}{x_j^\alpha x_k^\beta - x_k^\alpha x_j^\beta}. \tag{3.214}$$

The quantities Θ_{jk}^α and Θ_{jk}^β are the transition coefficients from an arbitrary initial system to the concrete reference system, or from absolute excesses to relative excesses; generally speaking, j and k may imply any components of the system. If a component is insoluble in a given phase, the corresponding concentration c_i is zero. In principle, it is quite admissible to restrict oneself to the expressions given above (3.213), but it often becomes necessary to pass from ion concentrations to concentrations of neutral salts, just as was done above in substituting the electrochemical potentials of ions by the chemical potentials of salts. In substituting the concentrations, the procedure used is somewhat different since unlike the chemical potentials, the salt concentrations in two phases are not the same. For this reason it is necessary to separately calculate the concentrations for α and β phases and the binary salts for different solutions must be prepared using different bases.

In the α phase we have selected the indicator salt $j' k'$. In addition, we selected other $r_c^\alpha + r_a^\alpha - 2$ salts that contain either the cation j', or anion k'. Their designations are $j' k$, $j' m$, jk', lk'. The concentrations of these salts are chosen such that after their dissociation the correct numbers of all ions are obtained. From equation (3.209) we find

$$c_{j'k}^\alpha = c_k^\alpha / v_{j'k}^-, \quad (k \neq k'); \qquad c_{j'm}^\alpha = c_m^\alpha / v_{j'm}^- \qquad (3.215)$$

$$c_{jk'}^\alpha = c_j^\alpha / v_{jk'}^-, \quad (j \neq j'); \qquad c_{lk'}^\alpha = c_l^\alpha / v_{lk'}^-. \qquad (3.216)$$

The concentration of the salt $j' k'$ is calculated differently. A certain portion of ions j' and k' is "introduced" into the solution with the salts indicated above. Therefore, the concentration $c_{j'k'}^\alpha$ should give the additional amount of substance $j' k'$ (per unit volume) to be introduced into the system to obtain the desired solution. It can be easily found that

$$c_{j'k'}^\alpha = \left(c_{j'}^\alpha - \sum_{k \neq k'} v_{j'k}^+ c_{j'k}^\alpha - \sum_m v_{j'm}^+ c_{j'm}^\alpha \right) \Big/ v_{j'k'}^+$$

$$= \left(c_{k'}^\alpha - \sum_{j \neq j'} v_{jk'}^- c_{jk'}^\alpha - \sum_l v_{lk'}^- c_{lk'}^\alpha \right) \Big/ v_{j'k'}^-. \qquad (3.217)$$

This double equation is the result of the condition of electroneutrality of the solution. Generally speaking, the quantity $c_{j'k'}^\alpha$ given by this equation may prove to be negative which is usually ignored in the literature, but we must not have any misgivings about this formal result. It simply means that the rest of the binary salts arbitrarily chosen in this procedure introduce into the solution more ions j' and k' than are needed and therefore one should not add the salt $j' k'$, but rather remove the excessive number of ions.

Solution β is composed in a similar manner. The indicator salt $1' m'$ is taken as the reference salt and on this basis the binary salts $1'k$, $l'm$, jm', lm' are made up. Their concentrations are

$$c_{l'k}^\beta = c_k^\beta / v_{l'k}^-, \qquad c_{l'm}^\beta = c_m^\beta / v_{l'm}^-, \qquad (m \neq m')$$

$$c_{jm'}^\beta = c_j^\beta / v_{jm'}^+, \qquad c_{lm'}^\beta = c_l^\beta / v_{lm'}^+, \qquad (l \neq l')$$

$$c^\beta_{l'm'} = \left(c^\beta_{l'} - \sum_k v^+_{l'k} c^\beta_{lk'} - \sum_{m \neq m'} v^+_{l'm} c^\beta_{l'm} \right) \Big/ v^+_{l'm}$$

$$= \left(c^\beta_{m'} - \sum_j v^+_{jm'} c^\beta_{j'm} - \sum_{l \neq l'} v^-_{lm'} c^\beta_{lm'} \right) \Big/ v^-_{l'm'}. \qquad (3.218)$$

From these equations the ion concentrations are found, which are then substituted into the expression for the ion excesses.

The final equations may be expressed either in terms of the concentrations of binary salts or as their mole fractions (Markin and Volkov, 1988a).

$$\Gamma_{j(\alpha,\beta)} = \Gamma_j + \Theta_{(\alpha,\beta)} v^+_{jk'} x^\alpha_{jk'} + \Theta_{(\alpha,\beta)} v^+_{jm'} x^\beta_{jm'}, \qquad (j \neq j');$$

$$\Gamma_{k(\alpha,\beta)} = \Gamma_k + \Theta_{(\alpha,\beta)} v^+_{j'k} x^\alpha_{j'k} + \Theta_{(\alpha,\beta)} v^-_{l'k} x^\beta_{l'k'}, \qquad (k \neq k');$$

$$\Gamma_{l(\alpha,\beta)} = \Gamma_l + \Theta_{(\alpha,\beta)} v^+_{lk'} x^\alpha_{lk'} + \Theta_{(\alpha,\beta)} v^+_{lm'} x^\beta_{lm''}, \qquad (l \neq l');$$

$$\Gamma_{m(\alpha,\beta)} = \Gamma_m + \Theta^\alpha_{(\alpha,\beta)} v^-_{j'm} x^\alpha_{j'm} + \Theta^\beta_{(\alpha,\beta)} v^-_{l'm'} x^\beta_{l'm'}, \qquad (m \neq m');$$

$$\Gamma_{j'(\alpha,\beta)} = \Gamma_{j'} + \Theta^\alpha_{(\alpha,\beta)} \left(\sum_k v^+_{j'k} x^\alpha_{j'k} + \sum_m v^+_{j'm} x^\alpha_{j'm} \right) + \Theta^\beta_{(\alpha,\beta)} v^-_{j'm} x^\beta_{j'm};$$

$$\Gamma_{k'(\alpha,\beta)} = \Gamma_{k'} + \Theta^\alpha_{(\alpha,\beta)} \left(\sum_j v^-_{jk'} x^\alpha_{jk'} + \sum_l v^-_{lk'} x^\alpha_{lk'} \right) + \Theta^\beta_{(\alpha,\beta)} v^-_{l'k'} x^\beta_{j'k'};$$

$$\Gamma_{l'(\alpha,\beta)} = \Gamma_{l'} + \Theta^\alpha_{(\alpha,\beta)} v^+_{l'k'} x^\alpha_{l'k'} + \Theta^\beta_{(\alpha,\beta)} \left(\sum_k v^+_{l'k'} x^\beta_{l'k} + \sum_m v^+_{l'm} x^\beta_{l'm} \right);$$

$$\Gamma_{m''(\alpha,\beta)} = \Gamma_{m''} + \Theta^\alpha_{(\alpha,\beta)} v^-_{j'm'} x^\alpha_{j'm'} + \Theta^\beta_{(\alpha,\beta)} \left(\sum_j v^-_{l'm'} x^\beta_{j'm'} + \sum_l v^-_{lm'} x^\beta_{lk'} \right). \qquad (3.219)$$

B

ELECTRIFIED INTERFACES

4

INTERFACIAL POTENTIALS

4.1. BOUNDARY POTENTIAL DIFFERENCE

An electric potential difference is established at the interface between two immiscible liquids. This is called the boundary potential difference or, briefly, the interfacial potential. Discussions of the nature of the potential drop at the interface between two immiscible electrolyte solutions (ITIES) dates back to the beginning of the century. The interfacial potentials were first classified by E. Lange and K. P. Miščenko (1930). Figure 4.1 shows the interfacial potentials at the interface between two immiscible phases α and β, as well as between these phases and a vacuum. The potential ψ^α produced by the surface charge of phase α and measured in vacuum in the vicinity of the surface is called the Volta potential or the external potential. The potential ϕ^α produced in the bulk phase α and calculated from infinity in a vacuum is called the Galvani potential or the internal potential. These two potentials differ by the surface potential drop χ^α.

The difference of Galvani potentials $\Delta_\beta^\alpha \phi = \phi^\alpha - \phi^\beta$ is of primary interest in the study of the interface between phases α and β. This difference is determined by the distribution of charged and dipolar particles near the interface. The dipole component of the interfacial potential difference $\Delta_\beta^\alpha \phi_i$ is sometimes considered as a separate term to obtain the quantity $\Delta_\beta^\alpha \phi - \Delta_\beta^\alpha \phi_i$ caused by the distribution of ions. This separation is useful in cases where the dipole potential drop does not depend on the total interfacial potential difference produced by ion distribution. One may therefore assume that changes in the interfacial potential difference are described only by the ionic contribution. However, it is uncertain whether the interfacial potential difference can be separated into two independent components caused by the ionic and dipole contributions, so that interpretation of results is to some extent approximate.

One should also bear in mind that, contrary to the claim frequently encountered in the literature (Hachisu, 1984), the potential jump $\Delta_\beta^\alpha \phi_h$ at the interface between two

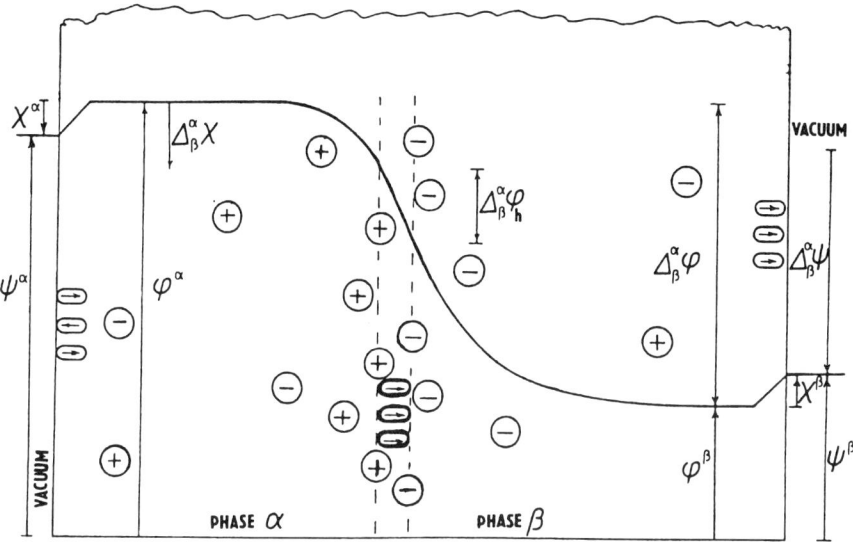

Figure 4.1. Interfacial potentials in a liquid–liquid system. Abbreviations: ψ^α—the external potential produced by the surface charge of phase α measured near the phase boundary; ϕ^α—the internal potential of phase α (these two potentials differ by the surface potential drop χ^α); $\Delta^\alpha_\beta \phi$—the Galvani potential difference; $\Delta^\alpha_\beta \psi$—the Volta potential difference; $\Delta^\alpha_\beta \phi_r$—the dipole component of the interfacial Galvani potential difference.

immiscible electrolyte solutions is not equal to the potential difference at the boundaries of these phases with a vacuum even in the case of specific adsorption:

$$\Delta^\alpha_\beta \phi_h \neq \Delta^\alpha_\beta \chi = \chi^\alpha - \chi^\beta.$$

The reason is the mutual influence of the phases on the orientation of the dipolar molecules in the boundary layers. Table 4.1 illustrates the thermodynamic feasibility of experimental measurement of boundary potentials.

The equilibrium potentials at a reversible or non-polarizable interface are divided into three types depending on which species pass between phases:

1. Potentials which arise in an equilibrium system where all the ions are capable of passing from one phase to a second are called *distribution potentials*.

2. If the phases in contact contain ions, some of which are capable of crossing the boundary while others are not, *Donnan potentials* or membrane potentials are generated (Donnan, 1911). The transfer of ions may be prevented by a semipermeable membrane or by chemical adsorption to ion exchangers. If ions of only one kind can cross the interface, then the membrane potential difference is called the *Nernst potential* (Nernst, 1888, 1889).

If two immiscible electrolyte solutions exchange electrons and are in equilibrium with respect to electrons, then an *oxidation–reduction potential* is established at the interface.

Table 4.1. Feasibility of measuring interfacial potentials in immiscible liquid phases α and β.

Boundary Potentials	Feasibility of Measurement
1. External potential of one phase ψ	Can be measured
2. Volta potential $\Delta_\beta^\alpha \psi$	Can be measured
3. Surface potential of one of the liquids χ	Cannot be measured
4. Difference between surface potentials $\Delta_\beta^\alpha \chi$	Cannot be measured, but can be determined when an additional non-thermodynamic hypothesis is used
5. Internal potential ϕ	Cannot be measured
6. Galvani potential, or difference between two inner potentials $\Delta_\beta^\alpha \phi$	Cannot be measured, but if both phases have the same composition, measurement is possible
7. Change of Galvani potential $\Delta_\beta^\alpha \phi$ during any process (adsorption, variation of temperature, pressure, concentration, etc.)	Can be measured
8. Potential difference in the cell between two phases of the same composition $E = \Sigma\Delta_\beta^\alpha \phi_i$	Can be measured
9. Distribution potential of a salt $\Delta_\beta^\alpha \phi_{ca}$ or of an individual ion $\Delta_\beta^\alpha \phi_{c+}$ or $\Delta_\beta^\alpha \phi_{a-}$	Cannot be measured, but can be found when an additional non-thermodynamic hypothesis is used
10. Change of $\Delta_\beta^\alpha \chi$ potential during any process (adsorption, variation of temperature, pressure, concentration, etc.)	Can be measured

Interfacial potentials at polarizable boundaries can be produced by adsorption of charged or dipolar species at the interface or by the charging of the interface from an external source. In an ideally polarizable interface, the potential is an additional independent variable characterizing the state of the boundary and is specified by the external potential source. Electrostatic rather than electrochemical equilibrium is maintained at an ideally polarizable interface, analogous to a charged condenser.

4.2. STANDARD POTENTIALS AND STANDARD GIBBS FREE ENERGIES

Suppose that the phases α and β in contact contain ions i_i with a charge number z_i and an activity a_i. It follows from the condition of electrochemical equilibrium that

$$\Delta_\beta^\alpha \phi = -\frac{\mu_i^{0,\alpha} - \mu_i^{0,\beta}}{z_i F} + \frac{RT}{z_i F} \ln \frac{a_i^\beta}{a_i^\alpha}, \tag{4.1}$$

where μ is the chemical potential and $F = 96\,500\ \text{C mol}^{-1}$. The difference between the standard chemical potentials of the ion i in the two phases

$$\mu_i^{0,\alpha} - \mu_i^{0,\beta} = \Delta_\beta^\alpha G_{tr}^0(i) \tag{4.2}$$

represents the energy of resolution or the standard Gibbs free energy $\Delta_\beta^\alpha G_{tr}^0(i)$ of the transfer of the ion i from the phase α to the phase β. According to the standard electrochemical nomenclature the distribution or partition coefficient of the ion i is:

$$P_i^{\alpha/\beta} = \exp\left(\frac{-\Delta_\beta^\alpha G_{tr}^0(i)}{RT}\right) \tag{4.3}$$

and the standard potential of the distribution of a specific ion between two immiscible liquid phases is:

$$\Delta_\beta^\alpha \phi^0 = -\frac{\Delta_\beta^\alpha G_{tr}^0(i)}{z_i F} \equiv \frac{RT}{z_i F}\ln P_i^{\alpha/\beta}. \tag{4.4}$$

According to equation (4.3),

$$P_i^{\alpha/\beta} = \frac{1}{P_i^{\beta/\alpha}}. \tag{4.5}$$

The interfacial potential difference can be represented in the form

$$\Delta_\beta^\alpha \phi = \Delta_\beta^\alpha \phi_i^0 + \frac{RT}{z_i F}\ln\frac{a_i^\beta}{a_i^\alpha} \equiv \frac{RT}{z_i F}\ln\left(P_i^{\alpha/\beta}\frac{a_i^\beta}{a_i^\alpha}\right). \tag{4.6}$$

The equilibrium ratios of the activities of the ions i in two phases are described by the expression

$$\frac{a_i^\alpha}{a_i^\beta} = \frac{\gamma_i^\alpha c_i^\alpha}{\gamma_i^\beta c_i^\beta} = \exp\left[-\frac{z_i F}{RT}(\Delta_\beta^\alpha \phi - \Delta_\beta^\alpha \phi_i^0)\right] = P_i^{\alpha/\beta}\exp\left(-\frac{z_i F}{RT}\Delta_\beta^\alpha \phi\right) \tag{4.7}$$

where γ_i is the activity coefficient and c is the concentration.

From this equation the physical meaning of the concepts introduced earlier becomes clear. First, the standard distribution potential of a given ion represents the interfacial potential difference which would be established at the interface if the equilibrium activities of this ion in the two phases were identical. The partition coefficient of the ions $P_i^{\alpha/\beta}$ corresponds to the equilibrium ratio of the activities of the ions in the two phases in the absence of an interfacial potential difference. Values of $\Delta_0^W \phi_i^0$ of interest for the study of ITIES have been compiled for certain solvent systems, and representative values are listed in Table 4.2.

If volumes of phases α and β are equal, the electroneutrality condition can be written as

$$\sum_i z_i c_i = 0. \tag{4.8}$$

The relation between the interfacial potential $\Delta_\beta^\alpha \phi$ and the concentration of ions in solution is described by equations (4.7) and (4.8). If there are n types of ions in the system, then the number of equations is $n + 2$ and the number of variables is $2n + 1$, so that the number of independent variables in the system is $n - 1$. In other words, we can specify the concentration of all species of ions, except one, in one phase. Then the concentration of the remaining species of ions in that phase, the concentrations of all ions in the other phase, and the interfacial potential are automatically determined.

Apart from the activities of individual ions, it is convenient to use the concept of the mean activity of a 1:1 electrolyte BA:

$$a_{BA} = \sqrt{a_B a_A}. \tag{4.9}$$

The mean activity coefficient of the electrolyte is then given by

$$\gamma_{BA} = \sqrt{\gamma_B \gamma_A}. \tag{4.10}$$

The partition coefficient of a 1:1 electrolyte can be introduced by analogy:

$$P_{BA}^{\alpha/\beta} = \sqrt{P_B^{\alpha/\beta} P_A^{\alpha/\beta}}. \tag{4.11}$$

It is easy to see that the electrolyte partition coefficient $P_{BA}^{\alpha/\beta}$ is the ratio of the mean activities of this electrolyte in the bulk phase

$$P_{BA}^{\alpha/\beta} = a_{BA}^\alpha / a_{BA}^\beta. \tag{4.12}$$

This ratio, in contrast to the analogous ratio for individual ions (equation (4.7)), is independent of the interfacial potential. It is also independent of the presence of other solute ions.

4.3. DISTRIBUTION POTENTIALS

Consider the simplest case where the interface is non-polarizable and both phases α and β contain a binary 1:1 electrolyte B^+A^-:

$$\alpha, B^+A^- \,|\, B^+A^-, \beta. \tag{4.13}$$

The interfacial potential can be readily found from equation (4.6) and the electroneutrality condition:

$$\Delta_\beta^\alpha \phi = \frac{\Delta_\beta^\alpha \phi_B^0 + \Delta_\beta^\alpha \phi_A^0}{2} + \frac{RT}{2F} \ln \frac{\gamma_B^\beta \gamma_A^\alpha}{\gamma_A^\beta \gamma_B^\alpha}. \tag{4.14}$$

The first term reflects the individual characteristics of the solute ions in the standard state, while the second depends on the activity coefficients and concentration effects. However, in practice, the activity coefficients of cations and anions can be assumed to be approximately equal, $\gamma_A^\alpha \approx \gamma_B^\alpha$ and $\gamma_A^\beta \approx \gamma_B^\beta$, so that a simplified expression for the interfacial potential can be written:

$$\Delta_\beta^\alpha \phi = \frac{\Delta_\beta^\alpha \phi_B^0 + \Delta_\beta^\alpha \phi_A^0}{2} \equiv \frac{RT}{2F} \ln \frac{P_B^{\alpha/\beta}}{P_A^{\alpha/\beta}}. \tag{4.15}$$

In this case the interfacial potential difference is equal to the arithmetic mean of the standard distribution potentials of individual ions. The potential difference between two phases with a binary electrolyte in both phases was called the distribution potential (Frumkin, 1919; Karpfen and Randles, 1953). In this approximation the distribution potential does not depend on the concentration of the electrolyte BA or the interface structure, but is determined only by the thermo-dynamic properties of the bulk phases. This has been confirmed by experimental studies (Gugeshashvili, 1974; Kalweit and Strelow, 1954; Karpfen and Randles, 1953; Koczorowski and Minc, 1963; Minc and Koczorowski, 1963), although under certain conditions deviations have been observed. At low concentrations of the salt BA, the interfacial potential measurements can be affected by HCO_3^- if they are performed in air rather than in an inert gas. Distribution potentials also depend on the concentration of ionic solutes when more than two ionic species are present (Markin and Volkov, 1987c, 1988f, 1989f, 1990f).

The structure of the interphase between two immiscible electrolyte solutions is determined by the bulk properties of the contacting phases. When a monolayer of surfactant that is practically insoluble in both phases is adsorbed at the interface, the surfactant should not alter the interfacial potential determined from equation (4.14). However, due to surface blocking the thermodynamic equilibrium can be affected, and as a result the measured potential difference will be a nonequilibrium quantity, which differs from the true distribution potential.

One may consider the general case of an asymmetric electrolyte $B_p^{k+} A_t^{l-}$ partitioning between phases α and β:

$$\alpha, B_p^{k+} A_t^{l-} \mid B_p^{k+} A_t^{l-}, \beta. \tag{4.16}$$

In this case, the interfacial potential, determined by the distribution of electrolytes, is described by the expression

$$\Delta_\beta^\alpha \phi = \frac{z_B \Delta_\beta^\alpha \phi_B^0 - z_A \Delta_\beta^\alpha \phi_A^0}{z_B - z_A} + \frac{RT}{(z_B - z_A)F} \ln \frac{\gamma_B^\beta \gamma_A^\alpha}{\gamma_B^\alpha \gamma_A^\beta}. \tag{4.17}$$

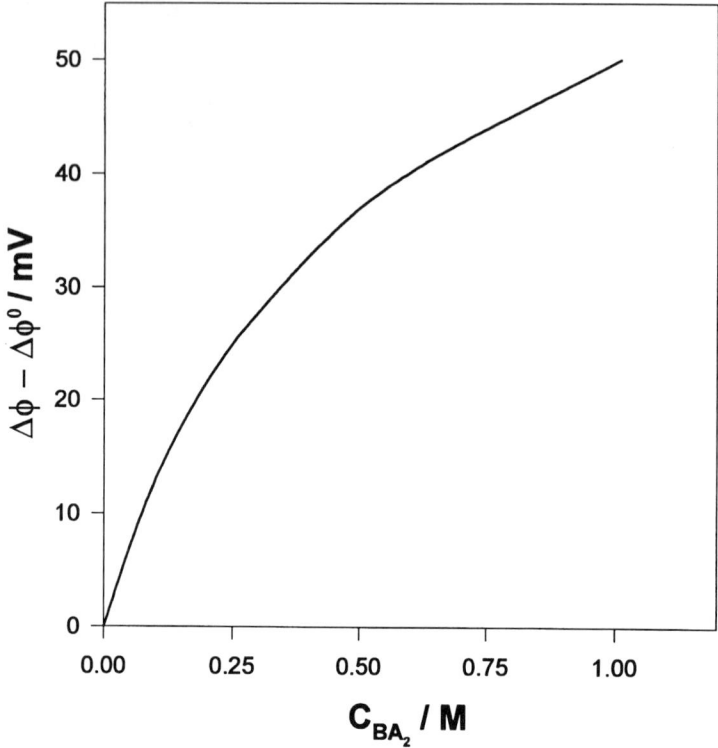

Figure 4.2. The distribution potential of a BA_2 electrolyte depends on concentration in the organic phase.

For an asymmetric electrolyte, the partition potential depends on the salt concentration. The second term on the right-hand side of equation (4.17) is given by the following expression according to the Debye–Huckel theory:

$$\frac{RT}{(z_B - z_A)F}\ln\frac{\gamma_B^\beta\gamma_A^\alpha}{\gamma_B^\alpha\gamma_A^\beta} = (z_B - z_A)\frac{e_0^3\sqrt{125N_A^2}}{2\pi F\varepsilon_0^{3/2}\sqrt{RT}}(J_\alpha^{1/2}\varepsilon_\alpha^{-3/2} - J_\beta^{1/2}\varepsilon_\beta^{-3/2}) \qquad (4.18)$$

where J_α is the ionic strength of the solution in the phase α, N_A is Avogadro's number, ε_0 is the electric constant, and ε_α is the dielectric constant of the phase α. In terms of the Debye approximation, the distribution potential is independent of concentration only for a symmetrical electrolyte, where $z_A = -z_B$.

In Fig. 4.2 the dependence of the distribution potential on concentration for a 2:1 electrolyte, for example, $CaCl_2$, is presented. For this illustration we chose $\varepsilon_\alpha = 10$ and $\varepsilon_\beta = 80$, which corresponds to the common experimental system of 1,2-dichloroethane–water. In this situation a hydrophilic electrolyte BA_2 is dissolved in a two-phase system, so that the term $J_\beta^{1/2}\varepsilon_\beta^{-3/2}$ can be neglected. As one can see from

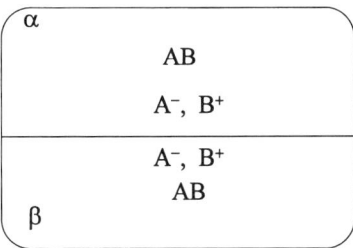

Figure 4.3. The distribution of incompletely dissociated binary electrolyte B^+A^- in a system of two immiscible liquids α and β.

Fig. 4.2, the distribution potential of a non-symmetrical electrolyte strongly depends on concentration.

4.4. INCOMPLETE DISSOCIATION OF THE SALT

We will next describe how incomplete dissociation of the electrolyte can affect the interfacial potential distribution (Fig. 4.3). Suppose that the reaction

$$B^+ + A^- \overset{K_{BA}^\alpha}{\Longleftrightarrow} BA \tag{4.19}$$

occurs in phase α, where K_{BA}^α denotes the dissociation constant of the BA molecule. This means that in equilibrium the equality

$$a_B^\alpha a_A^\alpha = K_{BA}^\alpha a_{BA}^\alpha \tag{4.20}$$

is fulfilled. The dissociation constant K_{BA}^α is related to the chemical potentials of the corresponding species by the equation

$$RT \ln K_{BA}^\alpha = \mu_{BA}^{0,\alpha} - \mu_{B+}^{0,\alpha} - \mu_{A-}^{0,\alpha}. \tag{4.21}$$

Similar equations can be written for dissociation of the electrolyte in the phase β. The undissociated BA molecules are distributed between the phases, and the partition coefficient $P_{BA}^{\alpha/\beta}$ is defined by the equation

$$RT \ln P_{BA}^{\alpha/\beta} = \mu_{BA}^{0,\beta} - \mu_{BA}^{0,\alpha}. \tag{4.22}$$

Incomplete dissociation of the molecules does not affect the value of an interfacial potential, and the distribution potential determined from equation (4.6) is established at the interface. The standard distribution potentials of the individual ions (see

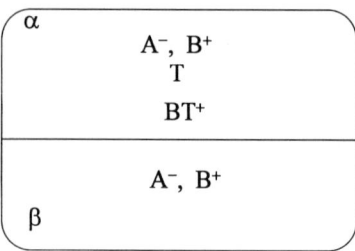

Figure 4.4. The distribution of a binary electrolyte B^+A^- in a system of two immiscible liquids α and β in the presence of a complexing agent T that selectively binds B^+ ions.

equation (4.5)) remain unchanged in exactly the same way. However, owing to the dissociation equilibrium and the interfacial partition of undissociated molecules, the constants of these processes are related to the difference between the Gibbs standard energies of the ions (and hence between the standard potentials):

$$RT\ln\frac{K_{BA}^{\beta}}{K_{BA}^{\alpha}P_{BA}^{\alpha/\beta}} = \Delta_{\beta}^{\alpha}G_{B+}^{0} + \Delta_{\beta}^{\alpha}G_{A-}^{0} = F\Delta_{\beta}^{\alpha}\phi_{A-}^{0} - F\Delta_{\beta}^{\alpha}\phi_{B+}^{0}. \qquad (4.23)$$

4.5. COMPLEX FORMATION IN ONE OF THE PHASES

Suppose that the binary electrolyte BA is distributed between the phases α and β and that phase α contains in addition a neutral compound T which can combine with the cation B^+ to form the neutral charged complex BT^+ (Fig. 4.4). The reaction

$$B^+(\alpha) + T(\alpha) \overset{K_{BT+}^{\alpha}}{\Longleftrightarrow} BT^+(\alpha) \qquad (4.24)$$

takes place in the phase α and the dissociation constant K_{BT+}^{α} is defined by the equation

$$RT\ln K_{BT+}^{\alpha} = \mu_{BT+}^{0,\alpha} - \mu_{B+}^{0,\alpha} - \mu_{T}^{0,\alpha}. \qquad (4.25)$$

Apart from the homogeneous dissociation constant K_{BT+}^{α} of the complex BT^+, it is also possible to introduce the constant of the heterogeneous or surface dissociation $K_{BT+}^{\alpha/\beta}$ referring to a process in which the complexing agent T and the complex BT^+ are in the phase α while the ion B^+ is in the phase β. This constant is defined by the equation

$$RT\ln K_{BT+}^{\alpha/\beta} = \mu_{BT+}^{0,\alpha} - \mu_{B+}^{0,\beta} - \mu_{T}^{0,\alpha}. \qquad (4.26)$$

The two dissociation constants and the partition coefficient of the ion B^+ are related by the following equation:

$$K_{BT+}^{\alpha/\beta} = K_{BT+}^{\alpha}/P_{B+}^{\alpha/\beta}. \tag{4.27}$$

The equation for the heterogeneous dissociation equilibrium is

$$c_T^{\alpha} c_{B+}^{\beta} = K_{BT+}^{\alpha/\beta} c_{BT+}^{\alpha} \exp\left(\frac{F\Delta_\beta^\alpha \phi}{RT}\right). \tag{4.28}$$

The case where the complex BT^+ is very stable (i.e. its homogeneous dissociation constant K_{BT+}^{α} is small) while the ion B^+ is only slightly soluble in the α-phase (the partition coefficient $P_{B+}^{\alpha/\beta}$ is small) is of special interest. One can then assume that, apart from the charged complex BT^+, the α-phase contains only one charged species—the counter-ion A^-. It is possible to find the interfacial potential difference for this case (Markin and Volkov, 1988f, 1989e):

$$\Delta_\beta^\alpha \phi = \frac{RT}{F} \ln\left[\left\{\left(\frac{c_{B+}^\beta}{2K_{B+}^{\alpha/\beta}}\right)^2 + \frac{c_{com}^\alpha}{P_{A-}^{\alpha/\beta} K_{BT+}^{\alpha/\beta}}\right\}^{1/2} - \frac{c_{B+}^\beta}{2K_{BT+}^{\alpha/\beta}}\right] \tag{4.29}$$

where c_{com}^α denotes the overall concentration of the free and bound complexing agent in phase α and the activity coefficients for simplicity are assumed to be unity.

We will consider two limiting cases arising from equation (4.29). Suppose that the concentration of the $A^- B^+$ electrolyte is low. Then, by expanding equation (4.29) in a series, we obtain

$$\Delta_\beta^\alpha \phi \approx \frac{RT}{2F} \ln\frac{c_{com}^\alpha}{P_{A-}^{\alpha/\beta} K_{BT+}^{\alpha/\beta}}. \tag{4.30}$$

Equation (4.30) can be written more conveniently by introducing the interphase partition coefficient of the ion B^+ taking into account complex formation:

$$\bar{P}_{B+}^{\alpha/\beta} = c_{com}^\alpha/K_{BT+}^{\alpha/\beta} \tag{4.31}$$

The expression for the interfacial potential is then

$$\Delta_\beta^\alpha \phi = \frac{RT}{2F} \ln\frac{\bar{P}_{B+}^{\alpha/\beta}}{P_{A-}^{\alpha/\beta}}. \tag{4.32}$$

This is a typical expression for the distribution potential which does not include the complexing agent concentration. The form of the relation is determined by the valences of the electrolyte, and equation (4.32) is valid only for a 1:1 electrolyte. For arbitrary valences, a more general expression is:

$$\frac{F}{RT}\Delta_\beta^\alpha \phi = \frac{1}{z_B - z_A} \ln\frac{c_{com}^\alpha}{P_A^{\alpha/\beta} K_{BT+}^{\alpha/\beta}}. \tag{4.33}$$

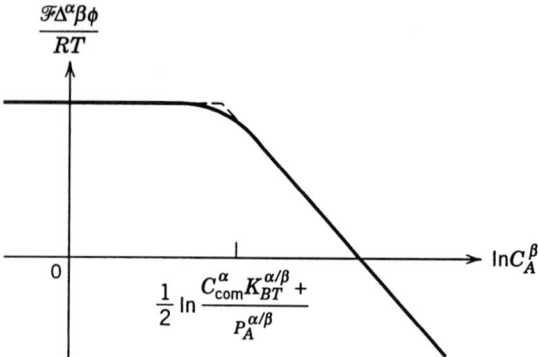

Figure 4.5. The distribution potential versus electrolyte concentration in phase β when complex formation occurs in phase α.

The dependence of $F\Delta_\beta^\alpha \phi/RT$ on $\ln c_{com}^\alpha$ is described by a straight line with a slope of 1/2 for a 1:1 electrolyte, 1/3 for 1:2 electrolyte, and 1/4 for 2:2 electrolyte (Markin and Volkov, 1988c, 1989f).

We will now proceed to another limiting case. Suppose that the electrolyte concentration is high so that the concentration of the complexing agent in the phase α is close to saturation. We then obtain for equation (4.31)

$$\Delta_\beta^\alpha \phi = \frac{RT}{F} \ln \frac{c_{com}^\alpha}{P_{A^-}^{\alpha/\beta} c_{A^-}^\beta}. \tag{4.34}$$

Figure 4.5 illustrates the dependence of $F\Delta_\beta^\alpha \phi/RT$ on $\ln c_{A^-}^\alpha$, which consists of two linear sections. The slope of the second section is -1 for a univalent anion A^- or $(z_A)^{-1}$ in the arbitrary case. The point of intersection (Fig. 4.5) of two linear parts has the abscissa $0.5 \ln (c_{com}^\alpha K_{BT^+}^{\alpha/\beta}/P_{A^-}^{\alpha/\beta})$, which makes it possible to determine the heterogeneous dissociation constant of the complex BT^+.

4.6. THE DONNAN POTENTIAL

The interfacial potential difference that arises when certain ionic solutes cannot cross the interface between two immiscible electrolyte solutions while the remaining ions are free to move reversibly from one phase to the other is called the Donnan potential. Suppose, for example, that an ion R with a charge number z_R has such a high partition coefficient $P_R^{\alpha/\beta}$ that it is almost totally concentrated in the phase α. Apart from the ions R, the system contains a binary electrolyte B^+A^- with a concentration in the β-phase of c_{BA}^β:

$$\alpha, R^{z_R}, B^+A^- \,|\, B^+A^-, \beta. \tag{4.35}$$

This type of equilibrium was first considered by Donnan (1911) for two aqueous (α and β) electrolyte solutions separated by a semipermeable membrane.

By definition, the partition coefficient of a salt is given by

$$P_{AB}^{\alpha/\beta} = \frac{a_{AB}^{\alpha}}{a_{AB}^{\beta}} \tag{4.36}$$

which can be expressed by the activities of ions B^+ and A^- in both phases:

$$a_{B+}^{\alpha} a_{A-}^{\alpha} = (P_{AB}^{\alpha/\beta})^2 a_{A-}^{\beta} a_{B+}^{\beta}. \tag{4.37}$$

Using this equation and the two electroneutrality conditions for the phases

$$z_R c_R^{\alpha} + c_B^{\alpha+} - c_{A-}^{\alpha} = 0, \tag{4.38}$$

$$c_B^{\beta} - c_A^{\beta} = 0, \tag{4.39}$$

one readily obtains the Donnan potential (Markin and Volkov, 1988c, 1989f)

$$\Delta_{\beta}^{\alpha}\phi = \frac{RT}{2F} \ln \left(\frac{P_{B+}^{\alpha/\beta} \gamma_{B+}^{\beta} \gamma_{A-}^{\alpha}}{P_{A-}^{\alpha/\beta} \gamma_{B+}^{\alpha} \gamma_{A-}^{\beta}} \right) + \frac{RT}{2F} \ln \left(1 + \frac{z_R^{\alpha} c_R^{\alpha}}{c_B^{\alpha+}} \right). \tag{4.40}$$

The Donnan potential can also be written as a function of the concentration of the dissolved electrolyte AB:

$$\Delta_{\beta}^{\alpha}\phi = \frac{RT}{2F} \ln \left(\frac{P_{B+}^{\alpha/\beta} \gamma_{B+}^{\beta} \gamma_{A-}^{\alpha}}{P_{A-}^{\alpha/\beta} \gamma_{B+}^{\alpha} \gamma_{A-}^{\beta}} \right) + \frac{RT}{2F} \ln \left[\sqrt{1 + \left(\frac{z_R c_R^{\alpha} \gamma_{BA}^{\alpha}}{2 P_{BA}^{\alpha/\beta} a_{BA}^{\beta}} \right)^2} \right.$$

$$\left. + \frac{z_R c_R \gamma_{BA}^{\alpha}}{2 P_{BA}^{\alpha/\beta} a_{BA}^{\beta}} \right]. \tag{4.41}$$

As shown in equations (4.40) and (4.41), the Donnan potential is the sum of the distribution potential and the term dependent on the concentration of the impermeable ion R. If this concentration is low, the second term vanishes, and equations (4.40) and (4.41) are reduced to the expression for the distribution potential.

If the concentration of R ions is high, the potential tends to another limit. If we set $z_R > 0$ then

$$\Delta_{\beta}^{\alpha}\phi = \frac{RT}{F} \ln \frac{z_R c_R^{\alpha} \gamma_{A-}^{\alpha}}{P_{A-}^{\alpha/\beta} a_{A-}^{\beta}}. \tag{4.42}$$

In this case the concentration of counter-ions A^- in phase α is no longer dependent on the partition coefficient and interfacial potential, and approaches the limiting value

$$c_{A-}^{\alpha} \approx z_R c_R^{\alpha}. \tag{4.43}$$

In this case the co-ions B^+ are almost completely expelled from phase α:

$$c_{B^+}^\alpha = \frac{1}{z_R\, c_R^\alpha}\left(\frac{P_{BA}^{\alpha/\beta}\, a_{BA}^\beta}{\gamma_{BA}^\alpha}\right)^2.$$

(4.44)

4.7. THE NERNST POTENTIAL

We now consider an especially important particular case of the Donnan potential where only one ion B^{z_B} can pass across the interface. The interfacial potential difference arising in such a system is called the Nernst potential

$$\Delta_\beta^\alpha \phi = \Delta_\beta^\alpha \phi_B^0 + \frac{RT}{z_B\, F}\ln\frac{a_B^\beta}{a_B^\alpha}.$$

(4.45)

If the phases α and β are identical but are separated by a membrane permeable to B^+, then the potential difference is given by

$$\Delta_\beta^\alpha \phi = \frac{RT}{z_B\, F}\ln\frac{a_B^\beta}{a_B^\alpha}.$$

(4.46)

Sometimes the potential difference determined from equation (4.45) is also referred to as the Nernst–Donnan potential, because this equation was first obtained by Nernst but corresponds to a special case of the Donnan equilibrium.

4.8. THE OXIDATION–REDUCTION INTERFACIAL POTENTIAL: GIBBS FREE ENERGY OF ELECTRON AND ION TRANSPORT COUPLING

Suppose that each of the two phases, α and β, contains its own oxidation–reduction couple Red/Ox (Fig. 4.6(a)). The exchange of n electrons across the interface results in the redox reaction

$$\nu_1\, Red_1 + \nu_2\, Ox_2 \longleftrightarrow \nu_3\, Red_2 + \nu_4\, Ox_1.$$

(4.47)

The following relation holds under equilibrium conditions:

$$\nu_1\, \bar\mu_{Red1}^\alpha + \nu_2\, \bar\mu_{Ox2}^\beta = \nu_3\, \bar\mu_{Red2}^\beta + \nu_4\, \bar\mu_{Ox1}^\alpha$$

(4.48)

from which the interfacial redox potential can be calculated:

$$\Delta_\beta^\alpha \phi = \Delta_\beta^\alpha \phi_{R/O}^0 + \frac{RT}{nF}\ln\frac{(a_{Red1}^\alpha)^{\nu1}(a_{Ox2}^\beta)^{\nu2}}{(a_{Ox1}^\alpha)^{\nu4}(a_{Red2}^\beta)^{\nu3}}.$$

(4.49)

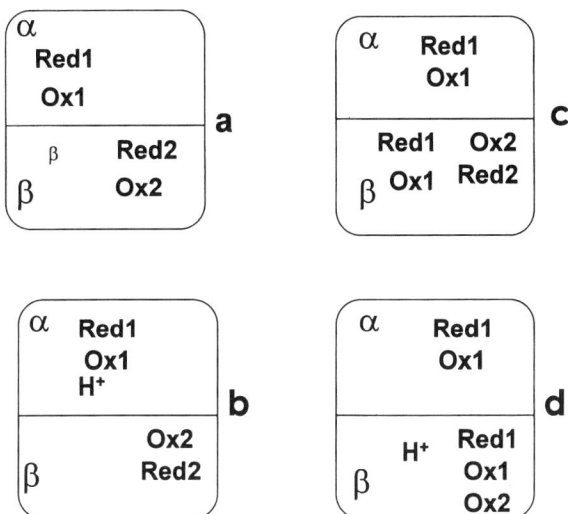

Figure 4.6. The distribution of redox couples in a system of two immiscible electrolyte solutions.

Here $\Delta_\beta^\alpha \phi_{R/O}^0$ denotes the standard interfacial redox potential of reaction (4.47), which is equal to

$$\Delta_\beta^\alpha \phi_{R/O}^0 = (\nu_1 \mu_{Red1}^{0,\alpha} + \nu_2 \mu_{Red2}^{0,\beta} - \nu_3 \mu_{Red_2}^{0,\beta} - \nu_4 \mu_{Ox1}^{\alpha})/nF \qquad (4.50)$$

where n is the number of electrons transferred from Red_1 to Ox_2. This is the interfacial potential for equal activities of the oxidized and reduced solutes in each solvent phase. It should be noted that the interfacial redox potential of the reaction (4.47) is defined by equations (4.49) and (4.50) even though one particle is neutral in each of the redox couples.

If the pH of one of the phases changes during reaction (4.47) (Fig. 4.6(b)),

$$\nu_1 Red_1 + \nu_2 Ox_2 \longleftrightarrow \nu_3 Red_2 + \nu_4 Ox_1 + mH_\alpha^+ \qquad (4.51)$$

then from the equilibrium condition

$$\nu_1 \bar{\mu}_{Red1}^{\alpha} + \nu_2 \bar{\mu}_{Ox2}^{\beta} = \nu_3 \bar{\mu}_{Red2}^{\beta} + \nu_4 \bar{\mu}_{Ox1}^{\alpha} + m\bar{\mu}_{H+}^{\alpha} \qquad (4.52)$$

one can find the interfacial redox potential

$$\Delta_\beta^\alpha \phi = \Delta_\beta^\alpha \phi_{R/O}^0 + \frac{RT}{nF} \ln \frac{(a_{Red1}^{\alpha})^{\nu 1}(a_{Ox2}^{\beta})^{\nu 2}}{(a_{Ox1}^{\alpha})^{\nu 4}(a_{Red2}^{\beta})^{\nu 3}(a_{H+}^{\alpha})^{m}} \qquad (4.53)$$

where

$$\Delta_\beta^\alpha \phi_{R/O}^0 = (\nu_1 \mu_{Red1}^{0,\alpha} + \nu_2 \mu_{Ox2}^{0,\beta} - \nu_3 \mu_{Red2}^{0,\beta} - \nu_4 \mu_{Ox1}^{0,\alpha} - m\mu_{H+}^{0,\alpha})/nF. \qquad (4.54)$$

The influence of the liquid interface on the standard redox potential of an electron-exchange reaction has been studied by Samec (1979a), Volkov (1984a,b, 1985a), Volkov and Kharkats (1986).

4.9. THE MIXED POTENTIAL

If each phase of two immiscible liquids contains common ions and redox couples, the interface acquires a mixed potential which is contributed by both the interfacial redox potential and the ion distribution between the phases.

A two-phase system allows various combinations of electron-exchange reactions. For example, if each of the phases α and β contain a redox couple and one of the couples is soluble in both phases (Fig. 4.6(c)) electron-exchange reactions can take place both at the interface

$$Red_1^\alpha + Ox_2^\beta \longleftrightarrow Red_2^\beta + Ox_1^\alpha \qquad (4.55)$$

and in the bulk phase β:

$$Red_1^\beta + Ox_2^\beta \longleftrightarrow Red_2^\beta + Ox_1^\beta. \qquad (4.56)$$

The equilibrium condition implies that

$$\bar{\mu}_{Red1}^\beta - \bar{\mu}_{Red1}^\alpha = 0 \qquad (4.57)$$

$$\bar{\mu}_{Ox1}^\beta - \bar{\mu}_{Ox1}^\alpha = 0, \qquad (4.58)$$

and it follows that

$$\Delta_\beta^\alpha \phi = \frac{RT}{nF} \ln \frac{P_{Ox1}^{\alpha/\beta}}{P_{Red1}^{\alpha/\beta}} + \frac{RT}{nF} \ln \frac{a_{Red1}^\alpha a_{Ox1}^\beta}{a_{Ox1}^\alpha a_{Red1}^\beta}. \qquad (4.59)$$

The standard mixed potential for reaction (4.53) is given by

$$\Delta_\beta^\alpha \phi_{R/O}^0 = \frac{RT}{nF} \ln \frac{P_{Ox1}^{\alpha/\beta}}{P_{Red1}^{\alpha/\beta}}. \qquad (4.60)$$

Equation (4.59) is equivalent to equation (4.11) for two-electron reactions if $z_{Red1} = -1$, $z_{Ox1} = +1$, and $a_{Red1}^\beta \cdot a_{Ox1}^\alpha = a_{Red1}^\alpha \cdot a_{Ox1}^\beta$. The last condition can be true only at a single point for specified activities of all the ingredients. This condition does not follow from phase electroneutrality because both phases contain not only the redox couples but also other ions.

Equation (4.59) does not explicitly contain the second redox couple Red_2/Ox_2, but it appears in this equation in the second term on the right-hand side. The second redox couple must match the first couple because equilibrium in the system is described by the four equations:

$$\bar{\mu}^{\beta}_{Red1} = \bar{\mu}^{\alpha}_{Red1} \tag{4.61}$$

$$\bar{\mu}^{\beta}_{Ox1} = \bar{\mu}^{\alpha}_{Ox1} \tag{4.62}$$

$$\bar{\mu}^{\alpha}_{Red1} + \bar{\mu}^{\beta}_{Ox2} = \bar{\mu}^{\alpha}_{Ox1} + \bar{\mu}^{\beta}_{Red1} \tag{4.63}$$

$$\bar{\mu}^{\beta}_{Red1} + \bar{\mu}^{\beta}_{Ox2} = \bar{\mu}^{\beta}_{Ox1} + \bar{\mu}^{\beta}_{Red2}. \tag{4.64}$$

If a common electrolyte $B_p A_t$ is introduced into both phases, this will not affect the general form of equation (4.59). However, the electroneutrality condition for each of the phases,

$$\sum_i z_i^{\alpha} c_i^{\alpha} = 0, \qquad \sum_j z_j^{\beta} c_j^{\beta} = 0, \tag{4.65}$$

will lead to a change in the distribution of the redox component concentrations in each phase, which in turn will contribute significantly to the second term on the right-hand side of equation (4.59).

Most redox reactions occurring at water–oil and water–biomembrane interfaces proceed with a change in pH, for example,

$$NADH \Leftrightarrow NAD^+ + 2e^- + H^+. \tag{4.66}$$

Consider a system of two immiscible liquids α and β (Fig. 4.6(d)). Let each of the phases have its own redox couple, and let one of the couples be common to the two phases:

$$Red_1^{\alpha} + Ox_2^{\beta} \leftrightarrow Red_2^{\beta} + Ox_1^{\alpha} + mH_{\beta}^+, \tag{4.67}$$

$$Red_1^{\beta} + Ox_2^{\beta} \leftrightarrow Red_2^{\beta} + Ox_1^{\alpha} + mH_{\beta}^+. \tag{4.68}$$

If in the course of the reaction n electrons are transferred from Red_1 to Ox_2 and if $m \neq n$, then from the equilibrium condition one obtains the expression for the mixed potential,

$$\Delta_{\beta}^{\alpha}\phi = \Delta_{\beta}^{\alpha}\phi_{R/O}^0 - \frac{RT}{(n-m)F} \ln \frac{a_{Red1}^{\beta} a_{Ox1}^{\alpha}}{a_{Red1}^{\alpha} a_{Ox1}^{\beta}}, \tag{4.69}$$

where the standard potential is given by

$$\Delta_{\beta}^{\alpha}\phi_{R/O}^0 = \frac{RT}{(n-m)F} \ln \frac{P_{Ox1}^{\alpha/\beta}}{P_{Red1}^{\alpha/\beta}}. \tag{4.70}$$

From the condition that the phases are in thermodynamic equilibrium and from the electroneutrality condition, one can write for reaction (4.63) 4 + 2 equations for two redox couples, which exceeds the number of independent variables in the system. If in the course of reaction (4.63) a proton equilibrium is established between phases α and β, the mixed potential can be written in the form

$$\Delta_\beta^\alpha \phi = \Delta_\beta^\alpha \phi_{R/O}^0 - \frac{RT}{nF} \ln \frac{a_{Red1}^\beta a_{Ox1}^\alpha}{a_{Red1}^\alpha a_{Ox1}^\beta} + \frac{2.3RT}{nF} \Delta_\beta^\alpha pH \tag{4.71}$$

where

$$\Delta_\beta^\alpha \phi_{R/O}^0 = \frac{RT}{nF} \ln \frac{P_{Ox1}^{\alpha/\beta}(P_{H^+}^{\alpha/\beta})^m}{P_{Red1}^{\alpha/\beta}}. \tag{4.72}$$

Another example of a redox reaction at the oil–water interface has the form

$$Red \leftrightarrow Ox + ne^- + mH^+. \tag{4.73}$$

The electrons which are the products of reaction (4.73) can be accepted at the interface by a second substance dissolved in one of the two phases.

The standard Gibbs energies of the reaction (4.73) for each phase, α and β, are:

$$\Delta G_\alpha^0 = {}_\alpha \mu_{Red}^0 - {}_\alpha \mu_{Ox}^0 - n\mu_e^0 - m_\alpha \mu_{H^+}^0, \tag{4.74}$$

$$\Delta G_\beta^0 = {}_\beta \mu_{Red}^0 - {}_\beta \mu_{Ox}^0 - n\mu_e^0 - m_\beta \mu_{H^+}^0. \tag{4.75}$$

Subtraction of equation (4.74) from equation (4.75) gives the change of the standard Gibbs energy at the interface if the electron acceptor is located in one phase only, or localized at the phase boundary:

$$\Delta G_\beta^0 - \Delta G_\alpha^0 = ({}_\beta \mu_{Red}^0 - {}_\alpha \mu_{Red}^0) - ({}_\beta \mu_{Ox}^0 - {}_\alpha \mu_{Ox}^0) - m({}_\beta \mu_{H^+}^0 - {}_\alpha \mu_{H^+}^0) \tag{4.76}$$

or

$$\Delta G^0 = RT \ln \frac{P_{Red}}{P_{Ox}(P_{H^+})^m}, \tag{4.77}$$

where P_i is the distribution coefficient of the i-th ion:

$$RT \ln P_i = {}_\beta \mu_i^0 - {}_\alpha \mu_i^0. \tag{4.78}$$

In the case of an n-electron reaction, the standard redox potential ΔE^0 at the interface is determined by:

$$\Delta E^0 = -\frac{RT}{nF} \ln \frac{P_{Red}}{P_{Ox}(P_{H^+})^m}. \tag{4.79}$$

It is possible to shift the redox potential scale in a desired direction by selecting appropriate solvents, thereby permitting reactions to occur that are highly unfavorable in a homogeneous phase. If the resolvation energies of substrates and products are very different, the interface between two immiscible liquids may act as a catalyst. The kinetic mechanism underlying the catalytic properties of the liquid–liquid interface was discussed by Kharkats and Volkov (1985, 1987, 1989, 1994).

4.10. THE ADSORPTION POTENTIAL

The adsorption potential is the interfacial difference of Galvani potentials caused by the adsorption of ions or dipoles at the interface. The concept of adsorption potential has been the subject of numerous discussions (Baur and Korman, 1917; Beutner, 1913a,b, 1918; Bonhoeffer *et al.*, 1953; Davies, 1951a; Davies and Rideal, 1955; Dean, 1939; Dean *et al.*, 1940; Ehresvärd and Sillén, 1938; Haber, 1908; Haber and Klemencsiewicz, 1909; Kalweit and Strelow, 1954; Karpfen and Randles, 1953; Koczorowski and Minc, 1963; Minc and Koczorowski, 1963; Nernst and Riesenfeld, 1902; Riesenfeld and Reinhold, 1902; Strehlow, 1952; Volkov, 1987b). If two immiscible liquids α and β are in equilibrium, an interfacial potential difference arises in an intermediate adsorption layer separating these phases. Formally, the electrolyte distribution potential may be calculated using only the ion solvation energies in each of the phases, and ignoring the properties of the interfacial adsorption layer. In equilibrium, however, the properties of ions in the adsorption layer are uniquely related to their properties in each of the contacting phases, so that the distribution potentials may be calculated either from the bulk properties of solvents α and β or from the properties of the adsorption layer.

At the beginning of the century Beutner (1913a,b) demonstrated that the formal thermodynamic theory of ion distribution can be applied to determine the potential difference between two immiscible electrolyte solutions, and that no auxiliary assumptions are needed for such calculations. The works of Beutner were criticized by Baur and Korman (1917). Dean (1939), Dean *et al.* (1940) and Ehrensvärd and Sillén (1938) considered the adsorption potential to be superimposed on the potential distribution drop, while Craxford *et al.* (1938) and Volkov (1987a) treated the adsorption potential as a part of the distribution potential.

Whereas the distribution potential is a purely thermodynamic concept, calculation of the adsorption potential goes beyond the scope of thermodynamics and requires the use of some structural assumptions. Nevertheless, the two approaches give similar results, provided the structure of the adsorption layer is a sufficiently exact replica of the real system. The adsorption approach is more general. For example, if surfactants perturb ionic equilibrium at the interface or if the non-aqueous solvent is a non-polar substance (such as octane, decane, or heptane) and does not contain ions, the theory of distribution potentials is inapplicable, while the adsorption approach permits a sufficiently rigorous description of the system (Dean *et al.*, 1940; Gugeshashvili, 1974; Volkov, 1987a,b).

4.11. VERIFICATION OF THE THERMODYNAMIC THEORY OF INTERFACIAL POTENTIALS

Karpfen and Randles (1953) proposed the following chain for measuring interfacial potentials:

$$\text{Hg, Hg}_2\text{Cl}_2 \left| \begin{matrix} \text{H}_2\text{O} \\ \text{KCl} \end{matrix} \right| \begin{matrix} \text{H}_2\text{O} \\ \text{M}^+\text{X}^- \end{matrix} \left| \begin{matrix} \text{Di-isopropyl} \\ \text{ketone or} \\ \text{nitrobenzene} \\ \text{M}^+\text{X}^- \end{matrix} \right| \begin{matrix} \text{Di-isopropyl} \\ \text{ketone} \\ \text{Et}_4\text{NPi} \end{matrix} \left| \begin{matrix} \text{H}_2\text{O} \\ \text{Et}_4\text{N}^+ \\ \text{Pi}^- \end{matrix} \right| \text{Hg, Hg}_2\text{Cl}_2 \qquad (4.\text{I})$$

$$\underbrace{\Delta^w_{RE1}\phi \qquad \Delta^w_w\phi}_{= \text{constant}} \qquad \Delta^{org}_w\phi_{M^+X^-} \qquad \Delta^{org}_{org}\phi \qquad \underbrace{\Delta^w_{org}\phi \qquad \Delta^{RE2}_w\phi}_{= \text{constant} \ = \text{constant}}$$

where M^+X^- is the salt under study which is equally distributed between the two solvents. Assuming zero diffusion potentials $\Delta^w_w\phi$ and $\Delta^{org}_{org}\phi$ and a constant distribution potential $\Delta^w_{org}\phi_{Et4NPi}$, if ions M^+ or X^- are replaced by other ions, it is possible to determine the change in the distribution potential $\Delta^{M1}_{M2}\Delta^{org}_w\phi_{MX}$ when passing from salt M_1X to salt M_2X:

$$\Delta^{M1}_{M2}\Delta^{org}_w\phi_{MX} = \frac{RT}{F}\ln\frac{P^{org/w}_{M_1X}}{P^{org/w}_{M_2X}} = \frac{RT}{2F}\ln\frac{P^{org/w}_{M_1}}{P^{org/w}_{M_2}}. \qquad (4.80)$$

Equation (4.80) was experimentally verified by Karpfen and Randles (1953) for the water–isopropylketone system, by Kalweit and Strelow (1954) for the water–acetonitrile and water–quinoline systems, and by Minc and Koczorowski (1963) and Gugeshashvili (1974) for the water–nitrobenzene system (Table 4.3).

Minc and Koczorowski (1963) measured $\Delta^{M1}_{M2}\Delta^{org}_w\phi_{MX}$ in the following chain:

$$\begin{matrix} \text{Vibrating} \\ \text{gold} \\ \text{electrode} \end{matrix} \left| \text{Gas} \right| \begin{matrix} \text{Nitrobenzene} \\ \text{M}^+\text{X}^- \end{matrix} \left| \begin{matrix} \text{Water} \\ \text{M}^+\text{X}^- \end{matrix} \right| \begin{matrix} \text{Water} \\ \text{KCl(saturated)} \end{matrix} \left| \text{Hg}_2\text{Cl}_2, \text{Hg} \right. \qquad (4.\text{II})$$

$$\Delta^{Gas}_{Au}\phi \quad \Delta^{org}_{Gas}\chi = \text{constant} \qquad \Delta^w_{org}\phi \qquad \Delta^w_w\phi_{dif} \qquad \Delta^{RE}_w\phi$$

The measured potential is the sum of potential drops across all the interfaces. If the salt M_2X is substituted for the salt M_1X, then $\Delta^{M1}_{M2}\Delta^{org}_w\phi_{MX}$ can be expressed as the difference of potentials in the chain (4.II) in the presence of M_1X and M_2X. The measuring technique is described in detail by Minc and Koczorowski (1963).

The results of verifying the thermodynamic theory of potential distributions are presented in Table 4.3, which shows that experimental changes of the distribution potential for a salt with one ion replaced are in satisfactory agreement with calculation by equation (4.80) (using the partition coefficient data).

Table 4.3. The distribution potential differences for two binary electrolytes M₁/M₂ with a common ion X in the water–nitrobenzene system at 298 K.

		$\Delta^{M1}_{M2}\Delta^{NB}_{H_2O}\phi$, mV			
M_1/M_2	X	Minc and Koczorowski, 1963	Karpfen and Randles, 1953	Gugeshashvili, 1974	Calculation by equation (4.71)
Na^+/K^+	Cl^-, Br^-, I^-	-52	-53	—	-50
Li^+/K^+	Br^-, I^-	-78	—	—	-73
Cl^-/I^-	$K^+, Na^+,$ $Et_4N^+,$ But_4N^+	$+98$	$+102$	$+100; +92$	$+95$
Br^-/I^-	$Li^+, Na^+, K^+,$ But_4N^+	$+79$	—	$+70; +75$	$+77$
Cl^-/NO_3^-	K^+	$+70$	—	—	—
Cs^+/K^+	Cl^-	$+40$	—	—	—
K^+/NH_4^+	I^-	$+30$	—	—	—
Et_4N^+/K^+	I^-, Cl^-	$+128$	$+126$	$+120$	$+124$
Me_4N^+/K^+		—	—	$+65$	—

4.12. MEASUREMENT OF INTERFACIAL POTENTIALS: THE TETRAPHENYLBORATE HYPOTHESIS

It is known from thermodynamics that the absolute value of a potential difference can be measured only in conductors of identical composition. Therefore, the difference of Galvani potentials between a point in an aqueous phase and a point in the bulk of an organic solvent cannot be measured experimentally, so that non-thermodynamic assumptions must be made. It is a simple matter to measure the interfacial potential difference by varying one ion of the dissolved salt. It is also possible to measure the partition coefficient for each salt, $P^{\alpha/\beta}_{salt}$. By virtue of equation (4.5), the interfacial potential can be represented in the form

$$\Delta^{\alpha}_{\beta}\phi = \Delta^{\alpha}_{\beta}\phi^0_{ion} - \frac{RT}{z_i F}\ln P^{\alpha/\beta}_{salt}. \qquad (4.81)$$

In this way a scale of interfacial potentials can be constructed, but this scale will include an unknown constant term, $\Delta^{\alpha}_{\beta}\phi^0_{ion}$, the standard distribution potential of the invariant ion. The problem, therefore, is to choose the origin of the scale correctly.

The origin cannot be chosen by thermodynamic methods, so that auxiliary methods are required. The most popular of these, using tetraphenylborate, was formulated by Grundwald et al. (1960), and later developed by Popovych (1970),

Parker (1969, 1976), and others (Hundhammer and Solomon, 1983; Kornyshev and Volkov, 1984; Volkov and Kornyshev, 1985). The historical aspects of the problem are considered in a review by Kolthoff (1971).

The idea of the tetraphenylborate method is to choose a cation and an anion with equal standard energies of transfer between two solvents, so that the interfacial distribution potential will be zero. In this case the ion partition coefficients will equal the salt partition coefficient which is easily determined experimentally. Once the latter coefficient is found, we can find the standard free transfer energy and the standard potential distributions of individual ions. These potentials are equal in magnitude but have opposite signs:

$$\Delta_\beta^\alpha \phi_+^0 = -\Delta_\beta^\alpha \phi_-^0. \tag{4.82}$$

In one application of this approach, Koczorowski worked with tetraethylammonium picrate $(Et)_4NPi$, while others used tetraphenylarsonium tetraphenylborate (Fig. 4.7). The ions comprising the latter salt, $TPhB^-$ and $TPhAs^+$, are very much alike. They are symmetric, almost spherical particles of sufficiently large radius, which can be assumed to have the same hydrophobic properties. Their charges are equal in magnitude, are located at the centers of the spheres, and are screened from the solvent, so that the electrostatic contributions to the free resolvation energy are equal. Such ions have a relatively small polarizability and a low surface charge density, and do not participate in a specific interaction with solvent molecules. For these reasons the standard transfer energies of these ions will be nearly equal for any two solvents.

The tetraphenylborate method allows one to determine the standard distribution potentials using cyclic voltametry, chronopotentiometry, polarography at a dropping electrolyte electrode, chronoamperometry, and also from solubility and extraction data (Czapkiewicz and Czapkiewicz-Tutaj, 1980; Hundhammer and Solomon, 1983; Hundhammer et al., 1982, 1984; Joos and Verburgh, 1978; Osakai et al., 1983; Rais, 1971; Rais et al., 1976). Figure 4.8 shows a typical current–voltage characteristic recorded at a scan rate of $25\,mV\,s^{-1}$ in a water–nitrobenzene system containing 10 mM TPhAsTPhB in the non-aqueous phase and $10\,mM\,Li_2SO_4$ in water. The potentials applied to the system range from the potential of the $TPhAs^+$ transfer from nitrobenzene into water in the negative region to the potential of the $TPHB^-$ transfer in the positive region. Since

$$\Delta_\beta^\alpha \phi_{TPhB^-}^0 = -\Delta_\beta^\alpha \phi_{TPhAs^+}^0 \tag{4.83}$$

we can choose a zero point on the potential scale and pass from the scale of applied potentials to the scale of Galvani potentials or distribution potentials.

The ion standard distribution potential depends both on the nature of solvent (primarily on the optical and static permittivities) and on the ionic radius (Figs. 4.9–4.12). Figures 4.10–4.12 show the distribution potentials versus the ion radius for water–nitrobenzene, water–1,2-dichloroethane, and water–dichloromethane systems. The radii of inorganic ions are taken in the Gourary–Adrian scale (Gourary and Adrian, 1960) and those of tetraalkylammonium salts are found according to

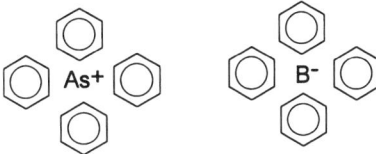

Figure 4.7. Structural formula of tetraphenylarsonium tetraphenylborate.

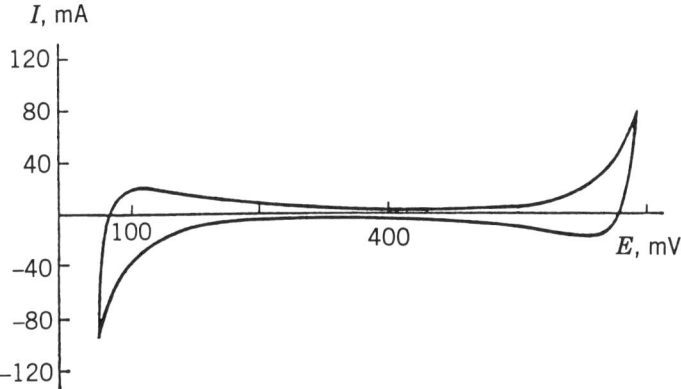

Figure 4.8. Cyclic current–voltage characteristic of a water–nitrobenzene system containing TPhBTPhAs in nitrobenzene and Li_2SO_4 in water.

Robinson and Stokes (1959). As can be seen from Figs. 4.10–4.12, the ion distribution potential drops sharply with increasing ion radius and the permittivity of the non-aqueous solvent. Solid curves in Figs. 4.10, 4.11, and 4.12 are from theoretical calculations (theoretical details will be discussed in Chapter 6). For ions of radius less than 0.3 nm the main contribution to $\Delta_\beta^\alpha \phi^0$ is due to the electrostatic portion of the free resolvation energy, and for ions of radius greater than 0.3 nm the main contribution is due to the solvophobic effect (Volkov and Kornyshev, 1985).

4.13. METHODS FOR MEASURING THE INTERFACIAL POTENTIAL DIFFERENCE

All the methods of measuring Volta potentials between phases α and β compensation are based on. This means that a potential difference is applied between the phases α and β which is equal in magnitude to the Volta potential but has an opposite sign (Bewig, 1964; Boguslavsky *et al.*, 1975a; Guyat, 1924; Kenrick, 1886; Knots and Dubovik, 1985; Randles, 1956).

Changes in Volta potentials and Galvani potentials can be measured by the dynamic capacitor method, the radioactive probe method, the jet electrode method,

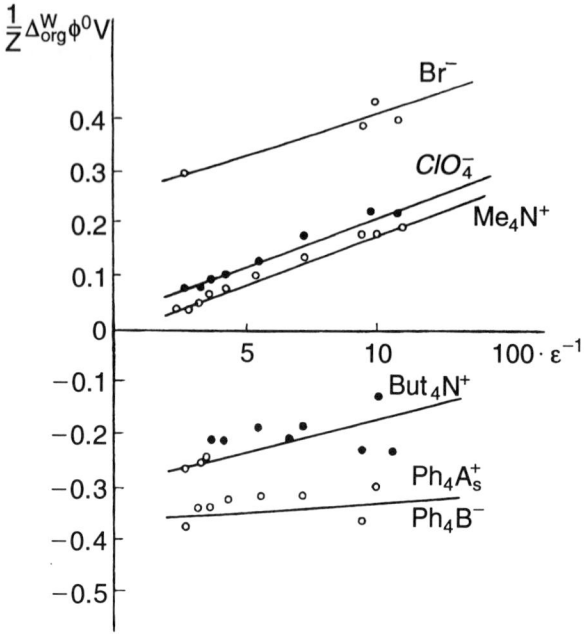

Figure 4.9. The standard ion distribution potential versus the static permittivity of non-aqueous solvents.

and also by using concentration chains. The dynamic capacitor method was proposed by Kelvin (1898) and later developed by Zisman (1932).

Two types of electrodes are usually employed to periodically alter the capacitance of the gap between the electrode surface and the surface under study: a vibrating electrode and a rotating one. Although the vibrating electrode is simpler than the rotating electrode, the former has a slightly lower stability.

A schematic diagram of the setup for measuring the Volta potential by the dynamic capacitor method is shown in Fig. 4.13. A generator of mechanical vibrations drives a gold electrode to vibrate at a distance of 0.2–1.0 mm from the surface of the liquid phase in question. To eliminate the effect of the diffusion potential, the aqueous solution is connected to a reversible reference electrode via a salt bridge filled with saturated KCl solution. The displacement current, which occurs due to the vibration of the gold electrode, is compensated by applying to the vibrating electrode a potential equal in magnitude to the measured Volta potential but having an opposite sign. The equivalent electrical circuit of the method is shown in Fig. 4.14.

The Volta potential $\Delta^w_{org} \psi$ at the interface between two immiscible balanced (mutually saturated) electrolyte solutions B^+A^- may be experimentally determined as the compensation potential in the chain

$$\text{Me} \left| \begin{array}{c} \text{Water,} \\ B^+A^- \end{array} \right| \text{Gas} \left| \begin{array}{c} \text{Nitrobenzene} \\ B^+A^- \end{array} \right| \text{Me,} \quad E_{\text{III}}. \qquad (4.\text{III})$$

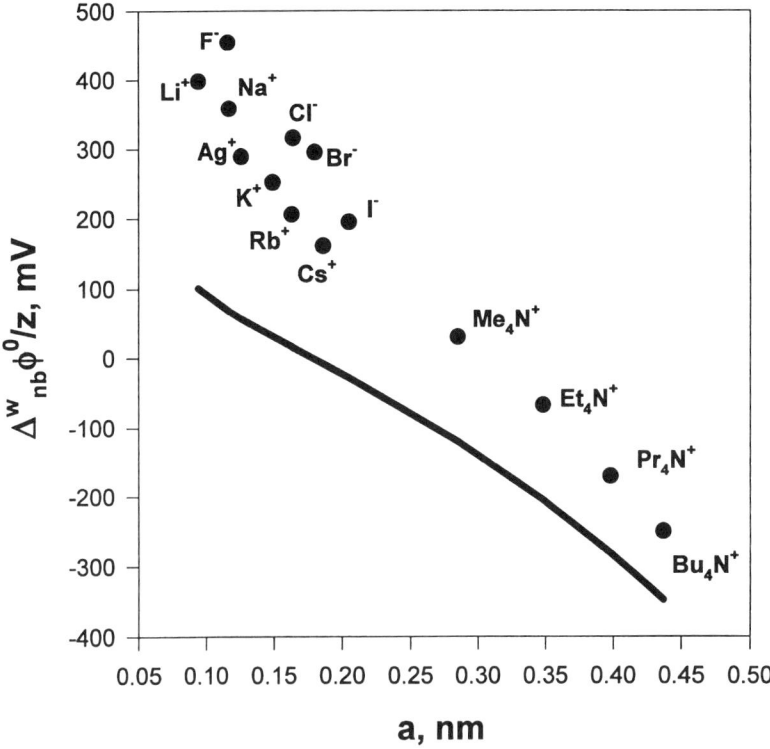

Figure 4.10. The standard ion distribution potential at the nitrobenzene–water interface versus the ion radius. The solid line was calculated using equations (6.1), (6.5), and (6.36).

The gas is a chemically inert gaseous phase, such as nitrogen, argon, or pure air; Me is a metal electrode; B^+A^- is the electrolyte.

The Volta potential can be measured by the vibrating electrode method as the difference between E_{II} and E_{III} for chains (4.II) and (4.III):

$$\text{Me} \left| \begin{matrix} \text{Vibrating} \\ \text{electrode} \end{matrix} \right| N_2 \left| \begin{matrix} \text{Nitrobenzene} \\ B^+A^- \end{matrix} \right| \begin{matrix} \text{Water} \\ B^+A^- \end{matrix} \left| \begin{matrix} \text{Saturated calomel} \\ \text{electrode} \end{matrix} \right| \text{Me}, \qquad E_{IV} \quad (4.IV)$$

and

$$\text{Me} \left| \begin{matrix} \text{Vibrating} \\ \text{electrode} \end{matrix} \right| N_2 \left| \begin{matrix} \text{Water} \\ B^+A^- \end{matrix} \right| \begin{matrix} \text{Saturated calomel} \\ \text{electrode} \end{matrix} \left| \text{Me}, \qquad E_V. \right. \quad (4.V)$$

Thus we have

$$\Delta_{org}^w \psi(BA) = E_{III} = E_{IV} - E_V \qquad (4.83)$$

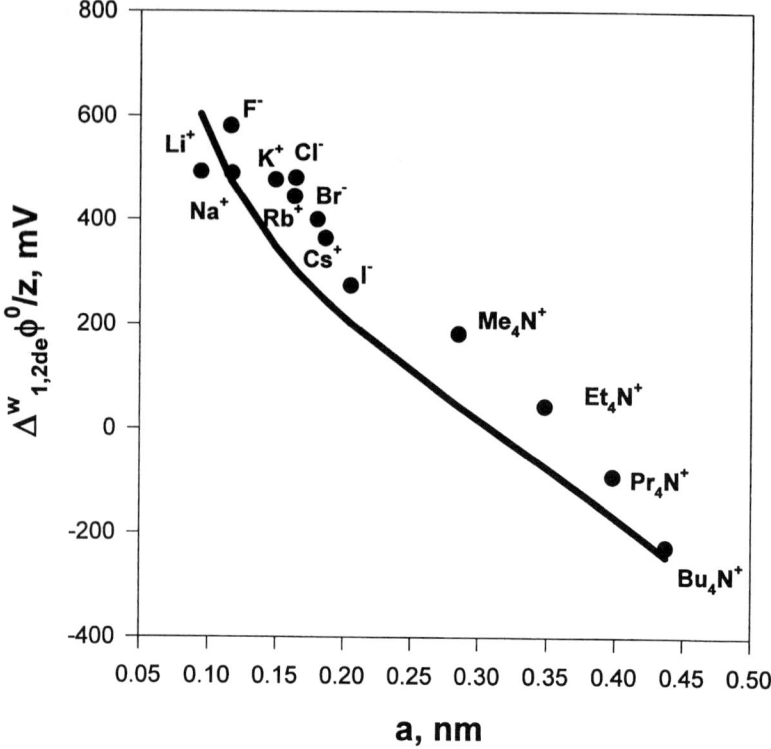

Figure 4.11. The standard ion distribution potential at the 1,2-dichloroethane–water interface versus the ion radius. The solid line was calculated using equations (6.1), (6.5), and (6.36).

provided the following conditions are satisfied:

1. The surface potentials of the vibrating electrode, the water-saturated nitrobenzene phase and the nitrobenzene-saturated aqueous phase are constant.
2. The calomel electrode is stable, and the contribution due to the diffusion potential between the aqueous phase and solution in the calomel electrode is constant.

The equivalent circuit of a rotating electrode is shown in Fig. 4.15. The rotating electrode is placed concentrically above a circular cell which is divided by a screen into two identical compartments filled with the liquids under study.

The radioactive probe method was proposed by Guyat (1924). The method uses ionizing α-particle irradiation of the gap between the electrode and the surface of the liquid in question to reduce the gap resistance, and the displacement current is measured by an electrometer. ^{210}Po (1 μCi), ^{241}Am (750 μCi), and ^{239}Pu are usually employed as the radioactive source (Costa *et al.*, 1976; Gugeshashvili, 1974;

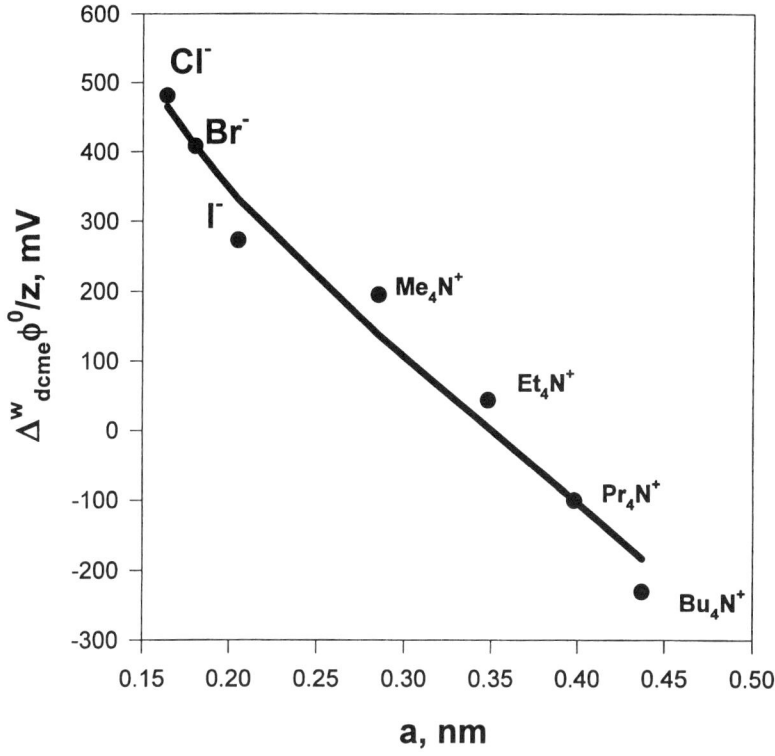

Figure 4.12. The standard ion distribution potential at the dichloromethane–water interface versus the ion radius. The solid line was calculated using equations (6.1), (6.5), and (6.36).

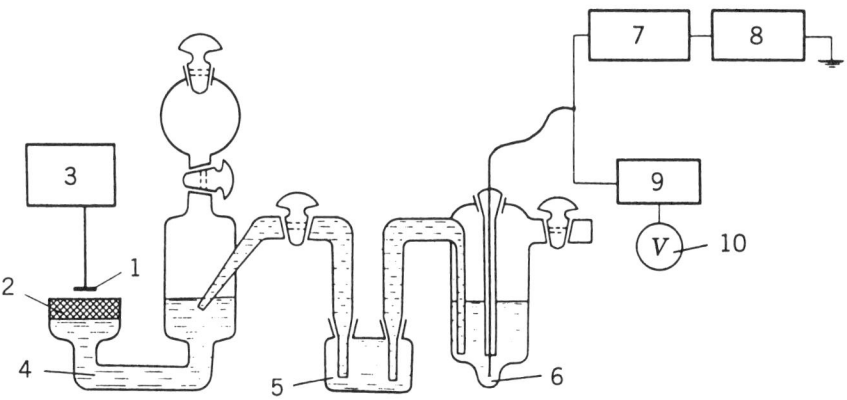

Figure 4.13. A schematic diagram of the setup for measuring Volta potentials between two immiscible liquids. Notation: 1—vibrating gold electrode; 2—layer of oil; 3—generator; 4—aqueous solution; 5—salt bridge; 6—calomel electrode; 7—amplifier; 8—oscillograph; 9—a compensating device. The compensating potential was measured by a voltmeter (10).

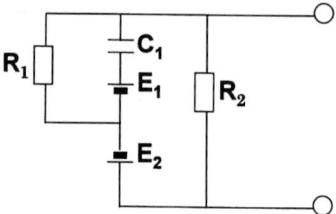

Figure 4.14. Equivalent electrical circuit for measuring potential difference by the vibrating-electrode method. Notation: R_1—electrode leakage resistance; C_1—vibrating electrode capacitance; R_2—resistor in the compensation circuit; E_1—potential difference between the vibrating electrode and the surface under study; E_2—EMF of the compensating source.

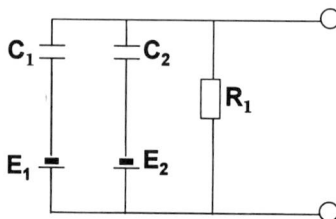

Figure 4.15. Equivalent electrical circuit for measuring potential difference by the rotating-electrode method. Notation: R_1—load resistor; C_1—capacitance between the cell compartments and the rotating electrode; E_1—compensating voltage.

Kamienski *et al.*, 1967a,b). The advantage of using [241]Am foil instead of [210]Po foil is twofold: the half-life period of the former source is much longer (458 years compared with 138.4 days) and it has a greater area, which can easily be isolated (Jaycock and Parfitt, 1981). The accuracy and reproducibility of the radioactive method are lower than for the vibrating-electrode method.

A necessary condition in measuring the Volta potential by the radioactive method is to shield the measuring chamber from the electrometric input. An overlarge gap between the electrode and the surface of the electrolyte solution is frequently a source of errors in measuring the potential difference, because of radiation scattering and parasitic leakage. A fairly narrow gap can also lead to erroneous results because of the interaction between α-particles and molecules of the liquid under study. The experimental setup was described by Bewig (1964), and an equivalent circuit of the radioactive probe method is shown in Fig. 4.16.

The jet electrode method was first proposed by F. Kenrick. A jet of the liquid in question flows along the axis of a tube whose inner surface is covered with a draining film of a second liquid (Fig. 4.17). Reversible reference electrodes connect the solutions to the measuring system consisting of a potentiometer and a null-detector

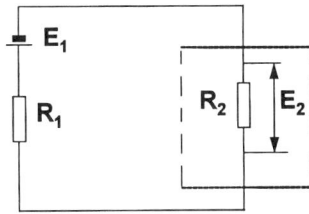

Figure 4.16. Equivalent electrical circuit for measuring potential difference by the radioactive probe method. Notation: E_1—potential difference to be measured; R_1—input electrometer resistance; R_2—air-gap resistance; E_2—voltage drop across the electrometer input.

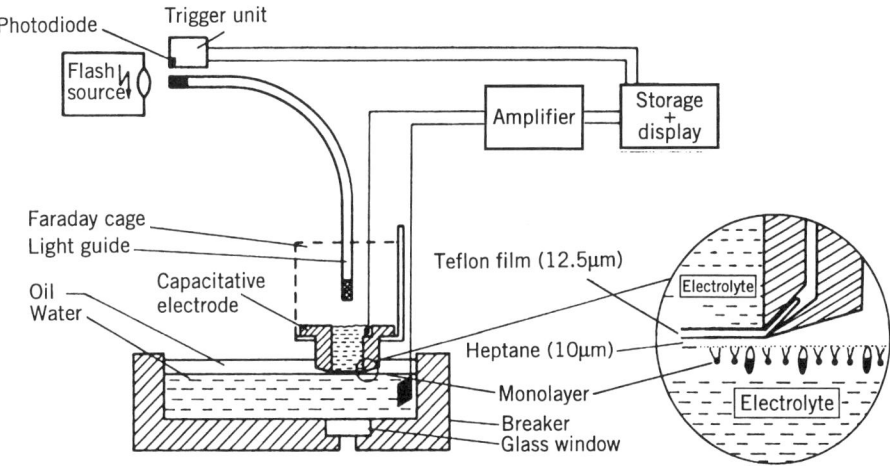

Figure 4.17. A schematic representation of the experimental setup to measure changes of electric potential differences at an oil–water interface with the capacitive electrode according to Trissl and Graber (1980). Reproduced with permission of Elsevier Sequoia S. A.

with a very high input resistance. When compensation is attained, the external potential of the jet is equal to the external potential of the solution draining along the inner surface of the tube. Because it is inconvenient to set up, the jet method is rarely used (Girault and Schiffrin, 1984b; Llopis, 1971).

When applied to measurement of the Volta potential, all three methods (except the jet method) give virtually identical results (Case and Parsons, 1974; Parsons and Rubin, 1974; Randles, 1956). The methods can give different results for nonpolar solvent–water interfaces containing ionogenic surface-active substances such as benzoic acid salts, sodium dodecylsulfate, liquid crystals and metallocompexes of valinomycin. This phenomenon was discovered by Boguslavsky and Gugeshashvili (1988, 1989) who studied the electret states of liquid hydrocarbons by vibrating-electrode and radioactive probes.

Another method for investigating electrical phenomena at oil–water interfaces was suggested by Trissl and Graber (1980), who described the construction and electrical properties of a novel capacitative electrode with a wide frequency range. The electrode can measure flash-induced photopotentials in monolayers at the oil–water interface with a sensitivity of 0.05–0.10 mV. The method is based on capacitative coupling of two aqueous compartments separated by a thin Teflon film, and the electrode is placed with its planar bottom about 0.01 mm above the interface. The aqueous subphase and the inner electrolyte are connected with Ag/AgCl electrodes to voltage amplifiers. A schematic representation of the experimental setup is shown in Fig. 4.17. The main advantage of this method is that it can measure rapid electrical signals with nanosecond resolution during photochemical processes at the oil–water interface. The main disadvantage is that it can only be used in the absence of ions, microemulsions or a diffuse double layer in the non-aqueous phase. Generally speaking, Trissl's method does not measure interfacial potentials between two phases, but instead monitors rapid light-induced electrochemical effects at the interphase boundary. The physical basis of such effects can vary between systems and must be carefully analyzed before drawing conclusions.

4.14. HUNG'S METHOD OF GALVANI-POTENTIALS CALCULATION: SMALL SYSTEMS AND THE EFFECT OF VOLUME

Interfacial potential values for systems with small volumes were analyzed by Hung (1980, 1983), who considered the distribution of charged particles between phases and the corresponding interfacial potentials. In particular, the finite volumes of contacting phases which affect the distribution of charged particles were taken into account. Such relationships are complex and generally do not permit analytical solutions. However, solutions can be obtained with the help of simplifying non-thermodynamic assumptions or by numerical methods.

Hung considered equilibrium distributions of ions I_i between two phases α and β:

$$
\begin{array}{c|c}
\alpha \quad I_i^{z_i} & I_i^{z_i} \quad \beta \\[4pt]
V_\alpha & V_\beta
\end{array}
\tag{4.85}
$$

where V_α and V_β are volumes of phases in contact.

From the mass conservation law it follows that

$$
V_\alpha c_i^\alpha + V_\beta c_i^\beta = m_i
\tag{4.86}
$$

where m_i is the amount of ion I_i in both phases. Hung also assumed electroneutrality of each phase

$$
\sum_i z_i c_i^\alpha = 0
\tag{4.87}
$$

$$
\sum_i z_i c_i^\beta = 0
\tag{4.88}
$$

and it follows from equations (4.85)–(4.87) that

$$\sum_i z_i m_i = 0. \tag{4.89}$$

Generally speaking, the total system consisting of phases α and β is electroneutral, but electroneutrality of a single phase (4.87) and (4.88) cannot be assumed if one of the phase volumes is small. According to Kakiuchi (1996a) if the diameter of a given phase is less than 10^{-6} m, the Poisson equation can be used to correlate local ionic concentrations with the interfacial potential, both of which vary with the distance from the phase boundary and depend on the shape of the interface.

Combining equations (4.86)–(4.89) and (4.7) gives Hung's equation:

$$\sum_{i=1}^{h} \frac{z_i c_i^{\alpha,0}}{1 + (\gamma_i^\alpha/v\gamma_i^\beta)\exp[(z_i F/RT)(\Delta_\beta^\alpha \phi - \Delta_\beta^\alpha \phi_i^0)]}$$
$$+ \sum_{i=h}^{j} \frac{z_i c_i^{\beta,0}}{1 + v(\gamma_i^\alpha/\gamma_i^\beta)\exp[(z_i F/RT)(\Delta_\beta^\alpha \phi - \Delta_\beta^\alpha \phi_i^0)]} = 0 \tag{4.90}$$

where $v = V^0/V^W$ and c^0 denote the initial concentration. Equation (4.90) has no analytical solution when the number of ions in the system is more than three and must be solved numerically. From Hung's equation (4.90) it follows that the value of $\Delta_\beta^\alpha \phi$ depends on temperature, the volume ratio of the two phases, initial concentrations of components, activity coefficients, and standard electrical potentials of ion transfer between phases α and β. If the system contains only one binary electrolyte A^+B^-, equation (4.90) gives the distribution potential (4.14). Detailed analysis of Hung's method can be found in reviews (Kakiuchi, 1996a; Koczorowski, 1987).

4.15. INTERFACIAL POTENTIAL MEASUREMENTS AS A TOOL FOR STUDYING MECHANISMS OF ENZYMATIC AND CATALYTIC REACTIONS

The vibrating-capacitor method has been used to measure electrical potential changes at the oil–water interface that result from catalyzed reactions accompanied by transfer of ions or electrons between phases (Boguslavsky and Volkov, 1975, 1977, 1987). The Volta potential is measured in the electric chain

Vibrating gold electrode	Gas	Oil, acceptor or donor of charges	Water, substrate buffer solution	Salt bridge	Reference electrode	(4.VI)
$\Delta_{Au}^{Gas}\phi$	$\Delta_{Gas}^{org}\chi = $ constant	$\Delta_{org}^{w}\phi$		$\Delta_w^w\phi_{dif}$	$\Delta_w^{RE}\phi$	

resulting in a sum of Galvani potentials $\phi = \Delta_{Au}^{Gas}\phi + \Delta_{Gas}^{org}\chi + \Delta_{org}^{w}\phi + \Delta_w^w\phi_{dif} + \Delta_w^{RE}\phi$. If all potentials except the investigated one are constant, the observed change of ϕ can be attributed to the changes of $\Delta_{org}^{w}\phi$. In general, some change of potential at the air–oil interface may occur, but its value depends on the thickness of the oil layer and tends to zero when this thickness increases. The measured Volta potential is independent of oil layer thicknesses between 1 to 5 mm.

The oil–water systems contained all necessary components except the catalyst and substrate, which were added to the cell immediately before measurements began. In the steady state, electrical currents between the oil and aqueous phases $\vec{I_{W}^{O}}$ and $\vec{I_{W}^{O}}$ are equal, so that the total charge transfer across the interphase boundary is zero. In reactions that alter the ionic concentration in the non-aqueous phase, the measured potential changed by an amount η, which corresponds to the overvoltage in chemical kinetics:

$$\eta = \Delta\phi_{W}^{O} - \Delta\phi_{W}^{O}(eq), \tag{4.91}$$

where "eq" denotes equilibrium, and "O" and "W" indicate the oil and water phases.

The rate of an electrochemical reaction involving charge transfer is characterized by the reaction current I. We can therefore use the kinetic relations for ordinary chemical reactions by replacing the rate of the catalytic reaction involving charge transfer by I/zF.

The simplest mechanism involving catalyzed charge transfer across the oil–water interface can be written as

$$S + K \underset{k_{-1}}{\overset{k_{1}}{\rightleftarrows}} SK,$$

$$SK \overset{k_{2}}{\longrightarrow} \sum_{i} P_{i}^{O} + \sum_{j} P_{j}^{W} + K,$$

$$P_{i}^{O} \overset{k_{3}}{\longrightarrow} P_{i}^{W}, \tag{4.92}$$

where S is the substrate, K is the catalyst and P_{i} and P_{j} are reaction products.

Two possible mechanisms of charge transfer from oil to water are diffusion of reaction products, or transfer by carriers when diffusion of carrier molecules from water to oil is not rate limiting. For a given sequence of reactions the form of the kinetic equation depends on the rate-limiting step. For instance, if the breakdown of the intermediate product SK is rate limiting the rate of accumulation of the product is given by the following expression:

$$\frac{d[P_{i}^{O}]}{dt} = -k_{3}[P_{i}^{O}] + k_{2}[SK]. \tag{4.93}$$

Taking into consideration the accumulation and decomposition of SK according to scheme (4.92) we obtain

$$\frac{d[KS]}{dt} = k_{1}[K][S] - k_{-1}[KS] - k_{2}[KS], \tag{4.94}$$

where the reagent concentrations are related by the material balance equations:

$$[K] = [K]_i - [KS], \qquad [S] = [S]_i - [KS]. \tag{4.95}$$

For the steady state established after the onset of the reaction,

$$\frac{d[KS]}{dt} = 0. \tag{4.96}$$

Taking into account that

$$[S] = [S]_i - [KS] \approx [S]_i \tag{4.97}$$

we obtain from (4.94) and (4.96)

$$[KS] = \frac{k_1[K]_i[S]_i}{k_{-1} + k_2 + k_1[S]_i}. \tag{4.98}$$

The initial steady reaction rate is then equal to

$$\frac{I_i}{zF} = k_2[SK] = \frac{k_2[K]_i[S]_i}{K_M + [S]_i} \tag{4.99}$$

where $K_M = (k_{-1} + k_2)/k_1$ is the Michaelis–Menten constant. In general the surface concentration of reactants and products c_s may differ from the bulk values ($0 < c_s < c_b$). In diffusion-controlled reactions (fast reactions) $c_s \approx 0$, while in kinetically controlled slow reactions $c_s \approx c_b$. The processes discussed here are kinetically controlled reactions. This is supported by the fact that the Michaelis–Menten's constants in some interfacial processes are equal to the bulk value (Kharkats et al., 1975, 1976, 1977).

Figure 4.18 shows a scheme illustrating catalyzed transfer of charged particles across the oil–water interface. Such reactions involving metalloporphyrins, ATPases and respiratory enzymes have often been investigated (Boguslavsky and Volkov, 1977, 1987; Kozlov and Skulachev, 1977), and can be theoretically described as follows.

The formula for the interfacial potential change

$$\eta = zF\frac{[P_i^O]}{C}, \tag{4.100}$$

where C is the integral capacitance of the electric double layer.

From equation (4.93) we obtain

$$\frac{k_2}{k_3}[SK] = [P_i^O]. \tag{4.101}$$

It follows from equations (4.100) and (4.101) that

$$\eta = \frac{zFk_2[SK]}{Ck_3} = \frac{I_i}{Ck_3}. \tag{4.102}$$

Therefore at steady state the change of interfacial potential η due to catalyzed charge

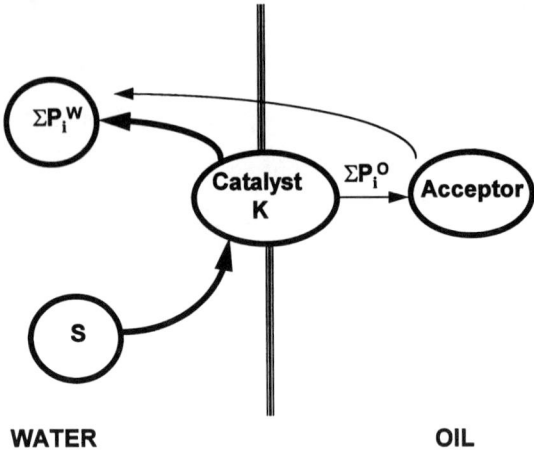

WATER   **OIL**

Figure 4.18. A simplified diagram of catalyst/enzyme functioning at the oil–water interface.

transfer across the oil–water interface is proportional to the rate of the reaction, as observed experimentally (Boguslavsky *et al.*, 1974b, Boguslavsky and Volkov, 1987). The vibrating-plate method is a simple way to obtain information and calculate the kinetics of such processes by the Michaelis–Menten equation. The values K_M and η_{max} can be found graphically by extrapolation of experimental data. Equation (4.98) can also be put in the form

$$\frac{1}{\eta} = \frac{K_M}{\eta_{max}} \frac{1}{[S_i]} + \frac{1}{\eta_{max}}. \tag{4.103}$$

Figure 4.19 shows the graph of equation (4.103), from which the Michaelis constant and η_{max} can be determined.

In intact cells many enzymes function simultaneously to catalyze reaction chains in which a product of one reaction serves as a substrate for the next. Such multi-enzyme systems are organized as complexes, bound to membranes or cell

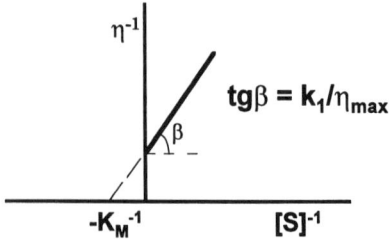

Figure 4.19. Graphic determination of the Michaelis–Menten constant from electrochemical data.

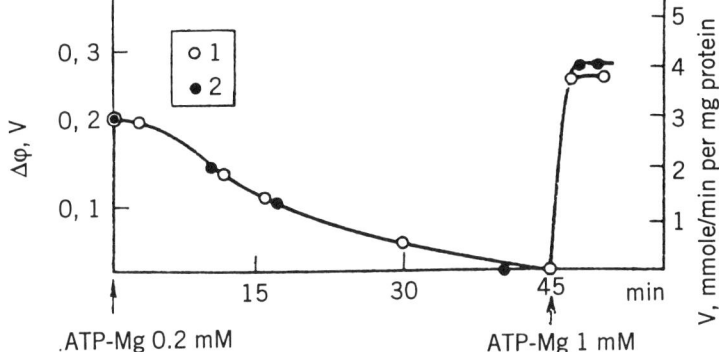

Figure 4.20. Volta potential (○) at the octane–water interface and enzyme activity V (●) as functions of incubation time of the reaction mixture. Medium: 0.2 mM MgATP, 10^{-8} M factor F_1, 1 mM 2,4-dinitrophenol, 10 mM Tris-HCl, pH 7.0. (Boguslavsky et al., 1974b).

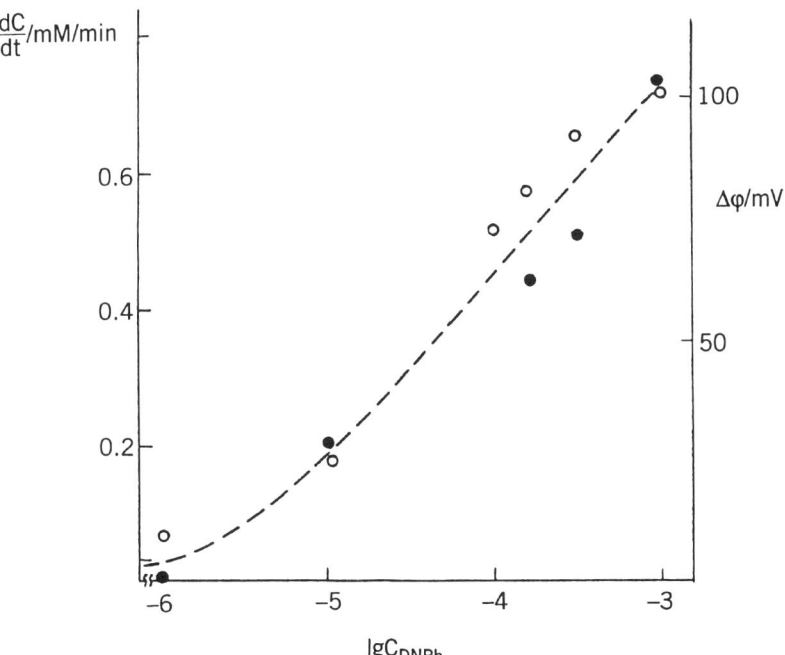

Figure 4.21. Dependence of the oxygen evolution rate (●) and Volta potential (○) on the concentration of electron acceptor $K_3Fe(CN)_6$ upon illumination. Medium: 2×10^{-6} M hydrated oligomer of chlorophyll a, 20 mM Tris-HCl, pH 7.4, 1 mM $NADP^+$ (Kandelaki et al., 1983b).

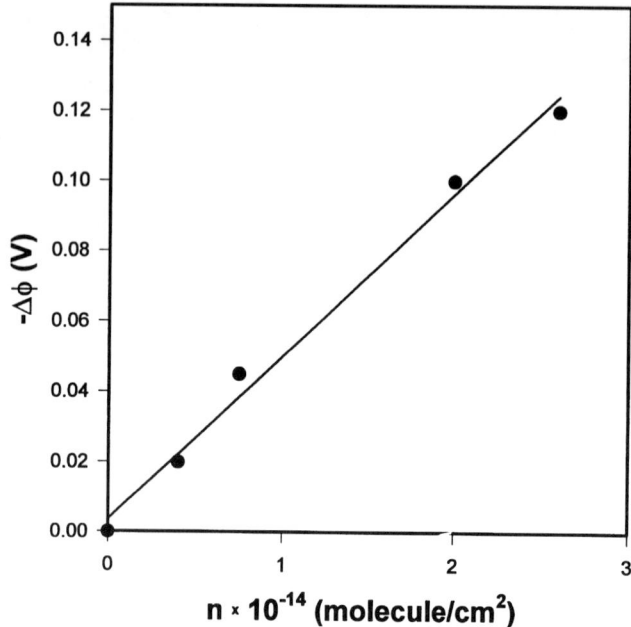

Figure 4.22. Dependence of Volta potential on the number of coproporphyrin molecules adsorbed at the water–octane interface during the interfacial electron transport reaction from NADH in water to 2N-methylamino-1,4-naphtoquinone (Volkov *et al.*, 1983a). Medium: 1 mM NADH, 0.1 mM 2N-methylamino-1,4-naphtoquinone, 10 mM Tris-HCl, pH 7.3.

organelles. It is possible to couple several enzymes at the octane–water interface so that the product of one becomes the substrate for another (Yaguzhinsky *et al.*, 1975a,b, 1976, 1977). Scheme (4.92) shows that there should be Michaelis-type dependence of η on the concentration of each of the substrates, and this was found to be the case experimentally (Kharkats *et al.*, 1975, 1976, 1977).

The usefulness of equations (4.102) and (4.103) is illustrated by Figs. 4.20–4.22, which show the correlation between interfacial potential and the rate of charge transfer reactions at the octane–water interface (Figs. 4.20 and 4.21) as well as a linear dependence of η on the interfacial concentration of adsorbed catalysts (Fig. 4.22).

5

ELECTROCAPILLARITY

5.1 THE ELECTROCAPILLARY EQUATION

Electrochemical processes occurring at the interface between two immiscible liquids are traditionally described based on Gibbs thermodynamics of surfaces. However, Hansen's method, by extending and generalizing Gibbs method, gives us a better understanding of the nature of interfacial phenomena and provides us with an improved method for describing them. Hansen's method is especially convenient for describing electrocapillary phenomena, and makes it possible to simplify otherwise complicated equations.

Surface Excesses of Ions. Consider two phases in contact, α and β, with r_h neutral components, whose chemical potentials are designated by μ_h, r_c types of cations and r_a types of anions with chemical potentials μ_i. Using Hansen's reference system and choosing the solvents α and β as reference substances one can write down Hansen's rendition of Gibbs adsorption equation:

$$d\gamma = -s^s_{(\alpha,\beta)}\,dT + \tau^s_{(\alpha,\beta)}\,dp - \sum_{h\neq\alpha,\beta}\Gamma_{h(\alpha,\beta)}\,d\mu_h - \sum_i \Gamma_{i(\alpha,\beta)}\,d\overline{\mu}_i. \qquad (5.1)$$

Here, as before, $s^s_{(\alpha,\beta)}$, $\tau^s_{(\alpha,\beta)}$ and $\Gamma_{h(\alpha,\beta)}$ are the surface excesses of the entropy, volume and substance h in the reference system, when the absolute excesses α and β are zero. The terms corresponding to different components of the system are divided into two summations, one for neutral and the other for charged particles, the reference substances α and β being excluded from the sum of neutral particles. Individual components of the system, both charged and neutral, can be present either in both phases, or in only one. It was shown in Chapter 3 that all charged particles can be divided into two groups α and β. The ionic sum in equation (5.1) can be divided into four parts:

$$\sum_i \Gamma_{i(\alpha,\beta)} \, d\bar{\mu}_i = \sum_j \Gamma_{j(\alpha,\beta)} \, d\bar{\mu}_j + \sum_k \Gamma_{k(\alpha,\beta)} \, d\bar{\mu}_k + \sum_l \Gamma_{l(\alpha,\beta)} \, d\bar{\mu}_l + \sum_m \Gamma_{m(\alpha,\beta)} \, d\bar{\mu}_m.$$

$$(5.2)$$

We shall substitute the electrochemical potentials of the ions in this equation with the chemical potentials of the salts of which solutions α and β are composed:

$$\sum_i \Gamma_{i(\alpha,\beta)} \, d\bar{\mu}_i = \sum_{j \neq j'} (1/\nu_{jk'}^{+}) \Gamma_{j(\alpha,\beta)} \, d\mu_{jk'} + \sum_{k \neq k'} (1/\nu_{j'k}^{-}) \Gamma_{k(\alpha,\beta)} \, d\mu_{j'k}$$

$$+ (1/z_{k'} \nu_{jk'}^{-}) \left(z_{k'} \Gamma_{k'(\alpha,\beta)} + \sum_{j \neq j'} z_j \Gamma_{j(\alpha,\beta)} \right) d\mu_{j'k'} + \sum_{m \neq m'} (1/\nu_{l'm}^{-}) \Gamma_{m(\alpha,\beta)} \, d\mu_{l'm}$$

$$+ \sum_{l \neq l'} (1/\nu_{lm'}^{+}) \Gamma_{l(\alpha,\beta)} \, d\mu_{lm'} + (1/z_{l'} \nu_{l'm'}^{+}) \left(z_{l'} \Gamma_{l'(\alpha,\beta)} + \sum_{m \neq m'} z_m \Gamma_{m(\alpha,\beta)} \right) d\mu_{l'm'}$$

$$+ (1/z_{j'}) \left(\sum_j z_j \Gamma_{j(\alpha,\beta)} + \sum_k z_k \Gamma_{k(\alpha,\beta)} \right) d\bar{\mu}_{j'} + (1/z_{m'}) \left(\sum_l z_l \Gamma_{l(\alpha,\beta)} + \sum_m z_m \Gamma_{m(\alpha,\beta)} \right) d\bar{\mu}_{m'}.$$

$$(5.3)$$

The right-hand side of this equation contains $r_c + r_a - 2$ differentials of the chemical potentials of the neutral salts and two differentials of the electrochemical potentials of the indicator ions, i.e., the total number of differentials in the ion sum is equal as before to $r_c + r_a$. The two last terms in this equation, however, can be transformed further. Note that the coefficients in parenthesis are the contributions to the interface charge (if multiplied by the Faraday constant F) made by ions of groups α and β, respectively. Designating these contributions by Q^{α} and Q^{β}, we have

$$Q^{\alpha} = F \left(\sum_j z_j \Gamma_{j(\alpha,\beta)} + \sum_k z_k \Gamma_{k(\alpha,\beta)} \right)$$

$$(5.4)$$

$$Q^{\beta} = F \left(\sum_l z_l \Gamma_{l(\alpha,\beta)} + \sum_m z_m \Gamma_{m(\alpha,\beta)} \right).$$

$$(5.5)$$

The interface as a whole is electrically neutral and therefore

$$Q^{\alpha} + Q^{\beta} = 0.$$

$$(5.6)$$

As a result, the last two terms in equation (5.3) can be written as

$$\left(\sum_j z_j \Gamma_{j(\alpha,\beta)} + \sum_k z_k \Gamma_{k(\alpha,\beta)} \right)(1/z_{j'}) \, d\bar{\mu}_{j'} + \left(\sum_l z_l \Gamma_l + \sum_m z_m \Gamma_m \right)(1/z_{m'}) \, d\bar{\mu}_{m'}$$

$$= Q^{\alpha} [(1/z_{j'} F) \, d\bar{\mu}_{j'} - (1/z_{m'} F) \, d\bar{\mu}_{m'}].$$

$$(5.7)$$

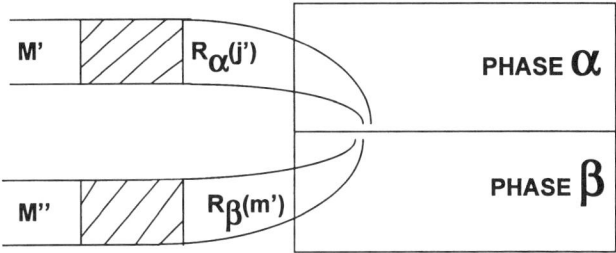

Figure 5.1. Electrochemical circuit for measuring $E_{\beta(m')}^{\alpha(j')}$ and electrocapillary phenomena at the interface between two immiscible electrolyte solutions.

The new parameters Q^α and Q^β that appeared in equations (5.4)–(5.6) represent thermodynamic, or total charges. It is tempting to ascribe them to the α and β phases; however, in general such an operation is incorrect (Frumkin *et al.*, 1980; Volkov, 1987b; Markin and Volkov, 1988a,b,c).

The electrochemical potentials of the indicator ions j' and m' act as a linear combination with constant factors. This means that the number of independent differentials in the electrocapillary equation has been reduced by one due to the assumption of electroneutrality of the interface. The number of independent variables in the ionic sum (5.3) becomes equal to $r_c + r_a - 1$.

EMF of the Circuit. Equation (5.7) allows a new parameter to be introduced—the electromotive force of a cell for electrocapillary measurements. The difference of electrical potential between the α and β solutions can be measured by the circuit shown in Fig. 5.1. The block diagram of the device for investigation of electrocapillary phenomena and measurement of potential difference at the interface between two immiscible electrolyte solutions is shown in Fig. 5.2. The experimental cell for measurements of electrocapillary curves is shown in Fig. 5.3.

From an external source a potential difference is applied between the α and β phases by means of polarizable electrodes. Electrode R_α, which is reversible with respect to the indicator ion j' is introduced into solution α. This electrode is connected through a series of intermediate phases with reversible interfaces to metal M'. Another electrode R_β reversible with respect to ion M', is connected in a similar manner to metal M'', which is identical to metal M'. Between terminals M' and M'' a potential difference $\Delta_{M''}^{M'}\phi$ is set up. This potential difference can be related to the electrochemical potentials of the indicator ions j' and m'.

Since electrode R_α is reversible with respect to ion j', the electrochemical potentials of this ion inside electrode R_α and in the α phase are identical: $\bar\mu_{j'}^{R_\alpha} = \bar\mu_{j'}^\alpha = \bar\mu_{j'}$. Hence

$$d\phi^{R_\alpha} = (1/Fz_{j'})\,d\bar\mu_{j'} - (1/Fz_{j'})\,d\mu_{j'}^{R_\alpha}. \tag{5.8}$$

When the electrode composition is kept constant,

$$d\bar\mu_{j'}^{R_\alpha} = -s_{j'}^{R_\alpha}\,dT + v_{j'}^{R_\alpha}\,dp \tag{5.9}$$

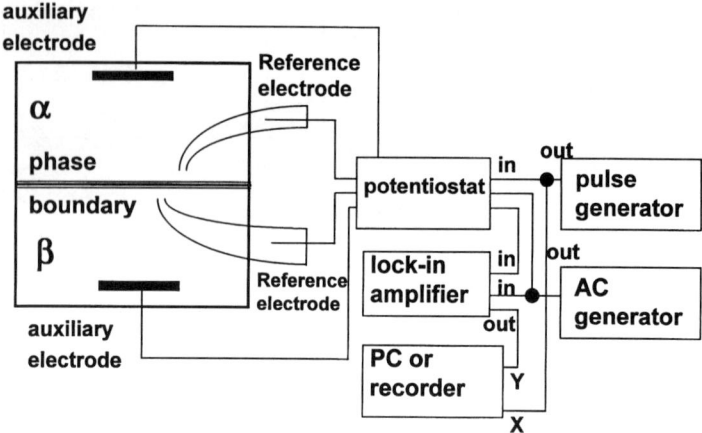

Figure 5.2. Block diagram of the device for measuring the potential difference and electrocapillary properties of the interface between two immiscible electrolyte solutions.

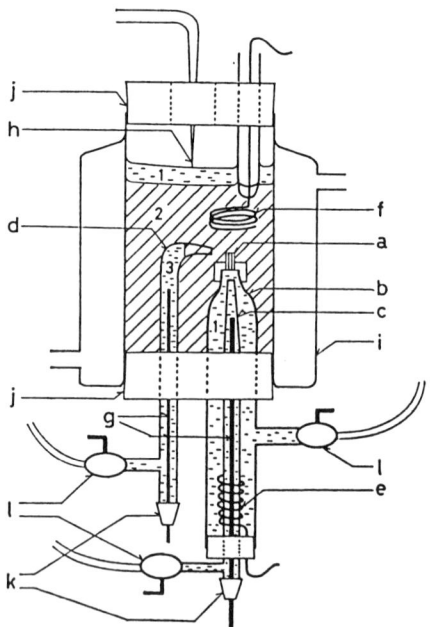

Figure 5.3. Electrochemical cell with dropping electrolyte solution electrode for measuring current–potential curves and electrocapillary curves at the interface between two immiscible electrolyte solutions. (1) LiCl aqueous solution; (2) TBATPhB nitrobenzene solution; (3) TBACl aqueous solution; (a) dropping electrolyte solution electrode tip; (b) glass tube; (c) and (d) reference electrodes with Luggin capillaries; (e) and (f) auxiliary electrodes; (g) Ag/AgCl electrodes; (h) fine glass capillary; (i) water-jacketted cell; (j) silicon rubber stopper; (k) silicone rubber cap; (l) stopcock (Kakiuchi and Senda, 1983a). Reproduced by permission of The Chemical Society of Japan.

where $s_{j'}^{R_\alpha}$ and $v_{j'}^{R_\alpha}$ are the molar entropy and the molar volume of component j' in electrode R_α.

An equation similar to (5.8) can be written for each interface. It is possible to reach terminal M' from the α phase, thus its Galvani potential can be related to the electrochemical potential of the indicator ion $\bar{\mu}_{j'}$:

$$d\phi^{M'} = (1/Fz_{j'})\,d\bar{\mu}_{j'} - (1/Fz_{j'})\,d\mu_{R_\alpha}^*. \tag{5.10}$$

As Mohilner (1966) proposed, we designated an appropriate linear combination of chemical potentials of pure components in the circuit from R_α to M' by $\mu_{R_\alpha}^*$, which appears in the transformations above. Similar to equation (5.9) one can write

$$d\mu_{R_\alpha}^* = -s_{R_\alpha}^*\,dT + v_{R_\alpha}^*\,dp \tag{5.11}$$

where $s_{R_\alpha}^*$ and $v_{R_\alpha}^*$ are the same linear combinations of the molar entropies and volumes of the components of intermediate phases. For the second part of the circuit from the β phase to terminal M'' similar equations are required

$$d\phi^{M''} = (1/z_{m'} F)\,d\bar{\mu}_{m'} - (1/z_{m'} F)\,d\bar{\mu}_{R_\beta}^* \tag{5.12}$$

$$d\bar{\mu}_{R_\beta}^* = -s_{R_\beta}^*\,dT + v_{R_\beta}^*\,dp. \tag{5.13}$$

We shall introduce the electromotive force of the circuit shown in Fig. 5.1 as the difference of Galvani potentials between two terminals made of identical metals:

$$E_{\beta(m')}^{\alpha(j')} = \phi^{M'} - \phi^{M''}. \tag{5.14}$$

The scripts $\alpha(j')$ and $\beta(m')$ indicate that the reference electrode in the α phase is reversible with respect to ion j', and in the β phase it is reversible with respect to ion m'. From equations (5.10)–(5.14), we obtain:

$$(1/z_{j'} F)\,d\bar{\mu}_{j'}^\alpha - (1/z_{m'} F)\,d\bar{\mu}_{m'}^\beta = dE_{\beta(m')}^{\alpha(j')} + (1/z_{j'} F)(v_{R_\alpha}^*\,dp - s_{B\alpha}^*\,dT) -$$
$$(1/z_{m'} F)(v_{R_\beta}^*\,dp - s_{R_\beta}^*\,dT). \tag{5.15}$$

The Final Form of the Electrocapillary Equation. We can now write the electrocapillary equation in its final form, which includes the electromotive force of the measuring cell:

$$d\gamma = -(s_{(\alpha,\beta)}^s - (1/z_{j'} F) Q^\alpha s_{R_\alpha}^* - (1/z_{m'} F) Q^\beta s_{R_\beta}^*)\,dT$$

$$+ (\tau_{(\alpha,\beta)}^s - (1/z_{j'} F) Q^\alpha v_{R_\alpha}^* - (1/z_{m'} F) Q^\beta v_{R_\beta}^*)\,dp - \sum_{h\neq\alpha,\beta} \Gamma_{h(\alpha,\beta)}\,d\mu_h$$

$$- \sum_{k\neq k'} (1/v_{j'k}^-)\Gamma_{k(\alpha,\beta)}\,d\mu_{j'k} - \sum_{j\neq j'} (1/v_{jk'}^+)\Gamma_{j(\alpha,\beta)}\,d\mu_{jk'} - \sum_{m\neq m'} (1/v_{l'm}^-)\Gamma_{m(\alpha,\beta)}\,d\mu_{l'm}$$

$$
- \sum_{l \neq l'} (1/v_{lm'}^{+}) \Gamma_{l(\alpha, \beta)} \, d\mu_{lm'} - (1/z_{k'} \, v_{j'k'}^{-}) \left(z_{k'} \Gamma_{k'(\alpha, \beta)} + \sum_{j \neq j'} z_j \Gamma_{j(\alpha, \beta)} \right) d\mu_{j'k'}
$$

$$
- (1/z_{lk'} \, v_{l'm'}^{+}) \left(z_{l'} \Gamma_{l'(\alpha, \beta)} + \sum_{m \neq m'} z_m \Gamma_{m(\alpha, \beta)} \right) d\mu_{l'm'} - Q^{\alpha} \, dE_{\beta(m')}^{\alpha(j')}. \tag{5.16}
$$

Note the symmetry of the last term with respect to the transposition of the α and β indexes:

$$
Q^{\alpha} \, dE_{\beta(m')}^{\alpha(j')} = Q^{\beta} \, dE_{\alpha(j')}^{\beta(m')}. \tag{5.17}
$$

Equation (5.16) contains $r_h + r_c + r_a - 1$ independent differentials of intensive variables, whose number is equal to the number of degrees of freedom of the system.

Similar equations were obtained by Kakiuchi and Senda (1983b) with the help of the surface phase method and by Mohilner (1962, 1966) with the help of the Gibbs method but are only applicable to an ideally polarized interphase boundary. Relation (5.16) turns out to be shorter and more convenient in many applications, as its components are simple and standardized. This is due to the fact that this derivation was performed with Hansen's surface excesses for extensive variables, rather than with traditional Gibbs excesses used by Mohilner, or with the surface concentrations used by Senda and Kakiuchi.

Other Combinations of Indicator Ions. In deriving equation (5.16), cation j' and anion m' are assumed to be indicator ions. This is not the only possibility since on the basis of the salts $j'k'$ and $l'm'$ it is possible to compose three more pairs of indicator ions: (j', l); (k', l'); (k', m'). The fact that the pair (j', l') is a combination of two cations, and the pair (k', m') is a combination of two anions does not complicate matters: in principle electrical measurements do not depend on this. We shall show how the electrocapillary equation changes when such substitutions are made. The changes are not very significant: they only concern certain coefficients before the differentials in equation (5.16).

Expressions (5.4)–(5.6) for the charges Q^{α} and Q^{β} do not change since they do not depend on the choice of indicator ions within the α and β groups. The coefficient in the term with dT contains the charge numbers of the indicator ions $z_{j'}$ and $z_{m'}$. When indicator ions change, they are substituted by new values. The coefficient before dp undergoes the same change. The terms with differentials $d\mu_{j'k}$, $d\mu_{jk'}$, $d\mu_{l'm}$, $d\mu_{lm'}$ do not change.

When the indicator ion changes from j' to k', the term with the differential $d\mu_{j'k'}$ changes to

$$
-(1/z_{j'} \, v_{j'k'}^{+}) \left(z_{j'} \Gamma_{j'(\alpha, \beta)} + \sum_{k \neq k'} z_k \Gamma_{k(\alpha, \beta)} \right) d\mu_{j'k'}. \tag{5.18}
$$

When the indicator ion changes from m' to l', the term with the differential $d\mu_{l'm'}$ changes to

$$-(1/z_{m'} v_{l'm'}^-)\left(z_{m'} \Gamma_{m'(\alpha,\beta)} + \sum_{l\neq l'} z_l \Gamma l\right) d\mu_{l'm'}. \tag{5.19}$$

Finally in the last term, $Q^\alpha \, dE_{\beta(m')}^{\alpha(j')}$, new indicator ions should only be substituted in the superscripts and subscripts: for example, $\alpha(k')$ and $\beta(l')$.

We shall write the electrocapillary equation for the indicator ions k', l' (which means that in both α and β phases the indicator ions are cations):

$$d\gamma = -(s_{(\alpha,\beta)}^s - (1/z_{k'} F) Q^\alpha s_{R_\alpha}^* - (1/z_{l'} F) Q^\beta s_{R_\beta}^*) \, dT$$

$$+ (\tau_{(\alpha,\beta)}^s - (1/z_{k'} F) Q^\alpha v_{R_\alpha}^* - (1/z_{l'} F) Q^\beta v_{R_\beta}^*) \, dp - \sum_{h\neq\alpha,\beta} \Gamma_{h(\alpha,\beta)} \, d\mu_h$$

$$- \sum_{k\neq k'} (1/v_{j'k}^-) \Gamma_{k(\alpha,\beta)} \, d\mu_{j'k} - \sum_{j\neq j'} (1/v_{jk'}^+) \Gamma_{j(\alpha,\beta)} \, d\mu_{jk'} - \sum_{m\neq m'} (1/v_{l'm}^-) \Gamma_{m(\alpha,\beta)} \, d\mu_{l'm}$$

$$- \sum_{l\neq l'} (1/v_{lm'}^+) \Gamma_{l(\alpha,\beta)} \, d\mu_{lm'} - (1/z_{j'} v_{j'k'}^-)\left(z_{j'} \Gamma_{j'(\alpha,\beta)} + \sum_{k\neq k'} z_k \Gamma_{k(\alpha,\beta)}\right) d\mu_{j'k'}$$

$$- (1/z_{m'} v_{l'm'}^+)\left(z_{m'} \Gamma_{m'(\alpha,\beta)} + \sum_{l\neq l'} z_l \Gamma_{l(\alpha,\beta)}\right) d\mu_{j'k} - Q^\alpha \, dE_{\beta(m')}^{\alpha(j')}. \tag{5.20}$$

Extension of this formula to the case where the indicator ion in the α phase is an anion and in the β phase is a cation, or both indicator ions are anions is trivial.

5.2. PARTICULAR CASES OF THE ELECTROCAPILLARY EQUATION

Incomplete Dissociation of Salts. In studying the interface between two immiscible electrolyte solutions, one is often faced with the situation where some salt $j''k''$ dissociates incompletely. When this happens one of the two solutions, apart from the ions j'' and k'', contains a neutral particle $j''k''$ and therefore there is an excess of neutral molecules $\Gamma_{j''k''(\alpha,\beta)}$ at the interface. How does this circumstance affect the electrocapillary equation?

An additional term $\Gamma_{j''k''(\alpha,\beta)}$ should appear in this equation, but the chemical potential of this incompletely dissociated salt is expressed in terms of the chemical potentials of its constituent ions:

$$d\mu_{j''k''} = v_{j''k''}^+ \, d\mu_{j''} + v_{j''k''}^- \, d\mu_{k''}. \tag{5.21}$$

Therefore the additional term can be easily transformed to the following form:

$$-\Gamma_{j''k''(\alpha,\beta)} \, d\mu_{j''k''} = -\Gamma_{j''k''(\alpha,\beta)}(v_{j''k''}^+ \, d\mu_{j''} + v_{j''k''}^- \, d\mu_{k''}) = -\Gamma_{j''k''(\alpha,\beta)}[(v_{j''k''}^+/v_{j'k'}^+) \, d\mu_{j''k'}$$

$$+ (v_{j''k''}^-/v_{j'k'}^-) \, d\mu_{j'k''} - v_{j''k''}^+ z_{j'}/v_{j'k'}^+ z_{j'}) \, d\mu_{j'k'}]. \tag{5.22}$$

The last equation was obtained by Kakiuchi and Senda (1983b); the only difference was that they used surface concentrations instead of Hansen's excesses in this derivation. If this term is now added to the electrocapillary equation, then additional terms such as Hansen's surface excess $\Gamma_{j''k''(\alpha,\beta)}$ will appear in the coefficient before the differentials $d\mu_{jk'}$, $d\mu_{j'k}$ and $d\mu_{j'k'}$.

Surface Reaction. On the boundary of two phases, reactions occur which result in the formation of insoluble compounds in either phase that remains in the interfacial region. If the reactions are reversible, the newly formed compounds are in equilibrium with all other compounds present. An example is formation of ionic couples from cations in one phase with anions of another (Koenig, 1934). Let us consider how this circumstance will manifest itself in the equation of electrocapillarity.

Formally one can take the surface reaction into consideration without changing the equation for electrocapillarity. For this it is necessary to return to Gibbs initial presentation of the equation of electrocapillarity (5.1). For simplicity we omit the term with the temperature differential, assuming temperature to be constant

$$d\gamma = -\sum_h \Gamma_h d\mu_h - \sum_i \Gamma_i d\bar{\mu}_i. \tag{5.23}$$

We designate the neutral components by index h and the charged ones by index i. Both soluble and insoluble components of this system are included in these sums. Therefore, analysis of the surface reaction will be completed in the most general form. Let us designate the neutral (h^s) and charged (i^s) components which are present only at the interface and result from arbitrary reactions between other components. The species h^s and i^s may be called surface components, and the species $h \neq h^s$ and $i \neq i^s$ the volume components. The corresponding reactions can be presented in the form of:

$$h^s = \sum_{h \neq h^s} v_{h^s}^{(h)} h + \sum_{i \neq i^s} v_{h^s}^{(i)} i \tag{5.24}$$

$$i^s = \sum_{h \neq h^s} v_{i^s}^{(h)} h + \sum_{i \neq i^s} v_{i^s}^{(i)} i. \tag{5.25}$$

The indices under the symbol of sum $h \neq h^s$ and $i \neq i^s$ show that summation is carried out over volume components. Symbols v are stoichiometric numbers for corresponding reactions. In order to derive the most general form of the electrocapillary equation, we include all the soluble components into these equations. However, if a certain component h or i does not occur in the present reaction its stoichiometric number should be made equal to zero. The first reaction describes neutral compounds such that

$$\sum_{i \neq i^s} v_{h^s}^{(i)} z_i = 0. \tag{5.26}$$

Now one can write the relationship between the chemical potentials as:

$$\mu_{h^s} = \sum_{h \neq h^s} \nu_{h^s}^{(h)} \mu_h + \sum_{i \neq i^s} \nu_{h^s}^{(i)} \mu_i$$

$$\bar{\mu}_{i^s} = \sum_{h \neq h^s} \nu_{i^s}^{(h)} \mu_h + \sum_{i \neq i^s} \nu_{i^s}^{(i)} \mu_i. \tag{5.27}$$

If these relations are substituted into the equation of electrocapillarity, we obtain

$$d\gamma = - \sum_{h \neq h^s} \left[\Gamma_h + \sum_{h^s} \nu_{h^s}^{(h)} \Gamma_{h^s} + \sum_{i^s} \nu_{i^s}^{(h)} \Gamma_{i^s} \right] d\mu_h$$

$$- \sum_{i \neq i^s} \left[\Gamma_i + \sum_{h^s} \nu_{h^s}^{(i)} \Gamma_{h^s} + \sum_{i^s} \nu_{i^s}^{(i)} \Gamma_{i^s} \right] d\bar{\mu}_i \tag{5.28}$$

$$d\gamma = - \sum_{h \neq h^s} \Gamma_h^\Phi \, d\mu_h - \sum_{i \neq i^s} \Gamma_i^\Phi \, d\bar{\mu}_i, \tag{5.29}$$

where

$$\Gamma_h^\Phi = \Gamma_h + \sum_{h^s} \nu_{h^s}^{(h)} \Gamma_{h^s} + \sum_{i^s} \nu_{i^s}^{(h)} \Gamma_{i^s}$$

$$\Gamma_i^\Phi = \Gamma_i + \sum_{h^s} \nu_{h^s}^{(i)} \Gamma_{h^s} + \sum_{i^s} \nu_{i^s}^{(i)} \Gamma_{i^s}. \tag{5.30}$$

Both Γ_h and Γ_h^Φ designate the surface excess of the h component. The difference between them is that the conventional term Γ_h describes the surface excess of h only; whereas, the new term Γ_h^Φ includes the combined excess of substance h present both as h and in any other form into which h is transformed as a result of the surface reaction.

Surface excesses in combination were often used by A. N. Frumkin. In his book Frumkin (1979) wrote: "There exist two different interpretations of the meaning of the quantity Γ in the case of adsorption leading to surface charging. One of these interpretations, which is quite common, treats the Gibbs adsorption of the components of a redox system Γ_O and Γ_R as measures of the surface excesses of these components, actually present in the surface layer. According to the second definition Γ_O and Γ_R are the amounts of the oxidizing and reducing agents to be introduced into the bulk phase, along with the Γ_i of other solution components undergoing adsorption, in order that the solution composition and hence the electrode potential, should remain unchanged with surface increase by a unity without electricity supply."

A. N. Frumkin considered the adsorption of hydrogen onto a platinum-promoted carbon electrode in hydrogen saturated aqueous KOH solution. Adsorbed hydrogen is ionized on the electrode and recombines with all ions forming water. As a result the electrode is charged up to the equilibrium potential. What is the value of the surface excess of hydrogen Γ_H in this case? It is conventional that $\Gamma_H = 0$; however, if Frumkin's interpretation is used, Γ_H is proportional to the electrode charge:

$\Gamma_H = -Q$. Actually, there is a reaction such that $H \rightarrow H^+ + e^-$ on the electrode and H^+ escapes into the bulk solution and e^- stays on the surface. Thus, this process is representative of the scheme for surface reactions given above.

When surface reactions are absent, conventional interpretations of excesses and Frumkin's interpretation coincide. They differ only when surface reactions occur. This is demonstrated by the relations given above where, in honor of Frumkin, the surface excesses are designated by Γ_h^Φ and Γ_i^Φ. When a new surface is formed, we need to add components h and i; however, at the interface they are partially transformed into new compounds. In order not to change the composition of the volume phases in this procedure, it is necessary to increase the masses of the components presented in the system by quantities Γ_h^Φ and Γ_i^Φ rather than the conventional Γ_h and Γ_i.

Thus, it is easy to take into consideration the presence of surface reactions in the equation of electrocapillarity. It is enough to summarize the soluble components in one phase and to understand that surface excesses Γ are the excesses Γ^Φ according to Frumkin's definition

$$d\gamma = -\sum_{h \neq h^s} \Gamma_h^\Phi \, d\mu_h - \sum_{i \neq i^s} \Gamma_i^\Phi \, d\bar{\mu}_i. \tag{5.31}$$

When using this transformation none of the derivations discussed above changes.

5.3. THERMODYNAMIC CHARGE AT THE INTERFACE

Definition. The electrocapillarity equation (5.16) relates the change in surface tension $d\gamma$ to the change in temperature, pressure, chemical potentials of neutral components and binary salts composed of the ions (constituents of the two indicator salts), and the potential of the total circuit. All these variables are independent and sufficient to describe the state of a system. For brevity, we shall designate this set of variables as $\{e\}$. The derivatives of γ with respect to the variables of this set determine the corresponding coefficients in equation (5.16), i.e., the surface excesses or their linear combinations. For example,

$$(\partial\gamma/\partial\mu_{j'k'})_{\{e\}} = -(1/z_{k'}\nu_{j'k'})\left(z_{k'}\Gamma_{k'(\alpha,\beta)} + \sum_{j \neq j'} z_j \Gamma_{j(\alpha,\beta)}\right). \tag{5.32}$$

The subscript $\{e\}$ of the derivative $\partial\gamma/\partial\mu_{j'k'}$ indicates that, besides $\mu_{j'k'}$, all the rest of the variables in set $\{e\}$ remain unchanged. The derivative with respect to temperature gives a combination of entropies, the derivative with respect to pressure gives a combination of volumes, etc.

Of particular interest is the derivative with respect to the electrical potential of the circuit $E_{\beta(m')}^{\alpha(j')}$ which gives the thermodynamic charge of the interface:

$$(\partial\gamma/\partial E_{\beta(m')}^{\alpha(j')})_{\{e\}} = -Q_{\{e\}}^\alpha \tag{5.33}$$

The right-hand side of equation (5.33) is the thermodynamic, or total, charge.

Historically, the concept of interface charge arose as a structural concept of the spatial separation of charges during formation of the electric double layer at the interface between two phases. It was formulated by Helmholtz (1879), who described the double layer as two layers of charges, equal in magnitude and opposite in sign, located at both sides of a certain plane. The charge of each of the sides of such a capacitor came to be called the free electrode charge q. It should be stressed that the free charge of the electrode is not a thermodynamic concept but a structural one, since it is essentially model based (Frumkin, 1919; Markin and Volkov, 1988a,d,e). Expressed in these terms, the concept of electrode charge appears to be simple and sufficiently well defined. The simple capacitor model underlying it, however, does not take into account all the possible details of electrical double-layer structure. For this reason, attempts at describing complex electrochemical phenomena in different systems by means of this concept have been fraught with difficulties. The best known example of these difficulties is the adsorption of ions from solution onto an electrode surface, in which case it is not clear where their charges should be ascribed: either to the metal or to the liquid side of the capacitor. The partial charge transfer concept introduced by Lorenz (1961, 1962, 1963, 1973) is an example of an unsuccessful attempt to overcome these difficulties. For this reason there is a need to use a strictly thermodynamic definition of the interface charge, and not one based on conceptual models.

Surface Excesses and the Interface Charge. The electrocapillary equation for reversible interfaces between two liquids was initially derived in 1962 by Mohilner based on Gibbs approach (Mohilner, 1962) and independently by Rusanov on the basis of the finite-thickness layer method (Rusanov, 1962, 1967). A drawback of many studies of electrocapillarity thermodynamics is that the thermodynamic charge Q is considered to be equal to the free surface charge. This assumption is only valid for an ideally polarizable electrode, but is not valid in the general case.

An essential shortcoming of Mohilner's (1962, 1966) approach is that in each particular case the equation has to be derived from scratch because of the lack of a general formula. This shortcoming was overcome by Kakiuchi and Senda (1983b). However, the electrocapillary equation (5.16) is more suitable for future advancements in the theoretical description of electrocapillary phenomena.

We will now use equation (5.16) to describe electrocapillary phenomena at polarizable and reversible interfaces between two immiscible electrolyte solutions.

5.4. POLARIZABLE INTERFACE

There are different types of polarizable interfaces. *A perfectly polarizable interface* is an interface whose state is completely determined at a given moment of time by the charge supplied to it. This definition does not specify how the charged particles coming from different phases are positioned in the transition layer. Their distributions may be of any type; for example, they may overlap (Fig. 5.4). A particular case of a perfectly polarizable interface is an *ideally polarizable* one, defined as follows (Fig.

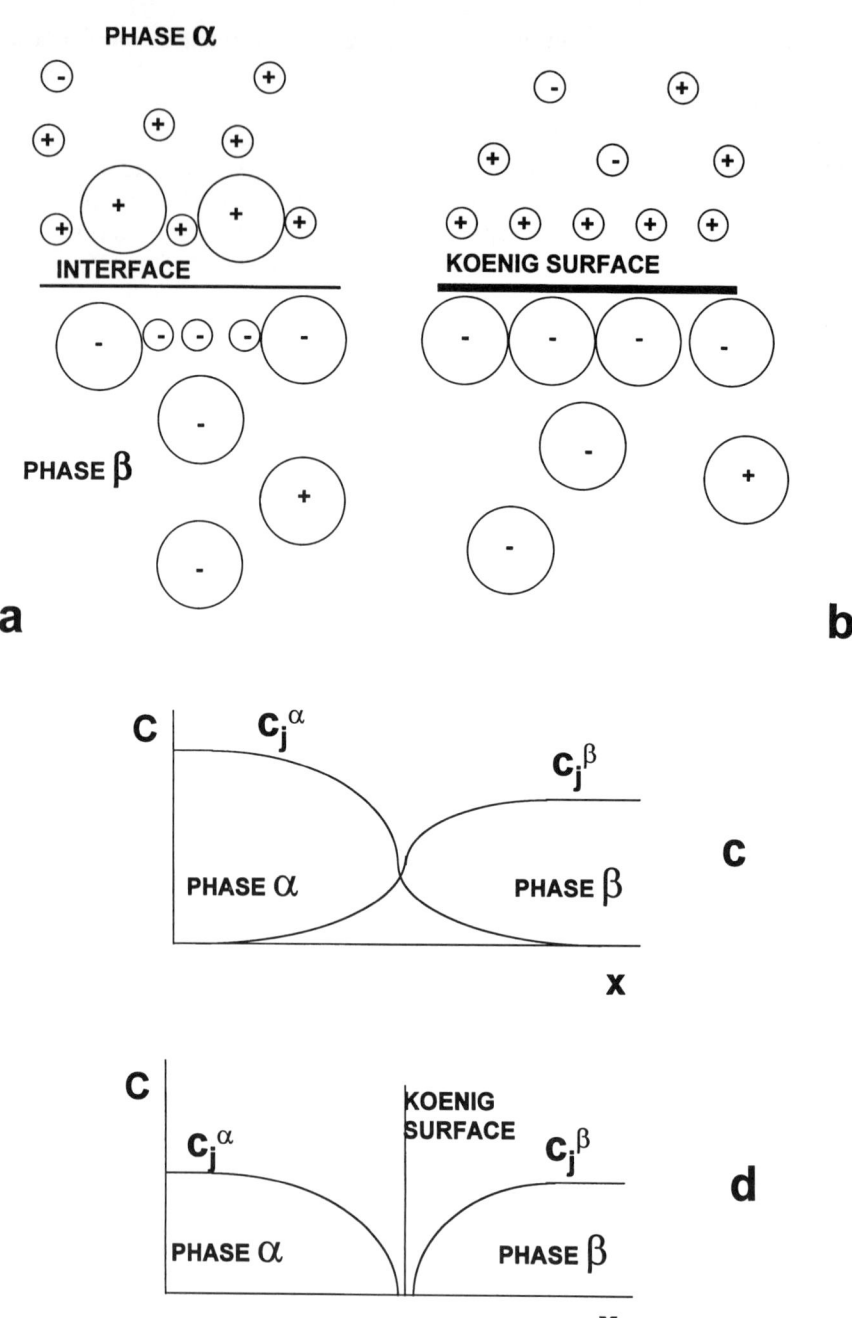

Figure 5.4. An arbitrary scheme of the structure of a perfectly polarizable interface between two immiscible electrolyte solutions (a, b, c, d) and an ideally polarizable interface (b, c).

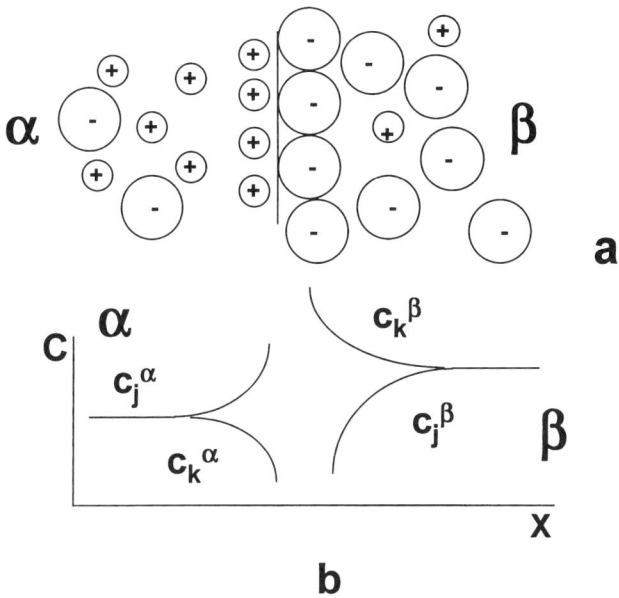

Figure 5.5. An arbitrary scheme of the structure of a reversible interface between two immiscible electrolyte solutions (*a*) and the distribution of the concentrations of ions (*b*) capable of passing from one phase to another.

5.4 (*b,d*). In this case a certain surface (Koenig surface) is assumed to exist in the transition layer, which is impermeable to all charged particles (both ions and electrons). This surface represents an infinitely high barrier for all charged particles. However, the definition of an ideally polarizable interface lies outside the scope of strict thermodynamics since it introduces a definite structure for the interface and makes use of non-thermodynamic assumptions for the existence of an infinitely high and thin barrier for all charged particles.

If at least one type of ion is present in a tangible amount in both the α and β phases and can cross the interface between these phases, such an interface may be called reversible or non-polarizable (Fig. 5.5). In reality, any boundary of two immiscible solutions of electrolytes is penetrable for any ion dissolved in the adjacent phases. For this reason one may speak about a polarized boundary only with a certain degree of approximation. This approximation is more accurate the lower the concentration of ions in one of the phases. The relationship of concentration, or more exactly the activity of ions, is defined by the value of the interfacial potential $\Delta_\beta^\alpha \phi$ and by the standard interfacial potential $\Delta_\beta^\alpha \phi_i^0$ according to:

$$(RT/z_i F) \ln a_i^\beta / a_i^\alpha = \Delta_\beta^\alpha \phi - \Delta_\beta^\alpha \phi_i^0. \tag{5.34}$$

The further the interfacial potential is from the value of the standard potential the

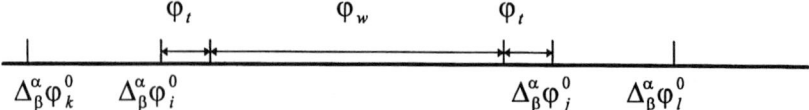

Figure 5.6. The useful potential window.

higher the concentration differential, and when it is large enough one can consider the ion to be soluble in only one phase. For example, Girault and Schiffrin (1984c) assumed that the presence of an ion in a phase can be neglected if its concentration does not exceed 10^{-5} M. If this criterion and the fact that the maximal concentration of ions in different solutions does not exceed 1 M are accepted, then the necessary deviation of interfacial potential from the standard value for a given ion ϕ_t, in the case of single-valence ions, is 300 mV. This is a maximal estimate. The corresponding value for bivalent ions is 150 mV.

Let us illustrate these relationships at the axis of potentials (Fig. 5.6) and mark on it the values of standard potentials for all ions present in system. The neighboring ones on this axis are separated by an interval $\Delta_\beta^\alpha \phi_j^0 - \Delta_\beta^\alpha \phi_i^0$. If this interval exceeds $2\phi_t$ then within this interval there is a sector from $\Delta_\beta^\alpha \phi_i^0 + \phi_t$ to $\Delta_\beta^\alpha \phi_j^0 - \phi_t$ for which the interface can be considered impermeable to ions. This sector with width $\phi_w = \Delta_\beta^\alpha \phi_j^0 - \Delta_\beta^\alpha \phi_i^0 - 2\phi_t$ is called "the useful potential window" (Girault and Schiffrin, 1984c) where each type of ion is soluble only in one of two phases, and hence the interface can be considered completely polarized. Let us recall that Girault and Schiffrin use the term "ideally polarized interface" though they point out that "the separation between phases is not as abrupt as for the metal-solution case. On the solution boundary, interfacial mixing and mixed ion solvation of adsorbed species is likely to occur. These effects will be complicated by interfacial ion pair formation and therefore, the definition of the surface charge density is not straightforward." This is absolutely true and that is why for the general case the term "completely (perfectly) polarized boundary" should be used instead of "ideally polarized boundary."

The size of the useful potential window ϕ_w at the interface of an Hg/aqueous solution is equal to 2.35 V in accordance with Girault and Schiffrin's estimation (1984c). It is much more difficult to realize conditions of complete polarization on ITIES. In the TBA$^+$TPhB$^-$(nitrobenzene)/LiCl(H$_2$O) system it turns out to be equal to only 0.02 V. However, if one uses low ionic concentrations of electrolytes, the above criteria can be relaxed and the useful window turns out to be much wider. If the interface is a perfectly polarizable one, then all the ions in the α group reside only in the phase α, and the ions in the β group reside only in the β phase. Therefore, no ion exchange reactions occur at the interface. This removes the uncertainty associated with the arbitrary division of all ions in the system into α and β groups during derivation of the electrocapillary equation. Neutral components may be distributed between the phases in any manner, in particular, the α and β solvents may be partially dissolved in each other, such that their mole fractions x_β^α and x_α^β may be greater than zero.

The α and β ion groups in two solutions with a perfectly polarizable interface are independent of one another. For this reason the amount of binary salts determining the ion composition of the α solution is $r_c^{\alpha} + r_a^{\alpha} - 1$, and that of the β solution is $r_c^{\beta} + r_a^{\beta} - 1$. These two amounts add up to $r_c + r_a - 2$ salts and hence there are this many differentials for their chemical potentials in equation (5.16). The ion components in the electrocapillary equation give $r_c + r_a - 1$ independent variables. This means that the "surplus" variable $dE_{\beta(m')}^{\alpha(j')}$ is not associated with the composition of the solutions and may be varied as desired by means of an external voltage source.

The quantities Q^{α} and Q^{β} are determined by equations (5.13)–(5.15) as the surface excesses of the charge carried by the α and β ion groups. If instead they were determined in terms of relative Hansen's excesses, then in separating the α and β ion group they would each become electroneutral. As a result, the surface excess of the charge of each of these groups no longer depends on the position of the dividing surface, i.e., on the choice of reference system. This means that in equations (4.13) and (4.15) both relative and arbitrary absolute excesses may be used, and the result remains invariant:

$$Q^{\alpha} = F\left(\sum_j z_j \Gamma_j + \sum_k z_k \Gamma_k\right),$$

$$Q^{\beta} = F\left(\sum_l z_l \Gamma_l + \sum_m z_m \Gamma_m\right). \tag{5.35}$$

The quantities Q^{α} and Q^{β} in equation (5.35) simply represent charges that should be applied to each side of the polarizable interface when its area increases by unity, so that the difference of the Galvani potentials should remain unchanged. At an ideally polarizable interface the ions of each group are separated by a Koenig surface, which precludes possible overlap (Fig. 5.4). In this case, one can think of the free charges of the α and β phases as being equal to thermodynamic charges:

$$q^{\alpha} = Q^{\alpha}, \qquad q^{\beta} = Q^{\beta}. \tag{5.36}$$

It is important to note that on the potential axis the point of zero thermodynamic charge is a thermodynamically measurable quantity, but it does not necessarily give the point of zero free charge and great care must be exercised in using this potential as a starting-point for using non-thermodynamic double-layer theories in the interpretation of interfacial electrocapillary data.

An Example of a System with a Polarizable Interface.

It was shown (Hajkova *et al.*, 1985; Kakiuchi *et al.*, 1987, 1988b; Kakiuchi and Senda, 1983a) that in a certain potential range the interface between a tetrabutylammonium tetraphenyl-borate solution in nitrobenzene (nb) and LiCl solution in water (w) can be considered to be polarizable. In these studies TBA$^+$ and Cl$^-$ ions were used as indicator ions in the electrochemical circuit. If we designate the TBATPhB solution in nitrobenzene as the α phase and the aqueous LiCl solution as the β phase, then $\alpha = nb$, $j' = TBA^+$, $k' = TPhB^-$, $\beta = w$, $l' = Li^+$, $m' = Cl^-$. There are no other ions and neutral compounds in the system. In this case equation (5.16) assumes the form

$$d\gamma = -(s^s_{(nb,w)} - q^{nb} s^*_{R_{nb}}/F + q^w s^*_{R_{nb}}/F) \, dT + (\tau^s_{(nb,w)} - q^{nb} v^*_{R_{nb}}/F + q^w v^*_{R_{nb}}/F) \, dp$$
$$- \Gamma_{TPhB^-(nb,w)} \, d\mu_{TBATPhB} - \Gamma_{Li^+(nb,w)} \, d\mu_{LiCi} - q^{nb} \, dE^{(TBA^+)}_{w(Cl^-)}. \qquad (5.37)$$

The last term in this equation can also be written as $-q^w \, dE^{w(Cl^-)}_{nb(TBA^+)}$. The charges of the nb and w phases are equal to

$$q^{nb} = F(\Gamma_{TBA^+} - \Gamma_{TPhB^-}), \qquad q^w = F(\Gamma_{Li^+} - \Gamma_{Cl^-}). \qquad (5.38)$$

The surface excesses of the $TPhB^-$ and Li^+ ions have been calculated with respect to the solvents nb and w. The transfer coefficients from absolute to relative excesses are

$$\Theta^{nb}_{(nb,w)} = \frac{x^w_w \Gamma_{nb} - x^w_{nb} \Gamma_w}{x^{nb}_{nb} x^w_w - x^w_{nb} x^{nb}_w}, \qquad \Theta^w_{(nb,w)} = \frac{x^{nb}_{nb} \Gamma_w - x^{nb}_w \Gamma_{nb}}{x^{nb}_{nb} x^w_w - x^w_{nb} x^{nb}_w}. \qquad (5.39)$$

Thus the relative excesses themselves can be represented as

$$\Gamma_{TPhB^-(nb,w)} = \Gamma_{TPhB^-} - (x^{nb}_{TBATPhB})(x^w_w \Gamma_{nb} - x^w_{nb} \Gamma_w)(x^{nb}_{nb} x^w_w - x^w_{nb} x^{nb}_w),$$
$$\Gamma_{Li^+(nb,w)} = \Gamma_{Li^+} - (x^{nb}_{LiCl})(x^{nb}_{nb} \Gamma_w - x^{nb}_w \Gamma_{nb})(x^{nb}_{nb} x^w_w - x^w_{nb} x^{nb}_w). \qquad (5.40)$$

The right-hand side of these equations, unlike the general formulas (5.15), contain only two terms each since the salts TBATPhB and LiCl are present only in different phases and hence $x^w_{TBATPhB} = x^{nb}_{LiCl} = 0$. These formulas appear in Kakiuchi and Senda (1983b) but for relative Hansen's surface excesses the inadequate symbols $\Gamma_{TPhB/nb}$ and $\Gamma_{Li^+/w}$ were used which can be easily confused with those for Gibbs relative excesses. As has been pointed out above, Hansen's excesses can change into Gibbs excesses only if the substance under consideration is present in only one phase and the solvents do not penetrate into one another, i.e., $x^{nb}_w = x^w_{nb} = 0$. Only then

$$\Gamma_{TPhB^-(nb,w)} = \Gamma_{TPhB^-(nb)} = \Gamma_{TPhB^-} - x^{nb}_{TBATPhB} \Gamma_{nb}/x^{nb}_{nb}, \qquad (5.41)$$

$$\Gamma_{Li^+(nb,w)} = \Gamma_{Li^+(nb)} = \Gamma_{Li^+} - x^w_{LiCl} \Gamma_w/x^w_w. \qquad (5.42)$$

Figures 5.7, 5.8, and 5.9 show the electrocapillary curves obtained for the interface between two immiscible electrolyte solutions. For the water–dichloroethane and water–nitrobenzene systems it is possible to charge the interface up to several hundred mV (Volkov, 1987b). It is inferred from equation (5.16) that in a relatively narrow ideal polarizability region at the interface between two immiscible electrolyte solutions, the surface state can be changed without altering the phase composition simply by changing the applied potential difference. In a state of thermodynamic

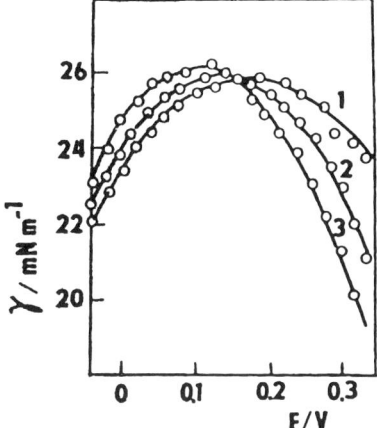

Figure 5.7. Electrocapillary curves for the interface between nitrobenzene solutions of 0.1 M TBATPhB and aqueous solutions of (1): 0.01 M, (2): 0.10 M, and (3): 1 M LiCl at 25 °C (Kakiuchi and Senda, 1983a). Reproduced by permission of The Chemical Society of Japan.

equilibrium the thermodynamic potential should be at a minimum. It follows from this condition and equation (5.1) that

$$[\partial^2 \gamma / (\partial E_{\beta(m')}^{\alpha(j')})_{T,p,\mu_i}^2] < 0. \tag{5.43}$$

It is evident from this inequality and equation (5.16) that when far from the critical point, the dependence of γ on the electromotive force (EMF) of the circuit resembles an inverted parabola with a maximum at the potential of zero thermodynamic charge, which is observed in the experiment shown in Figs. 5.7, 5.8, and 5.9.

Figures 5.10 and 5.11 show the dependencies of the relative surface excesses of components on the surface charge of the aqueous phase. The dependence of the volume excess on the interface charge is presented in Fig. 5.12. In this case, the water–liquid metal rather than the water–oil interface was investigated. Nevertheless, it does not matter because the electrocapillarity equation derived above (5.16) is applicable to all liquid interfaces.

5.5. NONPOLARIZABLE INTERFACE

An electrocapillary equation for a non-polarized, or reversible boundary was first proposed by Grahame and Whitney (1942). A simplified derivation of this equation was given by Overbeek (1952). Later, rigorous derivations of the equation of electrocapillarity for different types of reversible interfaces was made by Mohilner (1962) using Gibbs method and independently by Rusanov (1962) by the method of the surface phase. Subsequently, there have been numerous electrocapillary equations

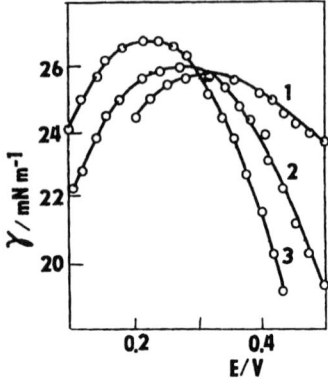

Figure 5.8. Electrocapillary curves for the interface between 0.1 M LiCl in water and a nitrobenzene solution of: (1) 0.02; (2) 0.05; (3) 0.17 M TBATPhB at 25 °C (Kakiuchi and Senda, 1983a). Reproduced by permission of The Chemical Society of Japan.

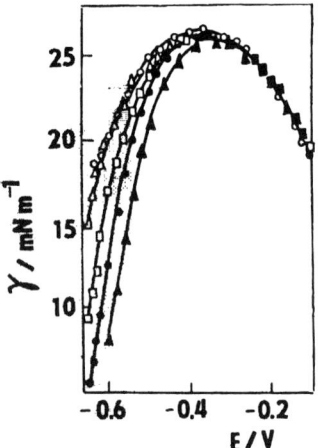

Figure 5.9. Electrocapillary curves for the interface between 0.05 M LiCl in water and $(0 - x)$ M TPnATPhB + x M HTMATPhB in nitrobenzene at 25 °C. $x = 0$ (○), 0.005 (△), 0.002 (□), 0.05 (●), and 0.1 (▲) (Kakiuchi *et al.*, 1987). Reproduced by permission of The Chemical Society of Japan.

for specific types of reversible ITIES (Dupeyrat and Nakache, 1980; Kakiuchi and Senda, 1983b; Reid *et al.*, 1983; Volkov, 1987b). To follow we shall apply general equations to the description of electrocapillary phenomena at reversible ITIES.

If the interface is a reversible one, then the component groups α and β are not to be considered independent, and if considered separately, are no longer electroneutral.

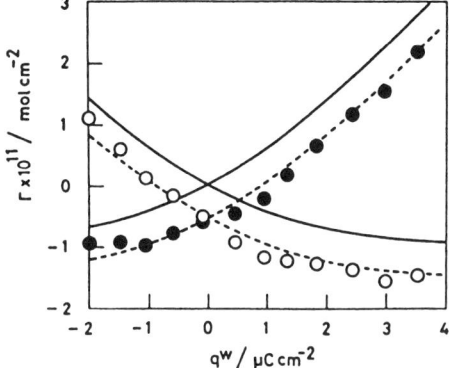

Figure 5.10. Relative surface excesses of lithium ions (●) and chloride ions (○) in the aqueous phase as a function of the surface charge density in the aqueous phase of 0.1 M LiCl. The solid lines are the theoretical curves based on Gouy–Chapman theory (Kakiuchi and Senda, 1983a). Reproduced by permission of The Chemical Society of Japan.

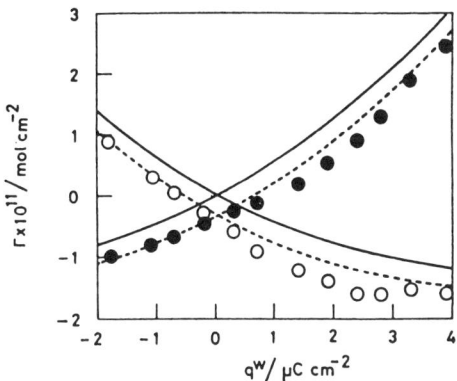

Figure 5.11. Relative surface excesses of tetrabutylammonium (○) and tetraphenylborate (●) ions in the nitrobenzene phase as a function of the surface charge density in the aqueous phase of 0.1 M TBATPhB (Kakiuchi and Senda, 1983). Reproduced by permission of The Chemical Society of Japan.

In this case the amount of independent chemical potential of the salts is equal to $r_c + r_a - 1$, and, therefore, the potential difference $E_{\beta(m')}^{\alpha(j')}$, which we introduced as a linear combination of the electrochemical potentials of indicator ions, cannot be considered as a variable independent of the solution's composition. This means that it is not possible to change the equilibrium potential difference between two phases without changing the chemical potential of at least one compound, or changing the temperature or pressure.

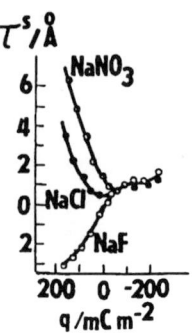

Figure 5.12. Surface excess volume as a function of a charge for 0.1 M solutions of the salts at 25 °C and 1 atm. Values are relative to the arbitrarily chosen zero at $q = 0$ for NaF system (Hills and Ovenden, 1966). Reproduced by permission of Interscience Publishers.

Changing these variables leads to a change of the potential difference $E_{\beta(m')}^{\alpha(j')}$, which can be readily recorded with a measuring circuit (Fig. 5.1). Simultaneously a change in the surface tension γ takes place. This enables one to estimate the derivative of γ with respect to E and to find the thermodynamic charge of the surface, Q. It is clear that the properties of the system can be changed in many ways. Accordingly, a large number of thermodynamic charges can be found. In order to distinguish these charges from one another it is necessary to indicate to which property change they correspond. When theoretically defining the thermodynamic charge by using the generalized electrocapillary equation, we specify the fixed variables and not those that change. Therefore, the change of potential is due to a variable that does not appear in the given set, i.e., the variable that in the given set replaces the electrochemical potential. The complete set of variables in the adsorption equation is designated by $\{e\}$. In calculating the partial derivative of surface tension with respect to potential, which (the derivative) we identify with the thermodynamic charge, the quantity Q should be provided with a corresponding subscript, indicating the system of variables used, i.e., to write the thermodynamic charge as $Q_{\{e\}}^{\alpha}$ or $Q_{\{e\}}^{\beta}$. It should be understood that $\{e\}$ is by no means a unique set of variables and, possibly, is far from being the most suitable one for practical use.

In set $\{e\}$ the potential E replaces the difference of electrochemical potentials of the indicator ions $\mu_{j'}/z_{j'} - \mu_{m'}/z_{m'}$. Formally, this difference may be used to form the chemical potential of the binary salt $\mu_{j'm'}/z_{j'}\nu_{j'm'}^{+}$. Thus, in the set of variables $\{e\}$ the thermodynamic charge is determined by the change of the chemical potential of the salt $j'm'$ at constant temperature, pressure, and chemical potential of the rest of the salts figuring in this set. Theoretically it is possible to achieve such a variation of the properties of the system, but its practical realization may prove to be very cumbersome, particularly for the case where each of the indicator ions j' and m' resides in different phases. In this version one may only speak conditionally of a salt, meaning that its chemical potential is merely a combination of the electrochemical potentials of the indicator ions. In such case it is not an easy combinatorial problem to change the chemical potential of "the salt $j'm'$," keeping constant the chemical potentials of the rest of the salts in the set $\{e\}$.

If a change of the interfacial potential is achieved by changing the chemical potential of a specific salt, and the chemical potentials of all the other ions not forming part of this chosen potential-determining salt, are kept constant, it would be reasonable to compose a new set of variables $\{g\}$. The potential-determining salt should contain at least one ion common to both phases. Let it be the salt $l'k'$ composed of the counter-ions l' and k' contained in the indicator salts $j'k'$ and $l'm'$. Since the choice of the counter-ions l' and k' was arbitrary, this assumption does not impose any constraints on the computations performed so far. Set $\{g\}$ will thus contain the following variables:

$$\{g\} = T, p, \mu_h(h \neq \alpha, \beta), \mu_j^\alpha, \mu_k^\alpha(k \neq k'), \mu_l^\beta(l \neq l'), \mu_{m'}, E_{\beta(m')}^{\alpha(j')}. \tag{5.44}$$

We will determine the chemical potentials of the ions for those phases to which the relevant group belongs. Therefore they are supplied with the indices α and β. The number of variables in the set $\{g\}$, just as in the set $\{e\}$, is equal to $r_h + r_c + r_a - 1$. The thermodynamic charge of set $\{g\}$ is determined by the generalized Lippmann equation

$$Q_{\{g\}}^\alpha = -(\partial\gamma/\partial E_{\beta(m')}^{\alpha(j')})_{\{g\}}. \tag{5.45}$$

Using $\mu_{l'k'}$ as an additional variable, we can represent this derivative as

$$(\partial\gamma/\partial E_{\beta(m')}^{\alpha(j')})_{\{g\}} = (\partial\gamma/\partial\mu_{l'k'})_{\{g\} \neq E}/(\partial E_{\beta(m')}^{\alpha(j')}/\partial\mu_{l'k'})_{\{g\}}. \tag{5.46}$$

Usually when annotating partial derivatives we indicate the set of variables to which these derivatives correspond, all others being kept constant. However, when using the additional variable $\mu_{l'k'}$ one has to define which basic variable it replaced. This explains the more detailed notation of the subscript $\{g\} \neq E$ used in the derivative $(\partial\gamma/\partial\mu_{l'k'})$.

For calculation of the parameters in (5.16), we shall use the electrocapillary equation with set $\{e\}$ variables e_i. We then obtain the following equation:

$$(\partial\gamma/\partial\mu_{l'k'})_{\{g\} \neq E} = \sum_i (\partial\gamma/\partial e_i)_{\{e\}} (\partial e_i/\partial\mu_{l'k'})_{\{g\} \neq E}. \tag{5.47}$$

The derivatives $\partial\gamma/\partial e_i$ are coefficients in equation (5.16) and to find them the derivatives $\partial e_i/\partial\mu_{l'k'}$ must be calculated. Some of these derivatives become zero, namely the derivatives of T, p, μ_h, $\mu_{j'k}(k \neq k')$, $\mu_{lm'}(l \neq l')$. This is because these variables (or their components) appear both in set $\{g\}$ and in set $\{e\}$ and therefore in calculating the derivative they are kept constant.

The derivatives of $\mu_{jk'}$, $\mu_{l'm}$ and $E_{\beta(m')}^{\alpha(j')}$ do not become zero: they can be transformed as follows:

$$(\partial\mu_{jk'}/\partial\mu_{l'k'})_{\{g\}\neq E} = v_{jk'}^-(\partial\mu_k^\alpha/\partial\mu_{l'k'})_{\{g\}\neq E}, \tag{5.48}$$

$$(\partial\mu_{l'm}/\partial\mu_{l'k'})_{\{g\}\neq E} = v_{l'm}^+(\partial\mu_l^\beta/\partial\mu_{l'k'})_{\{g\}\neq E}. \tag{5.49}$$

The derivative of $E_{\beta(m')}^{\alpha(j')}$ can be found in the following way. We transform the initial difference of electrochemical potentials of indicator ions, which is responsible for the appearance in the electrocapillary equation of the term with the electric potential:

$$\bar\mu_{j'}/z_{j'} - \bar\mu_{m'}/z_{m'} = \bar\mu_{j'}/z_{j'} - \bar\mu_{k'}/z_{k'} + \bar\mu_{l'}/z_{l'} - \bar\mu_{m'}/z_{m'} - (\bar\mu_{l'}/z_{l'} - \bar\mu_{k'}/z_{k'})$$

$$= \mu_{j'k'}/z_{j'} v_{j'k'}^+ - \bar\mu_{k'}/z_{k'} + \mu_{l'm'}/z_{l'} v_{l'm'}^+ - \mu_{l'k'}/z_{m'} v_{l'k'}^+. \tag{5.50}$$

Substituting this equation into the expression for the electrical potential and differentiating with respect to $\mu_{l'k'}$, we find

$$(\partial E_{\beta(m')}^{\alpha(j')}/\partial\mu_{l'k'})_{\{g\}} = (\partial\mu_l^\beta/\partial\mu_{l'k'})_{\{g\}\neq E}/z_{l'} F + 1/z_{k'} v_{l'k'}^+ F - (\partial\mu_k^\alpha/\partial\mu_{l'k'})_{\{g\}\neq E}/z_{k'} F. \tag{5.51}$$

Now expressions (5.48), (5.49), and (5.51) can be substituted into the equation for the derivative (5.47):

$$(\partial\gamma/\partial\mu_{l'k'})_{\{g\}\neq E} = -\Gamma_{k'}(\partial\mu_k^\alpha/\partial\mu_{l'k'})_{\{g\}\neq E} - \Gamma_{l'}(\partial\mu_l^\beta/\partial\mu_{l'k'})_{\{g\}\neq E} - Q_{\{e\}}^\alpha(\partial E_{\beta(m')}^{\alpha(j')}/\partial\mu_{l'k'})_{\{g\}}, \tag{5.52}$$

and finally we obtain the thermodynamic charge for the set of variables $\{g\}$:

$$Q_{\{g\}}^\alpha = Q_{\{e\}}^\alpha + \frac{F\Gamma_{k'}(\partial\mu_k^\alpha/\partial\mu_{l'k'})_{\{g\}\neq E} + F\Gamma_{lk}(\partial\mu_l^\beta/\partial\mu_{l'k'})_{\{g\}\neq E}}{1/z_{k'} v_{l'k'}^- - (\partial\mu_k^\alpha/\partial\mu_{l'k'})_{\{g\}\neq E}/z_{k'} - (\partial\mu_l^\beta/\partial\mu_{l'k'})_{\{g\}\neq E}/z_{l'}}. \tag{5.53}$$

The value of the thermodynamic charge $Q\{e\}$ in the initial set of variables $\{e\}$ is given by the equation

$$(\partial\gamma/\partial E_{\beta(m')}^{\alpha(j')})_{\{e\}} = -Q_{\{e\}}, \tag{5.54}$$

and the two remaining derivatives $\partial\mu_k^\alpha/\partial\mu_{l'k'}$ and $\partial\mu_l^\beta/\partial\mu_{l'k'}$ are determined by a specific electrochemical system. This completes the transformations of the derivatives in the general form.

Thus, we have arrived at a final expression for the thermodynamic charge (5.53). As has been already pointed out, there is a corresponding charge for each set of variables. The number of ways of compiling a particular set may be large. Even in the case under consideration, where we chose a potential-determining salt whose addition alters the interfacial potential, there are still many additional degrees of freedom associated with the arbitrary division of ions into the α and β groups.

Different versions of this division of ions give different sets of variables and hence different thermodynamic charges.

Approximate Calculation of the Derivatives $\partial\mu_{k'}^{\alpha}/\partial\mu_{l'k'}$ *and* $\partial\mu_{l}^{\beta}/\partial\mu_{l'k'}$.

These derivatives determine the final expression of the thermodynamic charge (5.53). To evaluate them we shall make a few simple transformations:

$$(\partial\mu_{k'}^{\alpha}/\partial\mu_{l'k'})_{\{g\}\neq E} = (\partial\mu_{l'k'}/\partial\mu_{k'}^{\alpha})_{\{g\}\neq E}^{-1} = [v_{l'k'}^{+}(\partial\mu_{l'}^{\alpha}/(\partial\mu_{k'}^{\alpha})_{\{g\}\neq E} + v_{l'k'}^{-}]^{-1} \quad (5.55)$$

$$(\partial\mu_{l'}^{\alpha}/\partial\mu_{k'}^{\alpha})_{\{g\}\neq E} = (\partial\ln a_{l'}^{\alpha}/\partial\ln a_{k'}^{\alpha})_{\{g\}\neq E}$$
$$= (\partial\ln a_{l'}^{\alpha}/\partial\ln c_{k'}^{\alpha})_{\{g\}\neq E}/(\partial\ln a_{k'}^{\alpha}/\partial\ln c_{k'}^{\alpha})_{\{g\}\neq E}$$
$$= [(c_{k'}^{\alpha}/c_{l'}^{\alpha})(\partial c_{l'}^{\alpha}/\partial c_{k'}^{\alpha})_{\{g\}\neq E}$$
$$+ (c_{k'}^{\alpha}/\gamma_{l'}^{\alpha})(\partial\gamma_{l'}^{\alpha}/\partial c_{k'}^{\alpha})_{\{g\}\neq E}][1 + (c_{k'}^{\alpha}/\gamma_{k'}^{\alpha})(\partial\gamma_{k'}^{\alpha}/\partial c_{k'}^{\alpha})_{\{g\}\neq E}]^{-1} \quad (5.56)$$

Here $a_{k'}^{\alpha}$ and $a_{l'}^{\alpha}$ are the activities of ions k' and l' in the α phase, $c_{k'}^{\alpha}$ and $c_{l'}^{\alpha}$ are the concentrations, and $\gamma_{k'}^{\alpha}$ and $\gamma_{l'}^{\alpha}$ are the activity coefficients of these ions. To proceed with the calculation we have to introduce two simplifying assumptions.

ASSUMPTION 1. Let the activity coefficients of ions k, and l' vary slowly with their concentration:

$$c_{k'}^{\alpha}|\partial\gamma_{k'}^{\alpha}/\partial c_{k'}^{\alpha}|_{\{g\}\neq E} \ll 1, \qquad c_{l'}^{\alpha}|\partial\gamma_{l'}^{\alpha}/\partial c_{l'}^{\alpha}|_{\{g\}\neq E} \ll 1. \quad (5.57)$$

These inequalities will permit us to neglect the derivatives of the activity coefficients: but not of the activity coefficients themselves.

ASSUMPTION 2: Let the change in the concentrations of ions k' and l' proceed stoichiometrically, i.e.

$$z_{k'} dc_{k'}^{\alpha} + z_{l'} dc_{l'}^{\alpha} = 0. \quad (5.58)$$

It should be noted that the condition (5.58) is by no means obvious even if an electrically neutral salt $l'k'$ is actually added to one of the solutions since its addition brings about a redistribution of all the ion components, including l' and k'. As a result, condition (5.58) may not be valid. Nevertheless, in practice it is realistic enough. For instance, it is definitely valid if there is only one common ion in the system. Then

$$(\partial c_{l'}^{\alpha}/\partial c_{k'}^{\alpha})_{\{g\}\neq E} = -z_{k'}/z_{l'}. \quad (5.59)$$

Using these assumptions, the derivatives sought after can be represented as follows:

$$(\partial\mu_{k'}^{\alpha}/\partial\mu_{l'k'})_{\{g\}\neq E} \approx z_{l'}^{2} c_{l'}^{\alpha}/v_{l'k'}^{-}(z_{k'}^{2} c_{k'}^{\alpha} + z_{l'}^{2} c_{l'}^{\alpha}), \quad (5.60)$$

$$(\partial\mu_{l}^{\beta}/\partial\mu_{l'k'})_{\{g\}\neq E} \approx z_{k'}^{2} c_{k'}^{\beta}/v_{l'k'}^{+}(z_{k'}^{2} c_{k'}^{\beta} + z_{l'}^{2} c_{l'}^{\beta}), \quad (5.61)$$

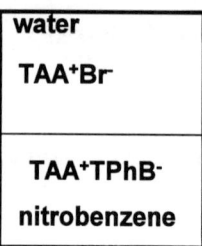

Figure 5.13. Scheme of the distribution of electrolytes between two immiscible liquids.

and we find a comparatively simple expression for the thermodynamic charge:

$$Q^\alpha_{\{g\}} = Q^\alpha_{\{e\}} + F[z_{l'} c^\alpha_{l'} \Gamma_{k'}(z^2_{l'} c^\beta_{l'} + z^2_{k'} c^\beta_{k'}) - z_{k'} c^\alpha_{k'} \Gamma_{l'}(z^2_{l'} c^\alpha_{l'} + z^2_{k'} c^\alpha_{k'})]/z_{l'} z_{k'}(c^\alpha_{k'} c^\beta_{l'} - c^\beta_{k'} c^\alpha_{l'}).$$
(5.62)

An Example: Calculation of Thermodynamic Charge. To illustrate the utility
of these equations with a concrete example, we shall consider the system studied by
Gros *et al.* (1978). A tetraalkylammonium tetraphenylborate (TAA$^+$TPB$^-$) solution in
a nitrobenzene phase (nb) was in contact with an aqueous solution of NaBr and
tetraakylammonium bromide (TAABr) (phase w). The interfacial potential was
controlled by addition of TAABr to the aqueous solution, its amount being small
compared with the amount of NaBr (Fig. 5.13). The characteristics of this system can
be readily obtained from the equations derived above.

Let the aqueous solution be the α phase and the solution in nitrobenzene the β
phase. Assume $j' = $ Na$^+$ and $m' = $ TPhB$^-$ to be the indicator ions. Accordingly,
$k' = $ Br$^-$ and $l' = $ TAA$^+$. The ions Na$^+$, Br$^-$, TAA$^+$ are present in the w phase and
the ions TAA$^+$ and TPhB$^-$ in the nb phase. The set of variables $\{e\}$ contains the
following quantities

$$\{e\} = \{T, p, \mu_{\text{NaBr}}, \mu_{\text{TAATPhB}}, E^{\text{w(Na}^+)}_{\text{nb(TPhB}^-)}\}.$$
(5.63)

The thermodynamic charge in this set of variables is equal to

$$Q^w_{\{e\}} = F(\Gamma_{\text{Na}^+\text{(w,nb)}} - \Gamma_{\text{Br}^-\text{(w,nb)}}).$$
(5.64)

The set of variables $\{g\}$ consists of the following quantities:

$$\{g\} = \{T, p, \mu^w_{\text{Na}^+}, \mu^{\text{nb}}_{\text{TPhB}^-}, E^{\text{w(Na}^+)}_{\text{nb(TPhB}^-)}\},$$
(5.65)

and the thermodynamic charge corresponding to this set is obtained from equation
(5.63):

$$Q^w_{\{g\}} = F(\Gamma_{\text{Na}^+\text{(w,nb)}} - F(1 + c^w_{\text{TAA}^+}/c^w_{\text{Br}^-})\Gamma_{\text{Br}^-\text{(w,nb)}}).$$
(5.66)

Gros *et al.* (1978) derived an approximate version of this equation which left out the term $c_{TAA^+}^w/c_{Br^-}^w$. With the use of the general equation (5.53) obtained by Hansen's method, these shortcomings can be easily overcome.

6

ENERGETICS OF EXTRACTION

In recent years there has been a great surge of interest in the experimental determination and theoretical calculation of the standard Gibbs free energy of ion transfer for individual ions moving between two solvents (Gibbs energy of resolvation). The Gibbs free energy of ion resolvation is a key concept intimately related to the electrochemistry of the interface between two immiscible electrolyte solutions (Markin and Volkov, 1987a,b,c,d, 1988f, 1989b,e,f; Volkov, 1984a, 1986a,b; Yaguzhinsky et al., 1976), ion transport across biological and artificial membranes (Deamer and Volkov, 1995; Paula et al., 1996), the mechanisms of interfacial and surface catalysis (Dehmlow and Dehmlow, 1993; Kharkats and Volkov, 1985, 1987, 1989, 1994; Starks et al., 1994), the kinetics of ion transfer across an oil–water interface (Samec, 1996), the coupling of heterogeneous reactions in bioenergetics (Yaguzhinsky et al., 1976), extraction (Hanna and Noble, 1985), and the design and manufacture of ion-selective electrodes (Koryta, 1979, 1987).

By solvation we mean the sum of all structural and energy changes occurring in a system when ions pass from the gaseous phase into solution. It has become customary to divide the interaction between ion and solvent and the corresponding energy into several components. To understand the energies involved in transferring an ion or dipole between two solvent phases, one must take the following effects into account:

    (i) electrostatic polarization of the medium;
    (ii) production in a medium of a cavity to accommodate the ion also known as the solvophobic, hydrophobic or neutral effect;
    (iii) changes in the structure of the solvent that involve the breakdown of the initial structure and the production of a new structure in the immediate vicinity of the ion;

(iv) specific interactions of ions with solvent molecules, such as hydrogen-bond formation, donor–acceptor and ion–dipole interactions;

(v) annihilation of defects: a small ion may be captured in a "statistical micro-cavity" within the local solvent structure so releasing energy of this defect;

(vi) existence different standard states.

This division is purely conditional since many of these different effects may overlap: for example, electric polarization of the medium may significantly influence its structure. Assigning such a division does permit analysis of the individual components. However, sometimes, components can be grouped into "blocks." For example, one may speak about the solvophobic effect which combines the formation of a cavity and structural changes of the solvent in the vicinity of the new particle. This is quite justifiable because the solvophobic effect, together with the electrostatic effect, provides the major contribution to the solvation energy (Deamer and Volkov, 1995; Kornyshev and Volkov, 1984; Markin and Volkov, 1989b). On the other hand, sometimes it becomes necessary to consider each component of a given effect.

For example, the total solvation (or resolvation) energy can be divided into electrostatic (el) and non-electrostatic parts (svph):

$$\Delta_\beta^\alpha G_i^0 = \Delta_\beta^\alpha G_i^0(\text{el}) + \Delta_\beta^\alpha G_i^0(\text{svph}). \tag{6.1}$$

To find the Gibbs standard free solvation energy, one needs to compare the Gibbs free energies for a given ion in each media. If one of the phases (β) is a vacuum (vac), the difference

$$\Delta_{\text{vac}}^\alpha G_i^0 = G_i^{0,\alpha} - G_i^{0,\text{vac}} \tag{6.2}$$

is called the solvation energy of the ion i in the α phase. The resolvation energy can be represented as the difference of two solvation energies:

$$\Delta_\beta^\alpha G_i^0 = -\frac{\Delta_\beta^\alpha \phi_i^0}{z_i F} = \Delta_{\text{vac}}^\alpha G_i^0 - \Delta_{\text{vac}}^\beta G_i^0, \tag{6.3}$$

where $\Delta_\beta^\alpha \phi_i^0$ is the standard potential for transfer of ion i from the β phase to the α phase.

Table 6.1 lists the standard free energies of ion transfer from water into different organic solvents obtained by cyclic voltammetry, chronopotentiometry, polarography and spin-lattice relaxation of quadrupole nuclei of ions, using solubility and extraction data (Abraham and Danil de Namor, 1976; Chapkiewich and Chapkiewich-Tutaj, 1980; Hundhammer and Solomon, 1983; Kihara *et al.*, 1989; Koczorowski and Geblewicz, 1983; Koczorowski *et al.*, 1984a,b; Marcus, 1996; Rais, 1971; Samec *et al.*, 1983, 1984a; Solomon *et al.*, 1984a,b). As can be seen from Table 6.1, the standard free energy of ion transfer from one solvent to another strongly depends on

Table 6.1. The standard Gibbs free energies (kJ mol^{-1}) of ions transfer from water into selected non-aqueous solvents at 298 K.

Ion	PC	FA	DMF	MeCN	PhNO$_2$	IBMK	ACPH	DCIMe	1,1-DCIE	1,2-DCIE
H$^+$	50	-10	-18	44	33					56
Li$^+$	24	-10	-10	25	38	20				56
Na$^+$	15	-8	-10	15	36					52
K$^+$	5	-4	-10	8	24				30	46
Rb$^+$	-1	-5	-10	6	20				29	37
Cs$^+$	-7	-6	-11	6	16					24
Sr^{2+}					67					
Ag$^+$	19	-15	-21	-31	30					5
Tl$^+$	11	-1	-12	8	17					26
NH$_4^+$					27					
Me$_4$N$^+$	-11	3	-5	3	4	9		18		16
Et$_4$N$^+$	-13		-8	-7	-6		4		18	5
Pr$_4$N$^+$	-22	-10	-17	-13	-16		-9		11	-9
Bu$_4$N$^+$	-31		-29	-32	-25	-19	-15	-22	-2	-18
Ph$_4$As$^+$	-36	-24	-39	-33	-36		-16		-27	-33
Mg^{2+}					69				29	
Ca^{2+}					68					
Ba^{2+}	46		-21	57	62					
Cu^{2+}	73	-4	-18	68	115					
Zn^{2+}	81		-30	69	140					
Cd^{2+}	70	-28	-34	42	93					
Hg^{2+}	80		-44	28	44					
Pb^{2+}	47	-12	-34	64	71					
F$^-$	56	25	85	71	70					65
Cl$^-$	40	14	48	42	31	50		46	58	54
Br$^-$	30	11	36	31	29			39	43	38
I$^-$	14	7	20	17	19		12	26	31	25
CN$^-$	36	13	40	35	38					41
N$_3^-$	27	11	36	37						
SCN$^-$	7	7	18	14	16		11			26
NO$_3^-$				21	24		20			34
ClO$_4^-$	-3	-12	4	2	8		2	21	22	17
IO$_4^-$					7		2			15
BF$_4^-$					11					18
DCC$^-$					-50					
2,4-DNP$^-$					7					
Pi$^-$		-7		-3	-5	12		-7		-3
BPh$_4^-$	-36	-24	-39	-33	-36		-16		-27	-33
SO$_4^{2-}$			78	88	141					

the nature of both ion and solvent. Figures 6.1 and 6.2 show experimental curves for the resolvation energy as a function of the static and optical permittivities of a non-aqueous solvent.

The standard ion hydration energy *versus* the ion radius is taken from Table 6.2 and presented in Fig. 6.3 (line 3). As can be seen from Figs. 6.1–6.3 and Table 6.1, the resolvation (hydration) energy decreases significantly with increasing ion size and static and optical permittivities. The standard free energies of ion resolvation when determined by different methods do not coincide exactly. The ion resolvation energy can be calculated or measured by two methods. For example, the ion solvation energy can be determined in each of two solvents, α or β, which then yield, by virtue of equation (6.3), the standard free energy of the transfer of ion i from solvent β to solvent α.

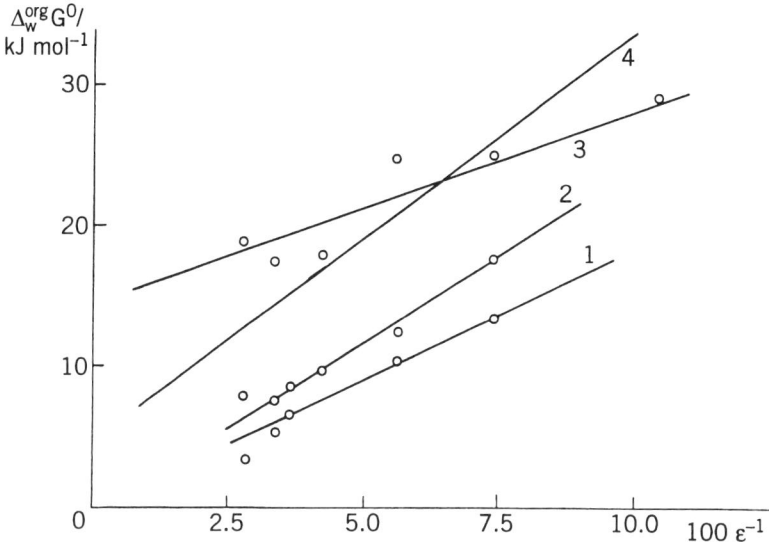

Figure 6.1. The standard free resolvation energy versus the reciprocal static permittivity of the non-aqueous phase: curves 1 (Me_4N^+), 2 (ClO_4^-), and 3 (I^-) are experimental data from Table 6.1. Curve 4 is calculated by the Born formula for Me_4N^+.

If two immiscible liquids α and β are mutually saturated, a certain amount of one solvent will be dissolved in the other solvent and *vice versa*. The corresponding energy of resolvation between two mutually balanced solvents is called the free partition energy. For most ions and solvents the standard free energies of transfer and partition coincide within experimental error. However, for some solvents and ions these quantities may differ significantly.

When considering the standard free resolvation energy, attention should be paid to whether one is dealing with ion transfer between pure or mutually saturated solvents.

6.1. ELECTROSTATIC CONTRIBUTION TO THE SOLVATION ENERGY

Two sets of models exist to calculate the electrostatic portion of the solvation energy. In the first set the medium is considered as a structureless continuum, while in the second set it is represented by a set of individual particles having either realistic or simplified properties. In early works the solvent structure was taken into account by directly calculating the energies of particular configurations of solvent molecules in the vicinity of the ion. The configuration and the number of molecules in it were chosen with a certain degree of arbitrariness, proceeding from some physical considerations, which ensured an excellent fit between experimental and theoretical data. Such models completely ignored the statistical properties of the solution which

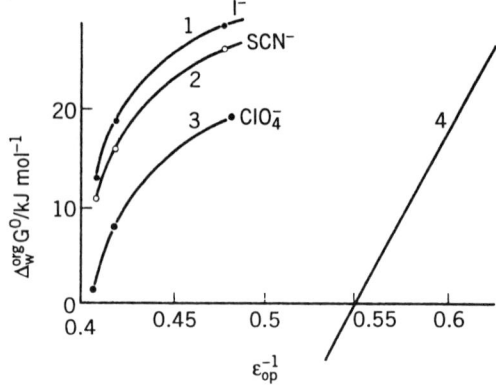

Figure 6.2. The standard free resolvation energy versus the optical permittivity of the non-aqueous phase. Curves 1 (I^-), 2 (SCN^-), and 3 (ClO_4^-) are experimental data from Table 6.1. Curve 4 is calculated by the Born formula for I^- under $\varepsilon = \varepsilon_{op}$. Optical permittivities are taken from Table 6.3.

Table 6.2. Bare and hydrated radii of ions.

Ion	Bare ion radius (nm)	Hydrated radius (nm)
H_3O^+	0.115	0.28
$H_9O_4^+$	0.395	0.395
Li^+	0.094	0.382
Na^+	0.117	0.358
Ag^+	0.126	0.341
Tl^+	0.144	0.330
NH_4^+	0.148	0.331
K^+	0.149	0.331
Rb^+	0.163	0.329
Cs^+	0.186	0.329
Be^{2+}	0.031	0.459
Mg^{2+}	0.072	0.428
Zn^{2+}	0.074	0.430
Cd^{2+}	0.097	0.426
Ca^{2+}	0.100	0.412
Pb^{2+}	0.132	0.401
Al^{3+}	0.053	0.480
OH^-	0.133	0.300
F^-	0.116	0.352
Cl^-	0.164	0.332
Br^-	0.180	0.330
I^-	0.205	0.331
NO_3^-	0.179	0.340
Me_4N^+	0.285	0.347
Et_4N^+	0.348	0.400
Pr_4N^+	0.398	0.452
Bu_4N^+	0.437	0.494

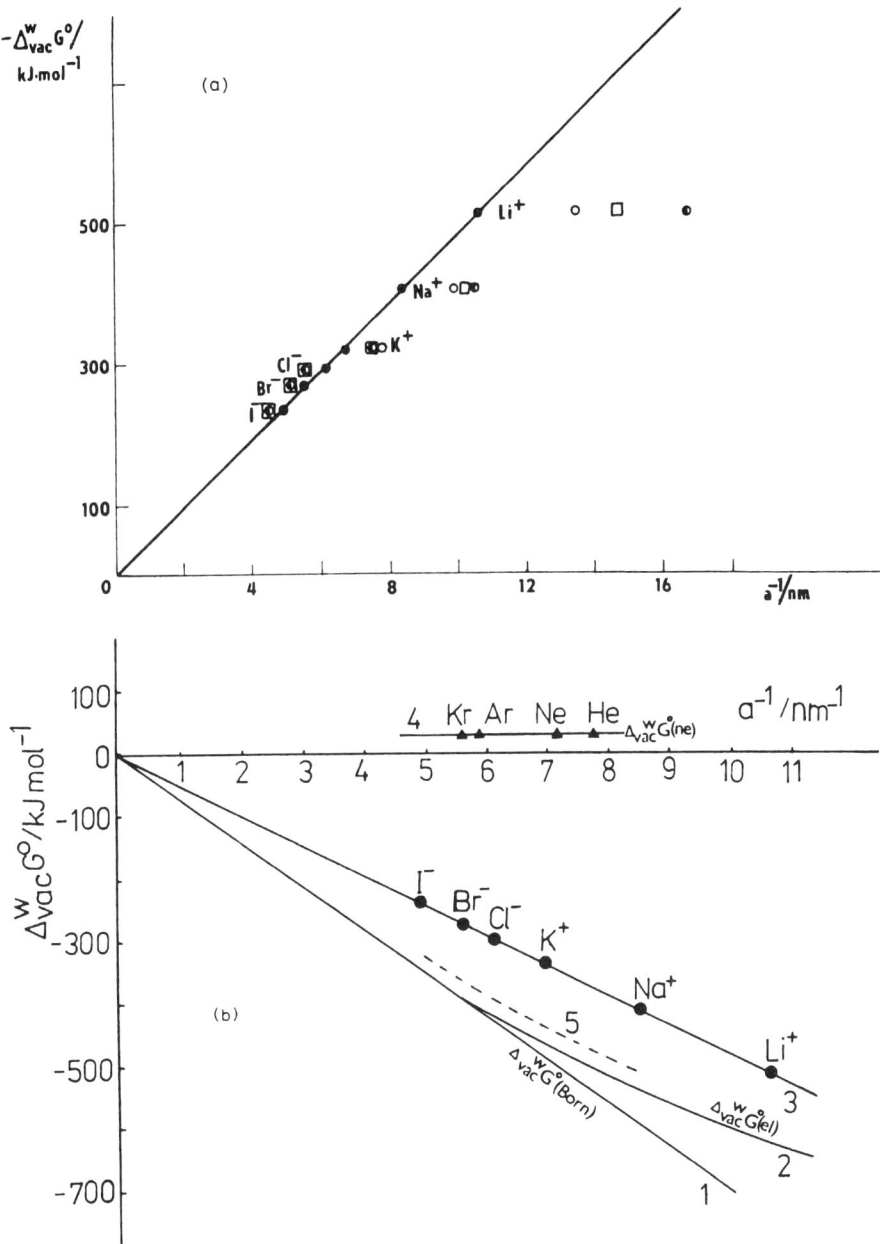

Figure 6.3. The standard free hydration energy for inorganic ions. The ion radii are taken according to the Gourary–Adrian (●), Waddington (○), Goldschmidt (□), and Pauling (◐) scales: (a) experimental points; (b) line 1—calculation by the Born formula; line 2— calculation with an allowance for dielectric saturation; line 3—experimental points; line 4—the standard free Gibbs energy of gas solutions (1 atm gas and unit mol fraction solution at 298 K); line 5—the sum of curves 2 and 4.

are very important for obtaining a correct description of solvation. The modern approach lies in developing a statistical theory of an ion–dipole plasma which describes both the energy and the statistical aspects of the phenomenon.

There is also an intermediate approach, in which, while remaining within the framework of continuum theories, attempts are made to take into account the discrete nature of the solvent. For this purpose, the non-linear dielectric effects are considered and the mutual correlation of the polarization vectors of the solvent molecules, situated at short distances from one another, is analyzed using the theory of non-local electrostatics. Each of these approaches, reflecting the role of different effects, has its own advantages.

6.2. THE BORN MODEL

The first continuum model, developed by Born (1920), considered the ion as a hard charged sphere of a given radius a immersed in a continuous medium α of constant dielectric permittivity ε^α (see Fig. 6.4(a)). This was a macroscopic theory, for it used macroscopic laws to describe the properties of a solvent at a small distance from the ion. In the Born theory, the solvation energy (its electrical part, to be exact) is given by the well-known relation:

$$\Delta^\alpha_{vac} G^0(\text{Born}) = -\frac{z^2 e_0^2}{8\pi\varepsilon_0 a}\left(1 - \frac{1}{\varepsilon^\alpha}\right) \tag{6.4}$$

where ε_0 is the electric constant. Equation (6.4) gives the energy per particle. Proceeding from this relation, the free energy of resolvation during the transfer of an ion from medium β to medium α or the difference between the energies in media α and β can be represented in the form:

$$\Delta^\alpha_\beta G^0(\text{Born}) = -\frac{z^2 e_0^2}{8\pi\varepsilon_0 a}\left(\frac{1}{\varepsilon^\beta} - \frac{1}{\varepsilon^\alpha}\right). \tag{6.5}$$

The solvation energies calculated by the Born formula differ noticeably from experimental values (Kornyshev and Volkov, 1984; Markin and Volkov, 1989b). Since, in most of the cases, the resolvation energy is the difference between two comparatively large solvation energies, even a relatively small error in calculation of each energy may give rise to considerable error in the resolvation energy, up to the wrong sign. For example, equation (6.5) implies that if $\varepsilon^\alpha > \varepsilon^\beta$ there is a higher probability for the ion to reside in solvent α, irrespective of ion size. In practice this is not always the case. It is known that small ions of radius $a < 0.2$ nm reside mainly in a polar solvent of high permittivity, while large organic ions reside preferably in the hydrophobic phase. Data on the partition coefficients, extraction, solubility, and current–voltage characteristics show that the standard free energy of ion transfer from water into a less polar solvent is positive for small radius ions and negative for large

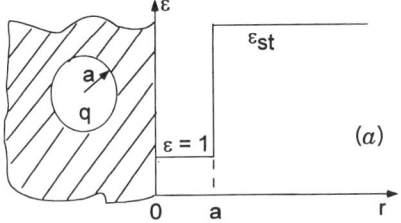 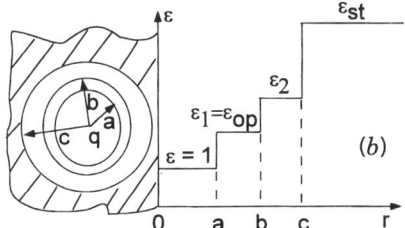

Figure 6.4. (a) The Born model. (b) The Abraham–Liszi model.

radius ions, while equation (6.5) implies that the sign of this energy does not depend on the ion radius.

The Born relation overestimates the solvation energy and provides values which are too negative (Figs. 6.1 and 6.3). However, when particular energy values are compared with experimental data, a certain radius must be assigned to each ion. This procedure is arbitrary to a certain extent because an ion is not a hard sphere of a definite radius. The ion radius is solely an effective parameter which enables a convenient description of different properties of an ionic system (Masterdon et al., 1971). It is frequently the case that different properties of a system are best described in terms of different ion radii. Although there are reasonable physical restrictions on the range within which an ion radius may change, it is meaningless to select one radius as being better than another.

The difficulty encountered in the selection of an ion radius is confirmed by the existence of at least four ion radius scales which have been calculated using different assumptions: namely the Pauling (1927) scale; the Goldschmidt (1926) scale; the Waddington (1966) scale; the Gourary–Adrian (1960) scale. The first three scales differ little; however, the fourth scale stands alone. Blandamer and Symons (1963) have attempted to choose the most relevant scales of ion radii by comparing the Gibbs free energy with the ion hydration entropy. They demonstrated that if any of these quantities is plotted as a function of $1/a$, all the points corresponding to alkali cations and anions of halide compounds fit the same smooth curve, provided the Gourary–Adrian scale is used. Other sets of ionic radii lead to two different curves: one for cations and the other for anions. It is difficult to say whether much attention should be paid to this fact since it does not have a theoretical explanation. Nevertheless, this unexplained circumstance does exist and makes one reflect on its meaning.

Since in the Born relation (6.4) the solvation energy is expressed in terms of a^{-1}, the same coordinates are used in Fig. 6.3. Figure 6.3(a) presents experimental data for six cations and anions for each of the four radius scales. For the Gourary–Adrian scale experimental data points ideally fit a straight line, thereby confirming the conclusion of Blandamer and Symons (1963). This fact undoubtedly deserves attention. For the other three scales, the fit is not very good, although deviation from a straight line is noticeable only for the smallest ions: Li^+; Na^+; K^+, but for Cl^-, Br^- and I^- deviation is small. Figure 6.3(b) compares the experimental results with theoretical predictions. To avoid encumbering the figure with a lot of points, it presents only experimental data for the Gourary–Adrian scale (curve 3). The non-electrostatic part of the free hydration energy is shown by curve 4. The hydration energy of inert gases may be determined by various independent methods. If the ion and atomic radii of inert gases coincide, the Gibbs free energy of the solution of gas atoms is just the non-electrostatic part of the free hydration energy.

The electrostatic part of the hydration energy, as calculated by the Born relation, is shown by curve 1. It can be seen that the calculated values are indeed lower (by approximately 50%) than experimental $\Delta_{\text{vac}}^{\text{w}} G^0$ values. It is the sum of curves 1 and 4 that should be compared with the experimental data and even this does not improve the correlation significantly.

Thus, the Born theory requires improvement. Equation (6.4) includes only two experimental parameters: the ion radius and dielectric permittivity ε. By appropriately changing one of these parameters, one can, in principle, achieve satisfactory agreement with experimental results. However, this formal "fitting" of the ion radius (Marcus, 1985) leads to unreasonable values which fall onto none of the four acceptable scales. Therefore, although such a fitting provides an empirical relation which is in good agreement with experiment, it still cannot constitute a physical explanation of the solvation phenomenon.

From this point of view, it seems more attractive to use the idea of dielectric saturation of the medium in the ion field (Goldman and Bates, 1972; Markin and Volkov, 1989b). This effect shifts the electrostatic part of the solvation energy in the proper direction (it decreases in magnitude). Again, agreement with experimental data may be very good if the size of the saturation region and dielectric permittivity are chosen properly. For example, ideal agreement was obtained by Abraham and Liszi (1978, 1980, 1981). Yet, a question remains as to whether these results offer a physical explanation of ion solvation or whether they only give a convenient empirical relation. Since the opinions of different authors on the role of saturation effects in ion solvation are contradictory, this topic deserves special consideration.

6.3. NON-LINEAR DIELECTRIC EFFECTS

One of the ways to improve the Born theory is to take into account non-linear dielectric effects. The permittivity of a medium containing polar molecules is a function of the electric field. This is due to both the perturbation of the intrinsic structure of associated dipoles and dielectric saturation (Bradley and Parry Jones, 1974; Vorotyntsev and Kornyshev, 1979). When analyzing the properties of water, Booth (1951a,b,c) showed that application of an external field reduced the water permittivity from the static value $\varepsilon = 80$ to the optical value ε_{op}, equal to the refractive index squared, i.e., 1.8. Saturation is attained in fields exceeding 10^9 V m^{-1}. However, because of technical problems the theory has not thus far been verified in fields higher than 10^7 V m^{-1} (Fig. 6.5) where the permittivity decreases by no more than 0.1%. This topic is considered at extent by Davies (1971, 1976), Malsch (1928, 1929), and Webb (1926).

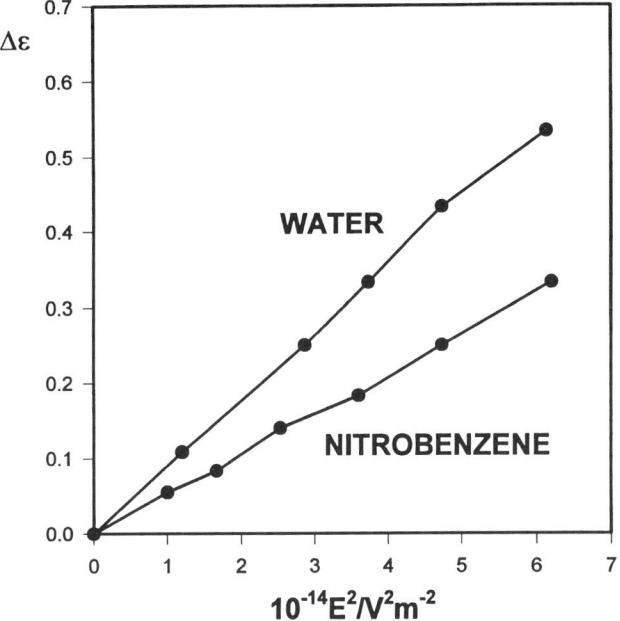

Figure 6.5. The effect of an electric field on water permittivity. Experimental data are taken from Malsch (1928, 1929); $\Delta\varepsilon = \varepsilon - \varepsilon(E)$; E is the electric field.

The first attempt at allowing for dielectric saturation of the medium near an ion (in hydration energy calculations) was made by Webb (1926). He found a correction to the Born theory which shifted the calculated values in the proper direction. Further interpretations, which attributed the discrepancy between the Born theory and experiment to dielectric saturation, faced serious objections (Kornyshev and Volkov, 1984; Volkov and Kornyshev, 1985). In recent years, however, there has been a new surge of interest in the problem (Abraham *et al.*, 1979, 1982; Markin and Volkov, 1989f; Stiles, 1980).

The free energy density for electric field in a dielectric is given, in general, by the integral:

$$g(r) = \int_0^D E \, dD \qquad (6.6)$$

where E is electric field and D, electric displacement. In weak fields the relation between D and E is linear, $D = \varepsilon_0 \varepsilon E$, and integration of free energy density (6.6) over the entire space around the ion yields the Born formula (6.4). In the general case (Booth, 1951a,b,c), electric displacement can be expressed as:

$$D = \varepsilon_0 \varepsilon E + N\mu L(E), \qquad (6.7)$$

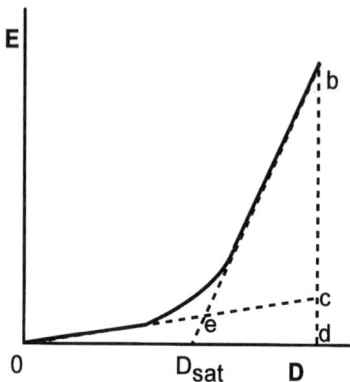

Figure 6.6. Electric field E versus electric displacement D.

with:

$$\mu = \frac{\varepsilon_{op} + 2}{3} \mu_0 \bar{g}^{1/2}, \tag{6.8}$$

where N is the density of molecules in the dielectric, μ_0 is their dipole moment in a vacuum, and $\bar{g}^{1/2}$ is the field dependent Kirkwood factor which takes into account the association of dipolar molecules. In a non-associated liquid and in high fields $\bar{g} = 1$. For the function $L(E)$ it is usual to employ the Langevin model function, but to estimate the effect it is sufficient to know the dependence of D on E in the most general form. It is clear from general physical considerations that the function $L(E)$ should increase with E and extend to a limiting value equal to unity. As a result, the D–E dependence changes from one limiting form to another. Therefore:

$$D \equiv [\varepsilon_0 \varepsilon_{op} + N\mu(L'|_{E=0})]E \approx \varepsilon_0 \varepsilon E, \tag{6.9}$$

for weak fields, and:

$$D \approx \varepsilon_0 \varepsilon_{op} E + N\mu, \tag{6.10}$$

for strong fields.

In Fig. 6.6 a plot of E versus D is shown. Straight line $0c$ corresponds to an ordinary linear dependence. When saturation occurs, the curve rises steeply and approaches another asymptote, eb. In the linear approximation, the free energy density is equal to the area of triangle $0cd$ and the total integral (equation (6.6)) is equal to a much larger area of the curvilinear triangle $0bd$. These quantities differ when the electric displacement D exceeds the characteristic value D_{sat}, which, as can easily be seen, is equal to:

$$D_{sat} = N\mu. \tag{6.11}$$

Since in a spherically symmetric field electric displacement does not depend on permittivity,

$$D = \frac{ze_0}{4\eta r^2}, \tag{6.12}$$

there exists a characteristic distance from the center of the ion,

$$r_{sat} = \sqrt{\frac{|z|e_0}{4\pi r^2}}, \tag{6.13}$$

within which the saturation effects become significant. The characteristic saturation radius does not depend on ion size and is only determined by ion charge and the dielectric properties.

Comparison of r_{sat} with the ion radius (a) shows to what extent saturation can contribute to the energy of the electric field. We now calculate r_{sat} for monovalent ions in some solvents.

For nitrobenzene $N = 6 \times 10^{27}$ m^{-3} and $\mu_0 = 3.9$ D, and we obtain $r_{sat}^{nb} = 0.31$ nm. For acetonitrile $N = 1.1 \times 10^{28}$ m^{-3}, $\mu_0 = 3.97$ D, and $r_{sat}^{ac} = 0.29$ nm. For water the corresponding parameters are equal to $N = 3.34 \times 10^{28}$ m^{-3}, and $\mu_0 = 1.84$ D, so the characteristic saturation radius amounts to $r_{sat}^{w} = 0.22$ nm.

These values exceed the radius of monoatomic inorganic ions, and for such ions dielectric saturation can contribute noticeably to the polarization energy. To determine this contribution, one should calculate the density of the electric energy g, i.e., equation (6.6). Approximating this integral by the area of two triangles in Fig. 6.6, we obtain:

$$g = \frac{D^2}{2\varepsilon_0 \varepsilon}, \qquad \text{if } D < D_{sat}, \tag{6.14}$$

$$g = \frac{D^2}{2\varepsilon_0 \varepsilon} + \frac{(D - D_{sat})^2}{2\varepsilon_0 \varepsilon_{op}}, \qquad \text{if } D > D_{sat}. \tag{6.15}$$

Integration of the energy density over the entire space outside the ion yields the electric part of the solvation energy:

$$\Delta_{vac}^{\alpha} G^0(\text{el}) = -\frac{z^2 e_0^2}{8\pi\varepsilon_0 a}\left[1 - \frac{1}{\varepsilon^{\alpha}} - \frac{1}{\varepsilon_{op}^{\alpha}}\left(1 - \frac{8}{3}x + 2x^2 - \frac{1}{3}x^4\right)\right], \tag{6.16}$$

where $x = a/r_{sat}$.

Besides the Born term $(1 - 1/\varepsilon^{\alpha})$, equation (6.16) contains a correction which depends on the ratio of the ion radius to the saturation radius. The correction vanishes at $a = r_{sat}$ but increases with decreasing a. This is quite clear because it is at small distances that the dielectric saturation becomes effective. To better illustrate this phenomenon let us consider some particular numerical examples. In Fig. 6.3 curve 2 describing the hydration energy is plotted according to equation (6.16). For different ion radii the values of the correction turn out to be different. For lithium ions the correction is equal to 16%. Consequently, lithium saturation reduces the discrepancy between the Born theory and experiment by more than 50%.

For potassium the correction is 4%, which amounts to 1/7 of the disagreement between theory and experiment. As the atomic number of the element increases, the contribution of saturation decreases. For rubidium it is equal to 2.9% or 0.1 of the required correction and for cesium, 1.3% or 0.04 of the correction.

The sum of the electrostatic and non-electrostatic parts of the free hydration energy is shown by the dashed line in Fig. 6.3, curve 5. It can be seen that this sum approaches the experimental data points, the difference between calculated and experimental $\Delta_{vac}^w G^0$ values being $10 \, kJ \, mol^{-1}$. Incidentally, the smallest discrepancy between theory and experiment is observed for the Gourary–Adrian ion radii, while for the other scales the discrepancy is greater.

Thus, dielectric saturation makes an appreciable contribution to the electrostatic solvation energy only for very small ions. Hence, any continuum theory must take into account dielectric saturation, at least when calculating the hydration energy for ions of radii less than 0.15 nm.

This estimate of the solvation energy with an allowance for dielectric saturation was made within the framework of a continuum model. All the quantities involved were calculated proceeding from the fundamental principles. The permittivity varied in space monotonically without experiencing discontinuities. Therefore, equation (6.16) does not contain fitting parameters. When deriving this relation, we did not attempt to describe in detail the displacement–field dependence, but only used the assumption that the true curve approaches very nearly the two asymptotes (Fig. 6.6). For the complete curve, as already noted, there is a model description in the form of the Langevin function. It can be shown that in this approximation, the function $E(D)$ rapidly reaches the second asymptote.

It should also be noted that the piecewise linear approximation considered above gives a somewhat lower E value (for a given D), so that the solvation energy obtained by this approach is an upper bound (absolute value) which is close to the true value. Approximation accuracy rises with decreasing ion radius. A detailed calculation performed with the use of the Langevin function has been reported by Glueckauf (1964a,b,c).

6.4. OTHER APPROACHES TO THE SATURATION PROBLEM

The most useful results to the saturation problem have been obtained by Abraham and Liszi (1978, 1980, 1981). They considered the field dependence of the permittivity and came to a conclusion that it is best to replace the continuous spatial distribution of permittivity by a stepwise one, i.e., to consider the solvent near an ion i as several concentric spherical layers of different permittivities (Fig. 6.4(b)).

Abraham and Liszi confined themselves to the analysis of two layers of radii b and c. Inside the first sphere the permittivity is equal to $\varepsilon_1 = \varepsilon_{op}$, but the calculated solvation energy strongly depends on the radius b where the saturation is maximum. Abraham et al. (1982) noticed that for acetonitrile the permittivity changes most abruptly at a distance (from the ion center) equal to the sum of the ion radius a and the solvent molecule radius a_{solv}.

This distance was used as the radius of the first sphere giving the result:

$$\Delta^{\alpha}_{vac} G^0(el) = -\frac{z^2 e_0^2}{8\pi\varepsilon_0}\left[\left(1-\frac{1}{\varepsilon_1}\right)\left(\frac{1}{a}-\frac{1}{b}\right) + \left(1-\frac{1}{\varepsilon_2}\right)\left(\frac{1}{b}-\frac{1}{c}\right) + \left(1-\frac{1}{\varepsilon}\right)\frac{1}{c}\right]. \quad (6.17)$$

It should be emphasized that such a choice of saturation radius is semi-empirical. It is interesting to compare this saturation radius with that obtained using equation (6.16). Since $N \sim a_{solv}^{-3}$,

$$r_{sat} \sim a_{solv}\sqrt{a_{solv}/\mu}. \quad (6.18)$$

Hence, for a constant ratio a_{solv}/μ the characteristic saturation radius r_{sat} is directly proportional to the solvent molecule radius. This proportionality is only in qualitative agreement with the empirical choice of the saturation radius by the Abraham–Liszi relation:

$$b = a + a_{solv}. \quad (6.19)$$

Nevertheless, equation (6.18) may serve as a "justification" of equation (6.19), although the transition from one solvent to another causes a change of both the radius a_{solv} and the dipole moment μ, such that the true dependence of the saturation radius on solvent nature is more complicated than that described by equation (6.19).

There is another, even more significant, circumstance associated with equation (6.17). When we set the permittivity of the first sheath equal to ε_{op}, we strongly overestimate the density of the electrostatic solvation energy, which should be calculated using integral equation (6.6). Such an *a priori* choice of the maximum permittivity in the first solvation sheath and calculation of the energy by equation (6.17) means that dielectric homogeneity is caused not by the electric field of the ion, but by some external forces ensuring complete orientation of the dipole moments. It is only in this case that equation (6.17) is valid. The difference between the exact approach and the approximate one, which leads to equation (6.17), becomes insignificant only in *very strong fields*, while in moderate fields, which are the ones usually dealt with in practice, this difference may be quite noticeable. Abraham et al. (1982) calculated the energy of Rb^+ and Br^- solvation in acetonitrile. For example, for a Rb^+ cation whose radius equals 0.143 nm (in accordance with the Pauling scale) the experimental value of the electrostatic part of the solvation energy is found to be -329 kJ mol^{-1}. The Born equation gives a value of -472 kJ mol^{-1}, so that there is a discrepancy of to 143 kJ mol^{-1}. The simplest Abraham–Liszi one-layer model with the parameters $\varepsilon_1 = 2.0$ and $b - a = 0.222$ nm or $b = 0.365$ nm calculates a value of

$-333 \, \text{kJ mol}^{-1}$, which differs from the experimental result by only $4 \, \text{kJ mol}^{-1}$. Improvement of the Abraham–Liszi model by fitting of ε_1 and $(b - a)$ leads to full agreement between the calculated and experimental data.

Now, let us compare these results with the predictions of equation (6.16). If one takes the radius of the rubidium ion to be equal to 0.143 nm (from Pauling's scale) and uses equation (6.16) then $\Delta_{vac}^{\alpha} G_{Rb^+}^{0}(\text{el}) = -446 \, \text{kJ mol}^{-1}$, i.e., the correction to the Born energy ($485 \, \text{kJ mol}^{-1}$) is as small as $39 \, \text{kJ mol}^{-1}$. This is much smaller than the value obtained by the Abraham–Liszi–Kristof (1982) semiempirical formula. Incidentally, the characteristic saturation radius in acetonitrile is 0.29 nm, which differs from the value $b = 0.365$ nm used by Abraham $et \, al.$ (1982). If the radius of Rb^+ is taken to be equal to 0.163 nm (the Gourary–Adrian scale), the Born electrostatic free energy amounts to $-414 \, \text{kJ mol}^{-1}$. If dielectric saturation is taken into account, calculation by equation (6.16) gives $-371 \, \text{kJ mol}^{-1}$, i.e., the correction to the Born formula is only $43 \, \text{kJ mol}^{-1}$. Let us now turn to the Br^- ion. If its radius is set equal to 0.18 nm (the Gourary–Adrian scale), the Born energy will be $-375 \, \text{kJ/mol}^{-1}$ and the energy calculated with an allowance for dielectric saturation, $-367 \, \text{kJ mol}^{-1}$. This means that for the Br^- ion the correction due to dielectric saturation is 11 times less than follows from Abraham–Liszi–Kristof calculations.

Consequently, our estimate of the contribution due to dielectric saturation differs significantly from the predictions of the Abraham–Liszi relation (equation (6.17)). It does not mean, of course, that this relation is wrong; on the contrary, the Abraham–Liszi relation proved to be very useful and convenient for practical applications. Its predictive power has been effectively used by Ford and Scribner (1983) to calculate the free solvation energy for a number of organic ions. Although the Abraham–Liszi relation gives a correct quantitative description of the solvation energy, it cannot explain physical principles underlying this phenomenon.

The works of Abraham and Liszi as well as those of Buckingham (1957) and Beveridge and Schnuelle (1975) were criticized by Stiles (1980) who called them physically unrealistic just because they used a stepwise permittivity profile. Instead, Stiles suggested describing the influence of dielectric inhomogeneity on the ion solvation free energy in terms of an arbitrary smooth function:

$$\varepsilon(r) = \varepsilon_a \left(\frac{\varepsilon_b}{\varepsilon_a} \right)^{f(r)} \qquad \text{for } a \leq r \leq b, \tag{6.20}$$

where:

$$f(r) = 1 - \left(\frac{b - r}{b - a} \right)^2. \tag{6.21}$$

This permittivity continuously varies from ε_α at point $r = a$ to ε_b at point $r = b$. Stiles (1980) wrote that although this profile was empirical, it could be more universal than the profiles obtained from particular mechanisms of dielectric inhomogeneity. Among such mechanisms we may note ligand solvent and solvent–solvent interactions, dielectric saturation, etc. In the case of dielectric saturation, Stiles has demonstrated that the electric solvation energy is of the form:

$$\Delta_{vac}^{\alpha} G^0(\text{el}) = \frac{z^2 e_0^2}{8\pi\varepsilon_0} \int_a^\infty \left[\frac{1}{\varepsilon(r)r^2} - \frac{1}{r^2} \right] dr$$

$$= \frac{z^2 e_0^2}{8\pi\varepsilon_0} \left[\frac{1}{b\varepsilon_b} - \frac{1}{a} + \frac{1}{\varepsilon_a} \int_a^b \left(\frac{\varepsilon_a}{\varepsilon_b} \right)^{f(r)} r^{-2} dr \right], \quad (6.22)$$

where integration can only be performed by numerical methods.

Stiles (1980) did not compare his calculation results with experimental data, but he came to the conclusion that although the correction to the Born formula due to dielectric inhomogeneity near the ion is small it still cannot be neglected.

Abraham *et al.* (1982) analyzed the Stiles relation and used it to calculate the solvation energy of Rb^+ in acetonitrile. They used the permittivity values $\varepsilon_a = 2$ and $\varepsilon_b = 36.02$, but retained the radius b as a fitting parameter because this quantity is not defined in the Stiles theory. To obtain the desired value of the electrostatic solvation energy, -329 kJ mol^{-1}, they had to assume that b is greater than 2.4 nm. They felt this value of the ion radius was too high, so they came to the conclusion that the Stiles profile (equation (6.20)) was unrealistic, at least for the Rb^+ acetonitrile system.

The Stiles profile is, of course, purely empirical and is not based on rigorous theoretical principles. However, the criticism by Abraham *et al.* (1982) is too strong. They proceeded from the erroneous assumption that an allowance for dielectric inhomogeneity will completely describe the electrical part of the solvation energy. When Abraham *et al.* (1979) obtained an unreasonably high value for the Rb^+ radius, $b = 2.4$ nm, they decided that the Stiles profile was unsuitable to describe dielectric saturation. However, in light of what has been said above, this result is the evidence that dielectric saturation cannot completely explain the discrepancy between the Born relation and experimental fact.

This result also points to the source of discrepancy between the true contribution of dielectric saturation to the solvation energy and the contribution due to the Abraham–Liszi relation (6.17). In the Abraham–Liszi model the solvent polarization energy is determined mainly by the layer $a \leq r \leq h$, and it follows from geometric considerations that the outer part of this layer gives the greatest contribution. Therefore, the solvent polarization energy is very sensitive to the permittivity in this region. By taking the minimum value $\varepsilon_1 = \varepsilon_{op}$ in that region, Abraham and Liszi considerably overestimated the contribution due to the saturation effect. In their papers this circumstance was somewhat masked, but the Stiles smoothed profile immediately revealed the discrepancy.

Thus, dielectric saturation of the medium gives a noticeable contribution to the solvation energy of small ions, yet this contribution is insufficient to correctly describe experimental data. Therefore, to develop a physical theory of solvation, one needs to go beyond the framework of the continuum model. Progress in this direction can be made by using non-local electrostatics methods (Kornyshev and Volkov, 1984; Volkov and Kornyshev, 1985).

6.5. THE NON-LOCAL ELECTROSTATICS METHOD

Another rapidly developing semi-macroscopic approach is to calculate the electro-static part of the solvation energy using the non-local electrostatics method. Non-local electrostatics takes into account that fluctuations of solvent polarization are correlated in space, since liquid has a structure determined by quantum interaction between its molecules. This means that the average polarization at a given point depends on the electric displacement at all other correlated points of the space correlated with a given point (Vorotyntsev and Kornyshev, 1993).

In non-local electrostatics the electric displacement D and electric field E are related by the tensor $\varepsilon_{mn}(r)$:

$$D^m(r) = \sum_n \int dr' \, \varepsilon_0 \varepsilon_{mn}(r - r') E^n(r'), \qquad (m, n = x, y, z). \qquad (6.23)$$

It should be noted that although this relation is spatially complicated, it is linear. Further calculations are carried out in terms of the Fourier transform of the tensor $\varepsilon_{mn}(r)$ which is called the static dielectric function $\varepsilon(k)$:

$$\varepsilon(k) = \sum_{m,n} \frac{k_m k_n}{k^2} \int d(r - r') \, e^{-ik(r-r')} \varepsilon_{mn}(r - r'). \qquad (6.24)$$

The potential produced in a medium by a charged sphere of radius a at a distance r from its center is given by:

$$\varphi(r) = \frac{ze_0}{2\pi^2 \varepsilon_0} \int_0^\infty \frac{dk}{\varepsilon(k)} \frac{\sin kr}{kr} \frac{\sin ka}{ka} \qquad (6.25)$$

hence we can easily find the electrostatic contribution to the solvation energy:

$$\Delta_{vac}^\alpha G^0(el) = \frac{ze_0}{4\pi^2 \varepsilon_0} \int_0^\infty dk \frac{\sin ka}{k^2 a^2} \left[1 - \frac{1}{\varepsilon(k)} \right]. \qquad (6.26)$$

Polarization of the medium can be divided into three main groups: optical, vibrational and orientational. To evaluate equation (6.26), one has to specify the function $\varepsilon(k)$. This can be done, for example, by subdividing fluctuations of the medium polarization into three modes relating to different degrees of freedom, namely: 1) electronic or optical; 2) vibrational or infrared; 3) orientational or Debye. If the radius of the correlation of the i-th mode of a fluctuation is λ_i, then

$$1 - \frac{1}{\varepsilon(k)} = 1 - \frac{1}{\varepsilon_{opt}} + \left(\frac{1}{\varepsilon_{opt}} - \frac{1}{\varepsilon_2} \right) \frac{1}{1 + k^2 \lambda_2^2} + \left(\frac{1}{\varepsilon_2} - \frac{1}{\varepsilon_3} \right) \frac{1}{1 + k^2 \lambda_3^2}. \qquad (6.27)$$

In this expression the correlation length of the electronic mode is set equal to zero.

The exact values of the correlation lengths λ_2 and λ_3 cannot be determined *a priori* but they can be approximated from physical considerations. In the infrared region, the length λ_3 depends on liquid type. In the case of non-associated liquids the correlation length for orientational vibration is approximately equal to the intermolecular distance, while for associated liquids (water for example) λ_3 is equal to the characteristic length of the hydrogen bond chain, i.e., 0.5–0.7 nm (Kornyshev and Volkov, 1984; Volkov and Kornyshev, 1985).

Integration of equation (6.26) with $(1 - 1/\varepsilon(k))$ expressed by equation (5.27) yields:

$$\Delta_w^{vac} G(el) = \frac{q^2}{8\pi\varepsilon_0 a}\left\{1 - \frac{1}{\varepsilon_{opt}} + \left(\frac{1}{\varepsilon_{opt}} - \frac{1}{\varepsilon_2}\right)\psi\left(\frac{2a}{\lambda_2}\right) + \left(\frac{1}{\varepsilon_2} - \frac{1}{\varepsilon_3}\right)\psi\left(\frac{2a}{\lambda_3}\right)\right\} \quad (6.28)$$

where:

$$\psi(x) = 1 - (1 - e^{-x})\frac{1}{x}. \quad (6.29)$$

The electric contribution to the hydration energy calculated by these relations is shown in Fig. 6.7, together with experimental data for small monovalent ions. The permittivities used in those calculations were taken from Table 6.3. The correlation lengths were chosen to provide the best fit between theoretical and experimental data: $\lambda_2 = 0.1$ nm and $\lambda_3 = 0.7$ nm. As can be seen from Fig. 6.7, the agreement between theory and experiment obtained by Kornyshev and Volkov (1984) is more than satisfactory. Yet, the question arises whether the results obtained are sensitive to the fitting parameters λ_i and whether such a choice of λ_i is justified. In addressing this question one can conclude that the results are most sensitive to parameter λ_2 and depend very little on λ_3 (Fig. 6.8). A detailed analysis of this situation is presented in Kornyshev and Volkov (1984). The non-local electrostatics method refers to the continuum models, but the effective parameters needed for calculation are chosen by analyzing the solvent structure. This method gives rather accurate values for the solvation energy for small ions, and also permits calculation, by virtue of equation (6.5), of the resolvation energy. For large ions there remains considerable discrepancy between theory and experiment, which makes it necessary to take into account other effects. In our case, these are the work done in the formation of a cavity in the solvent and the solvophobic effect. The largest value for the solvation energy is obtained in the Born limit $\lambda_2 = \lambda_3 = 0$, i.e., in a structureless solvent.

The major disadvantage of the continuum and semi-continuum approaches to the solvation problem lies in the solvent model itself. An allowance for dielectric saturation or dipole correlation is an attempt to partially describe the discrete properties of the solvent within the framework of the continuum model. Other analogous approaches attempt to take into account the unknown effect of the solvent molecular structure on the thermodynamic properties of a system (Ramanathan and Friedman, 1971). However, the problem can only be solved correctly using a statistical model of the solvent and taking into account its discrete structure.

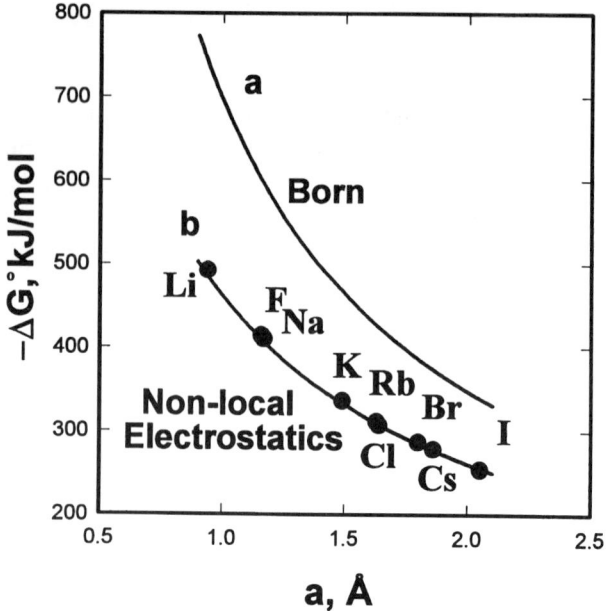

Figure 6.7. The hydration energy versus the ion radius. Points are experimental data from Randles (1956). The ion radii are taken from a paper by Gourary and Adrian (1960). Theoretical lines are calculated according to (*a*) Born and (*b*) using non-local electrostatics.

6.6. STATISTICAL SOLVENT MODELS

A consistent approach to the solvation problem is to develop a statistical theory that will consider both ions and solvent molecules as discrete species, though such an approach demands the introduction of some model potentials for particle interaction.

Unlike "primitive" electrolyte models which use continuum descriptions of the solvent, the modern statistical analysis employs a "non-primitive" or a so-called ion–dipole plasma model in which the electrolyte solution is considered as a system of hard spheres of radius a and a_0 each carrying at their center a point electric charge or a constant point dipole, respectively. Ordinary electrostatic interactions are assumed to exist between the charges and dipoles. The system is described in terms of many-particle correlation functions satisfying the corresponding equations (Henderson, 1983; Wertheim, 1979). To solve these equations, one has to introduce various simplifications, the mean-spherical approximation being the most widely used (Chan *et al.*, 1979).

The ion–dipole plasma model permits calculation of the ion solvation energy, and also the properties of the pure dipole liquid. Wertheim (1971, 1979) calculated the permittivity of the dipole liquid in the mean-spherical approximation:

$$\varepsilon = q(2\zeta)/q(-\zeta), \tag{6.30}$$

Table 6.3. Dielectric permittivities ε_l (Akhadov, 1972) in the three-mode polarization model ($T = 293$ K).

Solvent	ε_1	ε_2	ε_3
Water	1.8	4.9	78.8
Nitrobenzene	2.4	3.7	35.0
1,2-Dichloroethane	2.0		10.4
Acetophenone	2.45		17.4
Dichloromethane	2.04		9.1
Methylbutylketone	1.9		14.9

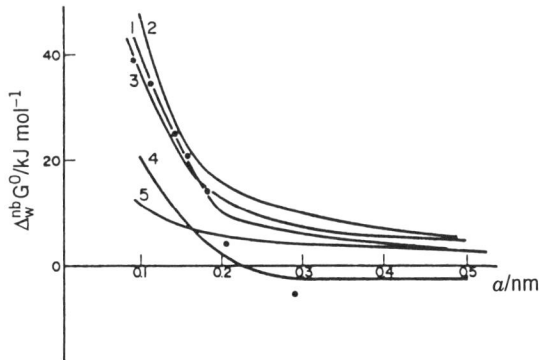

Figure 6.8. The effect of permittivity spatial dispersion in the aqueous and nitrobenzene phases on the electrostatic contribution to the free resolvation energy. Calculations are performed according to Kornyshev and Volkov (1984) with $\lambda_2^w = 0.1$ nm, $\lambda_3^w = 0.7$ nm and with permittivities taken from Table 6.3. The correlation radii were chosen as follows: (1) $\lambda_2^{nb} = 0.1$ nm, $\lambda_3^{nb} = 0.5$ nm; (2) $\lambda_2^{nb} = 0.1$ nm, $\lambda_3^{nb} = 0.3$ nm; (3) $\lambda_2^{nb} = 0.13$ nm, $\lambda_3^{nb} = 0.3$ nm; (4) $\lambda_2^{nb} = 0.1$ nm, $\lambda_3^{nb} = 0.7$ nm; (5) is calculated by the Born equation. Experimental data are taken from Table 6.1.

where:

$$q(x) = (1 + 2x)^2/(1 - x)^2, \tag{6.31}$$

and the parameter ζ is a solution of the equation:

$$q(2\zeta) - q(\zeta) = n_d \mu^2/3\varepsilon_0 kT. \tag{6.32}$$

Here n_d is the number of solvent particles per unit volume and μ is the solvent molecule dipole moment. This result is a considerable improvement over the well-known Onsager relation.

For highly associated liquids there exists a noticeable discrepancy between theory and experiment. This is not surprising because the model ignores a number of intricate forces which act between real solvent molecules. For example, for a dipole liquid with density and dipole moment (1.8 D) equal to the corresponding parameters of water, equation (6.30) yields, at room temperature, the permittivity value 48. To obtain a value of 80, the dipole moment should be set equal to 2.62 D.

Naturally, a system of hard, constant-dipole spheres can hardly seem a proper model for water. Yet, such a system may be a reasonable model of an organic solvent. In any case, it is the first step for constructing a relevant model of water. In particular, the application of the model to studying solvation and the interaction of ions has revealed a number of essential effects caused by solvent discreteness.

The energy of ion solvation (the "Born" energy) was calculated by Chan *et al.* (1979) using the mean spherical approximation:

$$\Delta_{vac}^{\alpha} G^0 = - \frac{(ze_0)^2}{8\pi\varepsilon_0(a + a_d/\lambda)} \left(1 - \frac{1}{\varepsilon}\right), \tag{6.33}$$

where λ is found using the equation:

$$\lambda^2(1 - \lambda)^4 = 16\varepsilon. \tag{6.34}$$

The parameter λ ranges from 1 to 3 (Fig. 6.9); for $\varepsilon = 80$ it is equal to 2.6. Together with the radius of the solvent dipole moment, this parameter determines a new characteristic length for the problem, a_d/λ which naturally arises in the solution of the equations. This characteristic length is solely determined by the properties of the dipole liquid and depends on two parameters: the molecule radius and permittivity.

Formally, equation (6.33) is analogous to the Born energy with the renormalized radius increased by a_d/λ. It is interesting to note that an empirical relation exists which is identical in form to equation (6.33), and which can be used to describe the solvation energy and the free resolvation energy (Govington and Newman, 1977). If the solvent molecule radius is set equal to 1.5 Å, the aforementioned relation gives the value 0.58 Å for a_d/λ.

An apparent increase of the ion radius in equation (6.33) is caused by solvent discreteness. It is obvious that the permittivity cannot have the bulk value immediately at the ion "surface." Equation (6.33) implies that this value is effectively attained at a distance of $a + a_d/\lambda$ from the ion center. One should, however, bear in mind that this idea of a stepwise change in the permittivity is only an approximate description of sufficiently rigorous physical results obtained in the discrete model. Furthermore, an analysis of the model shows that at small distances the concept of local permittivity is simply invalid and a more correct concept of medium polarization

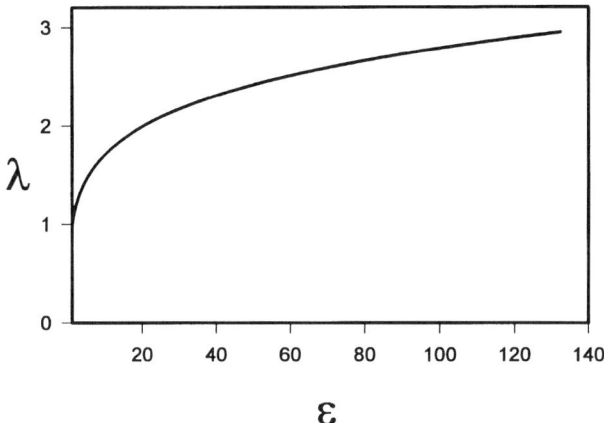

Figure 6.9. Dependence of parameter λ on the dielectric permittivity of the solvent.

must be used. If a dielectric is treated macroscopically, its polarization at a distance r from the ion is given by the well-known relation: $P_{mac}(r) = (\varepsilon - 1)z\,e_0/4\pi\varepsilon_0 r^2$.

Figure 6.10(a) shows the polarization of a medium (normalized to macroscopic polarization $P_{mac}(r)$ as calculated by Chan $et\ al.$ (1979). The main feature of the plot is its oscillating character, which means that because of strong ordering of dipoles adjacent to the ion the next-neighbor dipoles may have an opposite orientation. That is why the description of a medium in terms of local permittivity (e.g., in terms of a stepwise function), even though it is in good agreement with experiment, may lead to an erroneous physical picture at small distances.

An important consequence should be noted. The interaction of ions in a discrete medium differs fundamentally from that predicted by the macroscopic theory. This interaction is described by a potential of mean force W_{ij}, which, in addition to the Coulomb potential, includes the averaged interaction with all solvent molecules (Fig. 6.10(b)). As was shown by Chan $et\ al.$ (1979) at small distances the average force potential is also an oscillating function. For example, the curve for ions of opposite sign reveals a deep potential minimum W_{+-}, when the centers of the ions are at a distance of $4a$. This is a relatively stable configuration, in which the ions are separated by one dipole molecule; it can be regarded as the result of the formation of ion pairs in the solution. The potential for ions of the same sign W_{++} also has a deep minimum, at $r = 3a$. This configuration looks like a "double" ion, for which an increase in the Coulomb energy is more than compensated by the negative displacement of the Born solvation energy of the highly charged structure.

The results obtained in the mean spherical approximation model are mainly of a qualitative character, since the model is approximate. Nevertheless, they provide a more accurate physical representation of solvation and provide another way with which one can evaluate other theories and understand how these theories can agree with experimental results.

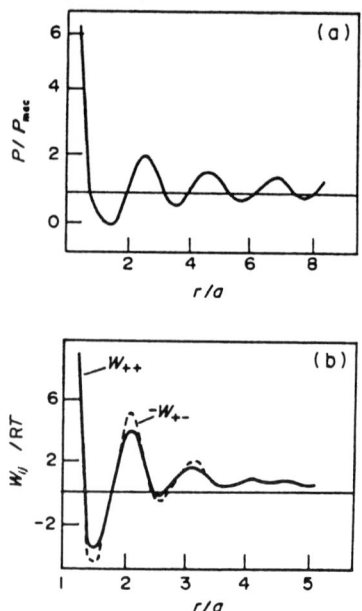

Figure 6.10. The normalized polarization density P/P_{max} (a) and the average-force potential W_{ij} for two monovalent ions i and j (b) as a function of the distance r from the ion center (Chan *et al.*, 1979).

6.7. CONTRIBUTION OF THE SOLVOPHOBIC EFFECT TO THE RESOLVATION ENERGY

Modern approaches to the calculation of the energy of the solvophobic ion–solvent interaction are based on phenomenological (Ben Naim, 1980; Tanford, 1980) or statistical mechanical molecular models (Pratt, 1985) in which the parameters describing both the ion and solvent are not always known. To avoid uncertainty, the solvophobic contribution to the resolvation energy is sometimes estimated using the semi-empirical solvophobic formula (Kornyshev and Volkov, 1984). The main idea is as follows (Kornyshev and Volkov, 1984; Markin and Volkov, 1989f; Sisskind and Kasarnowsky, 1933; Uhlig, 1937).

Surface energy, when expressed in terms of surface tension at the ion–solvent interface $\gamma_{0,\alpha}$, is equal to $4\pi a^2 \gamma_{0,\alpha}$, and the difference in the surface energies in media α and β is $4\pi a^2(\gamma_{0,\alpha} - \gamma_{0,\beta})$. According to Antonow's (1907, 1931) rule

$$\gamma_{0,\alpha} - \gamma_{0,\beta} = \gamma_{\alpha\beta}\, \mathrm{sgn}(\gamma_\alpha - \gamma_\beta) \tag{6.35}$$

where $\mathrm{sgn}(\gamma_\alpha - \gamma_\beta) = +1$ for $\gamma_\alpha > \gamma_\beta$ and $\mathrm{sgn}(\gamma_\alpha - \gamma_\beta) = -1$ for $\gamma_\alpha < \gamma_\beta$, $\gamma_{\alpha\beta}$ is the interfacial tension at a flat boundary between solvents, and γ_α, γ_β are the surface tensions at the boundaries of solvents α and β, respectively. With air under the same pressure and temperature, we obtain the formula for the solvophobic contribution to the resolvation energy:

$$\Delta_\beta^\alpha G^0(\text{svph}) = -4\pi a^2 \gamma_{\alpha\beta} \, \text{sgn}(\gamma_\alpha - \gamma_\beta). \tag{6.36}$$

Relation (6.36) infers that the molecules of the two solvents w and m do not mix or react chemically with each other. The hydrophobic contribution to the Gibbs free energy of the ion or molecule (dipole, multipole) transfer from water to hydrocarbon is negative and the absolute value is obviously greater for particles with larger radii, a.

The effect of curvature on the surface tension at a molecularly sized sphere was calculated by Tolman (1949). The surface tension at radius a can be written as

$$\gamma_a = \gamma \frac{1}{1 + 2\delta/a}, \tag{6.37}$$

where δ is a parameter, which according to Tolman is the distance from the surface of tension to the dividing surface for which the surface excess of fluid vanishes. Parameter δ can vary between 0 and a few angstroms (Tolman, 1949). Dependence of the change in surface tension with radius is shown in Fig. 6.11.

From the theoretical point of view, the limits of applicability of the solvophobic formula are not quite clear. Nonetheless, it works surprisingly well for calculation of the partition coefficients of a system of two immiscible liquids (Abramson, 1981). Some deviation from the solvophobic formula is observed if the interfacial tension of two pure immiscible solvents is less than 10 mN m^{-1} and if the size of the dissolved particle is less than 0.2 nm.

An important advantage of the solvophobic formula is that it does not involve fitting parameters because $\gamma_{\alpha\beta}$ can be determined experimentally. Figure 6.12 shows the solvophobic contribution to the resolvation energy versus the particle radius calculated by equation (6.36) for water–nitrobenzene and water–1,2-dichloroethane systems.

6.8. THE TOTAL RESOLVATION ENERGY

In the preceding sections we considered different effects which contribute to the energy of ion resolvation. However, for most solvents which are of practical interest the greatest contribution is due to the electrostatic and solvophobic effects. Therefore, the resolvation energy can be represented as a sum of all contributions (Volkov and Kornyshev, 1985):

$$\Delta G(\text{tr}) = \Delta G(\text{el}) + \Delta G(\text{solv}) + \Delta G(\text{si}), \tag{6.38}$$

where $\Delta G(\text{el})$ is electrostatic contribution, $\Delta G(\text{solv})$ is the hydrophobic effect and

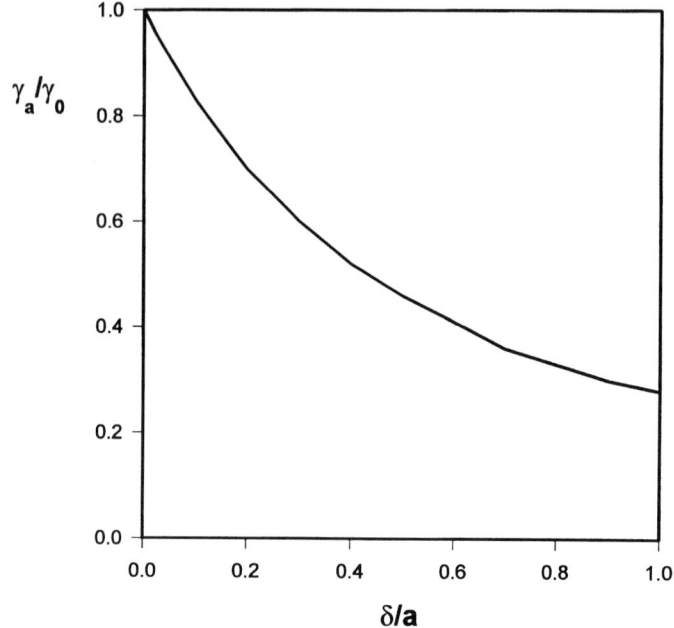

Figure 6.11. Dependence of surface tension on radius of curvature.

$\Delta G(\text{si})$ is caused by specific interactions of the transferred particle (ion, dipole) with solvent molecules, such as hydrogen-bond formation, donor–acceptor and ion–dipole interactions.

The electrostatic contribution was calculated by the non-local electrostatics method and the solvophobic contribution, by the solvophobic formula. Results of calculations for the water–nitrobenzene system, the most thoroughly studied pair, are presented in Fig. 6.13 (curve 1). For comparison, results of calculations using the Born formula with an allowance for the solvophobic effect (curve 2) are also shown. It can be seen that the solvophobic effect cannot explain the shortcomings of the Born formula.

Figure 6.13 shows that the solvophobic effect can make the resolvation energy (or the standard ion distribution potential) change sign as the ion radius varies. The calculation results are in excellent agreement with experimental data. The discrepancy observed for small radius anions may be attributed to the formation of hydrogen bonds between anion and solvent or to defect annihilation. The energy corresponding to this discrepancy is of the same order of magnitude as the energy gain obtained in the formation of a weak hydrogen bond between anion and solvent. It should, however, be noted that for small anions resolvation energies reported by different authors differ considerably.

The transfer of small inorganic ions with radii less than 0.3 nm through the interface between two immiscible liquids is impeded, because it requires considerable

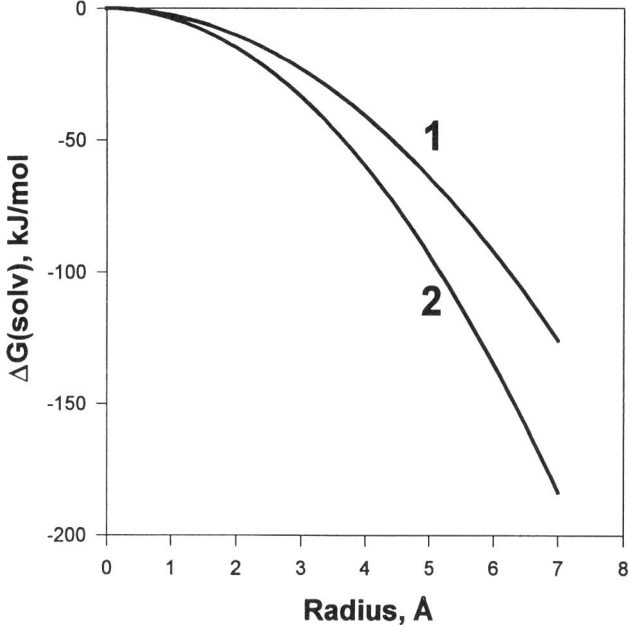

Figure 6.12. The solvophobic contribution to the energy of resolvation of particles of radius *a* from water into oil calculated by the solvophobic equation (6.36). (1) 1,2-dichloroethane; (2) *n*-octane.

energy for the resolvation and penetration of small charged particles into non-polar or low polar solvents. It is possible to facilitate the ion transfer from water to a non-polar solvent using complexing agents or chelators, which have relatively large sizes and can screen the charge of the ion. Chelators are well soluble in organic solvent due to a large hydrophobic effect and small electrostatic contribution and thereby facilitate the ion transfer between two contacting liquid phases.

6.9. DIPOLE RESOLVATION

Bell (1931) calculated the electrostatic part of Gibbs energy of dipole molecule transfer from a vacuum to a medium of dielectric constant ε_i:

$$\Delta_i^{\text{vac}} G(\text{el}) = -\frac{\mu^2}{12\pi\varepsilon_0 l^3}\left(\frac{\varepsilon_i - 1}{2\varepsilon_i + 1}\right), \tag{6.39}$$

where μ is the dipolar moment and l is the effective dipole size. Now one can calculate the electrostatic ("Born") Gibbs energy of dipole transfer from phase w to phase m

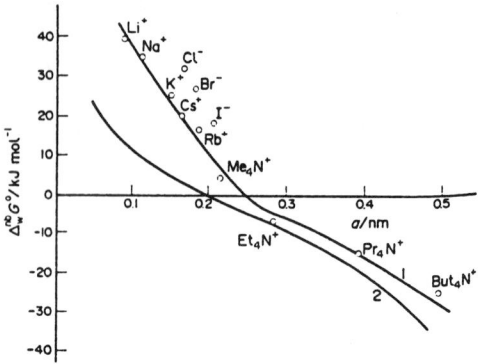

Figure 6.13. The resolvation energy versus the ion radius in the water–nitrobenzene system calculated by equation (6.38) with an allowance for the solvophobic effect: (1) the non-local electrostatics method (Kornyshev and Volkov, 1984; Volkov and Kornyshev, 1985); (2) the Born method. Points are experimental data from Table 6.1.

using a thermodynamic cycle (Volkov *et al.*, 1997):

$$\Delta_w^m G(\text{el}) = \frac{\mu^2}{12\pi\varepsilon_0 l^3}\left(\frac{\varepsilon_w - 1}{2\varepsilon_w + 1} - \frac{\varepsilon_m - 1}{2\varepsilon_m + 1}\right) = \frac{\mu^2}{4\pi\varepsilon_0 l^3}\left(\frac{\varepsilon_w - \varepsilon_m}{(2\varepsilon_w + 1)(2\varepsilon_m + 1)}\right). \quad (6.40)$$

Generally speaking, the dipole moment μ can vary from solvent to solvent.

The concentration of dipole molecules in the oil phase can be calculated (Paula *et al.*, 1996; Volkov *et al.*, 1997) from the estimated free Gibbs energy of a dipole molecule transfer from water to the hydrophobic phase:

$$c_{\text{oil}} = c_w \exp(-\Delta G_{\text{tr}}/RT). \quad (6.41)$$

The partition coefficient of an ion or dipole K_i can be also calculated from the free Gibbs energy of transfer :

$$K_i = \exp(-\Delta G_i^0(\text{tr})/RT). \quad (6.42)$$

Equations (6.40)–(6.42) are very useful in the theory of extraction, partitioning, and passive transport of small neutral molecules across liquid membranes. The partition coefficient K_i is also important in the formulation and delivery of pharmaceutical agents because it describes relative lipophilicity and solubility in lipid–water systems. For many drugs the inequality can be written:

$$0 < \log K_i < 4. \quad (6.43)$$

The choice of a given drug delivery method can be guided by the partition coefficient of the agent. For instance, a drug with a low partition coefficient will likely be given by injection, while a skin patch or ointment could be used for drugs with high partition coefficients. Oral consumption would be appropriate for drugs with coefficients in the intermediate range.

C

STRUCTURE OF INTERFACES

7

INTERFACIAL STRUCTURES AND ELECTRICAL DOUBLE LAYERS

The electrical double layer at the oil–water interface is a heterogeneous interfacial region that separates two bulk phases of polarized media and maintains a spatial separation of charges. Electrical double layers at such interfaces determine the kinetics of charge transfer across phase boundaries (Boguslavsky and Gugeshashvili, 1989; Kharkats and Volkov, 1987; Volkov and Deamer, 1996; Volkov et al., 1995a,b) stability and electrokinetic properties of lyophobic colloids (Kruyt, 1952; Verwey, 1950; Verwey and Niessen, 1939; Verwey and Overbeek, 1948), mechanisms of phase transfer or interfacial catalysis (Cunnane and Murtomaki, 1996; Kharkats and Volkov, 1985; Volkov et al., 1982a,b), effects in soil science (Yan and Masliyah, 1994) and bioelectrochemistry (Blank, 1991; Markin and Chizmadzhev, 1974; McLaughlin, 1989; Volkov et al., 1982a,b), charge separation in natural and artificial photosynthesis (Barber et al., 1977; Deamer and Volkov, 1996; Gugeshashvili et al., 1983a,b, 1991; Markin et al., 1992a,b,c; Volkov, 1984a,b, 1986a,b,c, 1989), and heterogeneous enzymatic catalysis (Boguslavsky and Volkov, 1987; Kakiuchi, 1996b). The elucidation of the structure of the interface between two immiscible liquids and the mechanism of charge separation is a fundamental problem in modern chemistry and chemical technology (Volkov and Deamer, 1996).

In classical electrochemistry the electrical double layer was generally considered to be the inhomogeneous region of an electrolyte near a charged metal surface (Delahay, 1965; Frumkin et al., 1980; Grafov and Damaskin, 1994; Parsons, 1954, 1961, 1980, 1990; Vorotyntsev and Kornyshev, 1984; Vorotyntsev and Mityushev, 1991). At the interface between two immiscible electrolyte solutions (ITIES) both plates of the capacitor are liquids. Verwey and Niessen (1939) first described the electrical double layer at ITIES as two non-interacting diffuse layers, one at each side of the interface. Both solvents were assumed to be structureless media with macroscopic dielectric permittivities, and potential distribution in the electrical double layer was defined by the Gouy–Chapman theory (Chapman, 1913; Gouy, 1910).

Gavach *et al.* (d'Epenoux *et al.*, 1979; Gavach *et al.*, 1977; Gros *et al.*, 1978; Seta and Gavach, 1972; Seta *et al.*, 1979) extended the Verwey–Niessen model by introducing an ion-free transition layer at the interface between two immiscible electrolyte solutions. This is directly analogous to the compact layer (or the inner Helmholtz layer) in classical electrochemistry. Stern theory (Stern, 1924) was extended to ITIES, and the final model is referred to as the modified Verwey–Niessen model (MVN). It should be noted that the MVN model does not completely describe all experimental data at ITIES (Watts and VanderNoot, 1996).

After discussing this model, which is commonly used to analyze electrical properties of ITIES, we will consider recent developments in the theory of the electrical double layer at liquid–liquid interfaces. The double layer at the oil–water interface consists of interacting electrical (ionic and dipolar) layers. Pertinent theory should therefore take into account ion penetration into adjacent phases, overlapping double layers, effects of ion size, dielectric saturation in compact and diffuse layers, discreteness of charges, ion–ion, ion–solvent and solvent–solvent interactions. Such double layer effects are very important in biological systems in which energy transducing processes take place at liquid interfaces between bilayer membranes and aqueous electrolyte solutions (Boguslavsky and Volkov, 1987; Deamer and Volkov, 1995; Volkov and Deamer, 1996; Yaguzhinsky *et al.*, 1975a,b, 1976, 1977).

7.1. THE MODIFIED VERWEY–NIESSEN (MVN) MODEL

In the MVN model the electrical double layer consists of two diffuse ion layers back to back, which produce a compact inner layer between the two phases (Fig. 7.1).

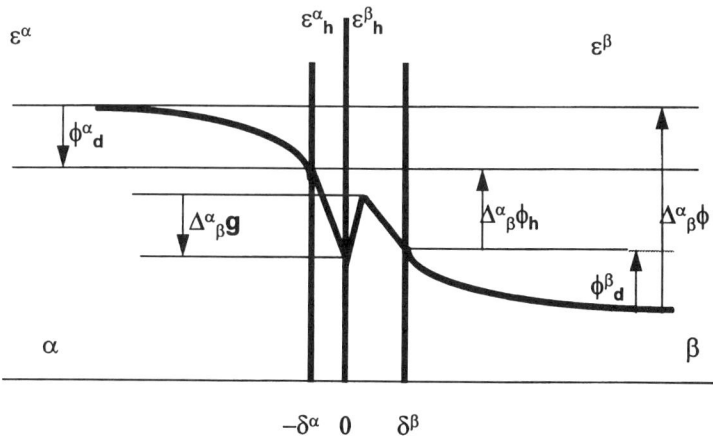

Figure 7.1. Distribution of the electric potential in the electric double layer at the interface between two immiscible electrolyte solutions.

Table 7.1. Relationship of the ion-free layer thickness at the oil–water and air–water interfaces to the thickness of a water monolayer assuming that surface concentration of water molecules in a monolayer corresponds to 1.72×10^{-9} mol cm^{-2}. Data from Girault and Schiffrin (1983).

Non-aqueous medium	LiCl	NaCl	KCl	MgSO$_4$
Nitrobenzene	0.4	0.7	0.4	3.0
1,2-dichloroethane	0.4	0.9	0.6	—
n-heptane	0.9	—	1.0	—
Air	0.9	1.0	1.0	2.2

Dielectric permittivity of the medium at any point in the diffuse layer is assumed to be constant and equal to the bulk phase value ε. The compact layer, or inner Helmholtz layer, is located between $-\delta^\alpha$ and $+\delta^\beta$. In a more detailed analysis the dielectric permittivities in both parts of the compact layer are ε_h^α and ε_h^β.

In the MVN model the boundary of the compact layer (called the outer Helmholtz plane) is at the distance of closest ion approach to the interface. For ions with different radii a few outer Helmholtz planes can be introduced as necessary. In the absence of specific adsorption the compact layer located between two nearest outer Helmholtz planes disposed on different sides of an ideal interface is usually called the ion-free layer.

Gavach *et al.* (1977) have introduced a compact layer at ITIES in analogy with a compact layer at metal electrodes in classical electrochemistry (Helmholtz, 1879). They defined the layer as an ion-free space between two planes of closest ion approach to the interface. However, applying definitions of classical metal–water interfaces to ITIES can be problematic. A plane of maximal ion approach to a metal electrode may not apply to ITIES since this plane can be deformed and ions can partially penetrate the interface. This phenomenon leads to an extremely thin compact layer that can have a thickness less than that of a solvent monolayer (Table 7.1) or even negative. Here the capacitance of the compact layer can be too high and will not contribute to the total interfacial capacitance.

The MVN model considers ions in the diffuse part of the electrical double layer as point charges and their dimensions are neglected. In aqueous electrolyte solutions this approximation is reasonable because inorganic ions have radii near 0.1 nm, but organic ions in a non-aqueous phase are usually considerably larger. For polarizability of the interface the aqueous phase must have small ions and the organic phase should have ions of maximal possible size. Small ions will be in the aqueous phase due to the electrostatic component of Gibbs free energy of solvation, and large ions will be in the organic phase due to the hydrophobic effect. The diameter of organic ions such as tetraphenylborate (TPhB$^-$), tetraphenylarsenate (TPhAs$^+$), dicarbollylcobaltate (DCC$^-$) is about 1 nm. The size of such adsorbed organic ions at ITIES usually exceeds the thickness of the compact layer and even that of the diffuse layer in

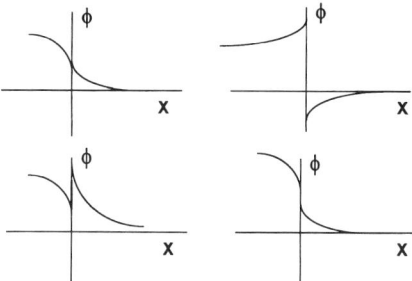

Figure 7.2. Different profiles of Galvani potential differences at the interface between two immiscible electrolyte solutions.

concentrated electrolyte solutions. This is why the theory of the electrical double layer at ITIES must take ion size into account.

We will now consider the potential distribution at the interface between two immiscible liquids (Fig. 7.1). The maximum electrical potential in the compact layer $\Delta_\beta^\alpha \phi_i$ includes a dipolar potential $\Delta_\beta^\alpha g$, which is shown schematically as a narrow region at the sharp interface. A dipolar layer can be located not only in the compact layer but can also occupy part of the diffuse layer. The amplitude and sign of $\Delta_\beta^\alpha g$ can differ from the total interfacial potential. Figure 7.2 illustrates four possibilities for potential distribution at ITIES. Generally speaking, the dipolar potential depends on the total interfacial potential $\Delta_\beta^\alpha \phi$ (Markin and Volkov, 1987c, 1988f, 1989f, 1990f).

The interfacial potential difference consists of the sum of the potential drops:

$$\Delta_\beta^\alpha \phi \equiv \phi^\alpha - \phi^\beta = \phi_d^\beta - \phi_d^\alpha + \Delta_\beta^\alpha \phi_h. \tag{7.1}$$

Charge density at each side of the electrical double layer is usually called the interfacial free charge density, and depends on the model chosen for the interphase. In an ideal polarized electrode the free charge equals the total thermodynamic charge in the Lippmann equation of electrocapillarity if all components of an interface have the same charges as they have in the bulk phases (Markin and Volkov, 1988e, 1989a).

We will denote surface charge density in the diffuse layers as q_d^α, q_d^β and in the compact layer as q_h. From electroneutrality of the interphase we have:

$$q_d^\alpha + q_h + q_d^\beta = 0. \tag{7.2}$$

At the potential of the free zero charge (fzcp) the compact layer has not adsorbed ions and the charges of the diffuse layers are equal to zero:

$$\phi_d^\alpha = \phi_d^\beta = 0 \quad \text{and} \quad \Delta_\beta^\alpha \phi_h = \Delta_\beta^\alpha \phi^{\text{fzcp}}. \tag{7.3}$$

The dipolar potential $\Delta_\beta^\alpha \phi_h$ also needs the index fzcp due to its dependence on total interfacial potential. Usually the dependence is weak and equation (7.3) is used to determine the dipolar potential.

From the Gouy–Chapman theory for a 1:1 electrolyte it follows that

$$q_d = -\frac{2\varepsilon_0 RT}{F}\varepsilon\kappa \sinh\frac{F\phi_d}{2RT}, \tag{7.4}$$

where κ is the Debye length,

$$\kappa = \sqrt{\frac{2F^2 c^0}{\varepsilon\varepsilon_0 RT}}, \tag{7.5}$$

and c^0 is the bulk electrolyte concentration in the corresponding phase.

If the Helmholtz layer does not have charges, potential drops in the two diffuse layers have a simple relation:

$$\varepsilon^\alpha \kappa^\alpha \sinh\frac{F\phi_d^\alpha}{2RT} + \varepsilon^\beta \kappa^\beta \sinh\frac{F\phi_d^\beta}{2RT} = 0. \tag{7.6}$$

The relation between potentials of two diffuse layers depends on the parameter

$$\eta^{\alpha/\beta} = \frac{\varepsilon^\beta \kappa^\beta}{\varepsilon^\alpha \kappa^\alpha} \equiv \sqrt{\frac{\varepsilon^\beta c^{0\beta}}{\varepsilon^\alpha c^{0\alpha}}}, \tag{7.7}$$

which is the reverse ratio of the diffuse layer capacitance. There is a direct proportional dependence between ϕ_d^α and ϕ_d^β with coefficient $\eta^{\alpha/\beta}$ when the potential drops are small:

$$\phi_d^\alpha / \phi_d^\beta = -\eta^{\alpha/\beta}. \tag{7.8}$$

From equations (7.7) and (7.8) it follows that the potential drop ϕ_d in the diffuse layer will be less if the dielectric permittivity ε or electrolyte concentrations c^0 are increased, or if the diffuse double layer capacitance C_d is increased. At the contact between aqueous and organic phases virtually the entire drop of the potential occurs in the organic phase. However, with increasing interfacial potential the situation changes dramatically. When potentials are large enough equation (7.6) can be presented in a different approximation:

$$\phi_d^\alpha = -\phi_d^\beta - \frac{RT}{F}\operatorname{sgn}\phi_d^\beta \ln\eta^{\alpha/\beta}. \tag{7.9}$$

Here the coefficient of proportionality is equal to one. The dependence between potential drops in the two diffuse layers is shown in Fig. 7.3. All curves in the region of high potentials have asymptotes with the same angle equal to 45°.

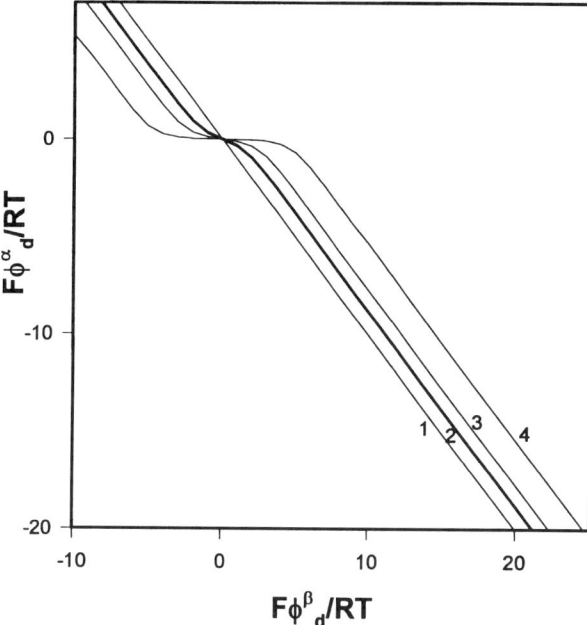

Figure 7.3. Dependencies between diffuse layer potentials at the interface between two immiscible electrolyte solutions.

From equations (7.1) and (7.6) it is possible to find the relationship between the potential drop in both diffuse layers $\Delta_\beta^\alpha \phi - \Delta_\beta^\alpha \phi_h$ and potential drops ϕ_d^α and ϕ_d^β in each diffuse layer:

$$\phi_d^\beta = \frac{RT}{F} \ln \frac{(\varepsilon^\beta \kappa^\beta / \varepsilon^\alpha \kappa^\alpha) + \exp\left[F(\Delta_\beta^\alpha \phi - \Delta_\beta^\alpha \phi_h)/2RT\right]}{(\varepsilon^\beta \kappa^\beta / \varepsilon^\alpha \kappa^\alpha) + \exp\left[-F(\Delta_\beta^\alpha \phi - \Delta_\beta^\alpha \phi_h)/2RT\right]} \tag{7.10}$$

$$\phi_d^\alpha = \frac{RT}{F} \ln \frac{(\varepsilon^\alpha \kappa^\alpha / \varepsilon^\beta \kappa^\beta) + \exp\left[F(\Delta_\beta^\alpha \phi - \Delta_\beta^\alpha \phi_h)/2RT\right]}{(\varepsilon^\alpha \kappa^\alpha / \varepsilon^\beta \kappa^\beta) + \exp\left[-F(\Delta_\beta^\alpha \phi - \Delta_\beta^\alpha \phi_h)/2RT\right]}$$

$$= -(\Delta_\beta^\alpha \phi - \Delta_\beta^\alpha \phi_h) + \frac{RT}{F} \ln \frac{(\varepsilon^\beta \kappa^\beta / \varepsilon^\alpha \kappa^\alpha) + \exp\left[F(\Delta_\beta^\alpha \phi - \Delta_\beta^\alpha \phi_h)/2RT\right]}{(\varepsilon^\beta \kappa^\beta / \varepsilon^\alpha \kappa^\alpha) + \exp\left[-F(\Delta_\beta^\alpha \phi - \Delta_\beta^\alpha \phi_h)/2RT\right]}. \tag{7.11}$$

These dependencies are shown in Fig. 7.4 for a system with $\varepsilon^\beta \kappa^\beta / \varepsilon^\alpha \kappa^\alpha = 0.1$. When interfacial potential is small, essentially the entire drop of the potential occurs in phase β. If the sum of two diffuse layer potentials exceeds the characteristic value $(2RT/F) \ln (\varepsilon^\alpha \kappa^\alpha / \varepsilon^\beta \kappa^\beta) = 118$ mV both potential drops in the diffuse layer change at the same rate.

The spatial distribution of potential near the interface is accompanied by changes in electrolyte concentrations that produce surface excesses of ions. In each phase the

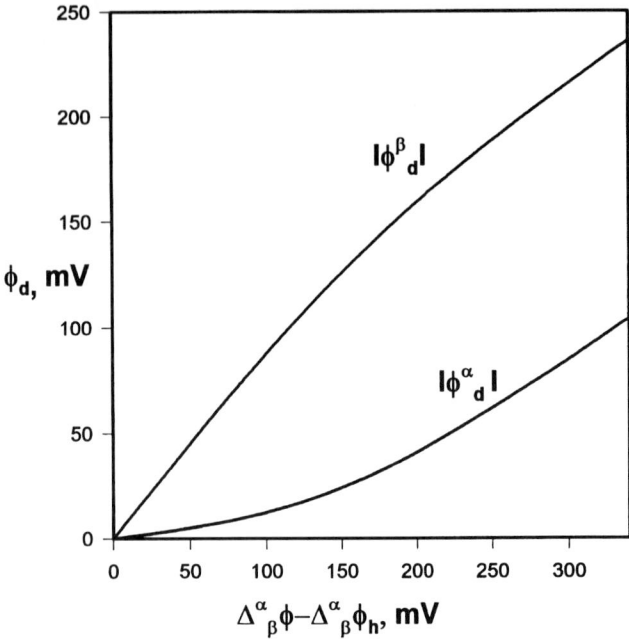

Figure 7.4. Dependence of diffuse layer potential drop from the sum of the potential drops in both diffuse layers.

surface excess Γ_i of an ion i can be divided into two components in the inner layer Γ_i^h and in the diffuse layer Γ_i^d:

$$\Gamma_i = \Gamma_i^h + \Gamma_i^d. \tag{7.12}$$

The charge of each contacting phases is equal to

$$q = \sum_i z_i F \Gamma_i. \tag{7.13}$$

For a binary electrolyte the ion excess in the diffuse layer can be written as

$$\Gamma_i^d = \frac{2c^0}{\kappa} \left[\exp\left(-\frac{z_i F \phi_d}{2RT} \right) - 1 \right] = \frac{2c^0}{\kappa^\alpha} \left[\sqrt{1 - \frac{q^2}{8\varepsilon_0 \varepsilon RT c^0}} + \frac{z_i q}{8\varepsilon_0 \varepsilon RT c^0} - 1 \right], \tag{7.14}$$

where c^0 is the electrolyte concentration in the bulk phase.

The interfacial capacitance C consists of the capacitance of the compact layer C_h and the capacitance of two diffuse layers C_d^α and C_d^β. Differentiating equation (7.13) with respect to charge q^α of phase α, and assuming that the drop of the potential in the compact layer $\Delta_\beta^\alpha \phi_h$ does not depend on an electrolyte concentration, we have

$$\frac{1}{C(q,c)} = \frac{1}{C_h(q)} + \frac{1}{C_d^\alpha(q,c)} + \frac{1}{C_d^\beta(q,c)}. \tag{7.15}$$

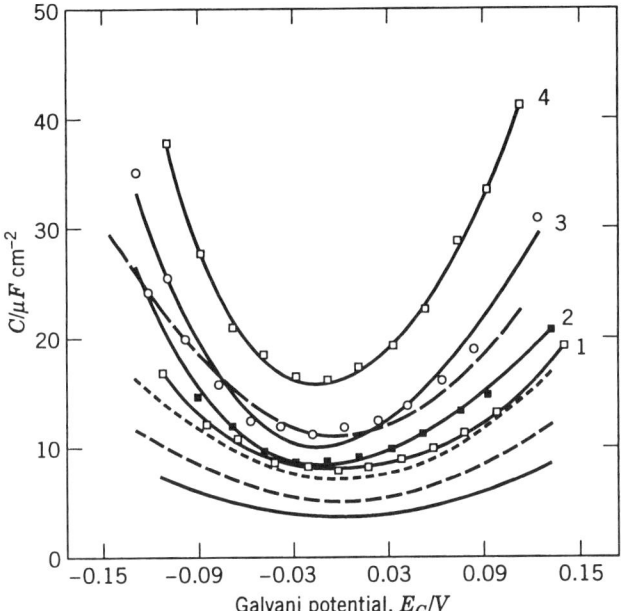

Figure 7.5. Experimental (E) and theoretical (Gouy–Chapman, GC) potential versus capacitance curves at the water–1,2-dichloroethane interface. KCl concentrations: 5 mM (1), E; long-dashed curve—GC; 10 mM (2), E; short-dashed curve—GC; 20 mM (3), E; dotted curve—GC; 50 mM (4), E (Yufei *et al.*, 1991). Reproduced by permission of The Chemical Society.

Diffuse layer capacitance depends on potential and electrolyte concentration (Fig. 7.5):

$$C_d = F\sqrt{\frac{2\varepsilon_0\varepsilon c}{RT}}\cosh\frac{F\phi_d}{2RT} = \frac{F}{2RT}\sqrt{8\varepsilon_0\varepsilon RTc + q^2}. \tag{7.16}$$

In classical electrochemistry the capacitance usually depends on surface charge q (Fig. 7.6).

7.2. POTENTIALS OF ZERO FREE CHARGE AND ZERO THERMODYNAMIC CHARGE

A comprehensive thermodynamic theory of the electrocapillary phenomena at polarizable and non-polarizable liquid interfaces was developed by Markin and Volkov (1988a,b,c,d, 1989a,b,c, 1990a,b,c) using Hansen's method (Hansen, 1962). Gibbs and Guggenheim methods were used by Blank (1966), Dupeyrat and Michel

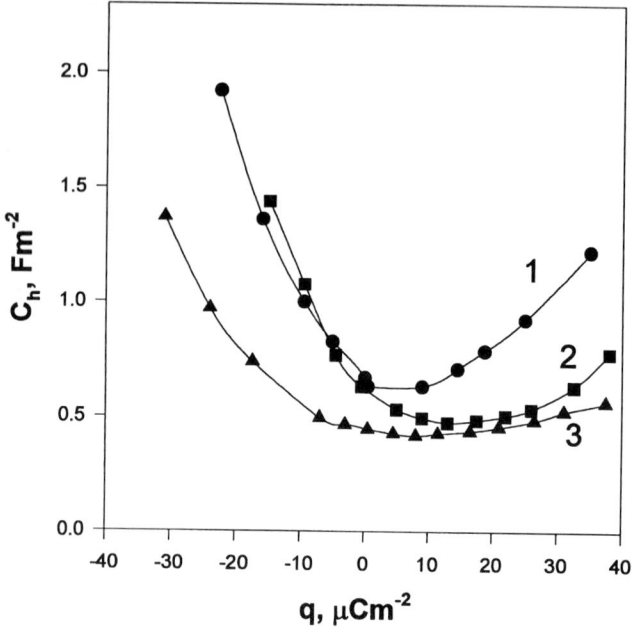

Figure 7.6. Capacitance of the inner layer C_h at the water–nitrobenzene interface as a function of the surface charge density q on the aqueous side of the interface evaluated from the experimental data using the non-iterative HCA results for the diffuse double layer (Samec *et al.*, 1985). Nitrobenzene phase: 0.1 M TBATPB and aqueous phase: (1) 0.1 M NaBr, (2) 0.1 M LiCl, (3) 0.05 M MgSO$_4$.

(1969), Dupeyrat and Nakache (1980), Gavach *et al.* (1977), Girault and Schiffrin (1983), Grahame and Whitney (1942), Joos and Vanden Bogaert (1976), Kakiuchi and Senda (1983b), Kandidow (1911, 1913), Mohilner (1962, 1966), Rusanov (1962, 1967), Silva (1984), Sparnay (1958), Verburgh and Joos (1980), Volkov (1987b; 1996), Watanabe (1984) and Watanabe *et al.* (1967a,b; 1968).

The impedance or electrocapillary properties of the interface between two immiscible electrolyte solutions can be measured with the following cell:

$$
\begin{array}{ccccc}
& 1 & 2 & 3 & \\
\text{RE1} & \left| \begin{array}{c} \text{H}_2\text{O } (\alpha) \\ \text{M}^+\text{Cl}^- \end{array} \right. & \left| \begin{array}{c} \text{A}^+\text{B}^- (\beta) \\ \text{oil} \end{array} \right. & \left| \begin{array}{c} \text{A}^+\text{Cl}^- (\alpha) \\ \text{H}_2\text{O} \end{array} \right. & \left| \; \text{RE2} \right. \\
& a_1 & a_2 & a_3 & \\
\Delta_\alpha^{Ag}\phi_{RE1} & \Delta_\beta^\alpha\phi & \Delta\phi_D & -\Delta_\alpha^{Ag}\phi_{RE2} &
\end{array}
\qquad (7.I)
$$

Silver/silver chloride electrodes are commonly used as the reference electrodes (RE), water is used as phase α, and nitrobenzene or 1,2-dichloroethane as phase β. The interface between phases 1 and 2 is the polarized oil–water interface serving as a working electrode (interface). The common cation A^+ is usually the terabutylammonium ion, and B^- is tetraphenylborate or dicarbollylcobaltate.

The potential difference measured in the cell (7.I) consists of the sum of two interfacial potential drops:

$$-E = \Delta_\beta^\alpha \phi + \Delta_\alpha^{Ag} \phi_{RE1} - \Delta_\alpha^{Ag} \phi_{RE2} + \Delta\phi_D. \tag{7.17}$$

$\Delta\phi_D$ is the Nernst–Donnan potential difference between the non-aqueous fraction and the aqueous solution of RE2 with the common ion A^+:

$$\Delta\phi_D = -\Delta_\beta^\alpha \phi_{A^+}^0 + \frac{RT}{F} \ln \frac{a_2}{a_3}, \tag{7.18}$$

and $\Delta_\beta^\alpha \phi_{A^+}^0$ is the standard potential difference for A^+ known from literature data (Marcus, 1996; Markin and Volkov, 1988f, 1989b,e, 1990f) or calculated theoretically (Kornyshev and Volkov, 1984; Markin and Volkov, 1987a,b, 1989f; Volkov and Kornyshev, 1985).

By suitable selection of the Cl^- concentrations and ionic strengths of the electrolytes into which Ag/AgCl electrodes are immersed, we obtain

$$\Delta_\alpha^{Ag} \phi_{RE1} - \Delta_\alpha^{Ag} \phi_{RE2} = \frac{RT}{F} \ln \frac{a_1}{a_3}. \tag{7.19}$$

Substituting (7.18) and (7.19) into (7.17) we obtain

$$d\Delta_\beta^\alpha \phi = -dE - d\Delta_\beta^\alpha \phi_D + \frac{RT}{F} d \ln \frac{a_3}{a_1}. \tag{7.20}$$

With equation (7.20) the electrocapillarity equation at the flat oil–water interface can be written as follows:

$$d\gamma = -\sum_i \Gamma_i d\mu_i - Q_\alpha \, dE + Q_\alpha \frac{RT}{F} d \ln \frac{a_2 a_1}{a_3^2} \tag{7.21}$$

where Q is the charge which must be supplied to each side (α or β, correspondingly) of the polarizable interface when expanding it by a unit area in order to maintain the potential difference between the phases. At the ideally polarizable interface, the ions of each group are separated by the Koenig barrier, which rules out a possible overlap between the groups (Fig. 5.4(b)). In this case the free charges q_α and q_β of the phases are equal to the thermodynamic charges: $q_\alpha = Q_\alpha = -q_\beta = -Q_\beta$. Historically the concept of the interface charge arose as a structural concept to describe the spatial separation of charges during formation of the electric double layer at the interface. It was first formulated by Helmholtz, who described the double layer as two layers of charges, equal in size and opposite in sign, located on both sides of a certain plane.

The charge on each of the interfacial capacitors came to be called the free electrode charge q. It should be stressed that the free charge of the electrode is a structural rather than thermodynamic concept, since it is based on a physical model. Expressed in these terms, the concept of the electrode charge appears to be simple and sufficiently well defined. The simple capacitor model underlying it, however, does not take into account all possible details of the electric double-layer structure. For this reason various attempts to describe complex electrochemical phenomena in different systems by free electrode charge met with fundamental difficulties. The best known example is adsorption of ions from solution on an electrode surface, in which case it is not clear whether their charges are on the metal or in the liquid side of the capacitor.

The equation of electrocapillarity (7.21) for ITIES resembles that for a mercury–water interface. The difference resides in the third term in equation (7.21), which arises from the contribution of the Nernst–Donnan potential in the diffusion bridge (Volkov, 1987b). For the mercury–water interface the difference arises from the chemical potential of the potential-determining ion of the reference electrode (Volkov, 1996).

Potentials of zero charge of the interface can be found reliably by the same independent methods (Frumkin, 1979; Markin and Volkov, 1988a,e, 1989a; Volkov, 1987b, 1996) that are used at the metal–water interface. These include finding the differential capacitance minimum of the electric double layer (Geblewicz *et al.*, 1984; Girault and Schiffrin, 1984a,b; Hajkova *et al.*, 1983; Homolka *et al.*, 1983; Kakiuchi and Senda, 1983a,c; Osakai *et al.*, 1984; Paleska *et al.*, 1990a,b; Samec *et al.*, 1981, 1984a,b; Volkov, 1996), from electrocapillary curves (Gros *et al.*, 1978; Kakiuchi and Senda, 1983a,c; Koczorowski *et al.*, 1987; Osakai *et al.*, 1984; Paleska *et al.*, 1990a,b; Volkov, 1987b), with a flowing-electrolyte electrode (Girault and Schiffrin, 1989), with the vibrating boundary method (Koczorowski *et al.*, 1987), with radiotracers (Alent'ev and Filatov, 1991), or by measuring the second-harmonic generation (Bell *et al.*, 1992; Brevet and Girault, 1996; Burbage and Wirth, 1992a,b; Conboy, 1996; Higgins and Corn, 1993). The promise of using nonlinear optical techniques to elucidate fundamental electrochemical phenomena at liquid interfaces has come to fruition in the past decade with the development of optical second-harmonic generation (SHG) as an extremely powerful analytical method (Conboy, 1996). SHG has the ability to examine the interface without interference from the surrounding bulk liquids. Under the electric-dipole approximation, SHG is forbidden in the bulk of centrosymmetric media such as liquids, but at interfaces this restriction is lifted (Bell *et al.*, 1992; Brevet and Girault, 1996; Burbage and Wirth, 1992a,b; Conboy, 1996; Higgins and Corn, 1993).

Potentials of the zero free charge at nitrobenzene–water and 1,2-dichloroethane–water interfaces, obtained from the differential capacitance minimum of the electric double layer in solutions of a surface-inactive electrolyte, do not necessarily correspond to the potentials of thermodynamic zero charge (Figs. 7.7 and 7.8, Table 7.2). They can depend on electrolyte concentration when the capacitance of the compact layer is affected by surface charge as a result of non-linear double-layer properties.

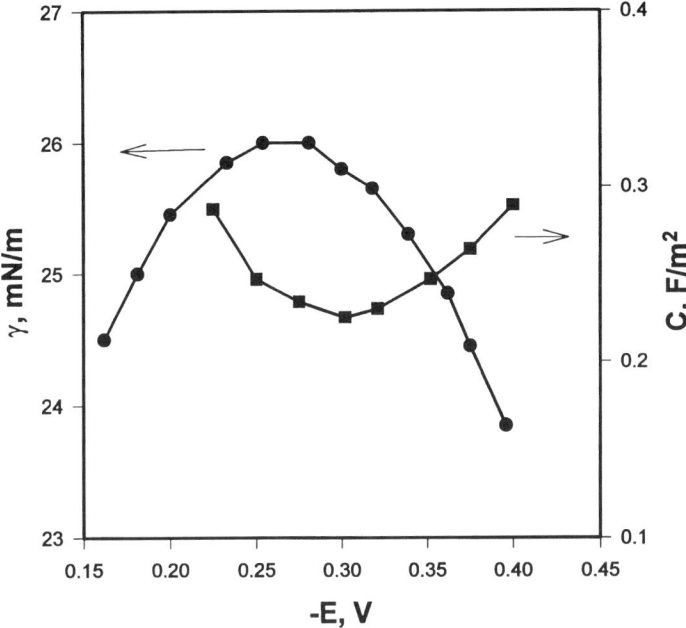

Figure 7.7. Interfacial tension and differential capacity as functions of applied potential difference. The medium: 0.1 M LiCl in water, 0.1 M tetrabutylammonium tetraphenylborate in the nitrobenzene at 25 °C. Experimental points are taken from Kakiuchi and Senda (1983) and Osakai *et al.* (1984).

The potential difference between two immiscible electrolyte solutions can be written as the sum:

$$\Delta_\beta^\alpha \phi = \Delta_\beta^\alpha \phi_h + \phi_d^\alpha + \phi_d^\beta, \tag{7.22}$$

where $\Delta_\beta^\alpha \phi$ and ϕ_d are the potential drops across the compact and diffuse layers, respectively. In the linear approximation and in the absence of specific adsorption, the electric double layer is equivalent to three capacitors in series:

$$\frac{d\Delta_\beta^\alpha \phi}{dq} = \frac{d\Delta_\beta^\alpha \phi_h}{dq} + \frac{d\phi_d^\alpha}{dq} + \frac{d\phi_d^\beta}{dq}. \tag{7.23}$$

One can use the following approximation when modeling the electric double layer:

$$\frac{d^2 \Delta_\beta^\alpha \phi}{dq^2} = \frac{d^2 \Delta_\beta^\alpha \phi_h}{dq^2} + \frac{d^2 \phi_d^\alpha}{dq^2} + \frac{d^2 \phi_d^\beta}{dq^2} \tag{7.24}$$

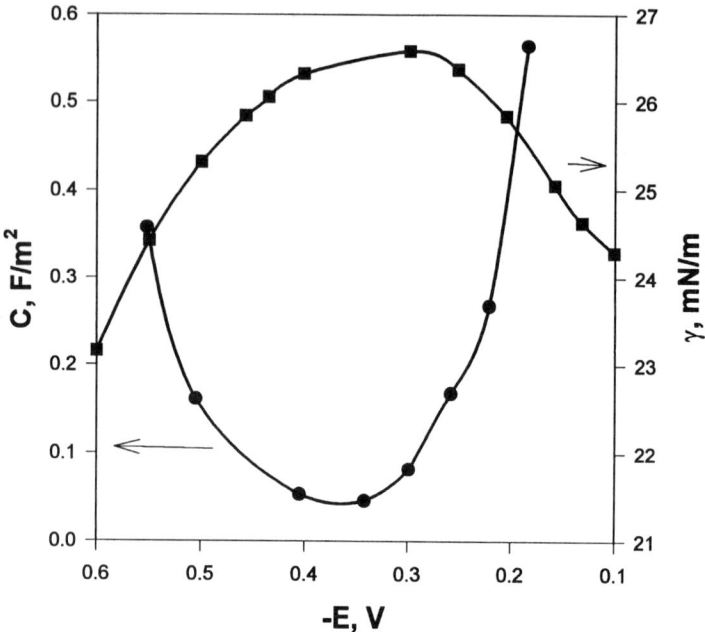

Figure 7.8. Interfacial tension and differential capacity as functions of applied potential difference. The medium: 0.01 M KCl in water, 0.01 M tetrabutylammonium tetraphenylborate in the 1,2-dichloroethane. Experimental points are taken from Girault and Schiffrin (1984a,b,c).

where

$$\Delta_\beta^\alpha \phi'' = -\frac{C'_\phi}{C^3}; \qquad \Delta_\beta^\alpha \phi''_h = -\frac{C'_{h\phi_h}}{C^3_h}; \qquad \phi''_\alpha = -\frac{C^{\alpha'}_{d\phi_\alpha}}{(C^\alpha_d)^3}; \qquad \phi''_\beta = -\frac{C^{\beta'}_{d\phi_\beta}}{(C^\beta_d)^3}. \quad (7.25)$$

We find the position of the minimum in the C versus $\Delta_\beta^\alpha \phi$ curve as

$$\Delta_\beta^\alpha \phi'' = \Delta_\beta^\alpha \phi''_h + \phi''_{d^\alpha} + \phi''_{d^\beta} = 0. \quad (7.26)$$

It follows from the Gouy–Chapman diffuse-layer theory that if $q_\alpha = q_\beta = 0$ then $(\phi^\alpha_d)'' = (\phi^\beta_d)'' = 0$. Therefore $\Delta_\beta^\alpha \phi'' = 0$ only when $\Delta_\beta^\alpha \phi''_h = 0$, that is, when the compact layer capacitance is independent of surface charge. It has been shown (Elkin et al., 1975; Mishuk et al., 1980) that the compact layer capacitance generally depends on potential and charge, and $\Delta_\beta^\alpha \phi''_i \neq 0$. As the value of $\Delta_\beta^\alpha \phi''_i$ in the region of the potential of zero free charge increases, the value of q corresponding to the minimum in the C versus $\Delta_\beta^\alpha \phi$ curve will also increase.

For the water–nitrobenzene system, one observes a strong dependence of capacitance of the compact double layer on surface charge, even in the absence of specific adsorption (Samec et al., 1982). This leads to a dependence of the potential

Table 7.2. $\Delta_{nb}^{w}\phi_h$ at the point of zero charge at the water–nitrobenzene interface. Abbreviations: TAABr—tetraalkylammonium bromide; TAATPhB—tetraalkylammonium tetraphenylborate.

Contents				
In water	In nitrobenzene	$\Delta_{nb}^{w}\phi_h$ (mV)	Method	References
3×10^{-2} M NaBr + TAABr	10^{-2} M TAATPhB	1 ± 5	γ	Gros et al., 1978
LiCl	TBATPhB	42, 15	C	Samec et al., 1981
LiCl	TBATPhB	20, −7	γ	Kakiuchi and Senda, 1983a,c
10^{-1} M LiCl	10^{-1} M TBATPhB	58	C	Kakiuchi and Senda, 1983a,c
5×10^{-3} M MgCl$_2$	10^{-2} M TBATPhB	27	C	Homolka et al., 1983
10^{-2} M LiCl	10^{-2} M TPhAsTPhB	46	C	Homolka et al., 1983
10^{-2} M LiCl	10^{-2} M TBATPhB	32	C	Homolka et al., 1983
3×10^{-2} M NaBr + TAABr	10^{-2} M TBATPhB	0 ± 5	γ	Hajkova et al., 1983
10^{-2} M NaBr	10^{-2} M TBATPhB	0	C	Samec et al., 1984a

minimum in the capacitance curve on electrolyte concentration (Fig. 7.9). Therefore the correct determination of the zero free charge potential for the nitrobenzene–water system in the presence of a binary surface-inactive electrolyte is possible only at base-electrolyte concentrations less than 0.01 M. For the water–nitrobenzene system, this quantity $\Delta_{nb}^{w}\phi_{pzfc} = -0.29$ V (Samec et al., 1982) and for the water–1,2-dichloroethane system $\Delta_{de}^{w}\phi_{pzfc} = -0.27$ V (Hajkova et al., 1983).

For many systems the maximum of the electrocapillary curve is located in the region of ideal polarizability of the electrode, and the maximum potential corresponds to the potential of zero total (thermodynamic) as well as zero free charge. At the mercury–water interface the region of ideal polarization is a few volts when soluble mercury salts are absent (2.2 V), but at the interface between two immiscible liquids the region is about a tenth of that value. Under conditions where current flow does not significantly alter the compositions of the liquid phases and where the equilibrium at the interface is not disturbed by the current, the electrocapillary curves yield the potential of zero total charge. However, under realistic conditions where the region of ideal polarization is narrow, the maximum of the electrocapillary curve corresponds to the potential of zero total charge, rather than zero free charge. This closely resembles systems consisting of amalgams and liquid electrolyte solutions that are described by Frumkin (1979).

Now we will consider ITIES that contain salts $B_i^+ A_i^-$ ($1 \le i \le k$) in phases α and β. The potential distribution in such systems has been analyzed by Markin and Volkov (1989f, 1990f) and the dependence of interfacial tension on solution composition and potential drop can be written as follows:

$$d\gamma = -\sum_{i=1}^{2k} \Gamma_i^\alpha d\mu_i^\alpha - \sum_{i=1}^{2k} \Gamma_i^\beta d\mu_i^\beta - Q^\alpha d\Delta_\beta^\alpha \phi \qquad (7.27)$$

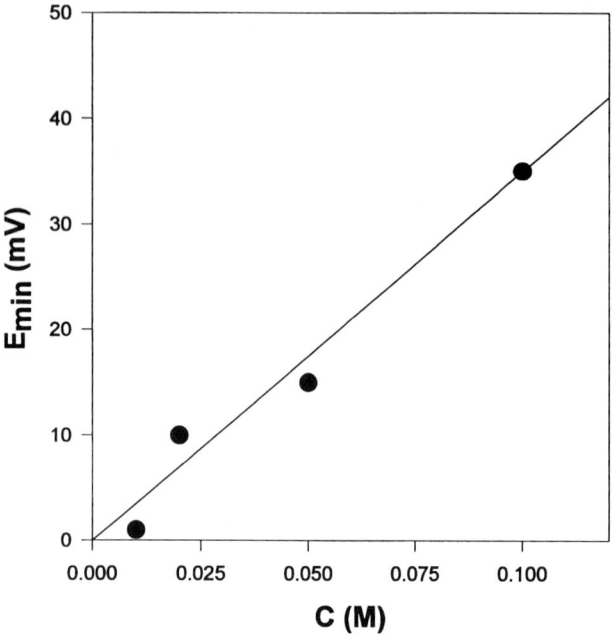

Figure 7.9. Displacement of the differential-capacity minimum on the potential axis as a function of electrolyte concentration in the water–nitrobenzene system. The medium: 0.1 M LiCl in water and 0.1 M tetrabutylammonium tetraphenylborate in nitrobenzene. Constructed from data of Samec *et al.* (1982).

where

$$Q^{\alpha} = -F \sum_{i=1}^{2k} \Gamma_i^{\beta} Z_i. \tag{7.28}$$

Suppose that ion j is the potential-determining ion. From the condition of phase equilibrium

$$d\bar{\mu}_j^{\alpha} = d\bar{\mu}_j^{\beta}, \tag{7.29}$$

or

$$d\Delta_{\beta}^{\alpha}\phi = \frac{d\mu_j^{\beta} - d\mu_j^{\alpha}}{Z_j F}. \tag{7.30}$$

Then

$$d\gamma = -\sum_{i=1}^{2k} \Gamma_i^{\alpha}\left(d\mu_i^{\alpha} + \frac{Z_i}{Z_j}d\mu_j^{\alpha}\right) - \sum_{i=1}^{2k}\left(\Gamma_i^{\beta}\,d\mu_i^{\beta} + \Gamma_i^{\alpha}\frac{Z_i}{Z_j}d\mu_j^{\beta}\right). \tag{7.31}$$

This expression had been discussed by Dupeyrat and Nakache (1980), and the assumption was made that interfacial tension in this case depends only on the composition of the phases. However, γ also depends on the potential drop:

$$
d\gamma = -\sum_{\substack{i=1 \\ i \neq j}}^{2k} \Gamma_i^\alpha \, d\mu_i^\alpha - \sum_{\substack{i=1 \\ i \neq j}}^{2k} \Gamma_i^\beta \, d\mu_i^\beta - \Gamma_j^\alpha \, d\mu_j^\alpha - \Gamma_j^\beta \, d\mu_j^\beta - q^\alpha \, d\Delta_\beta^\alpha \phi + \Gamma_j^\alpha \, d\mu_j^\beta - \Gamma_j^\alpha \, d\mu_j^\beta
$$

$$
\hspace{6cm} (7.32)
$$

$$
= -\sum_{\substack{i=1 \\ i \neq j}}^{2k} \Gamma_i^\beta \, d\mu_i^\beta - (q^\alpha - \Gamma_j^\alpha Z_j F) \, d\Delta_\beta^\alpha \phi - (\Gamma_j^\alpha + \Gamma_j^\beta) \, d\mu_j^\beta .
$$

If the polarized oil–water system contains n ionic impurities that are capable of reversibly crossing the interface, then

$$
\left(\frac{\partial \gamma}{\partial \Delta_\beta^\alpha \phi} \right)_{\mu_i, T, i \neq j} = -q^\alpha + \sum_{j=1}^{n} F Z_j \Gamma_j^\alpha - \sum_{j=1}^{n} (\Gamma_j^\alpha + \Gamma_j^\beta) \left(\frac{\partial \mu_j^\beta}{\partial \Delta_\beta^\alpha \phi} \right)_T = -q^\alpha + \Delta q. \quad (7.33)
$$

There are some limitations to this equation. For instance, consider the case in which a salt B_1L is poorly soluble in the other phase, and assume that $a_L \gg a_{1B}$ so that the salt B_1L keeps the chemical potential of ions B_1 in the phase β constant:

$$
\begin{array}{c|c}
\begin{array}{ll}
a_{1\alpha} & B_{1\alpha} A_{1\alpha} \\
a_{2\alpha} & B_{2\alpha} A_{2\alpha} \\
\end{array}
&
\begin{array}{ll}
B_{1\alpha} A_{1\beta} & a_{1\beta} \\
B_{2\beta} A_{2\beta} & a_{2\beta} \\
\end{array}
\\
\alpha \cdots\cdots\cdots\cdots\cdots\cdots &
\cdots\cdots\cdots\cdots\cdots\cdots\cdots\cdots\cdots\cdots \beta \\
\begin{array}{ll}
a_{k\alpha} & B_{k\alpha} A_{k\alpha} \\
\end{array}
&
\begin{array}{ll}
B_{k\beta} A_{k\beta} & a_{k\beta} \\
B_{1\beta} L & a_L \\
\end{array}
\end{array}
$$

Then

$$
\left(\frac{\partial \gamma}{\partial \Delta_\beta^\alpha \phi} \right)_{\mu_{i,i \neq j}, T} = -q^\alpha + Z_1 F \Gamma_1^\alpha = -F \sum_{i=2}^{2k+1} Z_i \Gamma_i^\alpha . \quad (7.34)
$$

It follows from equation (7.34) that at a polarizable interface where Faradaic processes are present, the potentials of zero charge obtained from measurements of differential capacitance of the electric double layer and from the position of the maximum in the electrocapillary curve may differ significantly from each other. The magnitude of the difference depends on the contribution of permeable ions to the surface charge (Volkov, 1996).

Figures 7.7 and 7.8 compare electrocapillary and differential double-layer capacitance curves measured experimentally by different authors. It is clear that at the nitrobenzene–water and dichloroethane–water interface the potential of the maximum in the electrocapillary curves (potential of zero thermodynamic charge) differs by several tens of millivolts from the potential of the minimum in the capacitance curves. The maximum of the electrocapillary curve in the water–nitrobenzene system with LiCl in the water and TBATPhB in the nitrobenzene is at -265 mV (Kakiuchi and Senda, 1983a,b), shifted away from the potential of zero free charge by 25 mV. The zero free charge differs from zero total charge by $0.85\ \mu C\ cm^{-2}$, which yields a value of adsorption for a penetrating ion of 9×10^{-12} M cm^{-2}. It was reported (Kakiuchi and Senda, 1983a,b) that the position of the maximum in the electrocapillary curve is a linear function of the logarithm of electrolyte concentration. The Esin–Markov plots thus obtained for zero total charge indicate that a diffuse layer is present, but that specific adsorption is absent, since the position of the zero free charge potential is independent of electrolyte concentration in the water–nitrobenzene system. When a correction for the $0.85\ \mu C\ cm^{-2}$ is introduced, the slope of the Esin–Markov plots for the potential of zero free charge coincides with the theoretical slope, while that for the potential of zero thermodynamic charge differs from the theoretical estimate by 3 mV.

Yufei *et al.* (1991) have measured the capacitance of the interface of immiscible electrolyte solutions for different cations: Li$^+$, Na$^+$, K$^+$, Rb$^+$, and Cs$^+$ chlorides in nitrobenzene and 1,2-dichloroethane. The position of the minimum of the (C, E) curve was shifted toward the positive potential range in the order Li $+$ $<$ Na$^+$ $<$ K$^+$ $<$ Rb$^+$ $<$ Cs$^+$, which corresponds to the order of the Gibbs energies for ion transfer across the interface. Such effects can be caused by diffusion-controlled cation permeation through the interface and non-ideal polarizability.

We will examine a few particular cases of equation (7.34). For $\mu_j^\beta = \text{constant}$, one has

$$\Delta q = \sum_{j=1}^{n} \Gamma_j^\alpha Z_j F. \tag{7.35}$$

For $\mu_j^\alpha = \text{constant}$, one has

$$\Delta q = \sum_{j=1}^{n} \Gamma_j^\beta Z_j F. \tag{7.36}$$

Maintaining constant chemical potentials of p ions in phase α and of $(n - p)$ ions in phase β,

$$\Delta q = \sum_{j=1}^{n-p} \Gamma_j^\alpha Z_j F - \sum_{m=1}^{p} \Gamma_m^\beta Z_m F. \tag{7.37}$$

Equation (7.34) shows that at the non-ideally polarizable interface between two immiscible electrolyte solutions a change in surface state can be caused by a change in applied potential difference without changes in phase composition other than the penetrating ion (Volkov, 1996).

7.3 MEASURING THE CAPACITANCE OF THE ELECTRICAL DOUBLE LAYER

The structure of the electric double layer at ITIES has been investigated using such common electrochemical methods of capacitance measurements as impedance, galvanostatic, and potentiodynamic techniques. More recently it became possible to measure electric double-layer capacitance at the interface between two immiscible electrolyte solutions using a four-electrode potentiostat (Figaszewski, 1982; Figaszewski *et al.*, 1982; Lhotsky *et al.*, 1996; Marecek *et al.*, 1995a,b; Samec, 1988; Samec *et al.*, 1981, 1984a,b; VanderNoot *et al.*, 1990) or two-electrode potentiostat (Yufei *et al.* (1991)). Commonly studied systems are nitrobenzene–water and 1,2-dichloroethane–water, in which the organic phase has a relatively high dielectric permittivity. These interfaces provide relatively simple models for biological membranes that can be used to investigate complicated enzymatic and photochemical reactions occurring in mitochondria, bacteria, and chloroplasts (Boguslavsky and Volkov, 1987; Deamer and Volkov, 1995, 1996; Volkov, 1984a,b, 1986a,b,c; Volkov and Deamer, 1996).

The EMF of the cell

$$Ag\,|\,AgCl\,|\,LiCl,\ H_2O\,|\,TBADCC,\ oil\,|\,TBACl,\ H_2O\,|\,AgCl\,|\,Ag \qquad (7.II)$$

is the sum of potential drops at all interfaces. The concentrations of the electrolytes used are the same in all phases, so that

$$E = \Delta_w^{org}\phi - \Delta_w^{org}\phi_{TBA^+}, \qquad (7.38)$$

where $\Delta_w^{org}\phi_{TBA^+} = -0.225\,V$ is the standard potential of TBA^+ transfer from 1,2-dichloroethane into water.

In typical experiments, cyclic current–potential curves are usually determined before impedance measurements in order to find the potential range where the contribution of Faradaic impedance is low. The electric double-layer capacitance is measured in the potential window between extremes in the cyclic voltamgram. Over this range of potentials, the water–dichloroethane interface has properties close to those of an ideally polarizable electrode. The equivalent circuit of the interface can be represented as shown in Fig. 7.10.

The equipment for recording current–voltage curves and impedance is illustrated in Figs. 5.2 and 7.12. Different types of electrochemical cells for measuring impedance have been described by Samec (1988). When current–potential curves are measured, the generator produces a triangular voltage (5 mV s^{-1}) which is fed to the potentiostat and the X-input of the recorder. The potentiostat senses the current through the cell in such a way that a potential difference equal to that produced by the generator is maintained between reference electrodes.

Impedance Z and phase shift δ can be calculated from the resistive and capacitive

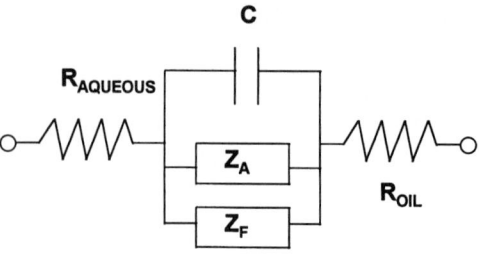

Figure 7.10. Electrical equivalent circuit for an ITIES. Z_A is the adsorption impedance, Z_F is the Faradaic impedance, C is the capacitance of the electric double layer, R is the solution resistance.

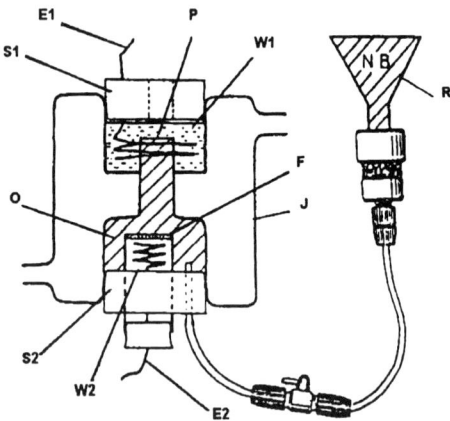

Figure 7.11. Cell design for impedance measurement: W1—aqueous solution; W2—aqueous solution of tetrabutylammonium chloride; O—nitrobenzene solution; E1 and E2—Ag/AgCl electrodes; F—glass frit; S1 and S2—silicone rubber stoppers; J—water jacket; R— nitrobenzene solution reservoir; P—polarized nitrobenzene/water interface (Kondo *et al.*, 1994). Reproduced by permission of Elsevier Sequoia S.A.

component of the sinusoidal current flowing across the interface, as follows:

$$|Z| = \frac{V_0}{\sqrt{J^2(0°) + J^2(90°)}}, \tag{7.39}$$

$$\delta = \tan^{-1} \frac{J(90°)}{J(0°)}, \tag{7.40}$$

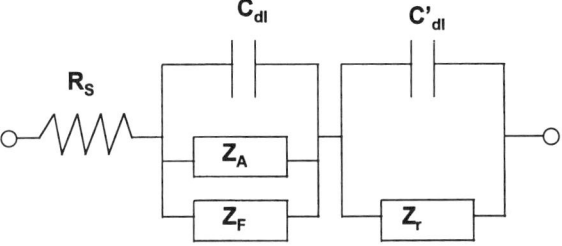

Figure 7.12. The equivalent circuit of a polarized oil–water interface in series with a non-polarized oil–water interface in the presence of ion transfer by a depolarizer ion (Kakiuchi and Senda, 1991). Abbreviations: R_s is solution resistance, Z_A, Z_t, and Z_t are the ion-transfer impedances for supporting electrolyte ions, a potential determining ion, and a depolarizer ion, respectively.

where V_0 is the amplitude of the applied sinusoidal voltage. Assuming that the Faradaic impedance coincides with the Warburg impedance Z_W, one has

$$Z' = Z \cos \delta = R_s + \frac{Z_c X}{(X+1)^2 + 1}, \tag{7.41}$$

$$Z'' = Z \sin \delta = \frac{Z_c X(X+1)}{(X+1)^2 + 1}, \tag{7.42}$$

where Z_c is the capacitive component of impedance:

$$Z_c = (\omega C)^{-1}. \tag{7.43}$$

C is the electric double-layer capacitance, ω is the angular frequency, and X is given by

$$X = \frac{Z_W}{Z_c} \sqrt{2} = 2\sigma c \omega^{1/2}. \tag{7.44}$$

The parameter σ is defined by the following expression:

$$\sigma = 4 \frac{RT}{F^2} S c_i^0 (dl) \sqrt{2D_i(dl)} \cosh^2 \frac{F(E - E'_{1/2}(i))}{2RT}, \tag{7.45}$$

where c_i^0 is the bulk concentration, $D_i(dl)$ is the diffusion coefficient of the ion being transferred, S is the interfacial area, and $E'_{1/2}$ is the reversible polarographic half-wave potential.

The use of a four-electrode potentiostat may cause problems in evaluating capacitance, and analysis of the frequency dependence must be carried out (VanderNoot and Schiffrin, 1990; VanderNoot et al., 1990; Watts and VanderNoot,

1996; Wiles *et al.*, 1990). The origin of experimental high-frequency artifacts in measuring of the a.c. impedance of ITIES was discussed by Wiles *et al.* (1990) and by Watts and VanderNoot (1996). The primary source of high-frequency artifact is the high resistance of the Luggin capillary in the organic phase, and an Ag/AgTPhB reference electrode was recommended to eliminate these problems.

Kakiuchi and Senda (1991) calculated the polarizability of a polarized oil–water interface commonly used as a working electrode, and also the non-polarizability of a non-polarized interface between phases 2 and 3 in the chain (7.I) used as a reference oil–water interface. Interfacial transport of ions due to the non-ideal polarizability of the working interface contributes to the total admittance of the interface (see the equivalent circuit in Fig. 7.12). The reference oil–water interface is not ideally reversible and can contribute to the measured admittance if the area of the reference oil–water interface is smaller than that of the working interface. For this reason Kakiuchi and Senda (1991) recommended a reference oil–water interface with a large effective area. They stated that it is important to check the frequency dependence of the experimentally observed double-layer capacitance values and to check the frequency dependence and depolarizer-concentration dependence of the kinetic parameters to estimate the degree of interference associated with the non-ideal behavior of working and reference interfaces.

Figure 7.5 shows experimental plots of electric double-layer capacitance at the dichloroethane–water interface, compared with calculated values from the Gouy–Chapman equations. It is clear that diffuse-layer theory does not fit the experimental data for KCl concentrations of 0.005 M, where the theoretical curve for capacitance as a function of applied potential is 40% below the experimental curve.

The galvanostatic and impedance techniques can only be used for salt concentrations of 0.001 to 0.1 M, so it is uncertain how much the capacitance of the compact layer depends on electrolyte concentration at higher ranges. The poten-tiodynamic method makes it possible to study capacitance at lower concentrations, because it does not require determination of the phase shift between oscillations of current and voltage that is necessary for impedance measurements. Instead the potentiodynamic method determines the current response to triangular pulses of small amplitude in an oscillogram. If the interface is equivalent to a capacitor, the entire current $I(t)$ crossing it is used to charge the interfacial capacitance:

$$I_C(t) = CA \frac{dU}{dt}, \tag{7.46}$$

where A is the surface area, U is voltage and t is the time. Due to linearity and periodicity of voltage variation with time it follows from (7.46) that

$$I_c(t) = CAv, \qquad n\tau < t \le (n + 1/2)\tau, \tag{7.47}$$

$$I_c(t) = -CAv, \qquad (n + 1/2)\tau < t \le (n + 1)\tau, \tag{7.48}$$

where v is the absolute magnitude of the rate of voltage variation in a pulse, τ is the period of voltage oscillations and n is the number of periods.

In the simplest case the dependence $I(t)$ will have the form of rectangular pulses and the capacitance is readily determined from equation (7.49):

$$C = \frac{1}{2Av}(I_+ - I_-),$$ (7.49)

where $I_+ = I(n + 1/2)\tau$ and $I_- = I(n\tau)$ are the absolute values of maximal and minimal currents in the oscillogram.

7.4. PARSONS–ZOBEL DEPENDENCIES

The classical test of Gouy–Chapman theory is the Parsons–Zobel (1965) or Grahame (1947) dependency. Grahame (1947) developed a phenomenological theory according to which the reverse capacitance is the reciprocal diffuse layer capacitance in the Gouy–Chapman approximation together with a contribution that depends on the interfacial charge but not on electrolyte concentration (see equation (7.15)). According to MVN theory the dependence of C^{-1} on C_d^{-1} must always be a straight line with unit slope that passes through the ordinate axis at a value equal to the reverse capacitance of the compact layer C_c^{-1}. A small deviation of the slope from unity is possible if an electrolyte does not dissociate completely. Using MVN theory and knowing the dissociation constant, it is easy to calculate the Parsons–Zobel dependency from the deviation of the slope.

Figures 7.13 and 7.14 show dependencies of the electric double-layer capacitance on diffuse layer capacitance in Parsons–Zobel coordinates for water–nitrobenzene and water–1,2-dichloroethane systems at the potential of zero free charge. At metal electrodes the slope of a Parsons–Zobel plot can be different from unity in some systems (Kornyshev and Ulstrup, 1985a,b; Partenskii and Feldman, 1983, 1988; Vorotyntsev and Kornyshev, 1979, 1993). To explain these experimental data by Gouy–Chapman theory we could use the equation

$$C_d^{-1} = \varepsilon_0 \varepsilon \kappa,$$ (7.50)

but this would require an absurd value for the dielectric permittivity of aqueous solution that is larger than that of water. The paradox was resolved using non-local electrostatics (Vorotyntsev and Kornyshev, 1993). It was shown that

$$C_d^{-1} = \varepsilon_0 \varepsilon \kappa_{eff}, \qquad \kappa_{eff} = \kappa[1 - M(\kappa \Lambda)^2].$$ (7.51)

Here the effective Debye screening length κ_{eff} depends on a radius of correlation L and a positive or negative parameter M, caused by the reflection of polarization waves. However, Parsons–Zobel dependencies at ITIES are noticeably different from dependencies at the metal–water interface (Figs. 7.13 and 7.14). At high electrolyte concentrations MVN theory satisfactorily describes the experimental data, although this theory was developed only for dilute electrolyte solutions. The complete agreement with experiment at high electrolyte concentrations was obtained with the

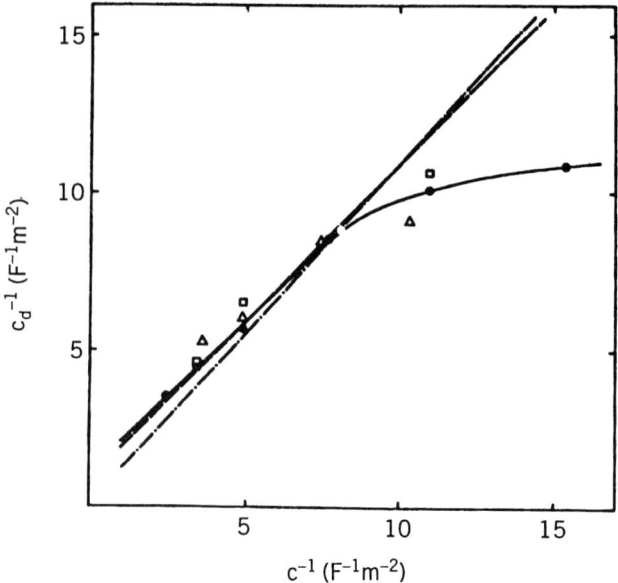

Figure 7.13. Parsons–Zobel dependencies for the nitrobenzene–water interface. Solid line—experimental data from Samec *et al.* (1985); dashed-dotted line—calculation according to Gouy–Chapman theory; dashed line—calculation using mean-spherical approximation. Reproduced by permission of Elsevier Sequoia S.A.

mean-spherical approximation (Fig. 7.13). Contrary to expectations, at low electrolyte concentrations the data show a very unusual deviation of capacitance (Figs. 7.13 and 7.14). If the model of two condensers in series is correct

$$C^{-1} = C_h^{-1} + C_d^{-1} < C_d^{-1},\qquad(7.52)$$

the differential capacitance of the compact layer should be negative and should depend on electrolyte concentrations in dilute solutions. It is possible that the simplest equivalent circuit of ITIES, corresponding to the model of two capacitors in series, is not applicable.

A number of papers on the negative differential capacitance in an electrical double layer discussed the possibility (Attard *et al.*, 1992; Feldman *et al.*, 1985, 1987; Halley and Price, 1987; Kornyshev, 1989; Partenskii *et al.*, 1996; Partenskii and Feldman, 1983; Torrie, 1992; Torrie and Valleau, 1982; Vorotyntsev and Kornyshev, 1993). In Monte Carlo simulations of an electrical double layer with a primitive model of 0.05 M 1:2 electrolyte, Torrie (1992) found that the total potential drop across the interface decreases as the surface charge increases.

Samec *et al.* (1985) proposed a model in which ions were allowed to penetrate into a compact layer to some extent. In this case,

$$C^{-1} = C_h^{-1} + C_d^{-1} + \Delta\qquad(7.53)$$

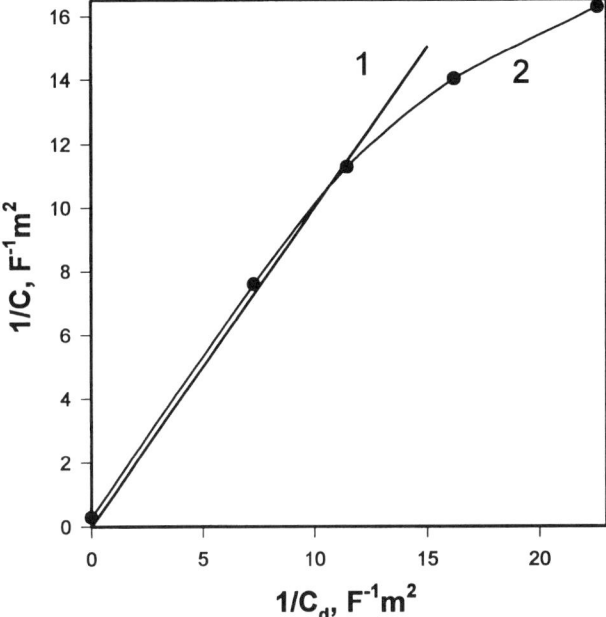

Figure 7.14. Parsons–Zobel dependencies for the 1,2-dichloroethane–water interface. Experimental points are taken from Hajkova *et al.* (1983) and Samec *et al.* (1987).

where

$$\Delta = (\varepsilon^h \varepsilon_0 \kappa^h)^{-1} \left[\frac{(1 - \eta^{o/h2}) \tanh (\kappa^h \delta)}{1 + \eta^{o/h} \tanh (\kappa^h/\delta)} - \kappa^h \delta \right] \qquad (7.54)$$

and $\eta^{o/h}$ is determined by equation (7.7). If $\kappa^h \neq 0$, the parameter Δ is negative and C^{-1} is reduced. In this case the equivalent circuit of ITIES can be presented as shown in Fig. 7.15.

7.5. POTENTIAL DISCONTINUITIES IN THE COMPACT LAYER

The structures of electric double layers at the metal–water and 1,2-dichloroethane–water interfaces are appreciably different. At the metal–water interface one has an "electronic capacitor" consisting of a negatively charged electron cloud extending about 1 Å into the aqueous solution, and a positively charged ionic metal core (Hajkova *et al.*, 1985; Vorotyntsev and Kornyshev, 1984; Vorotyntsev and Mityushev, 1991). This produces asymmetry in the capacitance curves relative to the point of zero charge. At the interface between two immiscible electrolyte solutions, the principal

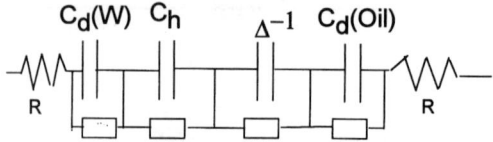

Figure 7.15. The equivalent circuit of a polarized oil–water interface with ion penetration of the inner layer.

potential drop occurs in two diffuse layers present in different phases. At potentials where the Faradaic impedance can be neglected, the base-electrolyte ions approach the interface to a distance defined by the position of the outer Helmholtz plane. A compact layer can be found between the two outer Helmholtz planes in the aqueous and organic phases, and the solvent dipoles or specifically adsorbing solutes can be localized at the inner Helmholtz planes. The potential drop across the compact layer formed by water and non-aqueous solvent dipoles is small (Table 7.3, Figs. 7.16 and 7.17).

A "diffuse" picture of the compact layer has been considered (Samec *et al.*, 1985). According to this hypothesis adsorbed ions can penetrate into the compact layer, which consists of solvent dipoles, analogous to non-localized electron gas penetration from a metal electrode to an aqueous electrolyte solution. Here a penetrating ion can act as a hydrophobic anion in the organic phase (Fig. 7.18) and potential distribution $\phi(x)$ near the point of zero charge can be calculated using the Poisson–Boltzmann equation:

$$\text{If}\quad x < x_2^w,\quad \varepsilon = \varepsilon_w: \qquad \phi'' = (\kappa_w)^2(\phi - \Delta_{org}^w \phi) \tag{7.55}$$

$$\text{If}\quad x_2^w < x < x_2^{org},\quad \varepsilon = \varepsilon_h: \qquad \phi'' = (\kappa_h)^2 \phi \tag{7.56}$$

$$\text{If}\quad x > x_2^{org},\quad \varepsilon = \varepsilon_{org}: \qquad \phi'' = (\kappa_{org})^2(\phi - \Delta_{org}^w \phi). \tag{7.57}$$

Table 7.3. Values of $\Delta_{nb}^w \phi$ corresponding to the minimum of the double-layer capacitance at the water–nitrobenzene interface in the presence of TBATPhB in nitrobenzene and NaBr (I), MgSO$_4$ (II) or LiCl (III) in water at 25 °C (Samec *et al.*, 1985).

$C(w)$ (M)*	I	II	III
0.005	11	—	—
0.010	5	3	18 (15)**
0.020	4	−8	— (25)
0.050	0	15	24 (29)
0.100	11	16	50 (45)

* Concentrations of electrolytes were equal in aqueous and organic phases if NaBr or LiCl was used. If MgSO$_4$ was used its concentration was half the electrolyte concentration in the organic phase.
** From capacitance measurements with the assumption that $\Delta_{org}^w \phi_{Bu_4N^+}^0 = -0.275$ V.

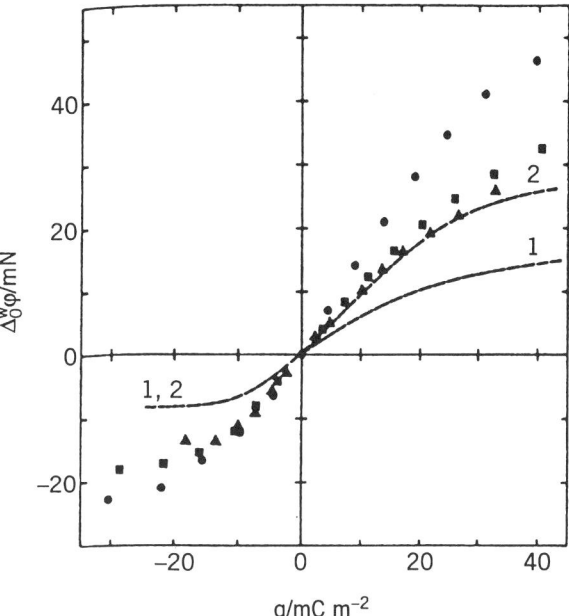

Figure 7.16. Potential difference across the inner layer at the water–nitrobenzene interface as a function of the surface charge density q^w on the aqueous side of the interface evaluated from the experimental data using non-iterative HNC results for the diffuse double layer at concentrations of NaBr in water and TBATPB in nitrobenzene: (▲) 0.02 M, (■) 0.05 M, and (●) 0.10 M. The dashed lines were obtained from experimental data using the Gouy–Chapman results for the diffuse double layer at concentrations (1) 0.05 M and (2) 0.10 M (Samec *et al.*, 1985). Reproduced by permission of Elsevier Sequoia S.A.

In each of these three areas the solution can be written as

$$\phi(x) = A_+ e^{\kappa x} + A_- e^{-\kappa x}, \tag{7.58}$$

where six coefficients A_+ and A_- can be determined from the six boundary conditions:

$$\phi(-\infty) = \Delta_{org}^w \phi$$
$$\phi(\infty) = 0$$
$$\phi(x_2^w - 0) = \phi(x_2^w + 0)$$
$$\phi(x_2^{org} - 0) = \phi(x_2^{org} + 0) \tag{7.59}$$
$$\varepsilon_w \phi'(x_2^w - 0) = \varepsilon_h \phi'(x_2^w + 0)$$
$$\varepsilon_h \phi'(x_2^{org} - 0) = \varepsilon_{org} \phi'(x_2^{org} + 0).$$

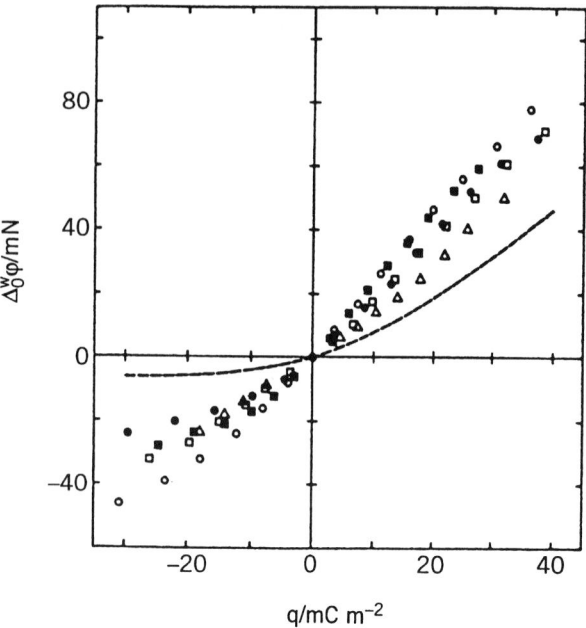

Figure 7.17. Potential difference across the compact layer at the nitrobenzene–water interface as a function of the surface charge density q^w on the aqueous side of the interface evaluated from the experimental data using the non-iterative HNC results for the diffuse double layer. Aqueous and nitrobenzene phases respectively: (■) 0.05 M LiCl and 0.05 M TBATPB; (●) 0.10 M LiCl and 0.10 M TBATPB; (△) 0.01 M MgSO$_4$ and 0.02 M TBATPB; (□) 0.025 M MgSO$_4$ and 0.05 M TBATPB; (○) 0.05 M MgSO$_4$ and 0.10 M TBATPB (Samec *et al.*, 1985). Reproduced by permission of Elsevier Sequoia S.A.

As a result the equation for the capacitance of the electric double layer is:

$$C = \frac{q^w}{\Delta_{org}^w \phi} = -\frac{(\kappa_w)^2 \, \varepsilon_w \, \varepsilon_{org}}{\Delta_{org}^w \phi} \int_{-\infty}^{x_2^w} (\phi - \Delta_{org}^w \phi) \, dx$$

$$= \kappa_w \varepsilon_w \varepsilon_{org} \tag{7.60}$$

$$\times \left[\frac{\left(1 + \dfrac{\varepsilon_h \kappa_h}{\varepsilon_{org} \kappa_{org}}\right) \exp(\kappa_h \delta) + \left(1 - \dfrac{\varepsilon_h \kappa_h}{\varepsilon_{org} \kappa_{org}}\right) \exp(-\kappa_h \delta)}{\left(1 + \dfrac{\varepsilon_w \kappa_w}{\varepsilon_h \kappa_h}\right)\left(1 + \dfrac{\varepsilon_h \kappa_h}{\varepsilon_{org} \kappa_{org}}\right) \exp(\kappa_h \delta) + \left(1 - \dfrac{\varepsilon_h \kappa_h}{\varepsilon_{org} \kappa_{org}}\right)\left(1 - \dfrac{\varepsilon_w \kappa_w}{\varepsilon_h \kappa_h}\right) \exp(-\kappa_h \delta)} \right]$$

where $\delta = x_2^{org} - x_2^w$. It should be noted that the size of a hydrophobic penetrating anion is large in relation to the compact part of the electrical double layer and it is not clear if the Poisson–Boltzmann equation can be used in this case.

Water   Oil

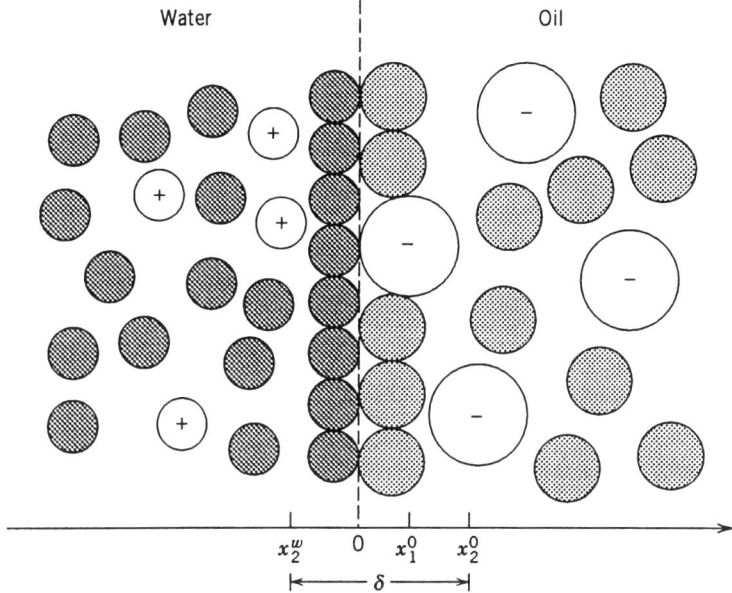

Figure 7.18. The structure of the electric double layer at ITIES according to Samec *et al.* (1985). Reproduced by permission of Elsevier Sequoia S.A.

7.6. SPECIFIC ADSORPTION: IONIC ASSOCIATION AND LIGAND BINDING

Ions can be adsorbed specifically if the main contribution to their interaction with the interface (ions, dipoles) is caused by non-Coulombic short-range forces. *Specific adsorption* cannot be explained using only the theory of diffuse double layers. Specifically adsorbed ions penetrate into the compact layer and form a compact or loose monolayer. The surface passing through the centers of specifically adsorbed ions is usually called the *inner Helmholtz plane.* If several kinds of specifically adsorbed ions are present, each ion can have its own inner Helmholtz plane.

At the polarized interface between two immiscible electrolyte solutions the specific adsorption of ions was discovered using impedance measurements (Hajkova *et al.*, 1985). Similar effects were observed earlier (Boguslavsky and Volkov, 1987; Volkov *et al.*, 1982a,b,c) at the octane–water interface using Volta-potential measurements in the chain:

$$Au \left| Air \right| Octane, FeP \left| Water, Salt\ KX \right| \begin{matrix} Saturated\ KCl \\ aqueous\ solution \end{matrix} \left| Hg_2Cl_2, Hg \right. \qquad (7.III)$$

$$\Delta_{water}^{octane}\, \phi$$

where KX indicates KCl, KBr, or KI, and FeP is the iron complex of 2,7,12,17-tetramethyl-3,13-octadecyl-8,18-bis(2-metoxycarbonylethyl)-porphyrin. If it is assumed that no specific adsorption of halogens occurs at the octane–water interface, then $\Delta_{water}^{octane}\phi$ should not depend on ionic species. It turned out, however, that $\Delta_{water}^{octane}\phi$ becomes negative in the series Cl^-, Br^-, I^- due to specific adsorption of halogens at the interface.

If dithionite is added to the aqueous phase, FeP undergoes reduction and the Volta potential shifts toward negative values. A more thorough investigation of the FeP reduction reaction showed that it occurs only at low electrolyte concentrations. Inhibition of porphyrin reduction depended not only on the ionic content of the aqueous solution, but also on the nature of the electrolyte anion. I^- anions, which are known to have a higher adsorptivity, inhibit the reaction at lower concentrations than Br^- anions do, the latter being more effective than Cl^- anions. To transfer the electron to the adsorbed porphyrin, the dithionite anion must approach it quite closely by occupying a position on the aqueous phase side. The cessation of porphyrin reduction with increasing halide concentration implies that halogen anions displace dithionite anions from the adsorption layer.

It was found (Boguslavsky and Volkov, 1987; Volkov et al., 1982a,b) while studying the mechanism of electron transfer across the interface between two immiscible liquids, that specific anion adsorption occurs at the octane–water interface in the presence of metalloporphyrins. This adsorption increases in the order of Cl^-, Br^-, I^-, and is caused by coordinative bonding of the anions as ligands of the porphyrin metal atoms.

When the iron complex of ethioporphyrin II (FeEP) or the iron complex of 2,7,12,17-tetramethyl-3,13-octadecyl-8,18-bis(2-carbometohyl)-porphyrin (FeP) is added to the system in concentrations much lower than that of the base electrolyte, the point of zero charge shifts in the negative direction and the capacitance of the electric double layer decreases. This indicates that the metalloporphyrins are adsorbed in the compact part of the electric double layer. The potential of free zero charge shifts by 50 mV for FeEP (Fig. 7.19) and by 80 mV for FeP (Fig. 7.20). The pzc shift depends on the pheophytin concentration, which is several orders of magnitude less than that of the primary electrolytes. The difference between the pzc shifts caused by the two different porphyrins is probably due to different dipole moments of the adsorbed particles. For FeEP, the adsorption spectra recorded in dry and "moist" (i.e., water-saturated) dichloroethane, coincide, but for the FeP the spectra are different. In dry non-aqueous solvent, the molecules of the porphyrin–iron complexes are present as undissociated salt. In contact with water the salt can dissociate, leading to a hydrolysis reaction that can be completed by μ-complex formation:

$$2FeP^+Cl^- \Leftrightarrow 2FeP^+OH^- \Leftrightarrow FeP—O—FeP. \qquad (7.61)$$

The absorption spectra indicate that FeEP is present as the salt of a hydroxide in dichloroethane, while FeP is present as the μ-complex FeP—O—FeP. This causes a marked long-wave shift in the absorption spectrum (Volkov et al., 1982a,b). When

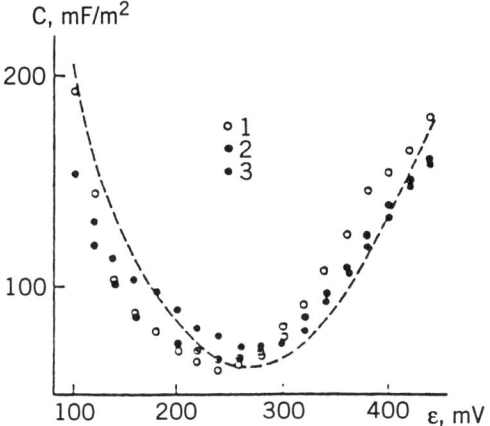

Figure 7.19. Potential dependence of electric double-layer capacity at the water–1,2-dichloroethane interface. The system contains 0.01 M LiCl in the water, 0.01 M TBADCC, and the iron complex of ethioporphyrin in the 1,2-dichloroethane, the latter in concentrations of: (1) 1 mM; (2) 0.1 mM; (3) 0. The dashed line is the theoretical relation calculated from the Gouy–Chapman equation in the absence of porphyrin (Hajkova *et al.*, 1985).

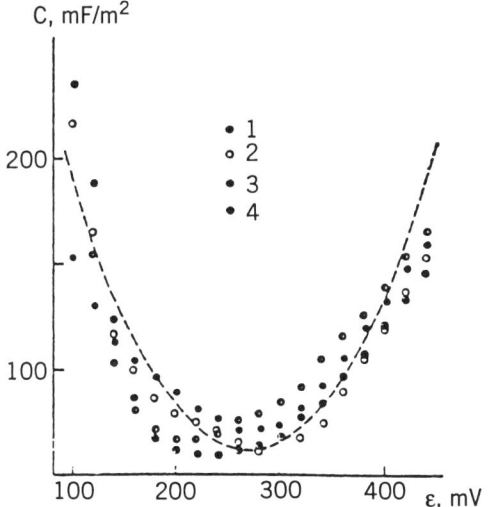

Figure 7.20. Potential dependence of electric double-layer capacitance at the water–1,2-dichloroethane interface. The system contains 0.01 M LiCl in the water, 0.01 M TBADCC and FeP in the 1,2-dichloroethane, the latter in concentrations of: (1) 0; (2) 5 μM; (3) 50 μM; (4) 100 μM. The dashed line is the theoretical relation calculated from the Gouy–Chapman equation in the absence of FeP (Hajkova *et al.*, 1985).

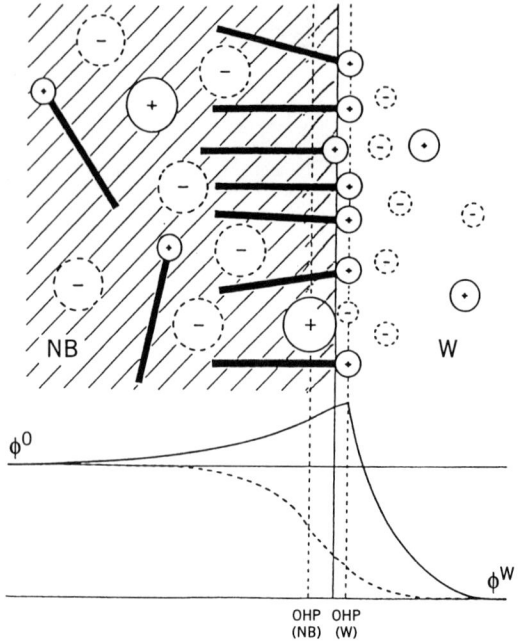

Figure 7.21. A schematic representation of the double-layer structure of the interface between nitrobenzene and aqueous solutions in the presence of the specific adsorption of hexadecyltrimethylammonium ions (Kakiuchi *et al.*, 1987). Reproduced by permission of the Chemical Society of Japan.

adsorbed at the oil–water interface, metallo-complexes of porphyrins are capable of ligand bonding to halide anions, which leads to specific adsorption in the compact layer.

Kakiuchi *et al.* (1987, 1988a,b, 1989, 1990b,c,d) studied specific ion adsorption at the polarizable nitrobenzene–water interface containing monolayers of phosphatidylcholine, phosphatidylserine, octaethylene glycol monodecyl ethers, tetraethylene glycol monodecyl ether and hexadecyltrimethylammonium (HTMA$^+$). HTMA$^+$ exhibited no specific adsorption in the potential range where the aqueous phase was positive, whereas a strong adsorption occurred in the potential range where the electric potential in the aqueous phase was negative with respect to the nitrobenzene (Fig. 7.21). Other examples of specific ion adsorption to phospholipids and surface-active compounds at the polarizable oil–water interface have been reported (Girault and Schiffrin, 1984a; Maeda *et al.*, 1987; Wandlowski *et al.*, 1987, 1988).

Specific adsorption of tetraalkylammonium salts and metallocomplexes of valinomycin was studied by Boguslavsky *et al.* (1976), Frumkin *et al.* (1971), Gugeshashvili *et al.* (1972, 1974a,b) and Krylov *et al.* (1977). Krylov (1964) proposed a theoretical model of specific adsorption of non-amphiphilic compounds

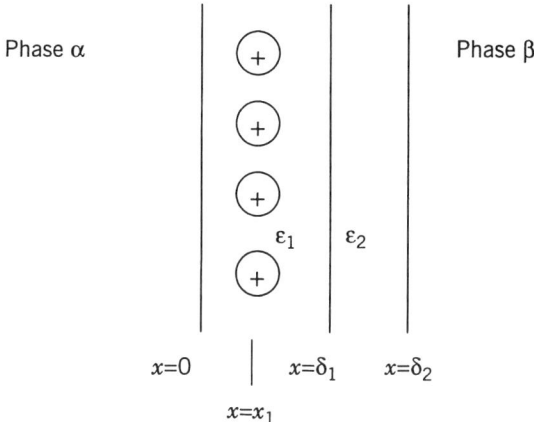

Phase α

Phase β

ε_1 ε_2

$x=0$ $x=\delta_1$ $x=\delta_2$

$x=x_1$

Figure 7.22. The double-layer structure at the oil–water interface in the presence of specific cation adsorption.

at the oil–water interface (Fig. 7.22) in which the inner Helmholtz plane is located at $x = 0$ in phase 1 and occupies a half-space $x \le \delta_1$. The inner Helmholtz plane in phase 2 is denoted by $x = \delta_2$, assuming that the effective dielectric permittivities of the Helmholtz layer are different and equal to ε_1 and ε_2, respectively. Using the Krylov (1964) method it is possible to write an equation for the electrochemical potential of the adsorbed ions:

$$\bar{\mu}_i^{\mathrm{ad}} = \mu^{0,\mathrm{ad}} + RT\ln\left[\frac{\Gamma}{\Gamma_\infty - \Gamma}\right] - \frac{2a}{RT}\frac{\Gamma}{\Gamma_\infty} + z_i F \int_0^1 \varphi_M(z_i F\mathfrak{s})\,d\mathfrak{s}. \qquad (7.62)$$

Here Γ is the Gibbs surface excess, Γ_∞ is the maximum limiting value, a is the attraction constant caused by the ion–ion short-range forces, and φ_M is the micropotential at the point where a given adsorbed ion is located. Krylov et al. (1977) obtained the following adsorption isotherm from the condition of thermodynamic equilibrium between the adsorbed layer and the volume of phase 2:

$$\frac{cf}{c_0}\exp\left(\frac{\Delta\mu^{\mathrm{ad}}}{RT}\right) = \frac{\Gamma}{\Gamma_\infty - \Gamma}\exp\left(-\frac{2a\Gamma}{\Gamma_\infty} + \frac{z_i F}{RT}\int_0^1 \varphi_M(z_i F\mathfrak{s})\,d\mathfrak{s}\right). \qquad (7.63)$$

Here f is the activity coefficient, $\Delta\mu^{\mathrm{ad}}$ is the partial standard Gibbs adsorption energy and c_0 is the total molar concentration of the solution including all the dissolved solutes and the solvent. Using the equation for micropotential, Krylov found the isotherm of specific ion adsorption at oil–water interfaces, which is similar to the Frumkin isotherm:

$$Bcf = \frac{\Theta}{1 - \Theta}\exp\left[-2a_{\mathrm{eff}}\Theta + G_i(\Theta)\right]. \qquad (7.64)$$

This equation was developed for non-amphiphilic compounds specifically adsorbed at the interface from an oil phase. According to Krylov's model this isotherm was obtained for specifically adsorbed ions in the absence of specific adsorption of counter-ions, without ion pair formation and in the complete absence of the diffuse layer. Although this theory is important from a historical perspective, there have been few if any applications to experimental data. Krylov's model predicts too thick a compact layer, which conflicts with experimental data.

The general isotherm of adsorption of amphiphilic and non-amphiphilic compounds was developed by Markin *et al.* (1992a,b,c), Markin and Volkov (1996), Volkov *et al.* (1994) and will be discussed in the next section. The structure of the oil–water interface can be deduced from measurements of adsorption of amphiphilic compounds, and the orientation of adsorbed solvent molecules at the interface can be determined. For example, the octane–water or decane–water interface is a sharp fractal structure with a monolayer of liquid hydrocarbon molecules oriented parallel to the interface (Markin and Volkov, 1996).

Another example of specific ion adsorption was discussed in terms of the formation of interfacial ion pairs between ions in the aqueous and the organic phase (Gugeshashvili, 1974; Yufei *et al.*, 1991). The contribution of specific ionic adsorption to the interfacial capacitance can be calculated using the Bjerrum theory of ion-pair formation. The results show that a phase boundary between two immiscible electrolyte solutions can be described as a mixed solvent region with varying penetration of ion pairs into it, depending on their ionic size. The capacitance increases with increasing ionic size in the order $Li^+ < Na^+ < K^+ < Rb^+ < Cs^+$ (Figs. 7.23 and 7.24). The MVN theory does not describe the experimental data (Fig. 7.24). Yufei *et al.* (1991) found that significant specific ion adsorption occurs at the interface between two immiscible electrolytes and the potential dependence of the capacitance is strongly influenced by the adsorption isotherms due to the interfacial ionic association.

When there is specific adsorption of ions dissolved in phase α the condition of electroneutrality can be written as

$$q^\beta = -q^\alpha = -(q_i^\alpha + q_d^\alpha), \tag{7.65}$$

where q_i is the charge of the inner Helmholtz plane. The division of q^α into q_i^α and q_d^α cannot be done without introducing a model of the interface.

Although capacitance of the interface can be calculated as before using the equation

$$\frac{d\Delta_\beta^\alpha \phi}{dq} = \frac{d\Delta_\beta^\alpha \phi_i}{dq} + \frac{d\phi_d^\alpha}{dq} - \frac{d\phi_d^\beta}{dq}, \tag{7.66}$$

the second term in the right side of the equation is not the diffuse layer capacitance since the charge of the diffuse layer in phase a is equal to

$$q_d^\alpha = -q_\beta - \sum_j q_i^{\alpha,j}. \tag{7.67}$$

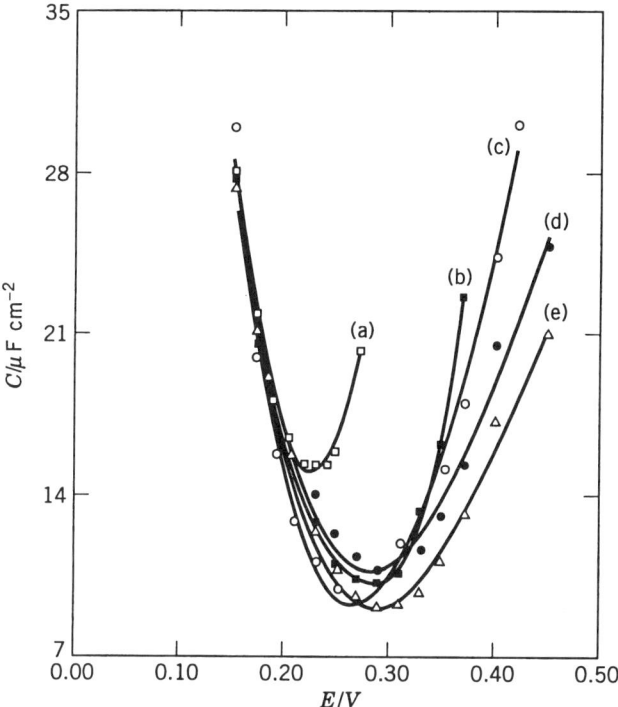

Figure 7.23. Potential–capacitance curves for the alkali-metal chlorides. The concentration of electrolytes in both phases was 10 mM: (a) CsCl; (b) RbCl; (c) KCl; (d) NaCl; (e) LiCl (Yufei *et al.*, 1991). Reproduced by permission of The Chemical Society.

The diffuse layer capacitance is equal to

$$C_{\mathrm{d}} = -\frac{\mathrm{d}q_{\alpha}}{\mathrm{d}\phi_{\mathrm{d}}^{\alpha}}, \qquad (7.68)$$

and one can write

$$C^{-1} = (C^{i})^{-1} + (C_{\mathrm{d}}^{\alpha})^{-1}\left(1 + \frac{\mathrm{d}\sum_{j} q_{i}^{\alpha,j}}{\mathrm{d}q^{\beta}}\right) + (C_{\mathrm{d}}^{\beta})^{-1}. \qquad (7.69)$$

The drop of potential in the compact layer depends on the surface charge q^{α} and the charge of the inner Helmholtz plane $q_i^{\alpha,j}$:

$$\mathrm{d}\phi_i = \left[\frac{\partial \phi_i}{\partial q^{\beta}}\right]_{q_i^{\alpha,j}} \mathrm{d}q^{\beta} + \sum \left[\frac{\partial \phi_i}{\partial q_i^{\alpha,j}}\right]_{q^{\beta}, q_i^{\alpha,k}} \mathrm{d}q_i^{\alpha,j}, \qquad (7.70)$$

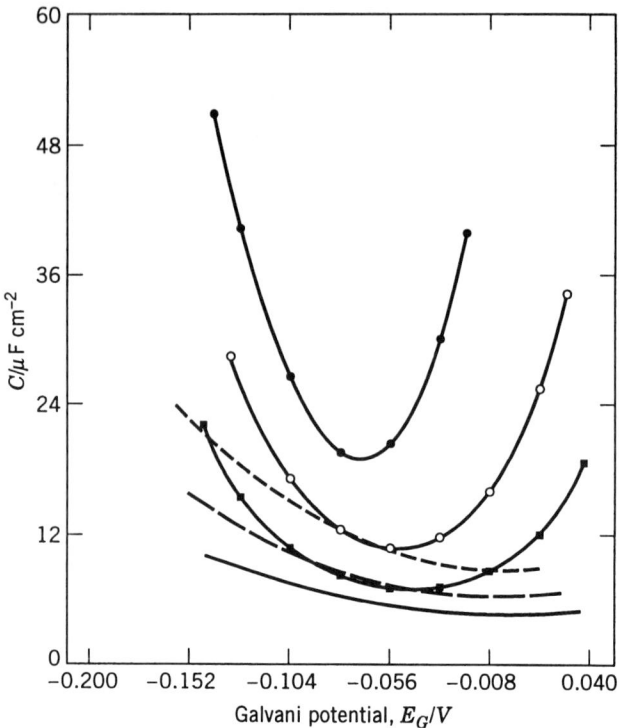

Figure 7.24. Experimental (E) and theoretical (Gouy–Chapman, GC) potential versus capacitance curves for CsCl in 1,2-dichloroethane at various concentrations: 5 mM, ■— E, (dotted curve—GC; 10 mM, ○— E, dashed curve—GC; 20 mM, ●—E, solid curve—GC (Yufei *et al.*, 1991). Reproduced by permission of The Chemical Society.

or

$$\frac{d\phi_i}{dq^\beta} \equiv \frac{1}{C_i} = \left[\frac{\partial\phi_i}{\partial q^\beta}\right]_{q_i^{\alpha,j}} + \sum \left[\frac{\partial\phi_i}{\partial q_i^{\alpha,j}}\right]_{q^\beta, q_i^{\alpha,k}} \frac{dq_i^{\alpha,j}}{dq^\beta}. \tag{7.71}$$

Here the index $q_i^{\alpha,j}$ means that all $q_i^{\alpha,j}$ are constant except one.

In the absence of specific adsorption the capacitances of the interface, compact and diffuse layers are always positive. The capacitance of the compact layer in the presence of specific adsorption can be either positive or negative.

7.7. ADSORPTION ISOTHERM AND STRUCTURE OF THE INTERFACE

Traditional models for calculation of adsorption isotherms are based on the assumption that surface active compounds at the interface can substitute for adsorbed molecules of one solvent but cannot penetrate the second phase (Adamson, 1990).

Although this approach is useful for metal–water interfaces, recent interest has focused on the surface chemistry of amphiphilic compounds which can penetrate both phases and replace adsorbed molecules of both solvents, for example water and oil (Markin *et al.*, 1992a,b,c; Markin and Volkov, 1996; Volkov *et al.*, 1994). Amphiphilic molecules consist of two moieties with opposing properties: a hydrophilic polar head and a hydrophobic hydrocarbon tail. We present here a theoretical analysis of the generalized Frumkin adsorption isotherm for amphiphilic compounds.

It is quite difficult in general to define the standard Gibbs free energy of the adsorption equilibrium, ΔG_{ads}^0, for any substance adsorbed at the interface between two immiscible liquids. For a neutral substance i at the metal–solution interface, ΔG_{ads}^0 is usually defined as the change of the Gibbs free energy for the adsorption process when all species are in their standard states. It can be calculated as a difference between the standard chemical potentials in the adsorbed state and in the bulk solution minus the difference between the chemical potentials of the solvent in the adsorbed state and in the bulk solution:

$$\Delta G_{ads}^0 = (\mu_i^{0,s} - \mu_i^0) - n(\mu_w^{0,s} - \mu_w^0), \tag{7.72}$$

where n is the number of solvent molecules which are displaced by one adsorbate molecule. For an amphiphilic substance that is adsorbed at the interface between two immiscible solvents, ΔG_{ads}^0 can be estimated from the value of $\Delta\mu^0$, both in water (w) and in the oil (o). Suppose that the solute has a partition coefficient P_i between these two phases. Then

$$\Delta_{ads}^w G_{ads}^0 = \mu_i^{0,s} - \mu_i^{0,w} - n_1\Delta\mu_w^0 - n_2\Delta\mu_o^0, \tag{7.73}$$

$$\Delta_{ads}^w G_{ads}^0 = \mu_i^{0,s} - \mu_i^{0,o} - n_1\Delta\mu_w^0 - n_2\Delta\mu_o^0, \tag{7.74}$$

and

$$\Delta_0^w G_{ads}^0 = \mu_i^{0,o} - \mu_i^{0,w} = RT\ln P_i. \tag{7.75}$$

Different definitions of Gibbs free energy of adsorption in the literature have resulted in different interpretations of adsorption behavior of amphiphilic molecules. The interface between two immiscible liquids may be considered to be a surface solution of surfactant in a special kind of solvent. In order to calculate the entropy of such a solution we will adopt a simplified lattice model and use lattice statistics, a widely used method for describing surface solutions. The transition from three-dimensional (3-d) to two-dimensional (2-d) geometry, may cause errors in statistical formulas, if some peculiarities of 2-d solutions are overlooked.

The main difficulty is that, when dealing with a monolayer of a surfactant, one can consider this monolayer as a 2-d system. The solvent molecules do not form a monolayer, but rather a multilayer. Therefore the transition from 3-d to 2-d geometry should be specified. Consider molecules of both solvents which are substituted with

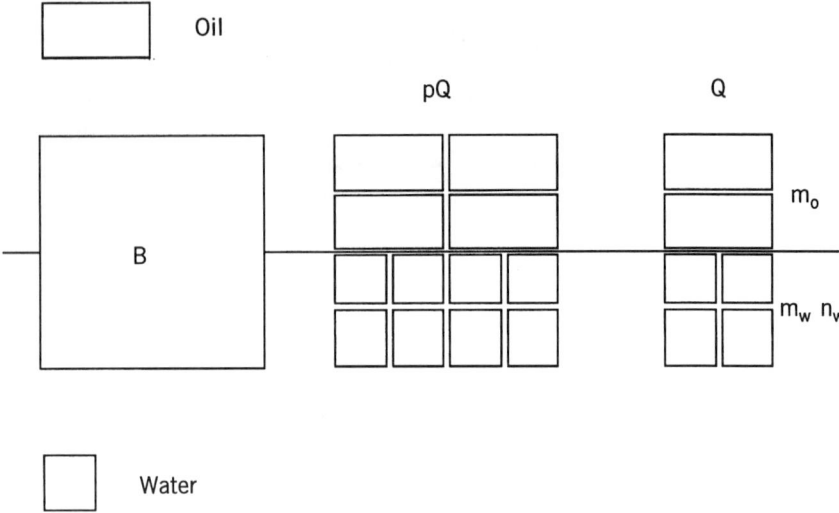

Figure 7.25. Structure of the oil–water interface with an adsorbed monolayer of amphiphilic surfactant B.

a surfactant (Fig. 7.25). Suppose that these molecules can be assembled into columns consisting of m_o molecules of oil and m_w molecules of water. Suppose that one column of oil molecules matches the n_w molecules of water. This match of one oil column and n_w water columns will be considered in what follows as a quasi-molecule of solvent Q. These quasi-molecules constitute a "monolayer" of solvent. They consist of m_o oil molecules and $n_w m_w$ water molecules.

Designate the molecules of surfactant in the bulk as A, and in the monolayer as B. At the interface aggregation of surfactant molecules can take place, $B \Leftrightarrow rA$, such as dimerization of porphyrin molecules or pheophytin at the octane–water interface. Let the surfactant B replace p quasi-molecules at the interface. Therefore one can write

$$pQ + rA = B + pm_o(\text{oil}) + pn_w m_w(\text{water}). \qquad (7.76)$$

The chemical potentials for (7.76) are

$$p\mu_Q^s + r\mu_A^b = \mu_B^s + pm_o \mu_o^b + pn_w m_w \mu_w^b. \qquad (7.77)$$

Taking the 2-d solution as ideal, we have

$$\mu_Q^s = \mu_Q^{0,s} + RT \ln X^s, \qquad (7.78)$$

$$\mu_B^s = \mu_B^{0,s} + RT \ln X_B^s. \qquad (7.79)$$

In the bulk phase we have

$$\mu_A^b = \mu_A^{0,b} + RT \ln X_A^b, \tag{7.80}$$

$$\mu_o^b = \mu_o^{0,b} + RT \ln X_o^b, \tag{7.81}$$

$$\mu_w^b = \mu_w^{0,b} + RT \ln X_w^b. \tag{7.82}$$

In all these equations X designates the mole ratio of corresponding substances. Substituting these equations into (7.77), one obtains:

$$p\mu_Q^{0,s} + r\mu_A^{0,b} - \mu_B^{0,s} - pm_o\mu_o^{0,b} - pn_wm_w^{0,b} + RT \ln \frac{(X_A^b)^r}{(X_o^b)^{pm_o}(X_w^b)^{pn_wm_w}} = RT \ln \frac{X_b^s}{(X_Q^s)^p}. \tag{7.83}$$

Using the standard Gibbs free energy of adsorption

$$\Delta_b^s G^0 = \mu_B^{0,s} - r\mu_A^{0,b} + pm_o\mu_o^{0,b} + pn_wm_w\mu_w^{0,b} - p\mu_Q^{0,s}, \tag{7.84}$$

one obtains the adsorption isotherm:

$$\frac{X_B^s}{(X_Q^s)^p} = \frac{(X_A^b)^r}{(X_o^b)^{pm_o}(X_w^b)^{pn_wm_w}} \exp\left(-\frac{\Delta_b^s G^0}{RT}\right). \tag{7.85}$$

The problem of choosing standard states of particles in the bulk solution and at the interface in these processes was analyzed by Mohilner et al. (1977).

We considered the 2-d solution of surfactant B in the solvent of quasi-particles Q, in which the mole ratios were defined as

$$X_B^s = \frac{N_B^s}{N_B^s + N_Q^s}, \qquad X_Q^s = \frac{N_Q^s}{N_B^s + N_Q^s}. \tag{7.86}$$

Some authors prefer another set of definitions when real particles in the interface are considered. The equation for this state with real particles A, O, W becomes:

$$X_A^s = \frac{N_A^s}{N_A^s + N_O^s + N_w^s}, \tag{7.87}$$

$$X_Q^s = \frac{N_Q^s}{N_A^s + N_O^s + N_w^s}, \tag{7.88}$$

$$X_w^s = \frac{N_w^s}{N_A^s + N_O^s + N_w^s}, \tag{7.89}$$

and we can obtain

$$X_B^s = \frac{m_o X_A^s}{m_o X_A^s + X_o^s}, \qquad X_Q^s = \frac{X_o^s}{m_o X_A^s + X_o^s}. \tag{7.90}$$

The adsorption isotherm can then be presented in the form

$$\frac{m_o X_A^s}{(X_O^s)^p}(m_o X_A^s + X_o^s)^{p-1} = \frac{(X_A^b)^r}{(X_o^b)^{p m_o}(X_w^b)^{p n_w m_w}}\exp\left(-\frac{\Delta_b^s G^0}{RT}\right). \quad (7.91)$$

In the past the adsorption isotherm was presented in terms of the fraction Θ of the surface actually covered by the adsorbed surfactant. The detailed description of the transition to these terms for the case of organic adsorption at the mercury–aqueous solution interface was presented by Mohilner et al. (1977). The main idea of this approach can be used in our case.

If we introduce η as the ratio of areas occupied in the interface by the molecules of surfactant and oil, the mole fractions in the surface solution can be presented as follows:

$$X_B^s = \frac{\Theta}{\Theta + \eta(1 - \Theta)}, \qquad X_o^s = \frac{\eta(1 - \Theta)}{\Theta + \eta(1 - \Theta)}. \quad (7.92)$$

The adsorption isotherm takes the form:

$$\frac{\Theta}{\eta^p(1 - \Theta)^p}[\Theta + \eta(1 - \Theta)]^{p-1} = \frac{(X_A^b)^r}{(X_o^b)^{p m_o}(X_w^b)^{p n_w m_w}}\exp\left(-\frac{\Delta_b^s G^0}{RT}\right). \quad (7.93)$$

In this isotherm the mole fractions X_A^b, X_o^b, X_w^b of the components in the bulk solution are presented. In the general case they must be substituted with activities:

$$\frac{\Theta}{\eta^p(1 - \Theta)^p}[\Theta + \eta(1 - \Theta)]^{p-1} = \frac{(a_A^b)^r}{(a_o^b)^{p m_o}(a_w^b)^{p n_w m_w}}\exp\left(-\frac{\Delta_b^s G^0}{RT}\right). \quad (7.94)$$

If the molecules B can interact as pairs in the adsorbed layer and the energy of each new particle is proportional to its concentration, then their chemical potential, μ_B^s, instead of equation (7.79), should be presented as

$$\mu_B^s = \mu_B^{s;0} + RT\ln X - 2aRTX, \quad (7.95)$$

where a is the so-called attraction constant (Damaskin et al., 1971; Frumkin, 1919). Then after some algebra we obtain, instead of equation (7.93), the isotherm:

$$\frac{\Theta[\eta - (\eta - 1)\Theta]^{p-1}}{\eta^p(1 - \Theta)^p}\exp(-2a\Theta) = \frac{(X_A^b)^r}{(X_o^b)^{p m_o}(X_w^b)^{p n_w m_w}}\exp\left(-\frac{\Delta_b^s G^0}{RT}\right). \quad (7.96)$$

Recall that η was introduced as the ratio of areas occupied in the interface by the molecule of surfactant to the same areas of oil and p was introduced as the number of columns of oil which could be supplanted with one molecule of surfactant (Fig. 7.25). Therefore p is a relative size of the surfactant molecule in the interfacial layer.

It is reasonable to suppose that

$$\eta = p. \tag{7.97}$$

If the concentration of surfactant in the solution is not high and the mutual solubility of oil and water is low, then we can use the approximation $X_o^b = X_w^b = 1$, so that the general equation (7.96) simplifies to

$$\frac{\Theta[p - (p - 1)\Theta]^{p-1}}{p^p(1 - \Theta)^p} \exp{(-2a\Theta)} = (X_A^b)^r \exp{\left(-\frac{\Delta_b^s G^0}{RT}\right)}. \tag{7.98}$$

This is the final expression for the isotherm that we will call the amphiphilic isotherm. It is straightforward to derive classical adsorption isotherms from the amphiphilic isotherm (7.98).

1. The Henry isotherm, when $a = 0$, $r = 1$, $p = 1$, $\Theta \ll 1$:

$$\Theta = X_a^b \exp{\left(-\frac{\Delta_b^s G^0}{RT}\right)}. \tag{7.99}$$

2. The Freundlich isotherm (Freundlich, 1926), when $a = 0$, $p = 1$, $\Theta \ll 1$:

$$\Theta = (X_a^b)^r \exp{\left(-\frac{\Delta_b^s G^0}{RT}\right)}. \tag{7.100}$$

3. The Langmuir isotherm (Langmuir, 1917), when $a = 0$, $r = 1$, $p = 1$:

$$\frac{\Theta}{1 - \Theta} = X_a^b \exp{\left(-\frac{\Delta_b^s G^0}{RT}\right)}. \tag{7.101}$$

4. The Frumkin isotherm (Damaskin et al., 1971; Frumkin, 1919), when $r = 1$, $p = 1$:

$$\frac{\Theta}{1 - \Theta} \exp(-2a\Theta) = X_a^b \exp{\left(-\frac{\Delta_b^s G^0}{RT}\right)}. \tag{7.102}$$

Therefore, the amphiphilic isotherm (7.98) could be considered as a generalization of the Frumkin isotherm, taking into account the replacement of some solvent molecules with larger molecules of surfactant. Of course, the amphiphilic isotherm includes all the features of the Frumkin isotherm and displays some additional ones. To elucidate them, it will be convenient to change the variable X_A^b to the relative concentration $y = X_A^b/X_A^b(0.5)$, where $X_A^b(0.5)$ is the concentration corresponding to the surface coverage $\Theta = 0.5$:

$$y = \frac{\Theta[p - (p - 1)\Theta]^{p-1}}{(p + 1)^{p-1}(1 - \Theta)^p} \exp{(a - 2a\Theta)}. \tag{7.103}$$

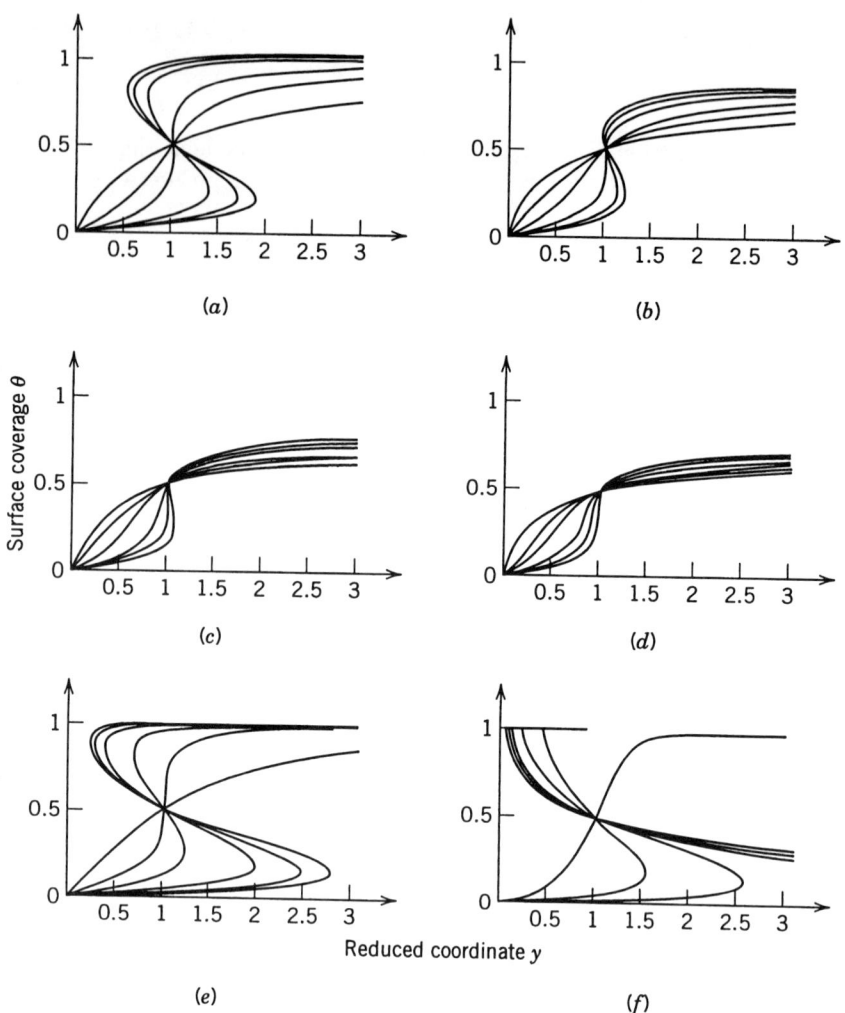

Figure 7.26. Surface coverage Θ as a function of relative concentration $y = c/c_{\Theta=0.5}$. Each panel presents a family of curves with a = 0.0; 1.299; 2.000; 2.840; 3.167; 3.333. (a) $p = 1$; (b) $p = 3$; (c) $p = 8$; (d) $p = 40$; (e) $p = 0.5$; (f) $p = 0.1$.

This equation gives the coverage fraction Θ as a function of relative concentration y, while a and p are the parameters of this isotherm, the first being the attraction constant and the second, the size of surfactant. These parameters play an important role because their effect on the shape of the amphiphilic isotherm is very strong.

The attraction constant in the Frumkin isotherm (7.99) or in the amphiphilic isotherm (7.98) with $p = 1$ determines the expanded-condensed transition in the adsorbed layer (Adamson, 1990). A family of amphiphilic isotherms (equation (7.98) with $p = 1$) with different parameters a is presented in Fig. 7.26. Because variables y and Θ were specially chosen, all the curves cross at the point with coordinates (1, 0.5), which is also an inflection point for those curves that have one. The first curve with $a = 0$ represents the well-known Langmuir isotherm. With increasing a, the curves tend to become S-shaped, which is a manifestation of the condensation in the adsorbed layer. The expanded-condensed transition occurs for the first time when $a = 2$. This case is presented by the third curve in Fig. 7.26(a), which goes vertical at the crossing-point. As a increases, the curves display a more pronounced S-shaped form. Therefore, increasing the attraction constant a leads to a more condensed adsorbed layer.

We can now consider the effects of surfactant size p. In Fig. 7.26(b), Fig. 7.26(c), and Fig. 7.26(d) the same family of isotherms is presented for p equal to 3, 8 and 40. The isotherms still cross at the same point, but increasing p suppresses the tendency of isotherms to be S-shaped. In Fig. 7.26(b), with $p = 3$, only the fourth curve with $a = 2.84$ displays the beginning of a phase transition. In Fig. 7.26(c), with $p = 8$, it occurs only at the fifth curve with $a = 3.168$. In Fig. 7.26(d), with $p = 40$, it happens only at the last isotherm with $a = 3.333$.

Therefore, increasing the surfactant size p prevents condensation of the interfacial layer. The reason is that the larger surfactant molecules must supplant more solvent molecules from the interfacial layer, which is unfavorable because the entropy of the interface thereby decreases. This becomes especially clear at higher concentrations. One can see in these figures that with increasing p the upper part of the curves is depressed, while the lower part deforms much less. Therefore larger surfactant molecules are less able to saturate the layer at high concentration than smaller molecules.

We will now consider the effect of decreasing surfactant size. Figure 7.26(e) presents the same family of isotherms with $p = 0.5$ and Fig. 7.26(f) with $p = 0.1$. One can see that smaller p values enhanced the tendency of curves to become S-shaped. In Fig. 7.26(e) the phase transition began at the second curve with $a = 1.299$ and in Fig. 7.26(f) even the Langmuir isotherm with $a = 0$ displays a concave portion of the curve.

We can also consider isotherms with parameter p correspondingly equal to 40, 1, 0.2, 0.1, and 0.01. Fig. 7.27(a) presents the family of isotherms for zero attraction constant ($a = 0$): there are no S-shaped curves and therefore no phase transitions to the condensed state. If we insert a very small attraction constant $a = 0.0336$, curve 5 in Fig. 7.27(b) displays the beginning of phase transition due to the very low value of parameter $p = 0.01$. Fig. 7.27(c) corresponds to $a = 0.321$. Here the isotherm with $p = 0.1$ displays the beginning of phase transition. The shift of the incipient phase transition can be followed in Fig. 7.27(d), Fig. 7.27(e) and Fig. 7.27(f) with attraction constants corresponding to 0.609, 2.000 and 3.333.

In the Frumkin isotherm (7.102) the inflection point coincides with the crossing point of all the curves (1, 0.5). This is not necessarily the case in the generalized Frumkin amphiphilic isotherm. Analysis shows that the coverage Θ_{cond}, at which the

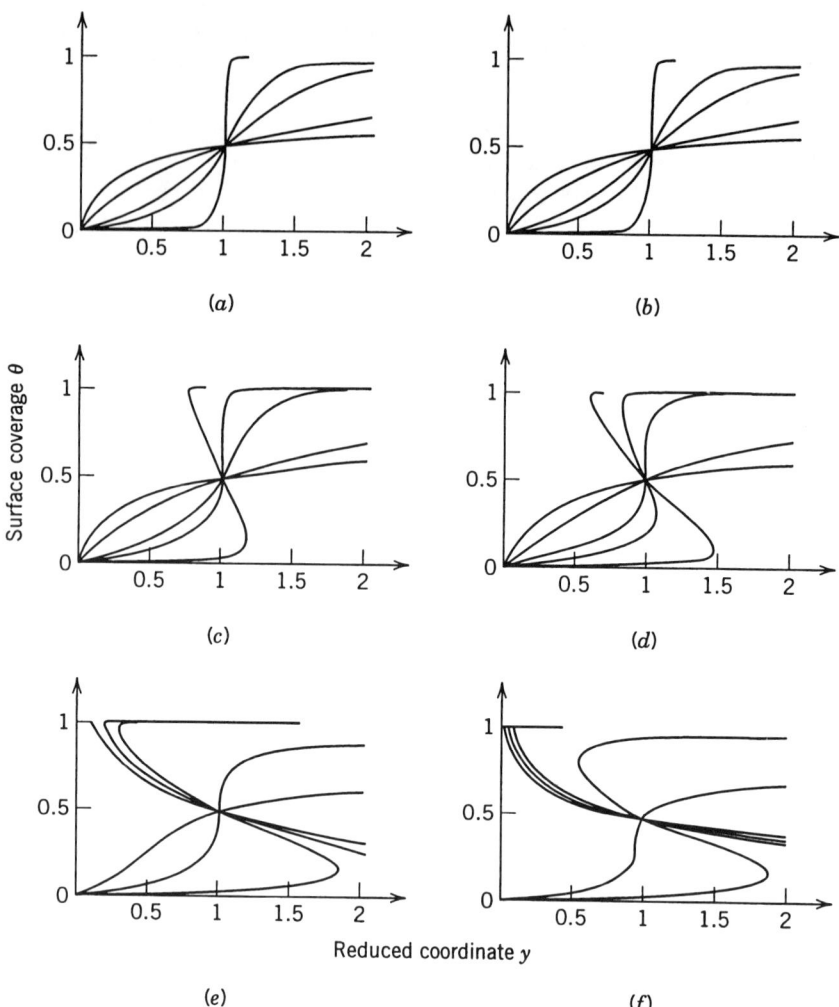

Figure 7.27. Surface coverage Θ as a function of relative concentration $y = c/c_{\Theta=0.5}$. Each panel presents a family of curves with $p = 40; 1.0; 2.000; 0.2; 0.1; 0.01$. (a) $a = 0$; (b) $a = 0.0336$; (c) $a = 0.321$; (d) $a = 0.609$; (e) $a = 2.000$; (f) $a = 3.333$.

condensation can begin, does not depend on attraction constant a but is a function of p as given by the equation below:

$$\Theta_{\text{cond}} = \frac{2p - 1 - \sqrt{p^2 - p + 1}}{3(p - 1)}. \tag{7.104}$$

When p varies from 0 to ∞ the value of Θ_{cond} changes from 2/3 to 1/3 as presented in Fig. 7.28, and when $p = 1$, critical covering becomes $\Theta_{\text{cond}} = 0.5$.

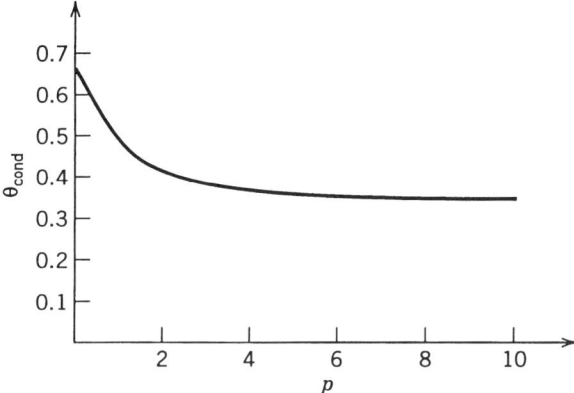

Figure 7.28. The dependence of the extent of the surface coverage Θ_{cond} on relative size of the surfactant molecule in the interfacial layer.

In the Frumkin isotherm condensation begins when $a = a_{cond} = 2$. With the amphiphilic isotherm, condensation depends on the size of surfactant p and begins when the attraction constant is

$$a_{cond} = \frac{1}{2\Theta_{cond}(1 - \Theta_{cond})[1 - (1 - 1/p)\Theta_{cond}]}, \qquad (7.105)$$

with Θ_{cond} given by equation (7.104). This dependence of a_{cond} on p is presented in Fig. 7.29. When p changes from 0 to infinity, parameter a_{cond} varies from 0 to 27/8, and the curve passes through the point $(1, 2)$.

Amphiphilic isotherm (7.98) can help in understanding the interfacial structure. An amphiphilic molecule, which consists of two moieties with opposing properties such as a hydrophilic polar head and a hydrophobic hydrocarbon tail, can be used as an analytical tool located at the interface. Chlorophyll a and pheophytin a are well-known surfactant molecules that contain a hydrophobic chain (phytol) and a hydrophilic head group. The amphiphilic isotherm yields adsorption parameters for amphiphilic compounds shown in Table 7.4. The value of p less than 1.0 indicates that adsorbed molecules of n-octane are parallel to the interface between octane and water (Fig. 7.30) Substitution of one adsorbed octane molecule requires about 4–5 adsorbed chlorophyll or pheophytin molecules. These are supported by molecular dynamic studies in the systems decane–water, nonane–water and hexane–water. The structure of both water and octane at the interface is different from the bulk. Adsorbed at the interface octane molecules (C_8H_{16}) have a lateral orientation at the interface, as shown in Fig. 7.30.

Experimental study of the amphiphilic molecule adsorption at the octane–water and benzene–water phase boundaries shows that adsorbed organic solvent molecules at the interfaces have lateral orientation (Fig. 7.31). Since parameter $p < 1$, only the

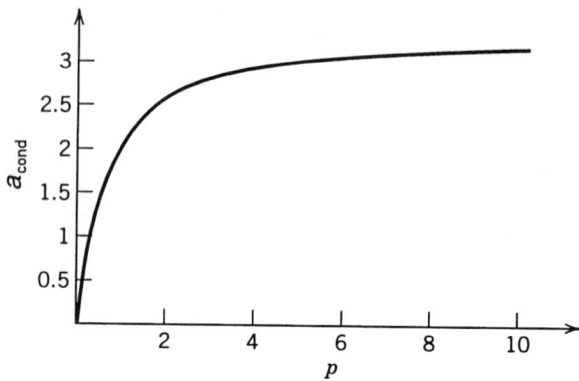

Figure 7.29. The dependence of attraction constant a_{cond}, corresponding to the beginning of condensation, on the relative size of the surfactant molecule in the interfacial layer.

Table 7.4. Adsorption parameters p, a, and Gibbs free energy of adsorption of amphiphilic molecules at the oil–water interface.

Amphiphilic compound	Non-aqueous solvent	p	a	ΔG^0	Reference
Chlorophyll a	n-octane	0.180	0.59	-27.4 kJ mol^{-1}	Markin et al., 1992a,b
Hydrated oligomer of chlorophyll a	n-octane	0.273	0.215	-35.2 kJ mol^{-1}	Markin et al., 1992a,b
Pheophytin a	n-octane	0.209	-0.36	-29.9 kJ mol^{-1}	Markin et al., 1992a,b
Pheophytin a	Benzene	0.200	0.643	-26.8 kJ mol^{-1}	Volkov et al., 1994

generalized adsorption isotherm (7.96) should be used for calculation and analysis of amphiphilic compound adsorption at the oil–water interface.

7.8. ROUGHNESS OF THE INTERFACE BETWEEN TWO IMMISCIBLE ELECTROLYTE SOLUTIONS

An original model of the electric double layer at ITIES, based on the assumption that ions affected by an external electric field distort the interface, was considered by Gugeshashvili et al. (1988), Indenbom (1985, 1995), Indenbom and Volkov (1986). According to Indenbom's model the smooth surface becomes grooved with increased voltage and the effective thickness of the compact layer decreases (Fig. 7.32). If the counter-ions on both sides of the interface react specifically or if the field is weak, dipoles may form at the phase boundary. These dipoles rotate under the action of an external field, also decreasing the effective thickness of the compact layer (Fig. 7.33).

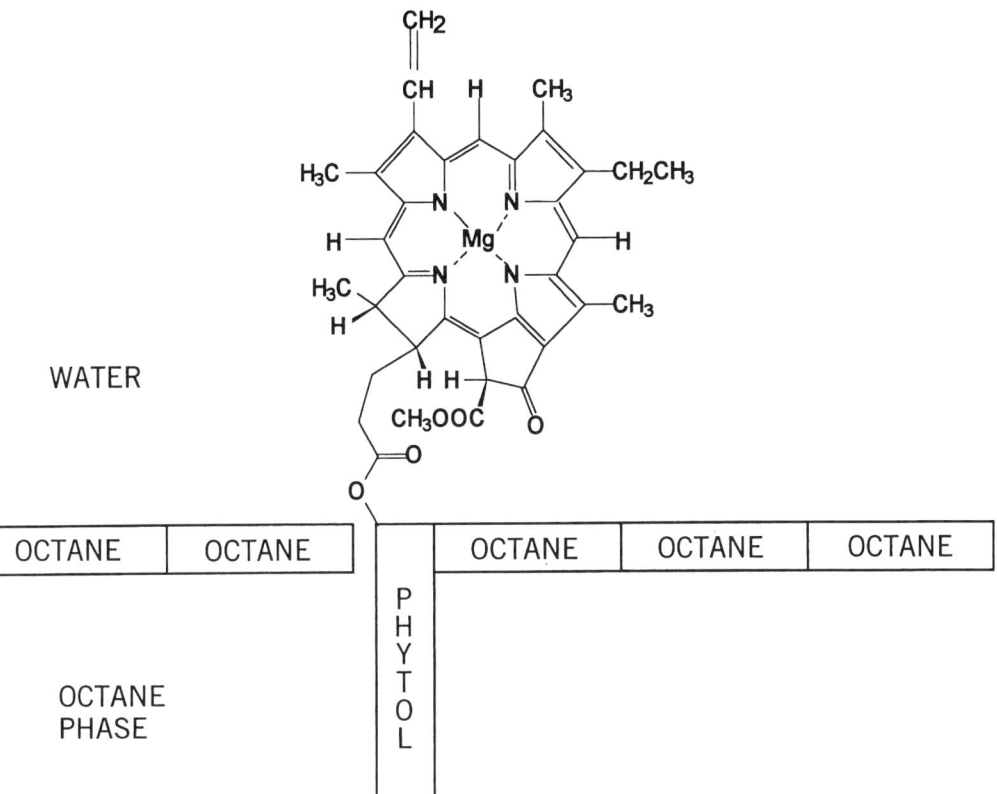

Figure 7.30. The amphiphilic compound chlorophyll *a*, which can substitute 1/6 of the lateral-oriented adsorbed molecule of octane, is a tool for the evaluation of the interfacial structure.

Such processes can explain unexpectedly high double-layer capacitance, or very low values of the potential difference across the compact layer.

According to the model proposed by Indenbom (1985), ions are retained at the interface due to the equilibrium of electric field force f and the force of the interfacial tension γ:

$$f = 2\pi r \gamma \sin \Theta, \qquad (7.106)$$

where Θ is the angle formed by the tangent to the surface and the axis r (Fig. 7.32).

This relation is easily transformed into a different form that describes the depression profile because

$$\frac{dz}{dr} = \tan \Theta = \frac{\sin \Theta}{\sqrt{1 - \sin^2 \Theta}} = \left[\left(\frac{r}{\zeta} \right)^2 - 1 \right]^{-1/2} \qquad (7.107)$$

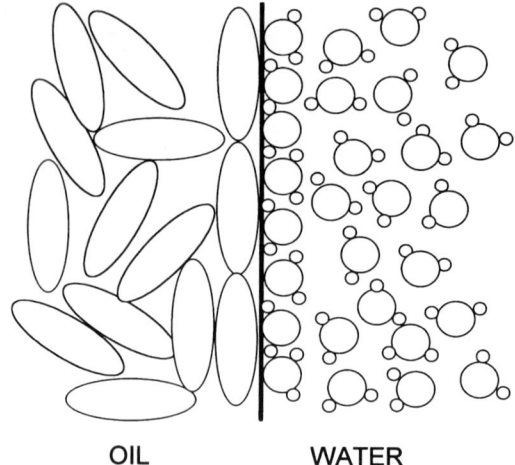

OIL   WATER

Figure 7.31. The lateral orientation of adsorbed solvent molecules at the oil–water interface.

The solution of the resulting equation becomes

$$z(r) = L - \sqrt{L(L-\zeta)} + \zeta \ln \frac{r/\zeta + \sqrt{(r/\zeta)^2 - 1}}{\sqrt{L/\zeta} + \sqrt{L/\zeta - 1}}, \qquad (7.108)$$

where

$$\zeta = f/2\pi\gamma. \qquad (7.109)$$

Equation (7.108) is meaningless when the electric field force is larger than the interfacial tension $\zeta \geq L$ and interfaces with adsorbed ions lose stability.

The maximal depth of an ionic pit (Fig. 7.34) can be determined when $r = d$, where d is the radius of the space occupied by a single adsorbed ion at the interface. Using Gouy–Chapman theory one can obtain:

$$d = \left(\frac{F}{N_A \pi}\right)^{1/2} (8RT\varepsilon_0 \varepsilon c)^{-1/4} \left(\sinh \frac{F\phi_d}{2RT} - \frac{2A}{1-A^2}\right)^{-1/2}, \qquad (7.110)$$

where

$$A = \exp\left(-\sqrt{\frac{2L^2 Fc}{RT\varepsilon_0 \varepsilon}}\right) \tanh \frac{F\phi_d}{4RT}. \qquad (7.111)$$

The electrical force f can be determined using the Ostrogradsky–Gauss theorem:

$$f = \frac{e_0 Q}{\varepsilon_0 \varepsilon}, \qquad (7.112)$$

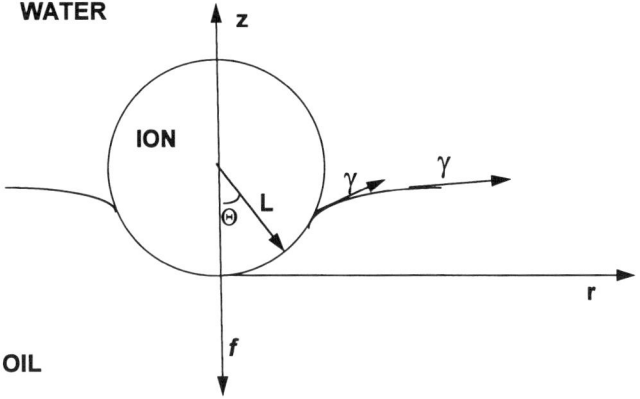

Figure 7.32. Ribbed model of the ITIES.

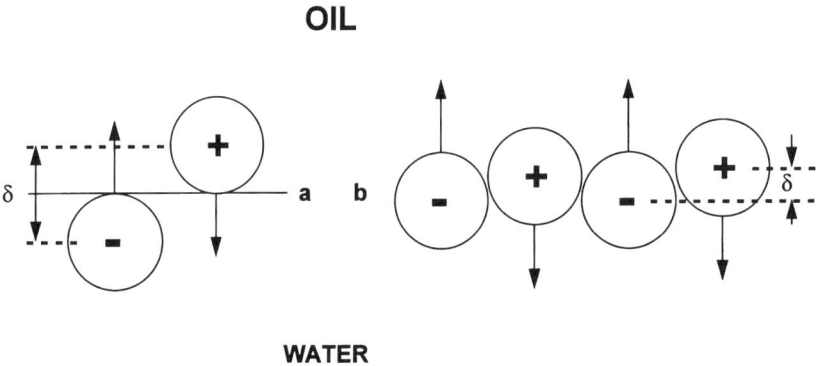

Figure 7.33. The structure of the compact part of the electric double layer in the absence of specific ion adsorption. The thickness of the ion-free layer is equal to the distance between dashed lines. (*a*) Traditional smooth interface; (*b*) ribbed interfacial model.

where Q is the surface charge density of the diffuse electric double layer:

$$Q = \sqrt{8RT\varepsilon_0\,\varepsilon c}\,\sinh\,(F\phi_d/RT). \tag{7.113}$$

When the formation energy is large the ionic depressions can form a 2-d lattice. Calculations performed for a water–nitrobenzene system containing equal concentrations of electrolytes show that the compact layer thickness can decrease to zero and the formation energy of the depression does not exceed 10 kT. Estimations, carried out by Indenbom (1985), show that depression depth is about a few angstroms and strongly depends on interfacial potential difference and ionic concentration.

The Indenbom (1985, 1995) model has a few limitations because it does not take

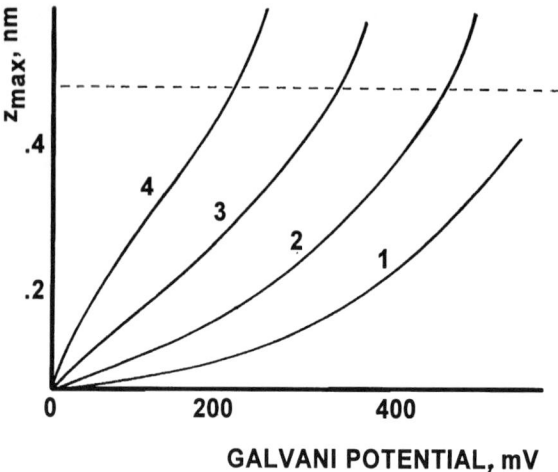

Figure 7.34. Dependencies of depression depths on the interfacial potential (Indenbom, 1985). Parameters: $R = 0.5$ nm, $T = 295$ K, $\varepsilon_1 = 36$, $\varepsilon_2 = 80$, $\gamma = 25$ mN m^{-1}. Curves 1, 2, 3, and 4 correspond to the solution concentrations 10^{-4}, 10^{-3}, 10^{-2}, and 10^{-1} M, respectively.

into account image forces, dielectric saturation in the compact layer, or the possibility that another ion can be symmetrically located in a second phase and compensate the ruffling effect. The theory of ionic ruffling of the oil–water interface should use non-local electrostatics rather than classical local electrostatics.

7.9. IMAGE FORCES

The image forces acting on charged particles near interfaces are defined as the forces of interaction between these particles and the "image" of free and bound charges induced by them in the region next to the interface, minus the analogous quantity in one bulk phase. As a rule, positive or negative adsorption of ions arises at the oil–water interface due to the effect of image forces. This effect is caused by different dielectric bulk properties of contact phases and by the inhomogeneous transition region where the ions and their solvation shells have different size. As a result, different planes of closest approaches arise where ions and dipolar molecules can interact specifically with the interfacial region.

Effects of image forces and electrostatic Gibbs free energy of an ion at the oil–water interface were considered by Gaevskii (1984), Kharkats (1976), Kharkats and Ulstrup (1991; 1993), Kharkats and Volkov (1985; 1987), Landau and Lifshitz (1984), Rehbinder (1927a,b), Torrie and Valleau (1986), Vorotyntsev and Ivanov (1987; 1989a,b), Vorotyntsev et al. (1993). The electrostatic Gibbs free energy of an ion in the vicinity of the boundary between two liquids with dielectric constants ε_1 and ε_2 (Fig. 7.35) is determined by the Born ion solvation energy and by the

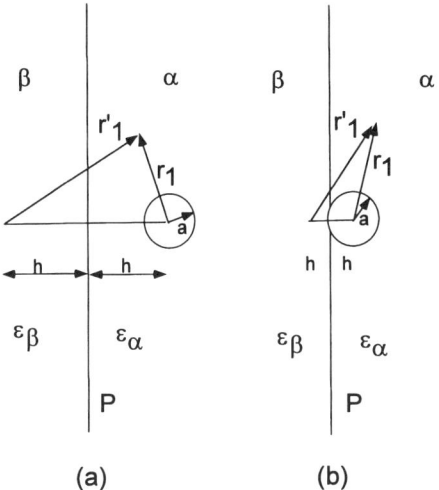

Figure 7.35. The relative arrangement of a spherical ion and a planar interface between two immiscible liquids with dielectric permittivities ε_α and ε_β: (a) the ion-center distance from the interfacial plane P exceeds the ion radius; (b) the ion partially penetrates the interface, $h < a$.

interaction with its image charge (Deamer and Volkov, 1995). In research dealing with the energy of image forces and with the interactions of charges at the oil–water interface, approximate models of the interface are often employed which are based on the traditional description of the interface between two local dielectrics. In the oil–water system the force of attraction (or repulsion) of charge in the organic (β) phase with its image in the aqueous phase (α) sitting on the same axis perpendicular to the dividing plane is given by

$$F(h) = -\frac{\varepsilon_\alpha - \varepsilon_\beta}{\varepsilon_\alpha + \varepsilon_\beta} \frac{(ze_0)^2}{16\pi\varepsilon_0\varepsilon_\beta h^2}, \tag{7.114}$$

where h is the distance from the interface. If $\varepsilon_\alpha > \varepsilon_\beta$, the charge in the non-polar phase is attracted to its image, but if $\varepsilon_\alpha < \varepsilon_\beta$ there is repulsion between the charge in an organic phase and its image. From equation (7.114) it follows that charges in the organic phase are attracted to the oil–water interface. Image forces attract the diffuse layer on the organic side, making it thinner, and repel and thicken the aqueous diffuse layer.

The Kharkats–Ulstrup model incorporates the finite radius of an ion a, which is assumed to have a fixed spherically uniform charge distribution and can continuously pass between the two phases. For a point charge at long distances from the interface, the electrostatic Gibbs free energy can be written as

$$\Delta G = \frac{(ze_0)^2}{32\pi\varepsilon_0\varepsilon_\alpha a}\left(4 + \frac{\varepsilon_\alpha - \varepsilon_\beta}{\varepsilon_\alpha + \varepsilon_\beta}\frac{2a}{h}\right), \tag{7.115}$$

where a is the ionic radius, h is the distance from the interface and ze is the charge of the ion.

Potential ϕ for the spherically symmetrical charge distribution is:

$$\phi = \begin{cases} \dfrac{ze_0}{4\pi\varepsilon_0\varepsilon_\alpha r_1} + \dfrac{ze_0}{4\pi\varepsilon_0\varepsilon_\alpha r_1'}\dfrac{\varepsilon_\alpha - \varepsilon_\beta}{\varepsilon_\alpha + \varepsilon_\beta} & \text{(in region } \alpha), \\[4mm] \dfrac{ze_0}{4\pi\varepsilon_0\varepsilon_\beta r_1}\dfrac{2\varepsilon_\beta}{\varepsilon_\alpha + \varepsilon_\beta} & \text{(in region } \beta). \end{cases} \tag{7.116}$$

Kharkats and Ulstrup (1991, 1993) obtained equations for the electrostatic part of the free Gibbs energy of a finite size ion.

When $h > a$,

$$\Delta G = \frac{(ze_0)^2}{32\pi\varepsilon_0\varepsilon_\alpha a}\left\{4 + \left(\frac{\varepsilon_\alpha - \varepsilon_\beta}{\varepsilon_\alpha + \varepsilon_\beta}\right)\frac{2a}{h} + \left(\frac{\varepsilon_\alpha - \varepsilon_\beta}{\varepsilon_\alpha + \varepsilon_\beta}\right)^2\left[\frac{2}{1 - (2h/a)^2} + \frac{a}{2h}\ln\frac{2h + a}{2h - a}\right]\right\}. \tag{7.117}$$

The first term in (7.117) is the Born solvation energy, and the second is the interaction with the image. For a charge located in region β, the electrostatic free energy is obtained from (7.117) by exchanging $\varepsilon_\alpha \Leftrightarrow \varepsilon_\beta$.

Kharkats and Ulstrup (1991) obtained the expression for the electrostatic free energy in the region $0 \leq h \leq a$:

$$\Delta G = \frac{(ze_0)^2}{32\pi\varepsilon_0\varepsilon_\alpha a}\left\{\left(2 + \frac{2h}{a}\right) + \left(\frac{\varepsilon_\alpha - \varepsilon_\beta}{\varepsilon_\alpha + \varepsilon_\beta}\right)\left(4 - \frac{2h}{a}\right) + \left(\frac{\varepsilon_\alpha - \varepsilon_\beta}{\varepsilon_\alpha + \varepsilon_\beta}\right)^2\right.$$
$$\left. \times \left[\frac{(1 + h/a)(1 - 2h/a)}{1 + 2h/a} + \frac{a}{2h}\ln\left(1 + \frac{2h}{a}\right)\right]\right\} + \frac{(ze_0)^2}{16\pi\varepsilon_0\varepsilon_\beta a}\left(\frac{2\varepsilon_\beta}{\varepsilon_\alpha + \varepsilon_\beta}\right)^2\left(1 - \frac{h}{a}\right). \tag{7.118}$$

If $h = 0$ and the center of the ion is at the interface, the electrostatic Gibbs energy is equal to

$$\Delta G(h = 0) = \frac{(ze_0)^2}{4\pi\varepsilon_0(\varepsilon_\alpha + \varepsilon_\beta)a}. \tag{7.119}$$

Figure 7.36 illustrates the effect of image forces on the electrostatic Gibbs energy of an ion at the oil–water interface.

Vorotyntsev and Ivanov (1987, 1989a,b) calculated the energy of the image forces and the interaction of an ion with a charged group at the electrolyte/dielectric solution interface using a non-local electrostatic approach. Their model is shown in Fig. 7.37. The dielectric was considered as local and homogeneous ($\varepsilon(x) = \varepsilon_d$ for $x < 0$, where $x = 0$ is the interface) while for the region of $x > 0$ the model of specular reflection

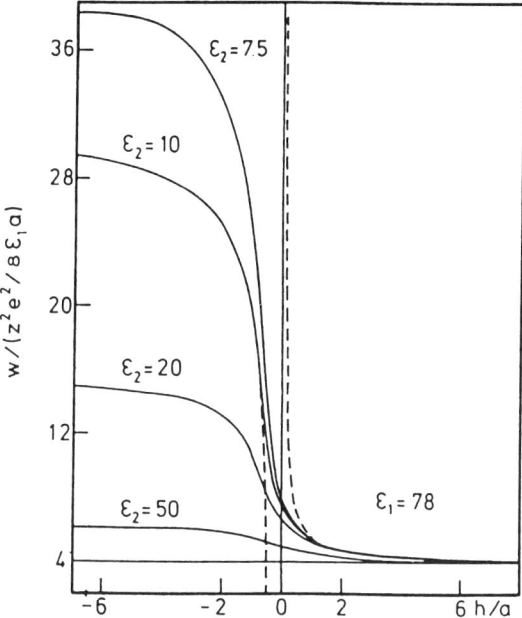

Figure 7.36. Electrostatic Gibbs energy profiles for ion transfer across the ITIES boundary. Solid lines: finite-size ion profiles in units of $(ze)^2/\varepsilon_1 a$, $\varepsilon_1 = 78$ and different values of ε_2. Dashed lines: profile for the point change model in the same units, $\varepsilon_1 = 78$, $\varepsilon_2 = 10$ (Kharkats and Ulstrup, 1991). Reproduced by permission of Elsevier Sequoia S.A.

of the polarization waves at the interface (Vorotyntsev and Kornyshev, 1993) was used. This allows the response of the solvent layer next to the interface to be expressed in terms of its bulk dielectric function $\varepsilon(k)$, which accounts for its spatial structure. Vorotyntsev and Kornyshev (1993) neglected screening the field by the electrolytes ionic plasma in the solution and described ions as point charges. Assuming $\varepsilon \gg \varepsilon^*$ and $\varepsilon \gg \varepsilon_d$, where ε^* is the short-wave dielectric permittivity, Vorotyntsev and Ivanov (1989a) expressed the image forces energy by the following equation:

$$\Delta G_{im}(a) \approx \Delta G_{im}^{local}(a) + \Delta G_{im}^{non}(a)\zeta(a), \qquad (7.120)$$

where $\Delta G_{im}^{local}(a)$ is the local classical energy of image forces,

$$\Delta G_{im}^{local}(a) = \frac{z_1^2 e_0^2}{16\pi\varepsilon_0\varepsilon} \frac{\varepsilon - \varepsilon_d}{\varepsilon + \varepsilon_d} \frac{1}{a}, \qquad (7.121)$$

and a non-local term can be written as

$$\Delta G_{im}^{non}(a)\zeta(a) = \frac{z_1^2 e_0^2}{16\pi\varepsilon_0\varepsilon^* a} \exp\left(-\frac{2a}{\Lambda}\right), \qquad (7.122)$$

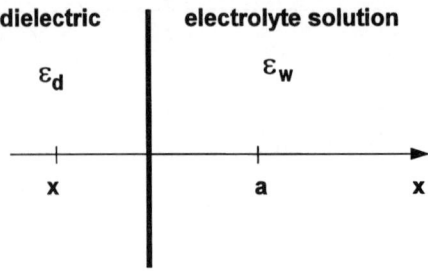

Figure 7.37. The sharp boundary model of the electrolyte–dielectric interface according to Vorotyntsev and Ivanov (1989a). ε_w and ε_d are the bulk dielectric permittivities of the media, a and x are the distances between the charges and interface.

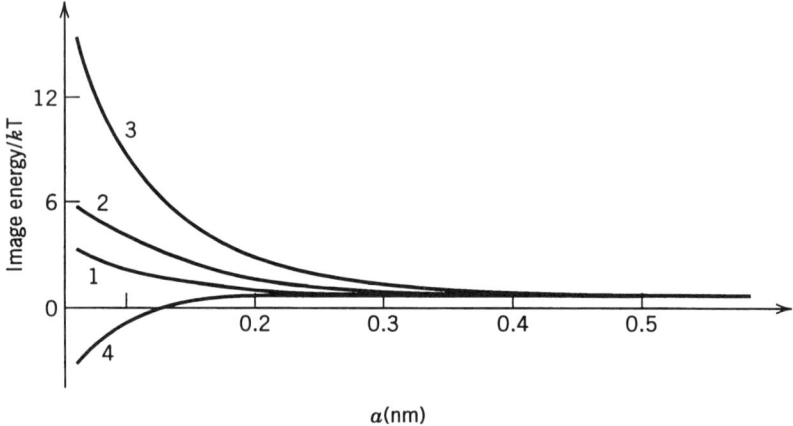

Figure 7.38. Dependence of the energy of image forces on the distance a from the interface for different values of the dielectric permittivity ε_d: 5 (2), 2.5 (3), and 9 (4); curve 1 was calculated according to the "classical" local model (Vorotyntsev and Ivanov, 1989a).

which is practically independent of the solvent's bulk dielectric constant ε. The function $\zeta(a)$ may be calculated from the relation:

$$\zeta(a) \approx \frac{\varepsilon^*\left(1 + \dfrac{2a}{p\Lambda}\right) - \varepsilon_d\sqrt{1 + \dfrac{4a}{p\Lambda}}}{\varepsilon^*\left(1 + \dfrac{2a}{p\Lambda}\right) + \varepsilon_d\sqrt{1 + \dfrac{4a}{p\Lambda}}}, \tag{7.123}$$

where p is a fitting parameter near unity, and Λ is the correlation radius. Figure 7.38 shows the dependence of image forces energy on the distance from the interface calculated using non-local electrostatics.

Image forces play a significant role in electric double-layer effects. The excess surface charge density is

$$q = ec_0 \int_{h_a}^{h_c} \{\exp[-\Delta g(h)/k_B T] - 1\} \, dh, \tag{7.124}$$

where

$$\Delta g(h) = \Delta G(h) - \frac{(ze_0)^2}{8\pi\varepsilon_0 \varepsilon_\beta a}. \tag{7.125}$$

$\Delta G(h)$ is the electrostatic contribution to the Gibbs energy of solvation, and h_a and h_c are the distances of closest approach of the anion and the cation. If image forces are taken into account, the diffuse layer capacitance can be calculated by the equation

$$C_d = C_d^{GC} \exp[\Delta g(h_0)/2k_B T] \tag{7.126}$$

where C_d^{GC} is the Gouy–Chapman diffuse layer capacitance without image terms. As noted by Kharkats and Ulstrup (1991), equation (7.126) always gives diffuse layer capacitance corrections toward higher values.

It is important to analyze the energetic profile of dipole molecules at the interface between two immiscible liquids (Arakelyan and Arakelyan, 1983a,b; Kirkwood, 1934, 1965; Kornyshev, 1988; Paula et al., 1996). Let us consider that a molecule has dipolar moment d_1 and it is located in the phase m with the dielectric permittivity ε_m. Its image d_2 is located in the phase w with the dielectric permittivity ε_m (Fig. 7.39). The dipole d_1 has angle Θ with an axis x. The interface is located at $x = 0$. The force of an interaction of a dipole with its image is:

$$F(x) = -\frac{3(\varepsilon_w - \varepsilon_m)d_1^2}{4\pi\varepsilon_0(\varepsilon_w + \varepsilon_m)\varepsilon_m(2x)^4}(1 + \cos^2\Theta). \tag{7.127}$$

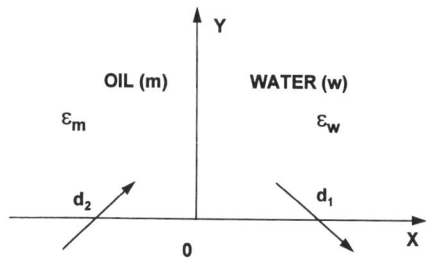

Figure 7.39. Dipole d_1 and its image d_2 at the interface between two immiscible liquids.

If $\varepsilon_w < \varepsilon_m$ the force $F(x)$ is negative and there is an attraction of a dipole to the interface. $F(x)$ is positive if $\varepsilon_w > \varepsilon_m$ and there exists a repulsion of the dipole from the interface.

Energy of the dipole interaction with its image can be calculated as

$$E(x, \Theta) = \int_x^\infty F(x)\,dx \qquad (7.128)$$

and substituting (7.92) in (7.93) gives the equation:

$$E(x, \Theta) = -\frac{(\varepsilon_w - \varepsilon_m)d_1^2}{8\pi\varepsilon_0(\varepsilon_w + \varepsilon_m)\varepsilon_m(2x)^3}(1 + \cos^2\Theta). \qquad (7.129)$$

There are two equilibrium dipole orientation with $\Theta = 0$ and $\Theta = \pi/2$ and a stable state will be achieved when a dipole is perpendicular to the interface. It is possible to determine the thickness of a layer in which dipoles are oriented perpendicular to the interface supposing that $E(x, 0) - E(x, \pi/2)$ is equal to kT:

$$x \approx \sqrt[3]{\frac{(\varepsilon_w - \varepsilon_m)d_1^2}{64\pi\varepsilon_0(\varepsilon_w + \varepsilon_m)\varepsilon_m kT}}. \qquad (7.130)$$

For the water–octane interface $x = 0.14$ nm and this result means that only one monolayer of water is oriented at the interface by image forces.

Bell (1931) calculated the Gibbs energy of a dipole molecule transferred from vacuum to a medium of dielectric constant ε_i:

$$\Delta_{vac}^i G = -\frac{d^2}{12\pi\varepsilon_0 a^3}\left(\frac{\varepsilon_i - 1}{2\varepsilon_i + 1}\right). \qquad (7.131)$$

The electrostatic ("Born") Gibbs energy of an ideal dipole transfer from the phase w to the solvent m can be calculated:

$$\Delta_w^m G_{dip} = \frac{d^2}{12\pi\varepsilon_0 a^3}\left(\frac{\varepsilon_w - 1}{2\varepsilon_w + 1} - \frac{\varepsilon_m - 1}{2\varepsilon_m + 1}\right) = \frac{d^2}{4\pi\varepsilon_0 a^3}\left(\frac{\varepsilon_w - \varepsilon_m}{(2\varepsilon_w + 1)(2\varepsilon_m + 1)}\right) \qquad (7.132)$$

where d is the dipole moment and a is the effective dipole size. The electrostatic Gibbs energy of the dipole profile in phase w can be evaluated from equations (7.129) and (7.131):

$$\Delta G_{dip} = -\frac{d^2}{4\pi\varepsilon_0 a^3}\left(\frac{\varepsilon_w - \varepsilon_m}{(2\varepsilon_w + 1)(2\varepsilon_m + 1)}\right) + \frac{(\varepsilon_w - \varepsilon_m)d^2}{6\pi\varepsilon_0(\varepsilon_w + \varepsilon_m)\varepsilon_m(2x)^3}. \qquad (7.133)$$

The electrostatic energy profile of water molecules at an octane–water interface is shown in Fig. 7.40.

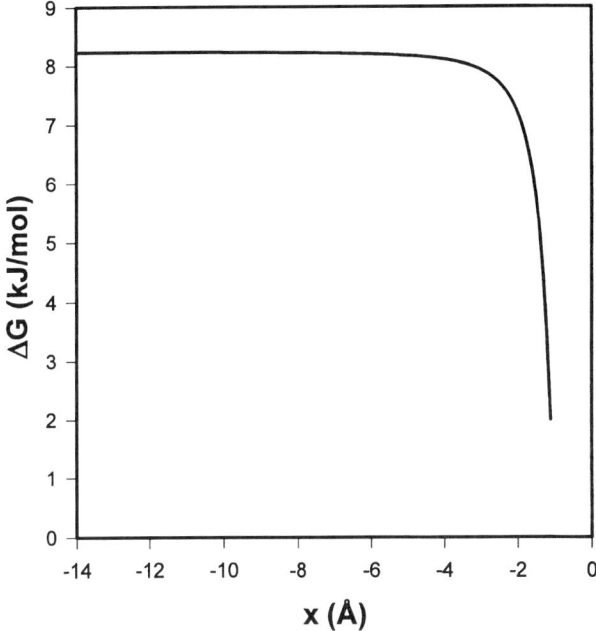

Figure 7.40. Electrostatics Gibbs energy profiles for a water dipole in the oil phase near the interface between water and an immiscible organic liquid. Parameters: $d_1 = 1.85$ D, $a = 0.135$ nm, $\varepsilon_w = 78$, $\varepsilon_m = 2$.

Dipolar molecules in the electric double layer have energetic characteristics different from the bulk phase. For example, amphiphilic compounds can produce self-organized molecular ensembles at the oil–water interface (Deamer and Volkov, 1995, 1996; Markin and Volkov, 1996). This is important for biotechnology and surface science. Recently Kornyshev (1988) presented a theory for non-local enhancement of the dipole–dipole interaction at the oil–water interface. Consider the interaction of two point dipoles oriented normally to the oil–water interface (Fig. 7.41). From classical electrostatics the energy of interaction between these two dipoles is:

$$\Delta G_{clas} = \frac{\mu\mu'}{4\pi\varepsilon_0\varepsilon}\left\{\begin{array}{c}\dfrac{1}{[R^2 + (l - l')^2]^{3/2}} - \dfrac{(\varepsilon_\beta - \varepsilon_\alpha)}{(\varepsilon_\beta + \varepsilon_\alpha)[R + (l - l')^2]^{3/2}} \\ -\dfrac{3(l - l')^2}{[R^2 + (l - l')^2]^{5/2}} + \dfrac{3(\varepsilon_\beta - \varepsilon_\alpha)(l - l')^2}{(\varepsilon_\beta + \varepsilon_\alpha)[R^2 + (l - l')^2]^{5/2}}\end{array}\right\}. \quad (7.134)$$

Taking into account the effect of spatial dispersion, Kornyshev (1988) also derived an equation for the energy between two dipoles at the interface ΔG. At $l/\Lambda \to 0$ a maximum enhancement of the dipole–dipole interactions takes place:

$$\Delta G(0, 0, R) = \beta^2 \Delta G_{clas}(R). \quad (7.135)$$

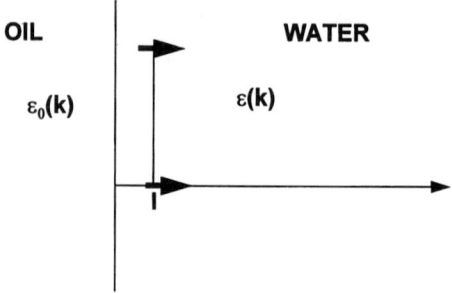

Figure 7.41. Two dipoles near the plane oil–water interface.

Assuming that $\beta = 16$ for water, enhancement of the dipole–dipole interactions is 256 times ($l = 0$) or 43 times ($l = \Lambda$) when compared to the classical approach.

7.10. DRAWBACKS AND DEVELOPMENT OF MVN AND GOUY–CHAPMAN–STERN THEORIES

The first statistical theory of the electric double layer at the metal–water interface was developed by Gouy and Chapman (Gouy, 1910; Chapman, 1913) and later applied to the interface at two electrolyte solutions (MVN). These theories provided the first important insights into the nature of the diffuse electrical layer and the first estimates of capacitance and potential distribution in the double layer. The MVN theory mechanically connected two Gouy–Chapman models, and therefore shares both the merits and drawbacks of the Gouy–Chapman theory. The major shortcomings of these models were that they seriously overestimated the values for double-layer capacitance and ionic concentration in the double layer at high electrode potentials, and their predictions for the Parsons–Zobel relationships were also unsatisfactory. In order to understand the future refinements of these models it is important to review their historical development.

Kirkwood (1934) conducted a thorough analysis of the Gouy–Chapman theory and found that it had three major drawbacks. It did not take into account the finite volume of ions, it incorrectly considered the ionic atmosphere around ions, and the average value of the electrical potential was used to calculate charged particle energy without justification. Elaborating on this list, the theory also neglects the correlation of ions within the diffuse layer because it does not take into consideration the proper volume of ions. The same is true for the MVN theory with one additional drawback of its own, that it neglects the correlation of ions in different layers (the between-layer correlation).

Numerous efforts have been made to correct the shortcoming of these models. The first attempt at improving the Gouy–Chapman model by Stern (1924) was to take into account the finite size of ions. Stern, in his correction, only considered the finite radius

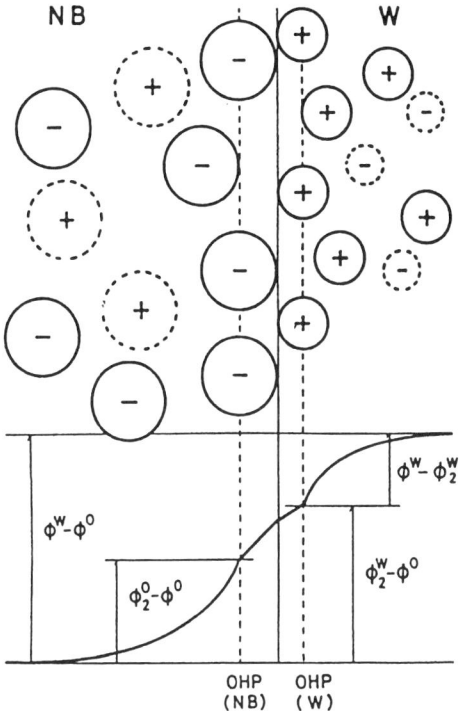

Figure 7.42. A schematic representation of the Gouy–Chapman–Stern model of the electric double layer at ITIES and a potential profile.

of ions in calculating their interactions with an electrode surface, but continued to use Gouy–Chapman's point ion assumption when considering ion–ion interactions in the diffuse layer. Nevertheless, his contribution constituted a very important advance. Since his work, the double layer has been considered to consist of two parts—an ion-free compact (or dense, or Helmholtz) layer and the Gouy–Chapman diffuse layer which extends from the compact layer into the bulk solution (Fig. 7.42).

Utilization of this approach immediately improved the description of the capacitance of the double layer because now the model included two capacitors in series:

$$\frac{1}{C} = \frac{1}{C_d} + \frac{1}{C_h}. \tag{7.136}$$

The first term in this equation is the capacitance contribution of the diffuse layer C_d and the second term is the capacitance contribution of the compact layer C_h. Therefore, if the compact layer has thickness δ and dielectric permittivity ε_h, and the diffuse layer capacitance is described by the Gouy–Chapman theory with the Debye parameter κ, then the capacitance of the total electric double layer can be presented in the form:

$$\frac{1}{C} = \frac{1}{\varepsilon \varepsilon_0 \kappa} + \frac{\delta}{\varepsilon_h \varepsilon_0}. \tag{7.137}$$

This correction by Stern dramatically changed the Gouy–Chapman results, which now predict a much smaller total capacitance at high potentials because the second capacitor is present in series. The thickness of the compact layer δ was assumed to be equal to the radius of adsorbed ions and the value of ε_h was used as a fitting parameter in order to obtain agreement between theory and experimental results.

Ions in the compact layer can be adsorbed either specifically or non-specifically. Grahame (1947) considered the specific adsorption of ions and modified the Gouy–Chapman–Stern model by introducing different degrees of ion approach to the electrode. It then became clear that ions in the compact layer do not "see" the average electrical potential that was employed in calculating their concentration, but rather they find the points with lowest potential energy. This phenomenon was named the discreteness-of-charge effect and the first approximate solution for it was derived by Esin and Shikhov (1943), who introduced the concept of micropotential at the point of ion adsorption as a correction for the previously used mean electrostatic potential. Unfortunately, none of these approaches was statistically based. Further progress would demand more realistic models of ions in the double layer and rigorous statistical analysis.

The first step to statistically correct the Gouy–Chapman theory for the diffuse double layer used a restricted primitive electrolyte model. This model considers ions to be charged hard balls of identical radii in a structureless dielectric continuum with constant dielectric permittivity. There are three main approaches to creating a statistical theory with this model. The first approach uses the Gouy–Chapman theory and the average electrical field in the diffuse layer is calculated from the Poisson equation. The structure and physical properties of the double layer were then calculated from the restricted primitive electrolyte model. As a result, the modified Poisson–Boltzmann theory (MPB) was developed (Andrietti et al., 1976; Bhuiyan and Outhwaite, 1994; Bhuiyan et al., 1991, 1992, 1994; Cui et al., 1994, 1995; Das et al., 1995; Gordillo et al., 1990; Outhwaite and Bhuiyan, 1991a,b; Outhwaite and Molero, 1989, 1990, 1991a,b; Outhwaite et al., 1980, 1991, 1993, 1994).

The second statistical approach incorporates correlation functions and integral equations from the theory of liquids. In this case, the well-known uniform Ornstein–Zernike equations were modified to calculate the ion distribution function near a charged interface. Modified in this way, the Ornstein–Zernike equations became non-uniform and were solved using hyperneted chain approximations (HCA) (Blum et al., 1984; Henderson, 1983), hyperneted chain approximations incorporating a mean spherical approximation for correlation functions (HCA/MSA) (Carnie and Chan, 1981; Carnie et al., 1981a,b; Greberg and Kjellander, 1994; Torrie et al., 1989), or mean spherical approximations (MSA) (Blum, 1977; Henderson, 1990; Henderson and Blum, 1978; Levine and Outhwaite, 1978; Outhwaite, 1970, 1981, 1983; Outhwaite and Bhuiyan, 1982, 1983). A third statistical approach to the electrical double layer was achieved by using the integral Born–Green–Ivon equations (Croxton and McQuarrie, 1979a,b, 1981).

These correlation function approaches resulted in the theory of ionic plasma in semi-infinite space. The next advance in describing the double layer was achieved by upgrading the restricted primitive electrolytes model to a "civilized" or "non-primitive" model. This was accomplished by addition of hard spheres with embedded point dipoles to the restricted "primitive" model (Gruen and Marcelja, 1983a,b; Miklavic and Attard, 1994; Valleau et al., 1980). This theory takes into account the presence of long-range Coulomb interactions, image effects, and external electrical fields. The new "non-primitive" model became the basis for the theory of ion–dipole plasma. The serious mathematical difficulties encountered in describing the "non-primitive" models necessitated the development of less statistically rigorous intermediate models. In these intermediate models, the first monolayer of solvent molecules at the charged interface was considered as a complex of discrete particles and an electrolyte outside this layer was described by the Gouy–Chapman theory. This intermediate approach was able to provide good theoretical agreement with experimental results.

The concept of constant dielectric permittivity in the double layer is also of concern. Different factors, such as high ionic concentration or strong electric fields can affect the dielectric permittivity (Chernenko, 1981a,b; Defay et al., 1966; Markin and Volkov, 1987a, 1989b; Sanfeld, 1968; Sparnay, 1958; Stilinger and Kirkwood, 1960). In addition, description of the dielectric properties of solutions by a *single parameter*—the dielectric permittivity—also came under criticism and led to the development of non-local electrostatics.

Non-local electrostatics takes into consideration discrete properties of solvent molecules (Feldman et al., 1985, 1987; Kornyshev et al., 1982; Kornyshev and Volkov, 1984; Partenskii and Feldman, 1983; Partenskii and Jordan, 1993; Patra and Ghosh, 1994a,b, 1995; Volkov and Kornyshev, 1985; Vorotyntsev and Kornyshev, 1984, 1993; Vorotyntsev and Mityushev, 1991). It assumes that fluctuations of solvent polarization are correlated in space. This means that the average polarization at each point is correlated with the electric displacement at all other points and therefore uses the solvent dielectric function which couples polarization vectors throughout the solvent.

Liquid structures near certain surfaces, especially near the surfaces of biological membranes, are very ordered and therefore generate hydration forces. These ordered layers extend into the diffuse layer and change its properties (Butt, 1991; Cevc, 1990a,b; Cevc et al., 1995; Grabe and Horn, 1993; Granfeldt and Jonsson, 1992; Israelachvili, 1992; Marsh, 1989; Ninham, 1989; Parsegian et al., 1991; Rand and Parsegian, 1989; Rutland and Christenson, 1990; Spitzer, 1992). Analysis of hydration forces reveals important features of the interface structure.

In the previous section we have discussed the shortcomings of the Gouy–Chapman theory and the subsequent modifications made to improve it. In the subsequent chapters we will discuss each modification in detail. It is important to realize that all theories are approximations of the real world and therefore one needs to be certain that a given theory appropriately reflects a given model before progressing to experimental correlations. The final chapters will define the criteria for success of these theoretical models.

7.11. EFFECTS OF VARIABLE DIELECTRIC PERMITTIVITY

Solvation of an ion in a polar liquid produces a new dynamic molecular structure around the ion which is different from its structure in a pure solvent. This makes the dielectric properties of the liquid near the ion different from the bulk properties. The spatial change in ion concentration in the double layer also changes the dielectric permittivity of this layer. Analysis of the dependence of ε on the electrolyte concentrations showed that the potential in the double layer can be changed by up to 10%, though deviations of ε from its bulk value occur only at the surface layer which has a thickness equivalent to the ion hydration shell radius (Chernenko, 1981a,b; Sparnay, 1958).

Electric fields can change the permittivity of a medium containing polar molecules due to perturbation of the intrinsic structure of associated dipoles and the dielectric saturation of the medium (Bradley and Parry-Jones, 1974; Cohen, 1970; Conway *et al.*, 1951; Davies, 1971, 1976; Grahame, 1950; Hurwitz and Steinchen-Sanfeld, 1965; Kornyshev and Volkov, 1984; Levine *et al.*, 1981; Markin and Volkov, 1987a,b, 1988f, 1989b; Volkov and Kornyshev, 1985; Vorotyntsev and Kornyshev, 1979; Wang and Bruner, 1978). Investigation of the dielectric properties of water by Booth (1951a,b,c) showed that an external field can reduce the permittivity of water from a static value $\varepsilon = 80$ to the optical value ε_{op}, which is equal to the refractive index squared, i.e., 1.8. Saturation occurs when field strength exceeds 10^9 V m^{-1}. However, because of technical limitations this concept has not been verified for fields exceeding 10^7 V m^{-1} (Fig. 7.43). Nevertheless, at 10^7 V m^{-1} the permittivity has been shown to decrease by less than 0.1%. This topic has been extensively reviewed by several authors (Booth, 1951a,b,c; Davies, 1971, 1976; Kolodziej *et al.*, 1975; Malsch, 1928, 1929).

Webb (1926) first considered the dielectric saturation of a solution near an ion for calculation of hydration energy. Incorporation of dielectric saturation into Born's theory of solvation (Born, 1920) improved calculation of hydration energy to a small degree. Other attempts to improve hydration energy calculation were met with much objection (Kornyshev and Volkov, 1984; Volkov and Kornyshev, 1985). However, recently there has been a new surge of interest in this problem and several new approaches have been put forward (Abraham *et al.*, 1979; Markin and Volkov, 1987b,d).

Now if we apply the same idea, which we discussed in section 6.3, to the double layer we will find that corrections to the Gouy–Chapman theory will become significant only when the interfacial potential deviates from that of zero charge (Chernenko, 1981a,b; Sanfeld, 1968; Sparnay, 1958). These corrections for a charged planar interface without specific adsorption of ions are due mainly to the dependence of the dielectric constant of the solution on the local concentration of the electrolyte. Under normal experimental conditions, corrections for the volume of ions and for the

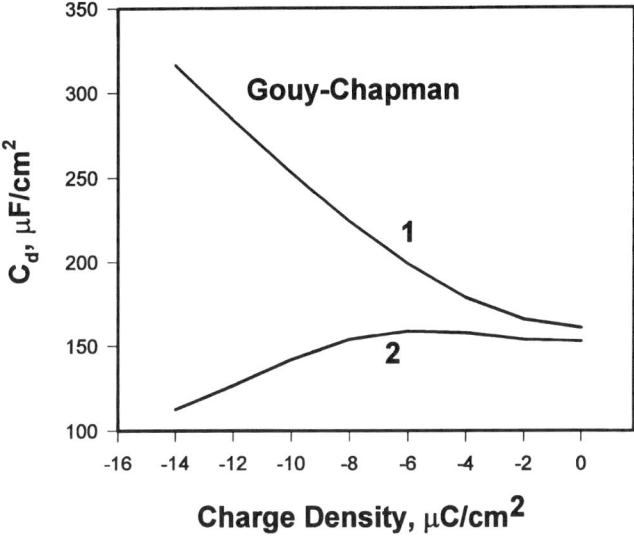

Figure 7.43. Capacitance of the diffuse double layer at a planar ideally polarized electrode without specific adsorption calculated according to the Gouy–Chapman theory without (1) and with (2) the effect of the dielectric saturation according to Chernenko (1981b).

dielectric saturation of the solvent by the interfacial field are small and can be disregarded.

The effects of dielectric saturation in the diffuse layer without specific adsorption of ions were analyzed in several papers (Chernenko, 1981a,b; Sanfeld, 1968; Sparnay, 1958). The Poisson equation for a planar interface can be written as

$$\frac{d}{dx}(\varepsilon E) = 4\pi e_0(c_+ - c_-), \tag{7.138}$$

where c_+ and c_- are the ion concentrations, e_0 is the elementary charge, x is the coordinate measured from the plane of closest approach of ions. Forces F_+ and F_- acting on the hydrated cation and anion can be represented by the expression:

$$F_{\pm} = \pm e_0 E - \frac{v}{8\pi}\frac{d}{dx}E^2, \tag{7.139}$$

where

$$v = \frac{4\pi\varepsilon_2(\varepsilon_2 - \varepsilon_1)}{2\varepsilon_2 + \varepsilon_1}r^3, \tag{7.140}$$

ε_2 is the dielectric constant of the pure solvent, ε_1 is the high-frequency value for the dielectric permittivity of the pure solvent, and r is the radius of the hydrated ion. Equilibrium concentrations can be calculated from the expression

$$c_\pm = c_0 \exp\left(-\frac{\upsilon}{8\pi RT}E^2 \mp \frac{zF\phi}{RT}\right), \tag{7.141}$$

where c_0 is the bulk ion concentration. Since the electrolyte solution can be treated as a two-component dielectric mixture of pure solvent and hydrated ions, Chernenko (1981a,b) used the Maxwell equation for electric permittivity:

$$\varepsilon = 3\varepsilon_2\left(1 + \tau\frac{\varepsilon_2 - \varepsilon_1}{2\varepsilon_2 + \varepsilon_1}\right)^{-1} - 2\varepsilon_2, \tag{7.142}$$

where τ is the volume fraction of the component with dielectric constant ε_1. Combining equations (7.140) and (7.142) yields the dependence of dielectric permittivity on the electrolyte concentration:

$$\varepsilon = 3\varepsilon_2\left[1 + \frac{\upsilon}{3\varepsilon_2}(c_+ + c_-)\right]^{-1} - 2\varepsilon_2. \tag{7.143}$$

It follows from equation (7.154) that if $\upsilon(c_+ + c_-)/3\varepsilon_2 \ll 1$ then

$$\varepsilon = \varepsilon_2 - \upsilon(c_+ + c_-). \tag{7.144}$$

It has been shown that equation (7.144) is a good approximation for ε up to an electrolyte concentration of 4 M. The complete solution of this problem is given by the following set of equations:

$$\phi_d = \frac{RT}{zF}\left\{ \upsilon c_0 \bar{E}^2 + \ln\left[\frac{1 + \bar{E}^2/2}{1 + 2\upsilon c_0 \bar{E}^2} + \mathrm{sgn}\,(\bar{E})\sqrt{\left(\frac{1 + \bar{E}^2/2}{1 + 2\upsilon c_0 \bar{E}^2}\right)^2 - \exp\left(-2\upsilon c_0 \bar{E}^2\right)}\right]\right\}, \tag{7.145}$$

$$C_d = \frac{\varepsilon_2}{4\pi\kappa}\sqrt{\frac{1}{\bar{E}^2}\left[\left(\frac{1 + \bar{E}^2/2}{1 + 2\upsilon c_0 \bar{E}^2}\right)^2 - \exp\left(2\upsilon c_0 \bar{E}^2\right)\right]}, \tag{7.146}$$

$$c_\pm = c_0\left[\frac{1 + \bar{E}^2/2}{1 + 2\upsilon c_0 \bar{E}^2} \mp \mathrm{sgn}\,(\bar{E})\sqrt{\left(\frac{1 + \bar{E}^2/2}{1 + 2\upsilon c_0 \bar{E}^2}\right)^2 - \exp\left(-2\upsilon c_0 \bar{E}^2\right)}\right], \tag{7.147}$$

where $\upsilon = (\upsilon/\varepsilon_0)N_A\,10^{-3}\,\mathrm{l\,mol^{-1}}$, N_A is Avogadro's number, and $\bar{E}$ is a dimensionless parameter. Figures 7.43 and 7.44 show the influence of dielectric saturation in the diffuse layer model of hydrated ions without specific adsorption according to Chernenko (1981b).

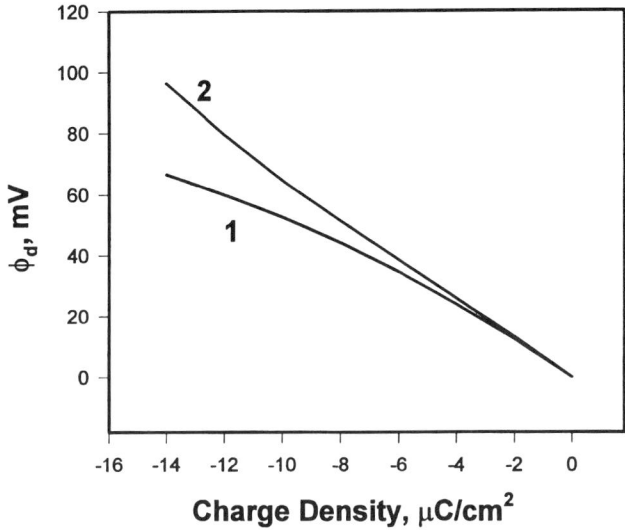

Figure 7.44. The potential of the diffuse layer at a planar ideally polarized electrode without specific adsorption calculated according to the Gouy–Chapman theory without (1) and with (2) the effect of the dielectric saturation according to Chernenko (1981b).

7.12. NON-LOCAL ELECTROSTATICS

Non-local electrostatics does not accept the classical description of a dielectric medium by the single parameter of dielectric permittivity ε. Instead, it assumes that fluctuations of solvent polarization in a liquid are correlated in space due to quantum interactions between molecules. This means that the average polarization at each point in the dielectric depends on the dielectric polarization at all other points in the medium (Kornyshev and Volkov, 1984; Vorotyntsev and Kornyshev, 1993). The method of non-local electrostatics was discussed in section 6.5.

The non-local electrostatic method belongs to the continuum models, but the effective parameters needed for calculations are chosen by analyzing the discrete solvent structure. This method gives accurate values for the solvation energy of small ions, and also permits calculation of the resolvation energy. For large ions there is a considerable discrepancy between theory and experiment, which makes it necessary to take into account some other effects. The most important contribution is given by the work of cavity formation in the solvent and the solvophobic effect.

Kornyshev and Ulstrup (1985a,b) developed a theory of the diffuse double layer at the metal–electrolyte interface which was described in terms of the dielectric function $\varepsilon(r, r')$. At low electrolyte concentrations and small electrical charges at an electrode the capacitance of the diffuse layer was obtained as a series expansion:

$$\frac{\varepsilon_0}{C_d} = \frac{1}{\varepsilon \kappa} + \frac{N\lambda^2}{\varepsilon} + M\frac{\kappa}{\varepsilon}\lambda^2. \tag{7.148}$$

The first term in this equation is the Gouy–Chapman diffuse layer capacitance, which depends on the electrolyte concentration. The second term represents the non-local structural effects of the solvent, which is concentration independent and can be interpreted as the inverse capacitance of the compact layer. Here, $N\lambda$ corresponds to the effective thickness of the Helmholtz layer, λ is the polarization correlation radius, and N is a coefficient equal to

$$N = \frac{1 - \varepsilon^*/\varepsilon}{\sqrt{\dfrac{\varepsilon^*}{\varepsilon}}\left(\sqrt{\dfrac{\varepsilon^*}{\varepsilon}} + \dfrac{1-\rho}{1+\rho}\right)}. \tag{7.149}$$

Parameter ρ is a fitting parameter that defines the reflection of polarization fluctuation waves from the surface (mirror reflection is represented when $\rho = 1$, and diffuse reflection when $\rho = 0$).

The third term in (7.148) is concentration dependent. The medium constant M is defined by the following equation:

$$M = \frac{1 - \dfrac{\varepsilon^*}{\varepsilon}\dfrac{1-\rho}{1+\rho} - 3\sqrt{\dfrac{\varepsilon^*}{\varepsilon}}}{2\dfrac{\varepsilon^*}{\varepsilon}\dfrac{1-\rho}{1+\rho} + \sqrt{\dfrac{\varepsilon^*}{\varepsilon}}}. \tag{7.150}$$

The sum of the first and third terms in equation (7.148) can be written as

$$\frac{\varepsilon_0}{\overline{C}_d} = \frac{1 + M\kappa^2\lambda^2}{\varepsilon\kappa}. \tag{7.151}$$

This quantity represents the concentration-dependent part of the diffuse layer capacitance, which at low concentrations reduces to the Gouy–Chapman expression. Thus, capacitance is weakly dependent on the type of electrolyte or on the value of the closest approach δ, since it is divided by a large value of ε. This is similar to the situation in the theory of ion–dipole plasma to be discussed later, where the ion radius a in the equations for capacitance is also divided by the dielectric permittivity. The solvent contribution to the double layer capacitance in non-local electrostatics and in ion–dipole plasma theories may look similar but the physical meaning of the corresponding terms is quite different.

Estimates of Parsons–Zobel relationships for metal electrodes at high 1:1 electrolyte concentrations show good agreement between theory and experimental data (Kornyshev and Ulstrup, 1985a,b). However, experimental data at nitrobenzene–water and 1,2-dichloroethane–water interfaces disagree with the theory at low electrolyte concentrations (see Figs. 7.13 and 7.14).

7.13. ELECTRIC DOUBLE LAYER WITH HYDRATION FORCES

Interactions between phospholipid membranes at very short distances, of the order 1– 3 nm, are controlled by hydration forces. Such hydration forces can be measured between various amphiphilic colloidal surfaces and phospholipid bilayers using hydrostatic or osmotic pressure techniques (Homola and Robertson, 1976; LeNeveu *et al.*, 1976), or the Surface Force Apparatus technique (Marra and Israelachvili, 1985). Although different manifestations of these forces have been observed for many years (Clunie *et al.*, 1967; Langmuir, 1938), they became the object of intensive investigations in the last decade (Israelachvili, 1992; Rand and Parsegian, 1989).

Figure 7.45 from Rand and Parsegian (1989) presents the pressure versus separation relationship for palmitoyloleoyl-phoshatidylethanolamine (POPE) and stearoyloleoyl-phoshatidylcholine (SOPC) bilayers. Pressure is plotted on a logarithmic scale, and at small separations d_w the dependence on separation distance is exponential:

$$P = P_0 e^{-d_w/\lambda}. \tag{7.152}$$

Other measurements of pressure versus separation also suggest an exponential relationship at high pressures with a decay distance (λ) between 0.1–0.3 nm (Rand and Parsegian, 1989).

There have been many theoretical attempts to rationalize this phenomenon. One approach to the problem was suggested by Marcelja and co-workers (Gruen and Marcelja, 1983a,b; Gruen *et al.*, 1984; Marcelja and Radic, 1976) which was based on non-local electrostatics. A polar surface can perturb and polarize an aqueous solvent immediately adjacent to it and this polarization exponentially diminishes as

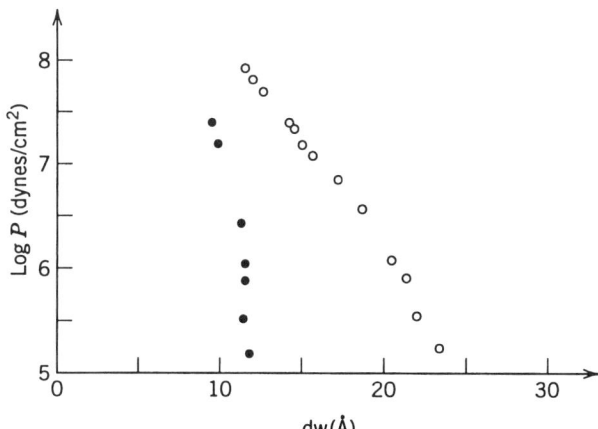

Figure 7.45. Pressure versus separation for POPE (□) and SOPC (■) at 30 °C (Rand and Parsegian, 1989). Reproduced by permission of Elsevier Science Publishers.

the distance from the polar surface increases. Polarization of the water layer will cause two adjacent polar surfaces (lipid membranes) to experience a strong repulsion force at small distances. The ordering of the solvent influences the charge and potential distribution near an electrified interface, and this in turn changes the Debye shielding length (Radic and Marcelja, 1978; Torrie *et al.*, 1989). A brief discussion of this phenomenon based on the concepts of non-local electrostatics (Gruen *et al.*, 1984) will now be presented.

Electrostatic induction $D(r)$ is the sum of the electrical field $E(r)$ and medium polarization $P(r)$:

$$D(r) = \varepsilon_0 E(r) + P(r). \tag{7.153}$$

In classical electrostatics the local response of a dielectric medium to an electrical field is proportional to this field, $P(r) = \chi \varepsilon_0 E(r)$, and the dielectric permittivity is $\varepsilon = 1 + \chi$. This is a very good approximation for bulk dielectrics but at dielectric interfaces and in the vicinity of charges the situation is more complicated.

An electrified surface perturbs the neighboring layers of the dielectric in two ways. First, the electrical field of the surface orients molecules of the dielectric according to classical electrostatics, but the presence of the surface *per se* also imposes constraints on potential molecular orientations. As a result, polarization may not be proportional to the local electrical field strength, although these parameters are closely interrelated.

The spatial variation of E and P can be found from the principle of minimal free energy of the entire system. To illustrate this approach, we will derive the classical Gouy–Chapman equation by this method. Generally speaking, the density of the Gibbs free energy $g(E)$ should contain terms proportional to E and its derivatives, and because of symmetry these quantities should be raised to the second power. Therefore, for the limiting case of small potentials, the linearized equation for the Gibbs free energy density of an electrical double layer can be written as follows:

$$g(E) = \frac{\varepsilon_0 \varepsilon E^2}{2} + \frac{\lambda_D^2 \varepsilon_0 \varepsilon}{2} \left(\frac{dE}{dz} \right)^2, \tag{7.154}$$

where λ_D stands for the Debye length.

At equilibrium the Gibbs free energy should be at its minimum, its variation with respect to E should be zero, and this results in the familiar equation of the linearized theory of Gouy–Chapman:

$$\lambda_D^2 \frac{d^2 E}{dx^2} = E. \tag{7.155}$$

It is well known that the solution for this equation decays exponentially with a characteristic length λ_D.

Implementation of this same methodology for non-local electrostatics and hydration forces is more complicated and therefore only the results of the theory are presented (Gruen and Marcelja, 1983b). At small distances the standard relationship $P(r) = \chi\varepsilon_0 E(r)$ no longer holds true; therefore one needs to include both $E(r)$ and $P(r)$ in the Gibbs free energy equation.

The overall polarization $P(r)$ of a medium is the sum of the orientational $P_c(r)$ and deformational $P_a(r)$ polarizations of its molecules:

$$P(r) = P_c(r) + P_a(r). \tag{7.156}$$

Since the effects of non-local interactions are caused by orientational polarization the electrostatic induction can be represented as

$$D = \varepsilon_0 E + P_a + P_c = \varepsilon_0 \varepsilon^{op} E + P_c, \tag{7.157}$$

where ε^{op} designates the optical dielectric permittivity. For approximation of small fields, the differential equations that minimize the integral of free energy in the one-dimensional case are derived as follows:

$$\lambda_D^2\left(\varepsilon_0 \varepsilon^{op} \frac{d^2 E}{dz^2} + \frac{d^2 P_c}{dz^2}\right) = \varepsilon_0 \varepsilon E, \tag{7.158}$$

$$\xi^2 \frac{d^2 P_c}{dz^2} = \frac{\varepsilon^{op}}{\varepsilon}(P_c - \varepsilon_0 \chi_c E), \tag{7.159}$$

where χ_c is the orientational susceptibility of the dielectric, and ξ is the characteristic radius of correlation between solvent molecules. It is very instructive to analyze a few limiting cases for these equations.

In classical electrostatics $P_c = \chi_c \varepsilon_0 E = \varepsilon_0(\varepsilon - \varepsilon^{op})E$; therefore, equation (7.158) transforms into equation (7.155) and the linearized case of the Gouy–Chapman theory is recovered.

If there is no external field and the solvent is polarized at a surface, we can write $\varepsilon_0 \varepsilon^{op} E = -P_c$ and it follows from equation (7.159) that $\xi^2(d^2 P_c/dz^2) = P_c$. This equation shows that solvent polarization decays with a characteristic length ξ.

From inspection of equations (7.158) and (7.159) one can conclude that a general solution for the functions $E(x)$ and $P(x)$ should contain terms with two exponentials that decay with characteristic lengths λ_1 and λ_2. In the case where $\xi \ll \lambda_D$ these characteristic lengths are:

$$\lambda_1 = \lambda_D\left[1 + \left(\frac{\varepsilon}{\varepsilon^{op}} - 1\right)\frac{\xi^2}{\lambda_D^2}\right]^{1/2}, \tag{7.160}$$

$$\lambda_2 = \xi\left[1 - \left(\frac{\varepsilon}{\varepsilon^{op}} - 1\right)\frac{\xi^2}{\lambda_D^2}\right]^{1/2}. \tag{7.161}$$

The first equation (7.160) gives the renormalized Debye length (λ_1) and equation (7.161) gives the corrected correlation length (λ_2). The new value for the Debye length (λ_1) is then slightly larger than its classical counterpart (λ_D) because the

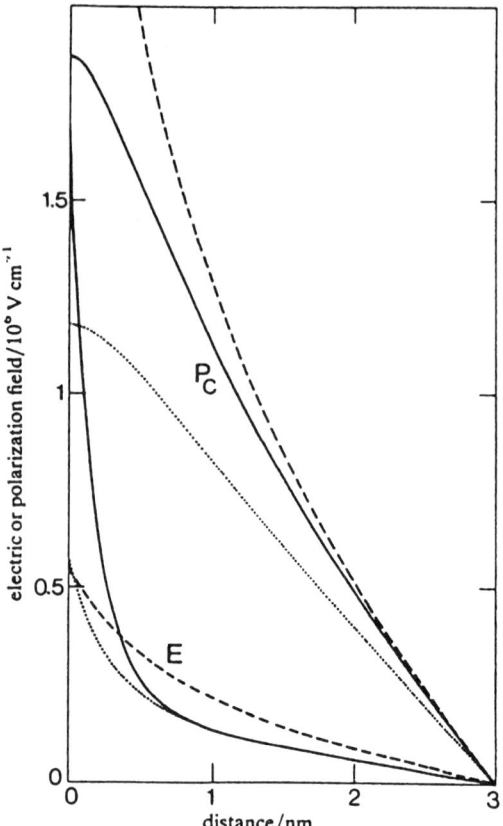

Figure 7.46. Behavior of electric and polarization fields in a system without specific surface–solvent interactions. The solutions were obtained for two identical parallel flat plates positioned 6 nm apart. The surface potential is 100 mV and the Debye length is $\lambda_D = 3$ nm. Solid curve—E and P_c in the Gruen and Marcelja model; dashed curve—the non-linear Gouy–Chapman theory; dotted curve—the linearized version of the Gruen and Marcelja (1983b) model. Reproduced by permission of The Royal Society of Chemistry.

dielectric medium resists the spatial variation of polarization. The solutions for functions P and E are both biexponential functions where λ_1 defines the long-range exponential component and λ_2 defines the short-range component. The short-range component is produced by rapid changes of the orientational polarization of solvent molecules. Because E and P_c are not locally related to one another, their values are determined by boundary conditions at the surface of the medium.

To define the boundary conditions, one can either stipulate the intensity of the electrical field $E(0)$ and the orientational polarization $P_c(0)$, or introduce a surface charge σ. These parameters are then related by the equation

$$\sigma = \varepsilon_0 \varepsilon^{op} E(0) + P_c(0). \tag{7.162}$$

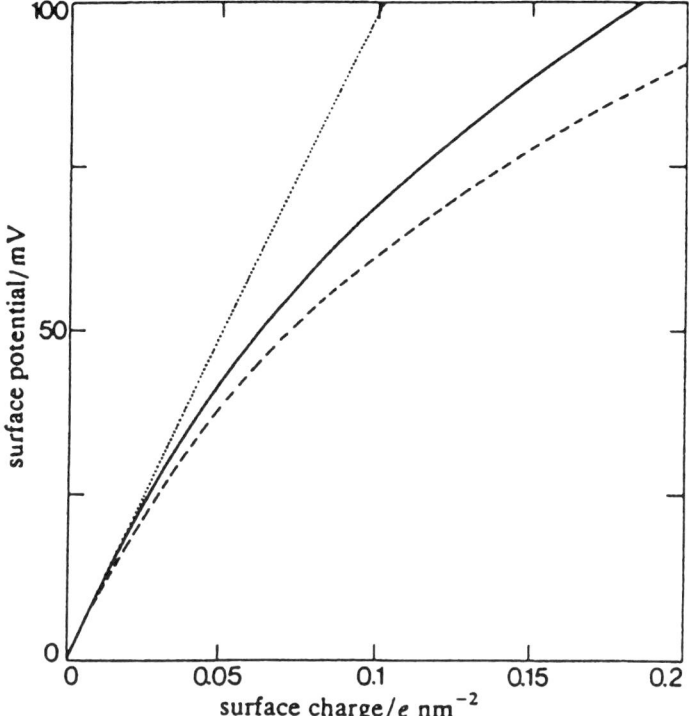

Figure 7.47. Relationship between the surface potential and the surface charge for two identical parallel plates positioned 6 nm apart, λ_D=3 nm. Solid curve—the Gruen–Marcelja model with no specific surface interactions; dashed curve—the Gouy–Chapman theory; dotted curve—the linearized version of the Gruen–Marcelja (1983b) model. Reproduced by permission of The Royal Society of Chemistry.

Gruen and Marcelja (1983a,b) derived the generalized Poisson–Boltzmann equation that includes hydration effects. It was numerically solved and we will present several examples from their paper (Gruen and Marcelja, 1983a,b). Figure 7.46 presents the distribution of electrical field and polarization between two identical charged plates. The parameters of this model were $\varepsilon = 80$, $\varepsilon^{op} = 6$, $\xi = 0.3$. The surface potential was 100 mV, and the Debye length was $\lambda_D = 3$ nm. The surface was not polarized. The difference in the Gruen–Marcelja model compared with the non-linear Gouy–Chapman theory is the lower value of polarization and correspondingly higher electric field at the surface. In the Gouy–Chapman model surface polarization would be 3.3×10^6 V cm^{-1} or approximately one eighth of the value for fully polarized water. For a surface polarization less than 25 mV, the non-linear and linearized theories of Gruen and Marcelja, and the Gouy–Chapman theory, all give nearly identical results.

Classical Gouy–Chapman theory describes the well-known relationship between

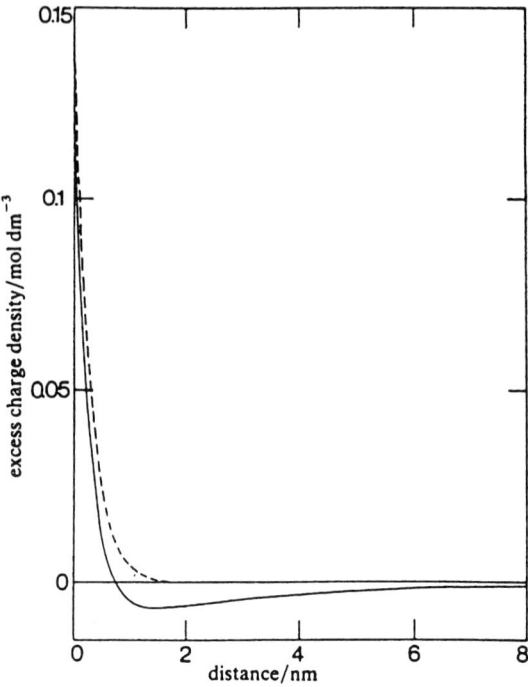

Figure 7.48. Screening of the solvent surface polarization by ions. In this example Gruen and Marcelja (1983b) have a single surface in which the polarization has been set at $0.1 P_0$ (pointing from the surface into the aqueous solution). P_0 is the maximal value corresponding to full orientation of water dipoles. The result has been derived with the linearized theory. The Debye length is $\lambda_D = 3$ nm corresponding to a bulk concentration of ions of 0.01 M. The curves show the excess charge density as a function of distance from surfaces with charge density $\sigma = 0$ (solid curve) and $\sigma = -0.024 e_0$ nm^{-2} (dashed curve). Reproduced by permission of The Royal Society of Chemistry.

surface charge and surface potential (Fig. 7.47). Taking non-local interactions into account changes this relationship, so that hydration forces give larger surface potentials for the same charge density. For example, even if the surface bears no charge, the surface layer of the solvent is oriented. The dipoles of the oriented layer create a drop of electrical potential. The solvent polarization, imposed at the surface, will gradually decay in the bulk of the solution. It will cause redistribution of charge, which will shield the polarization. Figure 7.48 shows the distribution of electrical charge near the surface with charge densities $\sigma = 0$ and $\sigma = -0.024 e_0$ nm^{-2}. The surface polarization was chosen to be $0.1 P_0$, where P_0 is the maximal polarization, which corresponds to complete orientation of all water dipoles. In the second case, where $\sigma = -0.024 e_0$ nm^{-2}, the charge corresponds to a polarization of $-0.013 P_0$. The surface polarization has a positive sign, caused by orientation of the positive poles of water molecules from the surface to the solvent volume. Note that the

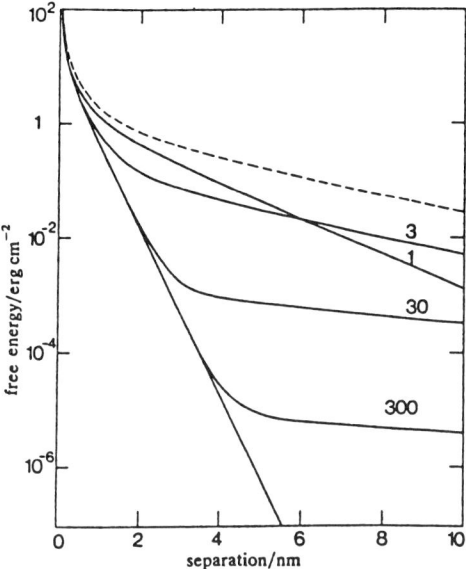

Figure 7.49. Interaction of two identical parallel planar surfaces within the linearized model (Gruen and Marcelja, 1983b). The polarization at the surface of the plates is assumed to have a constant value of $0.1P_0$, pointing toward the mid-plane. For all the solid lines there is no surface charge. The interaction free energy, shown for different values of λ_D (in nm), indicates the existence of strong, short-range hydration forces. The bottom curve corresponds to $\lambda_D = \infty$. A surface charge $\sigma = 0.033e\,nm^{-2}$ and a Debye length $\lambda_D = 3\,nm$ are shown as a dashed line. Reproduced by permission of The Royal Society of Chemistry.

positive charges are closer to the surface, while negative ones are shifted into the bulk solution. The charges are separated not by an external electrical field but rather by the field created by the oriented dipoles. The charge distribution is described by two exponentials with characteristic lengths $\lambda_2 \approx \xi$ and $\lambda_1 \approx \lambda_D$, representing short-range and long-range effects. The layer of positive charges near the surface is very thin and concentrated, while the layer of negative charges is rather extended with a smaller charge density.

Although the surface does not have a net electrical charge, from the point of view of the double-layer classical theory, it appears to carry a positive electrical charge. This effect shows up when distances exceed the Debye length. If a negative charge, corresponding to $-0.013P_0$, is placed at the surface, this dramatically changes the distribution of charges within the solution, as shown by the dashed line in Fig. 7.48. The long-range component of charge distribution (negative charge) disappears while the short-range component with positive charges remains and shields the surface charge.

This generalized theory can be applied to the interaction of two parallel planar surfaces. The interaction of free energy was defined as the free energy of the system minus the free energy of two surfaces at infinite separation. Figure 7.49 presents the results of calculations assuming that the surface–solvent polarization remains constant and equal to $0.1P_0$ and there is no surface charge. Gruen and Marcelja (1983a,b) estimated that this approximation should work at separations of more than 1 nm. Up to this distance the interaction free energy is small compared with kT per water molecule and cannot perturb the oriented layer.

As seen in Fig. 7.49, there are two regions of interaction between the surfaces, with a transition determined by the characteristic length ξ. For large separations, the interaction is given by classical double-layer repulsion of plates carrying an effective surface charge $0.1P_0$. Again, there is no actual charge at the surfaces and the double-layer interaction is generated by charge separation at the polarized surfaces. The apparent charge depends on the ionic strength of the solution, and increases as the Debye length decreases. When the distance between the surfaces decreases up to ξ, the interaction is dominated by the surface-induced polarization at the plates. The energy, and hence the repulsion, increases in this region exponentially with characteristic length $\lambda_2 \approx \xi$. These results qualitatively correspond to the experimental observations of Fig. 7.45, and with fitting parameters it is not very difficult to reach quantitative agreement.

The theory developed by Gruen and Marcelja (1983b) is based on a continuum model. As a result, it is not very accurate for describing phenomena where the effects of discreteness are important. For example, it smoothes oscillations in ion distribution caused by the finite size of molecules. In real liquids, attenuated oscillations caused by the finite size of molecules are inevitable and have been confirmed experimentally (Horn and Israelachvili, 1981a,b,c). This phenomenon will be discussed later. The non-local electrostatics approach to the problem of the double layer is used only for describing solvent dielectric properties, while the description of ion distribution follows the classical Gouy–Chapman approach. Radic and Marcelja (1978) made one attempt to take into account the finite size of ions. The theory of hydration forces is a very dynamic topic, and has been reviewed by Rand and Parsegian (1989) and Cevc *et al.* (1995).

7.14. MODIFIED POISSON–BOLTZMANN (MPB) MODEL

The popular Poisson–Boltzmann equation considers the mean electrostatic potential in a continuous dielectric with point charges and is therefore an approximation of the actual potential. An improved model and mathematical solution resulted in the modified Poisson–Boltzmann (MPB) equation (Outhwaite *et al.*, 1980). This equation is based on a restricted primitive electrolyte model that considers ions as charged hard spheres with diameter d in a continuous uniform structureless dielectric medium of constant dielectric permittivity ε. The sphere representing an ion has the same permittivity ε. The model initially was developed for an electrolyte at a hard wall with dielectric permittivity ε_w and surface charge density σ. The charge is distributed over the surface evenly and continuously.

This theory takes into account the finite size of ions, the fluctuation potential, and image forces in the electrolyte solution next to a rigid electrode, but it is still an approximation. The MPB theory begins with the Poisson equation for the mean electrostatic potential ψ in solution:

$$\frac{d^2\psi}{dx^2} = -\frac{1}{\varepsilon_0 \varepsilon} \sum_i z_i e_0 n_i^0 g_{0i}(x), \tag{7.163}$$

where ε_0 is the electrical constant, z_i is the charge number of ion i, e_0 is the elementary charge, n_i^0 is the bulk number density, and $g_{0i}(x)$ is the wall–ion distribution function representing the ratio of the local ion density to the ion density of the bulk electrolyte solution. The distance x is measured from the electrode so that the distance of closest approach of an ion i is $d/2$. According to Kirkwood (1934, 1965) the distribution function is given by the following equation:

$$g_{0i}(x) = \zeta_i(x) \exp\left[-\frac{1}{kT}z_i e_0 \psi(x) + \eta_i\right]. \tag{7.164}$$

The factor $\zeta_i(x)$ accounts for the excluded volume of the ion, η_i is the fluctuation potential that takes into consideration the ionic atmosphere around an ion created by other ions. The fluctuation potential can then be presented as

$$\eta_i = -\frac{1}{kT}\int_0^{z_i e_0} \lim_{r\to 0}(\phi_i^* - \phi_{ib}^*)\, d(z_i e_0), \tag{7.165}$$

where

$$\phi_i^* = \phi_i - \frac{z_i e_0}{4\pi\varepsilon\varepsilon_0 r}, \tag{7.166}$$

ϕ_i is the fluctuation potential created by the ion i and its atmosphere, and r is the distance from the center of the ion. The difference ϕ_i^* is the potential created by the ionic atmosphere only and ϕ_{ib}^* is the value of this potential in the bulk electrolyte solution. There are several methods for calculating ϕ_i and, if the image effect is taken into account, it can be presented as

$$\phi_i = B_0\left[\frac{1}{r}\exp(-\kappa r) + \frac{f}{r^*}\exp(-\kappa r^*)\right], \qquad \text{if } r>d, \quad x>0, \tag{7.167}$$

where f is the image factor determined by the difference between two dielectric constants of the electrolyte solution and the electrode:

$$f = \frac{\varepsilon - \varepsilon_w}{\varepsilon + \varepsilon_w}. \tag{7.168}$$

The local Debye–Huckel parameter at a given point is

$$\kappa^2 = \frac{e_0^2}{\varepsilon_0 \varepsilon kT} \sum_i z_i^2 n_i^0 g_{0i}(x). \tag{7.169}$$

In equation (7.167) B_0 is a constant, and r^* is the distance from the mirror image in the interface. Hence equation (7.167) takes into account both the ionic atmosphere and image forces.

If $r < d$ and $x > 0$, the equation for the fluctuation potential should be written as

$$\nabla^2 \phi_i = -\frac{d^2\psi}{dx^2}, \qquad r < d, \quad x > 0. \tag{7.170}$$

The term $\zeta(x)$ can be presented as a power series expansion in the bulk density:

$$\zeta_i(x) = 1 + B(x) + \frac{4\pi d^3}{3} \sum_j n_j^0, \tag{7.171}$$

where

$$B(x) = \pi \sum_j n_j^0 \begin{cases} \displaystyle\int_{d/2}^{x+d} [(x'-x)^2 - d^2] g_{0j}(x')\, dx', & \dfrac{d}{2} \le x \le \dfrac{3d}{2}, \\[4mm] \displaystyle\int_{x-d}^{x+d} [(x'-x)^2 - d^2] g_{0j}(x')\, dx', & x \ge \dfrac{d}{2}. \end{cases} \tag{7.172}$$

Now we can combine these equations and obtain the modified Poisson–Boltzmann equation for the diffuse layer:

$$\frac{d^2\psi}{dx^2} = -\frac{1}{\varepsilon_0 \varepsilon} \sum_i z_i e_0 n_i^0 \zeta_i(x) \exp\left\{ -\frac{z_i e_0}{kT}\left[L_1(\psi) + \frac{z_i e_0}{8\pi \varepsilon_0 \varepsilon d}(F - F_0) \right] \right\},$$

$$x > \frac{d}{2}. \tag{7.173}$$

Since the distance of the maximal approach of ions to the interface cannot be less than $d/2$, the usual linear dependence of potential in the compact layer can be introduced:

$$\psi(x) = \psi(0) + x\frac{d\psi(0)}{dx}. \tag{7.174}$$

Additional functions in equation (7.173) are:

$$\zeta_i(x) = 1 + \frac{4\pi d^3}{3}\sum_i n_i^0 + \pi \sum_i n_i^0 \int_{\max(d/2, x-d)}^{x+d} [(x'-x)^2 - a^2] g_{0i}(x')\, dx'. \tag{7.175}$$

$$L_1(\psi) = \frac{F}{2}[\psi(x+d) + \psi(x-d)] - \frac{F-1}{2d}\int_{x-d}^{x+d} \psi(x')\, dx'. \tag{7.176}$$

$$F = \cfrac{\left\{\begin{array}{l} 2(1+\kappa x)\exp(-\kappa d) + 2(1-\kappa d)\exp(-\kappa x) + 2\kappa^2\,dx \displaystyle\int_x^a r^{-1}\exp(-\kappa r)\,dr \\[2ex] + \dfrac{f}{\kappa x}[\exp(-\kappa d) + 2\kappa(d-x)\exp(-\kappa x) - \exp(-\kappa(2x+d))] \end{array}\right\}}{\left\{\begin{array}{l} 2[\exp(-\kappa d)(1+\kappa(x+d)) + \exp(-\kappa x)] \\[2ex] - \dfrac{f}{\kappa x}[2\kappa x\exp(-\kappa x) + (1+\kappa d)(\exp(-2\kappa x)-1)\exp(-\kappa d)] \end{array}\right\}},$$

(7.177)

$$\frac{d}{2} \le x \le d.$$

$$F = \frac{(1+f\delta_2)}{(1+\kappa d)(1-f\delta_1)}, \qquad \text{if } x \le d. \tag{7.178}$$

$$F_0 = \lim_{x\to\infty} F = \frac{1}{1+\kappa_0 d}. \tag{7.179}$$

$$\delta_2 = \frac{1}{2\kappa x}\exp[\kappa(d-2x)]\sinh\kappa d. \tag{7.180}$$

$$\delta_1 = \frac{(\kappa d\cosh\kappa d - \sinh\kappa d)\delta_2}{(1+\kappa d)\sinh\kappa d}. \tag{7.181}$$

$$\kappa_0^2 = \lim_{x\to\infty}\kappa^2 = \frac{e_0^2}{\varepsilon_0\,\varepsilon kT}\sum_i z_i^2 n_i^0. \tag{7.182}$$

The boundary conditions for these equations are:

1) $\psi(x)$, $\psi'(x) \to 0$, if $x \to \infty$,
2) $\psi'(0+) = -\delta/\varepsilon_0\varepsilon$, $\qquad\qquad\qquad\qquad$ (7.183)
3) $\psi(x)$, $\psi'(x)$ are continuous at $x > 0$.

This set of equations completes the mathematical formulation of the problem, but the equations are very cumbersome and can only be solved numerically. The major difficulty is related to the finite ion size. If we consider the limiting case of point charges, the equation for potential in the double layer would be

$$\frac{d^2\psi}{dx^2} = -\frac{1}{\varepsilon_0\varepsilon}\sum_i n_i^0 z_i e_0 \exp\left\{-\frac{z_i e_0}{kT}\left[\psi(x) + \frac{z_i e_0}{8\pi\varepsilon_0\varepsilon}\left\{\kappa_0 - \kappa + \frac{f}{2x}\exp(-2\kappa x)\right\}\right]\right\}.$$

(7.184)

In comparison with the Gouy–Chapman theory, the MPB equation for point charges contains two improvements in that the Debye–Huckel parameter κ depends on distance, and the ion image is screened.

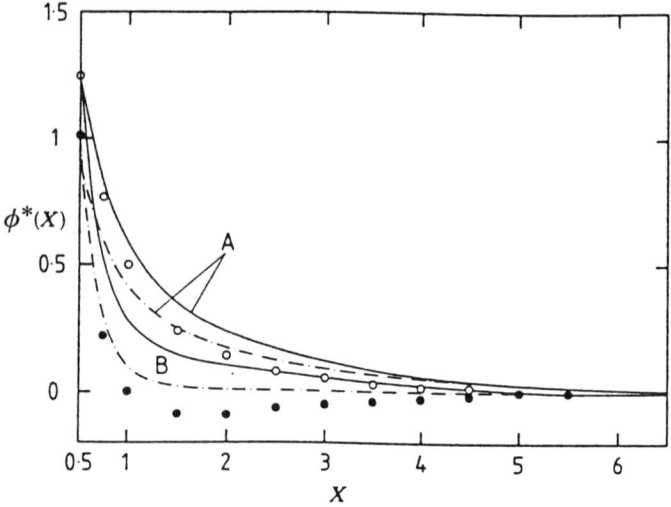

Figure 7.50. Mean electrostatic potential for a 1:2 electrolyte with $c = 0.05$ M. Continuous curves A and B are for $\sigma^* = 0.126$ and $\sigma^* = 0.284$, respectively. The Monte Carlo points ($\bigcirc$ and $\bullet$ for cases A and B, respectively) are those of Torrie and Valleau while the dashed-dotted line denotes MPB4 curve (Outhwaite and Bhuiyan, 1983). Reproduced by permission of The Royal Society of Chemistry.

In solving the MPB equations, different estimates of $\zeta_i(x)$ were considered. Depending on the type of $\zeta_i(x)$, the equations were called MPB1, MPB2, MPB3, and so on. Further details can be found in Outhwaite et al. (1980).

In the papers of Outhwaite et al. (1980) and Outhwaite and Bhuiyan (1983, 1991a,b) the MPB equation was solved numerically using a quasi-linearization iteration technique. It was shown (Gordillo et al., 1990) that the results of MPB theory coincide with Gouy–Chapman predictions at low electrolyte concentrations and small surface charges. Figure 7.50, obtained with the help of MPB equation (Outhwaite and Bhuiyan, 1983), shows the distribution of the mean electrostatic potential ψ^* in a double layer near a metal electrode for a 2:1 electrolyte with the following parameters: $d = 0.425$ nm, $T = 298$ K, $\varepsilon = 78.5$ and electrolyte concentration $c = 0.05$ M. The surface charge σ^*, potential ψ^* and coordinate X were presented in a dimensionless form:

$$\sigma^* = \sigma d^2/e_0, \qquad \psi^* = e_0 \psi/kT, \qquad X = x/d \qquad (7.185)$$

The Gouy–Chapman theory in this case predicts a much higher potential. The corresponding distribution of ions is shown in Fig. 7.51.

An even more interesting picture emerges at higher electrolyte concentrations. Figure 7.52 shows the potential distribution in the double layer at an ion concentration of 0.5 M. The difference between the MPB and Gouy–Chapman theories increases at higher ion concentrations and a new phenomenon appears in which the electrostatic

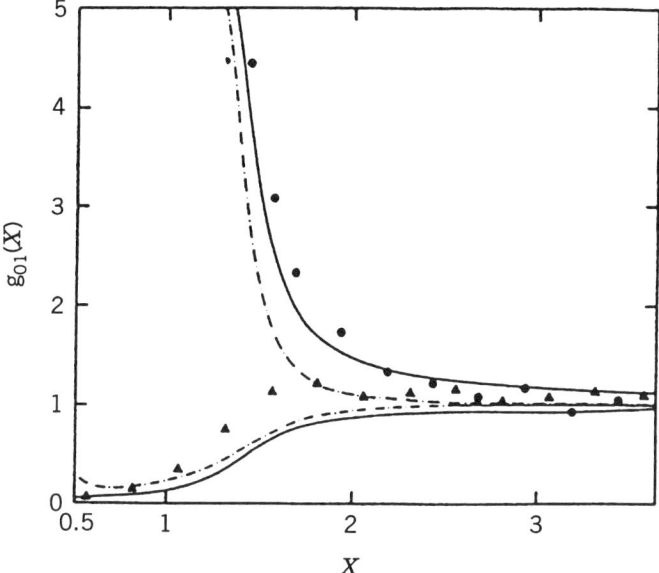

Figure 7.51. Co-ion and counter-ion profiles $g_{01}(X)$ for a 1:2 electrolyte with $c = 0.05$ M and $\sigma^* = 0.284$. The Monte Carlo points ($\bigcirc$, $\bullet$) are those of Torrie and Valleau while the dashed-dotted line denotes the MPB4 curve (Outhwaite and Bhuiyan, 1983). Reproduced by permission of The Royal Society of Chemistry.

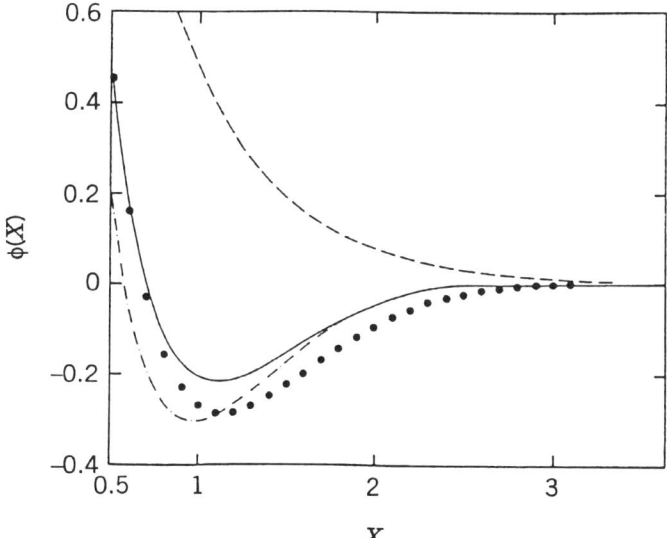

Figure 7.52. Mean electrostatic potential for a 1:2 electrolyte with $c = 0.5$ M and $\sigma^* = 0.1704$. The solid points are the MC results of Torrie and Valleau while the dashed-dotted curve and the dashed curve denote the MPB4 and modified Gouy–Chapman theories, respectively (Outhwaite and Bhuiyan, 1983). Reproduced by permission of The Royal Society of Chemistry.

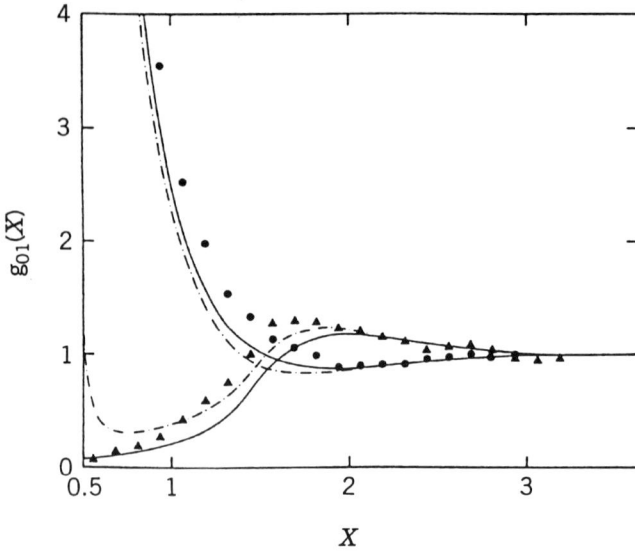

Figure 7.53. Co-ion and counter-ion profiles $g_{01}(X)$ for a 1:2 electrolyte at $c = 0.5$ M with $\sigma^* = 0.1704$. The MC points (▲, ●) are those of Torrie and Valleau while the dashed-dotted line denotes the MPB4 curve (Outhwaite and Bhuiyan, 1983). Reproduced by permission of The Royal Society of Chemistry.

potential in the double layer becomes non-monotonic. This phenomenon is in principal impossible in the Gouy–Chapman model. Corresponding ion concentration distributions are illustrated in Fig. 7.53, and a clear minimum for counter-ions and a maximum for co-ions are apparent. The fact that the ion density curves cross at some point means that there is an inversion of the double-layer charge in the region between $3d/2$ and $7d/2$, which produces the minimum in the electrical potential profile. The existence of oscillations implies charge stratification near a charged electrode and this depends on the electrolyte concentration.

The MPB theory overcomes certain limitations of the Gouy–Chapman model by taking into account ionic volume, the ionic atmosphere and image forces. Results of both theories coincide at low electrolyte concentrations and small surface charges, but beyond this limiting case there is a very significant qualitative and quantitative discrepancy. For instance, MPB theory predicts a thinner diffuse ion layer and higher capacitance of this layer than the Gouy–Chapman model. Near a charged interface MPB predicts considerable adsorption of co-ions. The major difference is that the decaying potential distribution of the Gouy–Chapman model transforms into the decaying-oscillating solution of the MPB theory, with oscillations beginning at high electrolyte concentrations when $\kappa_0 d > 1.24$. This oscillation was also predicted by Stilinger and Kirkwood (1960) who estimated a different value of $\kappa_0 d > 1.03$. Other modern theories predict oscillatory behavior of the electrostatic potential, but under different conditions. In the HCA theory the condition is $\kappa_0 d > 1.23$ (Blum, 1977) which is very close to MPB theory. The Bogolyubov–Born–Green–Ivon theory puts the condition at $\kappa_0 d > 1.72$ (Outhwaite and Bhuiyan, 1991a,b). Therefore, oscillatory behavior is a consistent finding of all theories. It is worthwhile to mention that damped oscillation in the MPB theory is due to a fluctuating potential and its existence does not depend on the excluded volume.

One of the major drawbacks of MPB theory is that it does not take into account the discreteness of a solvent. In the diffuse layer the discreteness of solvent molecules strongly influences ion–ion interactions at small distances. Furthermore, an understanding of the structure and properties of the compact layer is impossible without an adequate model of solvent molecules and their interactions with the electrode and between themselves.

Application of MPB model to ITIES was made first by Torrie and Valleau (1986). Cui *et al.* (1994, 1995) applied the modified Poisson–Boltzmann theory (MPB4) to the interface between two immiscible electrolyte solutions and found that MPB4 describes the experimental results at the nitrobenzene–water and dichloroethane–water phase boundaries and reproduces the Monte Carlo calculation more accurately than the MVN theory for 1:1 electrolytes over a wide range of conditions including variations in the image forces, ion size, the inner layer potential distribution, electrolyte concentration, surface charge density and solvent effects. They used the equation for electrostatic potential from the MPB4 model for the inner layer at ITIES when $x < a/2$:

$$\psi(x) = \psi(0) + x\psi'(0) - x\lambda c|q^{aq}|^{1/2} q^{aq}/\varepsilon_h, \qquad (7.186)$$

where $\psi(x)$ is the electrostatic potential at coordinate x, $\psi(0)$ is the electrostatic potential at the surface, $\psi'(0)$ is the derivative of $\psi(0)$, c is the electrolyte concentration, q^{aq} is the surface charge density of the aqueous phase, λ is the parameter determined by fitting the experimental values of the Galvani potential differences and ε_h is the dielectric permittivity of the compact layer. The third term in equation (7.186) represents the dependence on electrolyte concentration, surface charge density, and the solvent effect on the inner-layer potential distribution. These effects can be ascribed to ionic penetration into the opposite phase and ion–ion correlations across the interface. Cui *et al.* (1994, 1995) obtained good agreement between theoretical calculations and experimental data. Figures 7.54–7.57 show the most interesting results of their calculations. They came to the conclusion (Cui *et al.*, 1995) that the structure of the 1,2-dichloroethane–water interface is similar to that of the water–nitrobenzene interface, except that the effects of the image force and ion size are more pronounced and the inner-layer potential drop plays a more important role.

7.15. IONIC PLASMA IN A CONTINUOUS DIELECTRIC NEXT TO A CHARGED INTERFACE

Another approach to the theoretical description of the double layer is based on recent achievements in the theory of liquids. This theory is based upon multiparticle ionic distributions rather than on electrical potential. The fact is that the correct statistical description of a solution can be done only in terms of multiparticle ion and dipole distribution functions within a system. The interaction potential between these particles must be postulated and then equations for correlation functions of different orders need to be formulated. This results in an infinite chain of equations which should be uncoupled and interrupted at some point to obtain a closed set of equations. The most promising results using this technique are obtained with the so-called hyperneted chain approximation (HCA) or mean spherical approximation (MSA). A detailed description of these methods can be found in several reviews (Carnie and Torrie, 1984; Torrie and Patey, 1991, 1993; Watts and VanderNoot, 1996). We will provide a short introduction to these methods and their results.

In MVN and MPB theories an artificial division of a double layer into compact and diffuse moieties was made. Therefore, the goal of applying statistical mechanics to this model is to obtain a complete description of the double layer such that the division of the double layer, if it occurs at all, would be an automatic consequence of the general solution. We will show a few analytical results obtained with the MSA model. In this approach the potential is assumed to be small, which permits linearization of the equations. Of course, this assumption narrows the applicable potential range of MSA, but does not limit the electrolyte concentration.

One result of this theory is that the Debye screening length changes because it takes into account the finite ion radius. For example, in the expression for double-layer capacitance ($C = \varepsilon\varepsilon_0 \kappa$) the Debye parameter κ is changed to a new quantity 2Γ. Thus, the differential capacitance of the electric double layer becomes

$$C = \varepsilon\varepsilon_0 2\Gamma. \tag{7.187}$$

If the radii of all ions are identical, this new parameter is represented by the equation

$$2\Gamma = \frac{\sqrt{1 + 4\kappa a} - 1}{2a}. \tag{7.188}$$

For dilute solutions, where κ is very small, the difference between these two parameters disappears:

$$2\Gamma \cong \kappa. \tag{7.189}$$

Therefore, at low electrolyte concentrations, the equation for capacitance can be written as

$$\frac{1}{C} = \frac{1}{\varepsilon\varepsilon_0 \kappa} + \frac{a}{\varepsilon\varepsilon_0}. \tag{7.190}$$

This is a significant result. The theory, in a self-consistent way, came to the formal division of capacitance, *but not of space*, into two terms corresponding to the diffuse

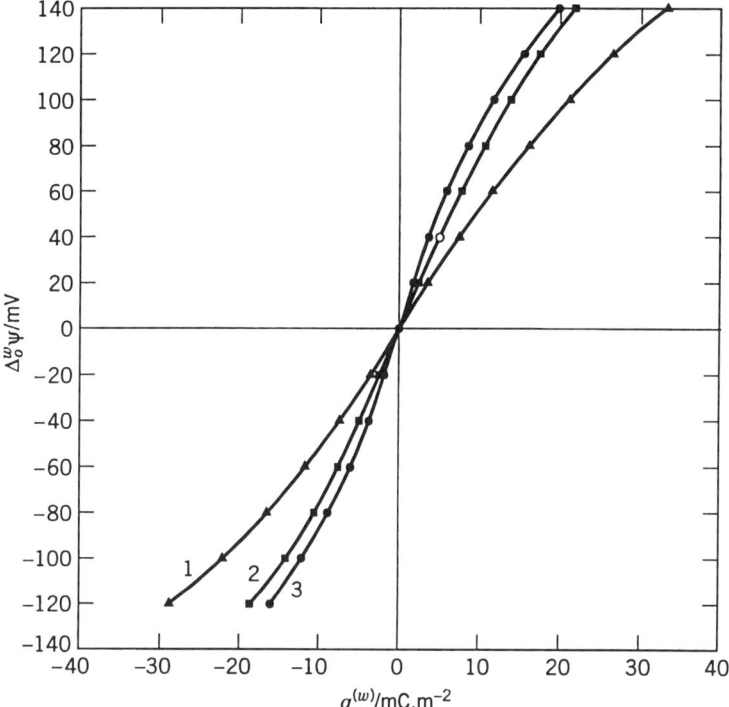

Figure 7.54. Plot of the Galvani potential difference across the water–nitrobenzene interface evaluated using the MPB4 theory as a function of the surface charge density of the aqueous phase at different electrolyte concentrations: (1) 0.01 M NaBr + 0.01 M TBATPB; (2) 0.02 M NaBr + 0.02 M TBATPB; (3) 0.05 M NaBr + 0.05 M TBATPB. The points represent experimental data (Cui *et al.*, 1994). Reproduced with permission from Elsevier Sequoia S.A.

$(1/\varepsilon\varepsilon_0\kappa)$ and compact $(a/\varepsilon\varepsilon_0)$ parts of the electric double layer. However, the diffuse layer can only be regarded as an independent component of the double layer if the polarization fluctuation of the solvent molecules in the diffuse part are not correlated with the polarization fluctuations in the compact layer next to the interface.

Another drawback of this solution is that it implies that dielectric permittivity in the compact layer is equal to the bulk value. This is the result of neglecting the solvent structure and substituting in its place a simple dielectric continuum. The situation was improved by taking into account the discrete solvent structure, which led to the development of the ion–dipole plasma theory.

7.16. ION–DIPOLE PLASMA AT A CHARGED INTERFACE

The ion–dipole plasma theory uses the principles of statistical mechanics to describe a system of ions and solvent molecules without involving the concept of a continuous

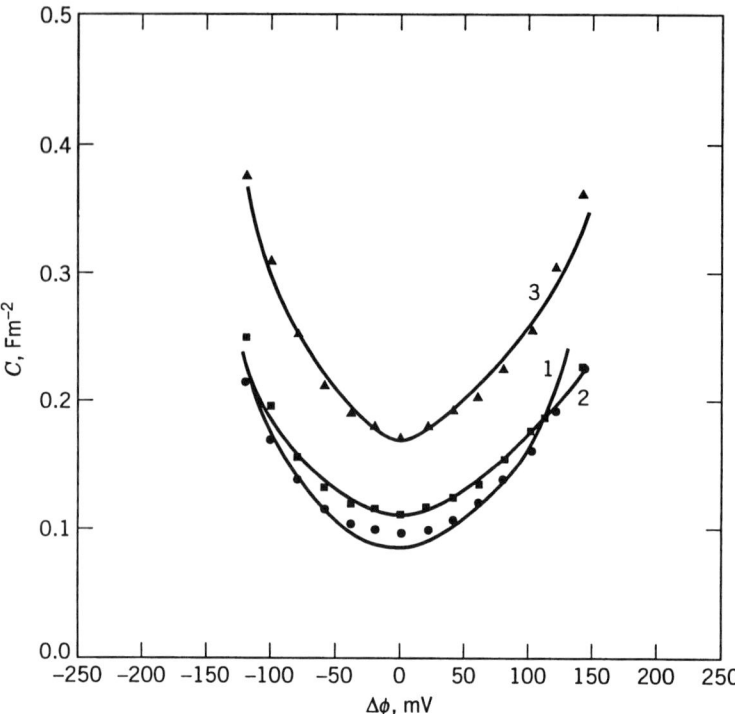

Figure 7.55. The differential capacitance of the nitrobenzene–water interface as a function of the Galvani potential difference at different electrolyte concentrations: (1) 0.01 M NaBr + 0.0.1 M TBATPB; (2) 0.02 M NaBr + 0.02 M TBATPB; (3) 0.05 M NaBr + 0.05 M TBATPB (Cui *et al.*, 1994). Reproduced by permission of Elsevier Sequoia S.A.

dielectric medium. This theory is used for the analysis of bulk properties of electrolyte solutions (Blum, 1974; Carnie *et al.*, 1981a,b; Levesque *et al.*, 1980) and for studying the structure of the electric double layer (Attard *et al.*, 1992; Carnie and Chan, 1980; Henderson, 1983; McCormack *et al.*, 1995; Outhwaite, 1981; Patey and Torrie, 1989; Patra and Ghosh, 1995; Torrie and Patey, 1993; Torrie *et al.*, 1989; Wei *et al.*, 1990, 1993).

Description of the ion–dipole plasma is based on a non-primitive model, consisting of hard spheres with point charges and point dipoles. The ionic and dipolar balls are non-polarizable, and their dielectric permittivity is assumed to be equal to one. The interface represents a charged wall of the same non-polarizable material to avoid considering the image forces. The usual electrostatic interactions are assumed between ions and dipoles. The only other force included in this model is the repulsion of hard spheres at close distances. This model is again strongly idealized, even though it is a "non-primitive" model. Nonetheless, its analysis has helped to define important qualitative properties of the ion–dipole plasma and of the double layer. We will present some results obtained by the MSA model.

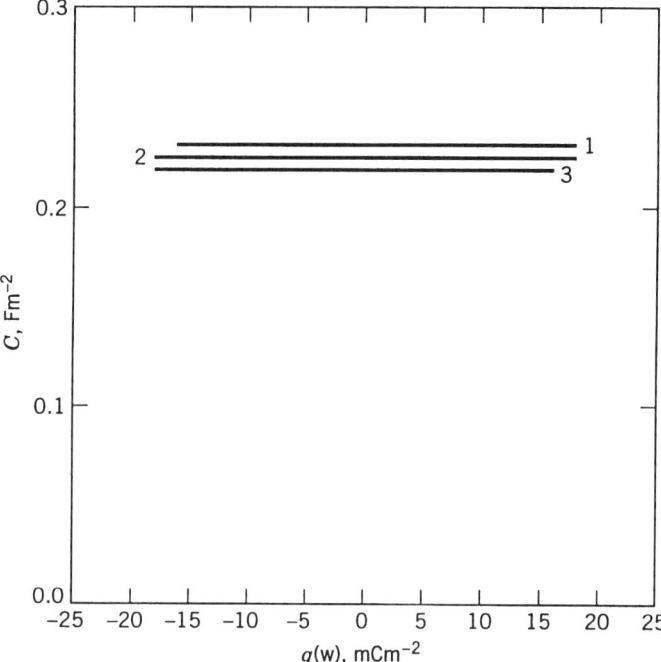

Figure 7.56. Plot of the inner-layer capacitance as a function of the surface charge density of the aqueous phase at different electrolyte concentrations: (1) 0.005 M LiCl +0.005 M TBATPB; (2) 0.01 M LiCl + 0.01 M TBATPB; (3) 0.02 M LiCl + 0.02 M TBATPB (Cui *et al.*, 1995). Reproduced by permission of Elsevier Sequoia S.A.

The MSA is a linear theory, similar to the Debye–Huckel theory, which includes the case of finite molecular size in a self-consistent way. As mentioned earlier, it is applicable only at small potentials, but provides a solution for any electrolyte concentration. Analytical solutions have been obtained for low concentrations, while for high concentrations numerical integration is used.

Suppose ions of a binary electrolyte solution have a radius a and solvent molecules have the radius a_d and the wall is charged with a surface density σ. The bulk dielectric permittivity of the pure solvent is equal to ε. Then, within the limits of low potentials and low concentrations ($\kappa a \ll 1$), the theory gives the following expression for the potential of the wall relative to the potential in the bulk electrolyte solution:

$$\phi_0 = \frac{\sigma}{\varepsilon \varepsilon_0 \kappa} \left[1 + \kappa a \left(1 + \frac{\varepsilon - 1}{\lambda} \frac{a_d}{a} \right) \right], \tag{7.191}$$

where κ is the usual Debye parameter, the definition of which remains unaltered at low electrolyte concentrations, and λ is determined by the equation

$$\lambda^2 (1 + \lambda)^4 = 16\varepsilon. \tag{7.192}$$

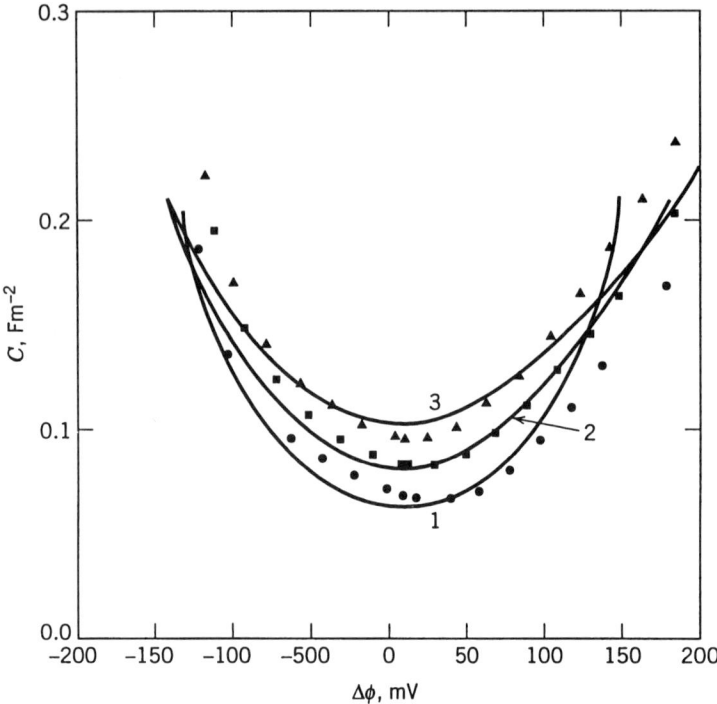

Figure 7.57. The differential capacitance of the 1,2-dichloroethane–water interface as a function of the Galvani potential difference at different electrolyte concentrations: (1) 0.005 M LiCl + 0.005 M TBATPB; (2) 0.01 M LiCl + 0.01 M TBATPB; (3) 0.02 M LiCl + 0.02 M TBATPB (Cui *et al.*, 1995). The points represent the experimental values obtained by Samec *et al.* (1987). Reproduced by permission of Elsevier Sequoia S.A.

This parameter λ is always larger than one and, at all reasonable values of ε, it falls within the interval $1 < \lambda < 3$ (Fig. 7.58). For example, if $\varepsilon = 78.4$ then $\lambda = 2.65$.

By using equation (7.122) we obtain the following expression for capacitance:

$$\frac{1}{C} = \frac{1}{\varepsilon \varepsilon_0 \kappa} + \frac{a}{\varepsilon \varepsilon_0} + \frac{\varepsilon - 1}{\varepsilon \varepsilon_0} \frac{a_d}{\lambda}. \tag{7.192}$$

The inverse capacitance now consists of three terms. The first term is the inverse capacitance of the Gouy–Chapman diffuse layer, and the sum of the second and third terms can be interpreted as the reverse capacitance of the compact layer. Again, rigorous analysis of the ion–dipole plasma results in automatic division of the capacitance of the double layer. The second and the third capacitance terms are separate contributions from ions,

$$C_{\text{ion}} = \frac{\varepsilon \varepsilon_0}{a}, \tag{7.194}$$

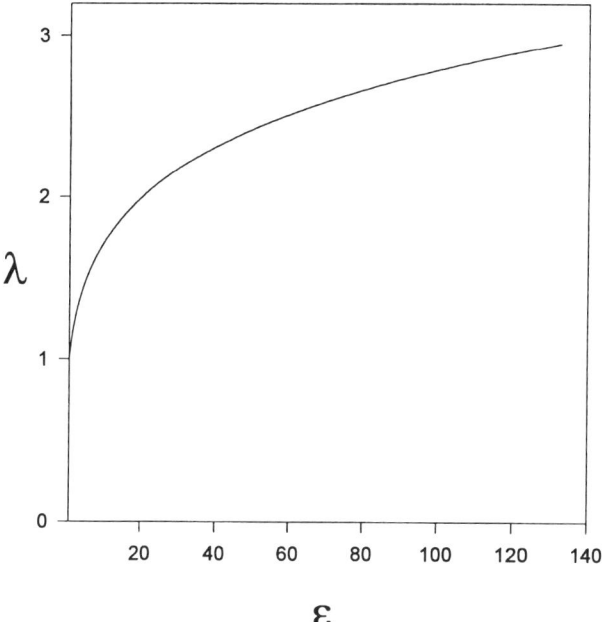

Figure 7.58. Dependence of parameter λ on the dielectric permittivity of the solvent.

and from solvent dipoles,

$$C_{\text{solv}} = \frac{\varepsilon \varepsilon_0 \lambda}{(\varepsilon - 1) a_d}. \qquad (7.195)$$

Note that in equation (7.193) the dependence on electrolyte concentration resides in the first term only, while the second and third terms are concentration independent. This corresponds to the conventional understanding of the properties of diffuse and compact parts of double layers. Formula (7.193) gives correct transitions to the limiting cases. If one neglects the finite size of ions and solvent dipoles ($a, a_d \to 0$), the Gouy–Chapman capacitance is recovered. However if only $a_d \to 0$, the main result of the "primitive" model (7.190) is recovered.

The dipolar terms in (7.193) can be formally compared to the capacitance of a Stern layer having thickness δ and dielectric permittivity ε_h:

$$\frac{\delta}{\varepsilon_h} = \frac{a}{\varepsilon} \left(1 + \frac{\varepsilon - 1}{\lambda} \right). \qquad (7.196)$$

For the sake of simplicity, $a_d = a$. This formula defines only the ratio δ/ε_h. Additional assumptions are required to determine the thickness of the compact layer and its dielectric permittivity.

Suppose that the thickness of the compact layer is equal to the radius of ions and dipoles, $\delta = a$. At first glance, it is natural to expect that the dielectric permittivity of the compact layer ε_h will be equal to one. This is because the center of ions and dipoles cannot enter the compact layer, and, by convention, ions and dipoles are not polarizable. However, formula (7.196) gives a different result and demonstrates that the dielectric permittivity is not unity:

$$\varepsilon_h = \frac{\varepsilon}{1 + (\varepsilon - 1)/\lambda}. \tag{7.197}$$

For example, if $\varepsilon = 78.4$, then $\varepsilon_h = 2.6$.

Another situation arises if we assume that the dielectric permittivity in a compact layer is equal to unity, $\varepsilon_h = 1$. Then the thickness of this layer will be found to be equal to

$$\delta = \frac{a}{\varepsilon}\left[1 + \frac{(\varepsilon - 1)}{\lambda}\right]. \tag{7.198}$$

Using the same value $\varepsilon = 78.4$, the thickness of the compact layer is equal to $\delta = 0.38a$. Taking solvent structure into account results in a compact layer which is thinner than the radius of the particles. Both results follow from attempts to visualize a complex dielectric response in terms of two different regions with constant dielectric permittivities. It is clear that such an interpretation is not entirely adequate. The theory *does not predict* the physical division of the space occupied by an electrolyte solution into two separate zones. However, the capacitance of the double layer can be divided into a sum of independent terms and this can be interpreted in terms of the Stern model. The capacitive contributions to the second and the third terms that we attributed to the Stern layer come from the particles in the entire double layer and not just those next to the surface.

The discrepancy between the ion–dipole plasma model and the Gouy–Chapman model of a diffuse layer becomes especially clear when one looks at the solution polarization distribution in the double layer calculated by Carnie and Chan (1980). Recall that the Gouy–Chapman theory operates with the macroscopic concept of the dielectric permittivity. For polarization of the diffuse layer the Gouy–Chapman theory gives the expression:

$$P_{\text{mac}}(x) = \varepsilon_0(\varepsilon - 1)E(x) = \frac{\varepsilon - 1}{\varepsilon}\sigma\exp[-\kappa(x - \delta)]. \tag{7.199}$$

The polarization $P(x)$ found by Carnie and Chan using the MSA, normalized by $P_{\text{mac}}(x)$, is presented in Fig. 7.59. If the results of the two models were identical, we would have only one curve in this figure. However, in the range of several molecular layers, $P(x)$ essentially deviates from $P_{\text{mac}}(x)$. It is especially interesting that it oscillates in this range. This is a consequence of the discrete nature of the solvent

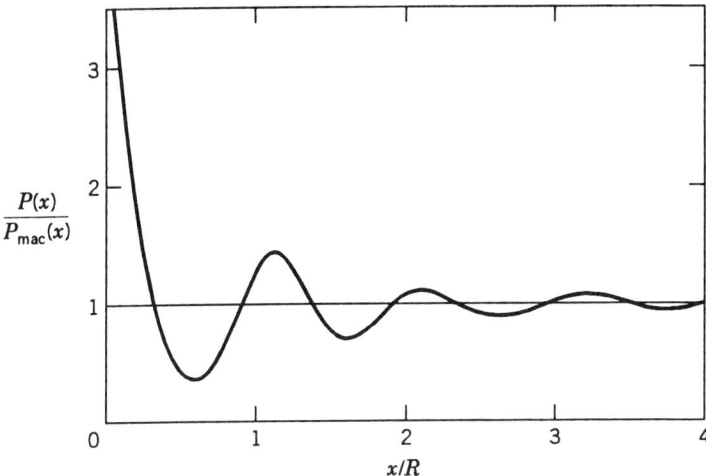

Figure 7.59. Polarization density at the surface (Carnie and Chan, 1980). Reproduced by permission of the American Institute of Physics.

which manifests itself by a strong cooperative ordering of dipoles next to the surface. Far from the surface, the polarization calculated by Carnie and Chan coincides with that of the macroscopic theory. Thus, the physical picture of dipolar particle distribution and orientation at a charged wall appears to differ from the predictions of the macroscopic theory. In the theory of Carnie and Chan total polarization of the double layer $\int_a^\infty P(x)\,dx$ slightly exceeds the value derived using the macroscopic theory. This results in a somewhat greater dielectric screening of the charged surface and shows that at distances of a few molecular diameters, the concept of local dielectric permittivity makes no sense. At small distances, it is therefore more appropriate to operate with the concept of medium polarization.

The "non-primitive" model analyzed by Carnie and Chan (1980) and referred to in other publications is very difficult to solve analytically. Still it is a simplified model because it oversimplifies short-distance interactions, and does not include image forces, quadruple moments, or molecular polarization. However, a rigorous derivation of Stern capacitance demonstrates that this quantity originates from the discrete solvent structure rather than dielectric saturation or hydrogen bonds. This result represents an important achievement of the ion–dipole plasma theory.

7.17. THE MONTE CARLO METHOD AND THE DOUBLE LAYER

If one is to determine whether a statistical-mechanical theory is an accurate representation of the real world is it enough to say that the experimental data fit the theory? Unfortunately, the answer is no. Comparison of existing theories with experimental data does not give an unambiguous answer about the accuracy of the theory, since any theory is built upon a particular model. In addition, a theory is usually an approximation, as is the case for the double layer. Therefore, its predictions have two different sources of error—the inadequacy of the model and the inaccuracy of its mathematical analysis.

The correct question to ask regarding the merits of a given theory should be how well it describes *the model* it was built upon, rather than the experimental system. This means that theoretical predictions should be compared with experimental data, but the experiment should not be conducted in the real world *but within the model itself.* The development of computational experiments, and in particular the development of the Monte Carlo (MC) method, solved this problem (Carnie *et al.*, 1981a,b; Torrie and Valleau, 1979, 1980, 1982; Torrie *et al.*, 1989). The results of the MC experiment can be considered to be a decisive criterion for judging the accuracy of existing theories of the double layer at the interface between two immiscible electrolyte solutions. We will present here the essential conclusions of this research.

A series of such computational experiments on a double layer was carried out by Torrie and Valleau (1979, 1980, 1982, 1986). The authors addressed mainly the properties of the diffuse layer and compared them with predictions of the Gouy–Chapman (GC) and MPB theories. These "experiments" were carried out with ions of finite size in the dielectric continuum with dielectric permittivity ε. Some of their data have been already included in Figs. 7.50–7.53.

In their paper, Torrie and Valleau (1982) presented a computational experiment for 2:1 and 2:2 electrolytes in an "aqueous" solution. The authors stressed that their data for 2:2 electrolyte correspond exactly to a 1:1 electrolyte solution when $\varepsilon = 20$. The following parameters were used: the diameter of hard balls $d = 0.425$ nm, the dielectric permittivity $\varepsilon = 78.5$ and the temperature $T = 298$ K. Results were presented in terms of the dimensionless potential $\psi^* = \psi e_0/kT$, dimensionless surface charge density $\sigma^* = \sigma d^2/e_0$ and dimensionless length $X = x/d$. Hence, the dimensional values are equal to $\psi = 25.7\psi^*$ mV, $\sigma = 88.7\sigma^*$ μC cm^{-2} and $x = 0.425X$ nm. The distance of closest approach of the ions to the charged surface was equal to the ionic radius $d/2$. Therefore, the diffuse layer began at the point $x = d/2$, and the potential of the diffuse layer was equal to $\psi(d/2)$. These results were compared to the GC and MPB theories.

Figure 7.60 shows the dependence of the potential of the diffuse layer calculated by all three methods for 2:1 electrolyte solutions. At positive σ the counter-ions had a single charge, and at negative σ they had two charges. Obviously, the curves were non-symmetrical with respect to the origin of coordinates. As one can see in this figure, the GC theory on the right side of the plot overestimates the thickness of the diffuse layer and the potential drop in it. This is because the GC theory does not take into account the correlation between ions in the diffuse layer accurately enough, and overestimates their mutual repulsion. The MPB theory gives a satisfactory agreement with the computational experiment in the right-hand portion of Fig. 7.60, and in the case of a 1:1 electrolyte in an aqueous solution (Torrie and Valleau, 1980).

The left-hand portion of Fig. 7.60, where the counter-ions are doubly charged, is more interesting. First, the MC results show a definite minimum on the curve of $\psi^*(d/2)$ versus σ^*. The classical GC theory strongly deviates from MC and is

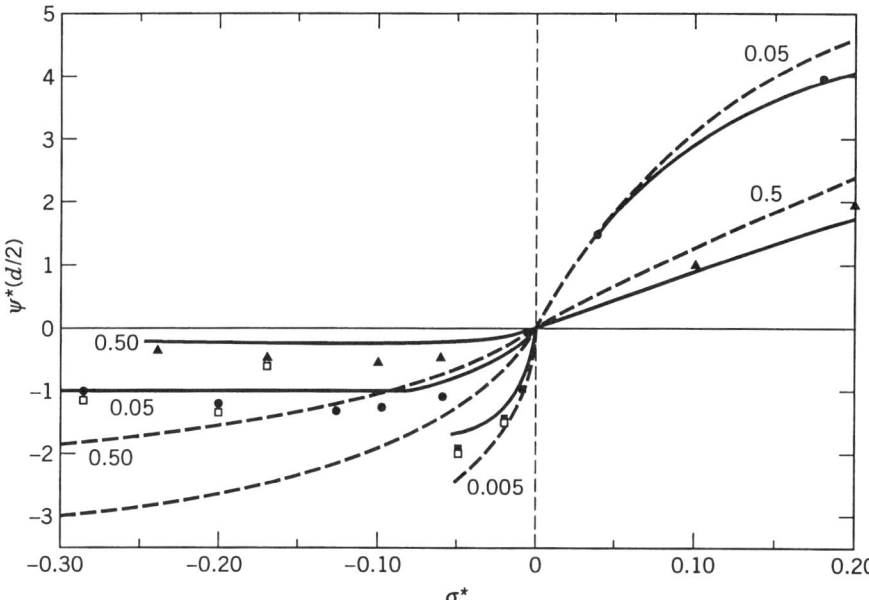

Figure 7.60. Surface charge dependence of the diffuse layer potential. The counter-ions are singly charged for $\sigma^* > 0$, doubly charged for $\sigma^* < 0$. ($\square$, $\blacksquare$, $\bullet$, $\blacktriangle$) are MC results for 2:1 electrolytes at 0.005, 0.05 and 0.5 M, respectively. Open symbols are MC results for 2:2 electrolytes. Dashed curve—Gouy–Chapman theory for 2:1; solid curve—MPB theory for 2:1 (Torrie and Valleau, 1982). Reproduced by permission of the American Chemical Society.

completely inaccurate at $\sigma < 0$. It cannot describe the qualitative features of the computational experiments, and also gives large quantitative errors even at rather low electrolyte concentrations and surface charges. At a concentration 0.005 M and $\sigma = -0.9 \,\mu\text{C cm}^{-2}$ the MVN theory overestimates the potential of the diffuse layer by 5%, but at the same concentration and moderate surface charge of $-4.4 \,\mu\text{C cm}^{-2}$ the error grows to 36%. At the same time, curves calculated with MPB, in contrast to the classical theory, are more consistent with MC data and also demonstrate a minimum for the ψ^*–σ^* relationship. However, we should note that the quantitative agreement between MPB and MC is not as good as in 1:1 electrolyte or 2:1 electrolyte solutions where the counter-ion carries a single charge.

A more detailed understanding of the structure of the double layer and the reasons for the discrepancies of different theories is provided by the analysis of ionic charge density and mean electrostatic potential in the double layer. Figure 7.61 presents these functions for a 2:1 electrolyte solution with an ion concentration 0.5 M and surface charge density $\sigma = 25.2 \,\mu\text{C cm}^{-2}$. These results are very interesting, and clearly show a fundamental difference between the GC and MC results. The GC theory again predicts an overly extended diffuse layer. The concentration of the counter-ions in the MC data displays a shallow minimum, while the co-ion concentration gives a rather

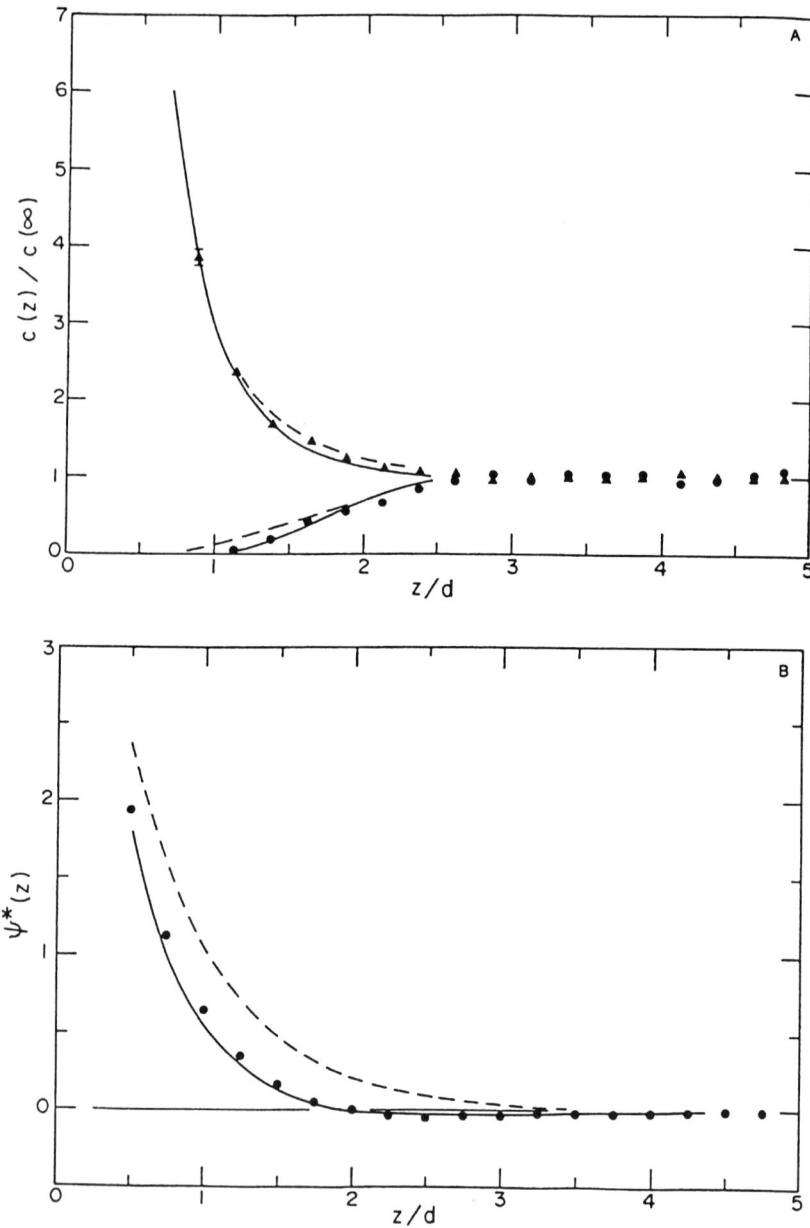

Figure 7.61. (A) Local concentration divided by the bulk concentration for counter-ions and co-ions in 2:1 systems at 0.5 M and a surface charge $\sigma^* = 0.20$: ($\bullet$, $\blacktriangle$) are MC results; dashed curve—Gouy–Chapman theory; solid curve—MPB theory. Some typical statistical uncertainties in the MC data are indicated. (B) Mean electrostatic potential $\psi^*(z)$ for a 2:1 electrolyte at 0.5 M and a surface charge $\sigma^* = 0.20$. All symbols as in (A) (Torrie and Valleau, 1982). Reproduced by permission of the American Chemical Society.

distinct maximum, which is in principal impossible in the MVN theory. The MPB theory demonstrates monotonic behavior of the ionic concentration.

The oscillatory properties of ionic density and potential in the electrical double layer become more obvious as the electrolyte concentration increases up to 0.5 M (Fig. 7.62). The oscillations are seen even at a smaller surface charge $\sigma = -15.1 \, \mu C \, cm^{-2}$ in this figure. The MPB predictions qualitatively agree with the results of the MC computational experiment, yet there is a rather pronounced quantitative discrepancy between them. Note that the agreement is much better for a 1:1 electrolyte.

The dependency of the potential of the diffuse layer on surface charge for 2:1 and 1:1 electrolytes at $\sigma > 0$ is very similar. This proves that the failure of the classical theory is not due to the charge asymmetry between the co- and counter-ions, but rather to electrical interactions between counter-ions. To study this possibility, the MC experiment was performed with a 2:2 electrolyte.

In this case, the predictions of the classical theory for negative charge of the surface in Fig. 7.60 do not differ from the predictions for a 2:1 electrolyte. The results of the computer simulation experiment are shown in Fig. 7.62 by open symbols. They confirm the expectation that the potential of the diffuse layer does not depend on the co-ion charges except at the smallest values of surface charges. In Fig. 7.63 the ion density and electrical potential distribution in the diffuse layer for a 2:2 electrolyte are presented at the same electrolyte concentration and surface charge density. These curves demonstrate charge inversion and potential oscillation. Thus, when the counter-ion is multivalent, the MVN theory is completely inadequate even at rather low concentrations and surface charges. The potential of the diffuse layer, obtained in the computational experiment, very quickly ceases to depend on surface charge (Fig. 7.60), although it displays a shallow minimum when $\sigma = -10 \, \mu C \, cm^{-2}$. Oscillations of ionic density arise at a concentration of 0.05 M and are observed at all surface charges when the concentration is equal to 0.5 M. The MPB theory is capable of qualitatively predicting the majority of these phenomena. However, it slightly overestimates the compression of the diffuse layer in comparison with MC data.

7.18. COMPUTER SIMULATION OF ITIES

At the interface between two immiscible electrolyte solutions (ITIES) there are two ionic diffuse layers and probably two compact layers. The MVN model considers these two diffuse layers as completely independent and linked only by the mean electrical field between them. In that sense the MVN theory actually deals with the model of independent diffuse layers (IDL). The legitimacy of such an approach to the interface between two immiscible electrolyte solutions was questioned by Torrie and Valleau (1986) who analyzed three factors that distinguish ITIES from the IDL:

1. *Interaction of Discrete Charges between Different Diffuse Layers Located in Adjacent Phases.* It was shown above that ionic correlation, even within one

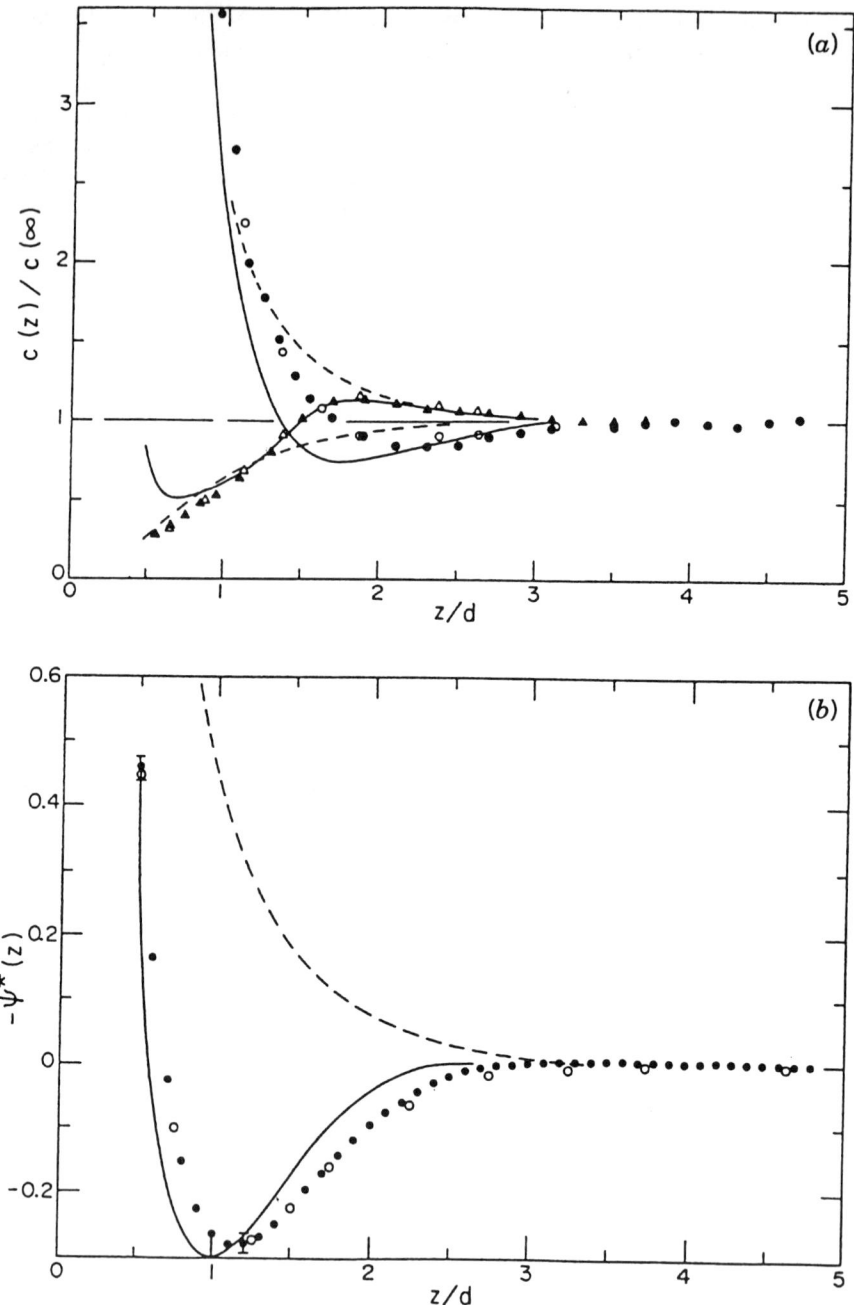

Figure 7.62. (a) Local ion concentrations for a 2:1 electrolyte at 0.5 M, $\sigma^* = -0.1704$: (O, △, ●, ▲) MC results for a cell with $L = 15d$, $W = 14.9d$; (O, ■, ●, ▲) are MC results for a cell with $L = 12d$, $W = 19.1d$; dashed curve—Gouy–Chapman theory; solid curve—MPB theory. (b) $\psi^*(z)$ for 2:1 electrolyte at 0.5 M, $\sigma^* = -0.1704$. All symbols as in (a) (Torrie and Valleau, 1982). Reproduced by permission of the American Chemical Society.

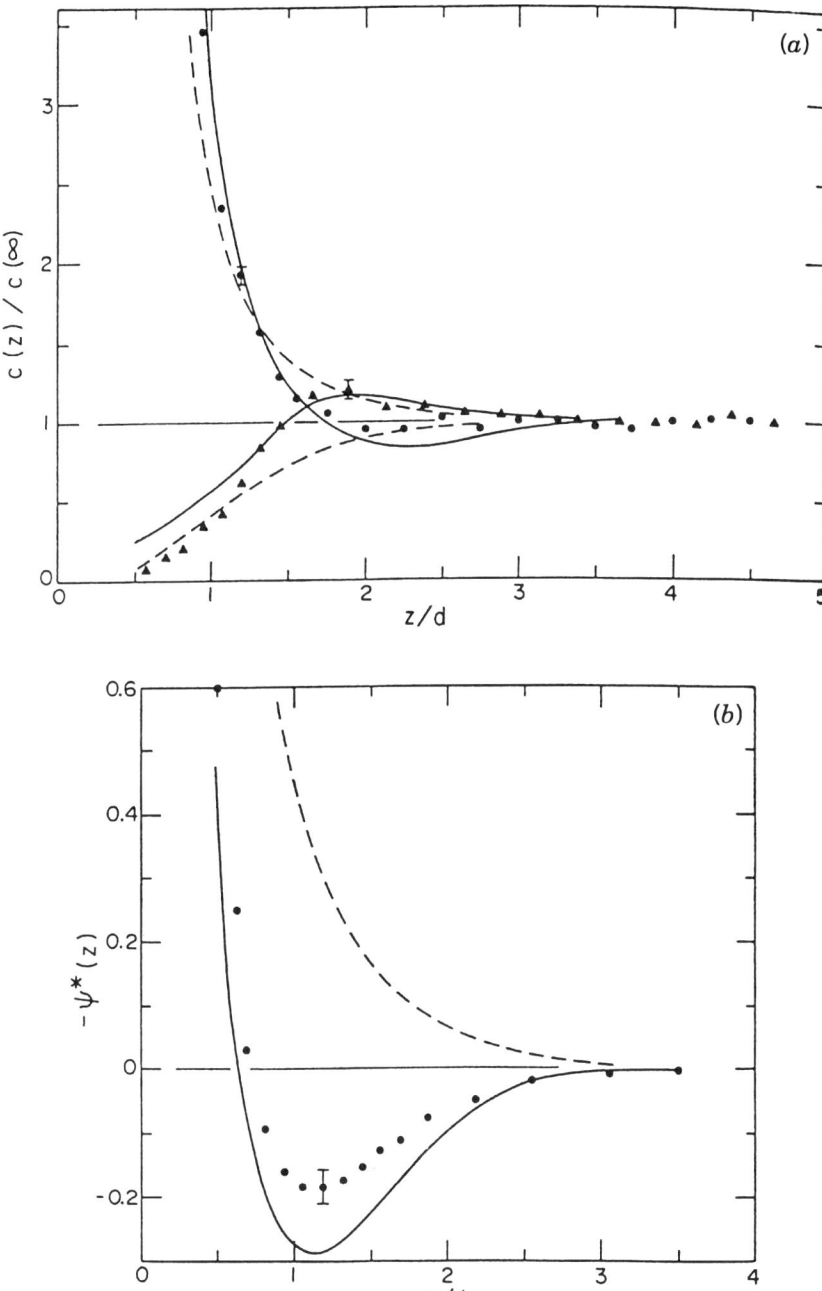

Figure 7.63. (*a*) Local ion concentrations for a 2:2 electrolyte at 0.5 M, $\sigma^* = -0.1704$: (●, ▲) MC results from a two-plate experiment; dashed curve—a Gouy–Chapman theory for the 2:2 system. (*b*) $\psi^*(z)$ for a 2:2 electrolyte at 0.5 M, $\sigma^* = -0.1704$ (Torrie and Valleau, 1982). Reproduced by permission of Elsevier Sequoia S.A.

diffuse layer, results in compression of this layer and reduction of its potential drop. One might expect that taking into account the correlations of ions between layers would enhance this tendency.

2. *Interpenetration of Ions.* In the MVN and MPB theories the distance of closest approach of the ions to a smooth electrode is usually assumed equal to the radius of the ions, and therefore penetration of ions through the interface is not allowed. However, at the interface between two liquid solvents, where the centers of ions can appear much closer to the imaginary sharp interface, this may be incorrect (Koczorowski *et al.*, 1987). The partial mutual penetration of two charged layers can have two consequences: a thinner inner layer and a reduction in the interfacial potential difference.

3. *Image Forces.* These forces should undoubtedly manifest themselves at the interface between phases with different dielectric permittivities and should have their major effect on the properties of the non-aqueous phase.

Torrie and Valleau (1980) have analyzed these factors for both ITIES and IDL by MC simulation of the restricted primitive model of an electrolyte, with the diameter of balls (d) equal to 0.425 nm at temperature $T = 298$ K. The effect of inter-layer ion correlations is shown in Table 7.5, which gives the data for the diffuse layer potential in two adjacent phases ψ_1^* and ψ_2^*. To single out this effect, the dielectric permittivities of two phases were assumed identical with $\varepsilon = 19.625$ (one quarter that of water). Each layer bears the surface charge density $|\sigma| = 4.324 \ \mu C \ cm^{-2}$. The distance of closest approach of the ions to the interface was assumed equal to $z_{dca} = d/2 = 0.2125$ nm. The calculations were performed for different electrolyte concentrations in each phase as specified in Table 7.5, and results of the MC experiment for ITIES and IDL were compared with predictions of the MVN theory. As shown in Table 7.5, the MVN theory strongly overestimates the potential of diffuse layers in each phase. For example, at a concentration 0.5 M the overestimation reaches 60%, as graphically presented in Fig. 7.64.

Comparison of these models helps to clarify the role of different correlations. The difference between the MVN model and IDL is due to the inter-layer ion correlations

Table 7.5. Diffuse layer electrical potential drops at both sides of the interface between two immiscible electrolyte solutions. Parameters: $\varepsilon_1 = \varepsilon_2 = 19.625$, $d = 2z_{dca} = 0.325$ nm, $\sigma = 0.04324$ C m^{-2}. IDL correspond to two independent diffuse layers, MVN—modified Verwey–Niessen (1939) model.

ϕ_1^*					ϕ_2^*		
MVN	IDL	ITIES	c_1 (M)	c_2 (M)	ITIES	IDL	MVN
3.77	2.81(0.03)	2.62(0.06)	0.05	0.05	2.62(0.06)	2.81(0.03)	3.77
1.8	1.22(0.05)	1.10(0.05)	0.50	0.50	1.10(0.05)	1.22(0.05)	1.8
3.77	2.81(0.03)	2.56(0.04)	0.05	0.50	1.10(0.04)	1.22(0.05)	1.8

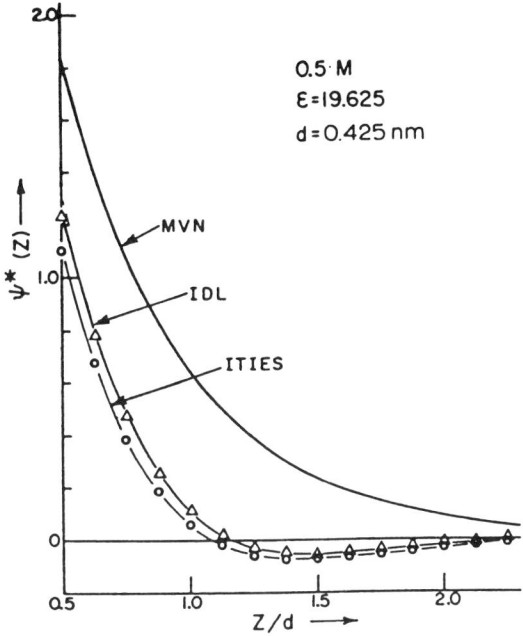

Figure 7.64. Profiles of the mean electrostatic potential in the diffuse layer. ITIES refers to a symmetrical ITIES of a 0.5 M 1:1 electrolyte, with $\varepsilon = 19.625$ on both sides, $z_{dca} = d/2 = 0.2125$ nm, and with a surface charge density $\sigma = 0.04324$ C m^{-2}. IDL shows Monte Carlo results for an isolated independent diffuse layer, and MVN gives the profile in that theory. The curves through the Monte Carlo points are merely visual aids. One sees (from the IDL curve) the effect of the within-layer ion correlations in giving a much thinner diffuse layer than predicted by MVN. The between-layer correlations give (compare ITIES and IDL curves) a further contraction of the overall potential drop. In both sets of Monte Carlo results one sees also a small oscillation of $\psi(z)$ forbidden in MVN theory (Torrie and Valleau, 1986). Reproduced by permission of Elsevier Sequoia S.A.

and this difference appears rather large. The difference between IDL and ITIES is due to the within-layer ion correlations and is quite appreciable, up to 11%. Therefore, these findings demonstrate the existence and importance of inter-layer interactions. Obviously, the MVN theory ignores both within- and inter-layer correlations.

The result of these interactions is the compression of diffuse layers and the reduction of interfacial potential difference. It should also be expected that these effects would increase with decreasing solvent dielectric permittivity.

The partial penetration of ions into the boundary layer of another phase was investigated by varying the distance of closest approach z_{dca}. Figure 7.65 shows the dependence of the potential drop in the diffuse layer on the distance of closest approach. It is clear that the reduction of z_{dca} (up to 0.5d) leads to the decrease of absolute values of potentials ψ_1^* and ψ_2^*. However, at further reduction of z_{dca} the

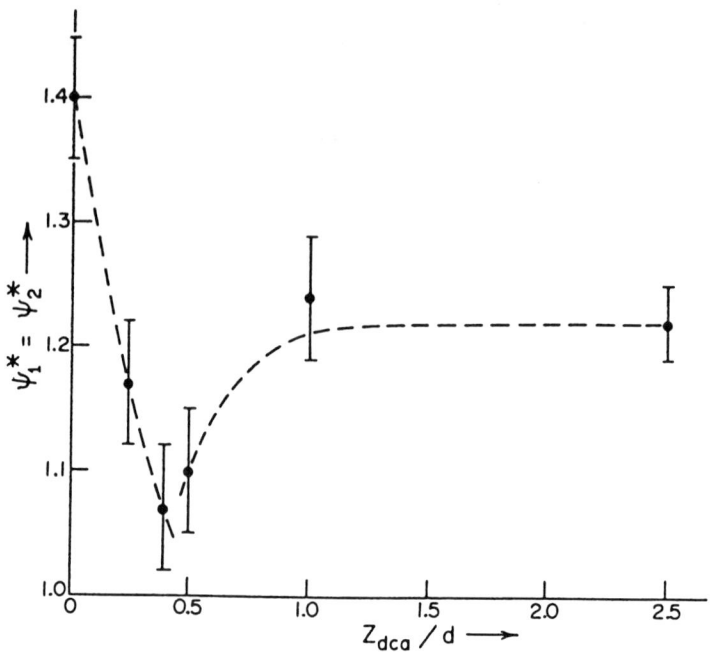

Figure 7.65. Potential drop across each of the diffuse layers of a symmetrical ITIES, as a function of the distance of closest approach z_{dca} of the ions to the surface. The dashed curve is only a visual aid. One sees that the between-layer correlation effect, reducing ψ_i^*, falls off sharply as z_{dca} is decreased, as expected. For $z_{dca} < d/2$, hard-sphere interactions between the two layers become possible; these counteract the Coulombic correlation effect, eventually becoming dominant for very small z_{dca} (Torrie and Valleau, 1986). Reproduced by permission of Elsevier Sequoia S.A.

diffuse layer potential begins to increase due to steric repulsion of ions (Torrie and Valleau, 1986).

The image forces also need to be considered. The computer "experiment" was carried out for phases with $\varepsilon_1 = 78.5$ and $\varepsilon_2 = 19.625$ containing a 1:1 electrolyte. The distance of closest approach was assumed equal to the ion radius. The results of the "experiment" are presented in Table 7.6, which shows the data for ITIES and MVN as well as the potential drops in IDL both with and without the image forces. To compare these data, moving from the center of the table to its sides means moving from ITIES to a system that ignores inter-layer correlations (IDL with image forces), then to a system that additionally ignores the image forces (IDL without image forces) and eventually to a system that ignores both of these and the within-layer correlations. It is clear that in the aqueous phase the image forces tend to increase the thickness of the diffuse layer and hence its potential (ψ_1^*), while this effect is opposed by the within-layer and inter-layer correlations. In the non-aqueous phase all three factors tend to reduce the thickness and potential. Because changes in the non-aqueous

Table 7.6. Diffuse layer electrical potential drops at two sides of the interface between two immiscible electrolyte solutions. Parameters: $\varepsilon_1 = 78.5$, $\varepsilon_2 = 19.625$, $d = 2z_{dca} = 0.325$ nm.

	ψ_1^*							ψ_2^*		
	IDL							IDL		
	Without image	With forces	ITIES	c_1 (M)	σ (C m^{-2})	c_2 (M)	ITIES	Without image	With forces	MVN
MVN										
1.40	1.38[a]	1.43[a]	1.41(0.02)	0.01	0.00887	0.005	2.38(0.05)	3.39	2.94	3.02
4.08	—	3.99	3.89(0.02)	0.01	0.04435	0.005	3.71(0.06)	3.82	4.29	6.14
0.47	—	0.50	0.48(0.01)	0.10	0.00887	0.05	1.03(0.02)	1.05	1.49	1.25
2.03	1.94	2.01	1.95(0.01)	0.10	0.04435	0.05	2.24(0.05)	2.27	2.76	3.84
1.21	—	0.95	0.93(0.03)	1.0	0.07557	0.50	0.85(0.06)	0.97	1.21	2.71

[a] These results were obtained from MPB5 theory (Torrie and Valleau, 1986).

phases are larger, the overall effect is to reduce the total interfacial potential difference. This computer "experiment" shows that the MVN theory is misleading because it is based on a structure that predicts an unrealistic potential drop.

7.19. MOLECULAR DYNAMICS AND THE STRUCTURE OF INTERFACES

In recent years our understanding of the structure of liquid–liquid interfaces has benefited from the application of new theoretical approaches to this problem. The use of molecular dynamics computer simulations has enhanced our understanding of both the equilibrium and transport properties of interfaces. We will present a general outline of this technique and its results; the interested reader can find details in a number of excellent reviews (Allen and Tildesley, 1987; Benjamin, 1996a; Blum, 1990; Carpenter and Hehre, 1990; Egberts *et al.*, 1994; Hayoun *et al.*, 1987, 1994; Torrie *et al.*, 1989).

Molecular dynamic simulation of interfaces began with liquid–vapor interfaces (see for example Lee and Scott, 1980) and later progressed to liquid–liquid interfaces (Linse, 1987). The idea of molecular dynamic simulation is actually very simple. It is based on classical mechanics that can calculate the movement of particles within the liquid system provided that the forces between these particles are known. The quantum-mechanical nature of particles and their interaction are included in the description of forces between the particles, although in a simplified way. By using this approach one can simultaneously analyze the position and movement of many particles over time intervals from picoseconds to nanoseconds. In this way, one can actually visualize the structure of a liquid interface (actually, of its model) and its evolution over time, and thereby calculate its thermodynamic and kinetic properties.

From a thermodynamic point of view the most important characteristics of any system are its free energy and we will describe how this quantity can be calculated by molecular dynamics methods (Benjamin, 1996a,b). Interaction between molecules in a system is addressed in a pairwise manner. In addition, the many-body nature of this interaction is partially taken into account by consideration of the mutual potential energy between two molecules. Each atom in the molecule is modeled as a soft sphere with a central charge, and the total intermolecular energy of interaction is represented by the sum of the Coulomb and Lennard-Jones potentials:

$$U_{\text{pair}} = \sum_{i,j} \left\{ \frac{Q_i Q_j}{r_{ij}} + 4\varepsilon_{ij} \left[\left(\frac{\sigma_{ij}}{r_{ij}} \right)^{12} - \left(\frac{\sigma_{ij}}{r_{ij}} \right)^{6} \right] \right\}. \tag{7.200}$$

Here Q_i and Q_j are the charges of atoms in a molecule. Considerable latitude is employed to select them in such a way as to produce the correct value of the dipole moment of the molecule in a condensed phase. The term ε_{ij} is the van der Waals binding energy between atoms i and j, and σ_{ij} is the distance at which the attractive dispersion force (the first term) is balanced by the repulsive force (the second term). The parameters of this equation are adjusted in such a way that the thermodynamic properties of the liquids can reflect experimental results. Numerical values for the parameters of different liquids have been published by Benjamin (1996a).

The internal energy of the system also includes an intramolecular component consisting of bond stretching, bending vibrations, and torsional motion. For computational reasons, it is highly desirable to maximize the integration time step. For this purpose the internal energy is usually simplified by omitting the high-frequency vibration modes. This allows simulations over longer time periods, but unfortunately makes the calculation of the dielectric constant less reliable. This drawback can be compensated by adjusting the parameters of the pairwise interaction in equation (7.200).

When liquids and their interfaces are modeled analytically, they can be considered in infinite space, which is impossible for computer simulation. Numerical calculations can be done only for a finite set of molecules that are contained within a set of periodic boundary conditions. One can think about these conditions as an infinite chain of identical units in contact with each other. When interfaces are included in the system, periodic boundary conditions are also imposed, which represent two different types of units (two different fluids) in the chain. When construction of the model is complete, the next step is to calculate the free energy of the system using classical statistical-mechanical equations and to follow the pathway of molecules in the system.

Several liquid–liquid interfaces have been studied by molecular dynamics methods. However, the results strongly depend on the model selected and on the numerous model parameters. Molecular dynamics simulation of interfaces has two goals: to appropriately select the surface tension and solubility parameters consistent with experimental data, and to predict the actual behavior of liquid–liquid interfaces.

These preliminary results can provide interesting insights into the structure of interfaces.

The decane–water interface studied by Vanbuuren *et al.* (1993) consisted of 50 decane molecules and 389 water molecules in a two-phase system. An interesting question concerns the difference in the properties of molecules in the bulk solution versus those at the interface. In bulk solution, water has an unlimited ability to form hydrogen bonds without any special orientation. However, at the interfacial layer where there are fewer neighboring water molecules, a given water molecule is limited in its molecular orientation. Vanbuuren *et al.* (1993) demonstrated that water molecules tend to orient normal to the interface, while decane assumes a more lateral orientation with respect to the interface. Immediately adjacent to the interface this orientation is reversed with hydrogen atoms of the water pointing toward the decane phase.

Another popular interface in electrochemistry is water–1,2-dichloroethane (DCE), which has been examined in detail by Benjamin (1992a,b). His model included 216 DCE molecules and 500 H_2O molecules, and the molecular pathways were followed for periods up to 100 ps. It was found that the thickness of the transition boundary between two liquids was about 1.2 nm and the density of each liquid changed gradually in this interfacial region. The interfacial layer was defined as the region where the density of each liquid was between 10% and 90% of the bulk density. Density oscillations were found in bulk water and DCE. However, it was unclear whether these oscillations were artifacts of using a small system size or due to short simulation periods. With proper selection of parameters, the two liquids were immiscible during the chosen simulation period.

An important characteristic of the molecular structure of liquids is the pair correlation function which gives the probability of finding a given pair of molecules at two given points. While in bulk liquids the pair correlation function depends only on the distance between two molecules, at the interface it also depends on the position of the atoms normal to the interface. The pair correlation function is characterized by a series of peaks. These were analyzed by Benjamin, who found that the interaction of water molecules at close range does not change much in the interfacial region, and that most of the water molecules have the same first coordination shell as they do in bulk solution. The situation is different with DCE molecules, which have weaker self-solvation at the interface. Substantial interactions between water and DCE molecules were found, although they were not strong enough to produce solvation of water molecules by a shell of DCE molecules or vice versa. This is consistent with the poor mutual solubility of the two liquids.

The molecular orientation at the liquid–liquid interface is of special interest. It was found that water dipoles are distributed in a certain range of angular orientation with maximal distribution being parallel to the interface. However, a significant fraction of dipoles was also perpendicular to the interface. The orientation of water molecules parallel to the surface declined very rapidly as their distance from the interface increased, with a characteristic length of 0.7 nm.

The dynamic properties of molecules at a surface, such as dipole reorientation time and the diffusion constant, did not differ from bulk values. The mechanical fluctuations of this surface were also studied and were found to have a characteristic time in the range of a few tens of picoseconds.

In summary, Benjamin concluded that the water–DCE interface is rather sharply defined with no mixed solvent region, but on a molecular scale the interface is very rough. This roughness manifests itself as capillaries or fingers of one liquid protruding into another. These protrusions can extend as far as 0.8 nm within a few nanoseconds, and may play an important role in transport processes. For instance, Benjamin (1993), and Schweighofer and Benjamin (1995) proposed that aqueous capillaries or water fingers can be the precursors for the transfer of ions across the interface between two immiscible electrolyte solutions.

D

CHEMISTRY AT LIQUID INTERFACES

8

INTERFACIAL CATALYSIS

8.1. OIL–WATER INTERFACE AS A MODEL OF MEMBRANES

Michael Faraday first studied electron transfer reactions at oil–water interfaces to prepare colloidal metals by reducing metal salts at ether–water or carbon disulfide–water interfaces (Faraday, 1857). As the field progressed after Faraday's pioneering observations, it became clear that vectorial charge transfer at the interface between two dielectric media is an important stage in many bioelectrochemical processes such as those mediated by energy-transducing membranes (Blank, 1991; Blank and Feig, 1963; Boguslavsky and Volkov, 1977, 1987; Kharkats and Volkov, 1985, 1987; Rehbinder, 1927a,b). Boundary membranes play a key role in the cells of all contemporary organisms, and adequate models of membrane function are therefore of considerable interest. The interface of two immiscible liquids has been widely used for this purpose. For example, the fundamental processes of photosynthesis (Volkov, 1984a,b, 1985a,b, 1986a,b,c, 1987c, 1988, 1989; Volkov *et al.*, 1996), biocatalysis (Aliev *et al.*, 1976; Kharkats and Volkov, 1987), membrane fusion (Gingell and Fornes, 1975; Gingell *et al.*, 1977), ion pumping (Volkov *et al.*, 1985; Volkov and Deamer, 1996; Yaguzhinsky *et al.*, 1975), and electron transport (Bell, 1928; Samec, 1996; Volkov *et al.*, 1975a) have all been investigated in such interfacial systems. The multi-electron redox reaction at the interface between two immiscible liquids was first investigated by Bell (1928). Studies have been made on redox and hydrolysis reactions catalyzed by enzymes (Boguslavsky *et al.*, 1975a,b; Goodall and Sachs, 1977), photosynthetic pigments (Kandelaki and Volkov, 1991), metal complexes of porphyrins (Volkov *et al.*, 1976a,b, 1983a,b), bacteria (Rosenberg, 1991), submitochondrial particles (Boguslavsky *et al.*, 1976g), as well as in systems with an extended surface—in microemulsions (Deamer and Volkov, 1996; Gugeshashvili *et al.*, 1991), vesicles and reversed micelles (Zamaraev and Parmon, 1983). Enzymes and pigments embedded in a hydrophilic–hydrophobic interface have properties similar to their functional state in a membrane. For instance, certain enzymes can be highly active at the interface, but virtually inactive in a homogeneous medium. The interface between two immiscible liquids with immobilized photosynthetic pigments can also serve as a simple model for investigating photoprocesses accompanied by spatial separation of charges across a membrane (Volkov, 1989). Such light-dependent redox reactions at the oil–water interface have been discussed in recent reviews and books (Deamer and Volkov, 1996; De Armond and De Armond, 1996; Gratzel, 1989; Kotov and Kuzmin, 1996; Volkov and Deamer, 1996). Here we will first present theoretical concepts of interfacial redox reactions, then show how the theory has been applied to experimental results.

8.2. MULTI-ELECTRON REACTIONS AT INTERFACES

Synchronous multi-electron reactions in membranes have recently drawn the attention of both chemists and biologists. In a synchronous multi-electron reaction the energy can be utilized very economically. Furthermore, the biotechnological application of multi-electron reactions makes it possible to drive redox reactions in relatively mild conditions under the action of weak oxidants or reductants. Synchronous multi-electron reactions may proceed without forming highly reactive intermediate radicals which have the potential to damage the catalytic complex. Since multi-electron reactions do not produce significant toxic intermediates, they can be used by living organisms for biochemical energy conversion in respiration and photosynthesis. In multi-electron reactions that occur as consecutive one-electron stages, the Gibbs energy necessary per single electron transfer obviously cannot be uniformly distributed over the stages. The energy needs for various stages will be different and the excess energy in the lower energy stages will be converted into heat.

The term *synchronous multi-electron* reaction does not mean that all *n* electron transfer occurs simultaneously. Instead, each electron is transferred from donor to acceptors individually, but the time required for "intermediate" formation is much less than the time of the reorganization of the medium, so that "intermediates" as individual chemical compounds do not exist.

From thermodynamic and kinetic principles the interface between two immiscible liquids can have catalytic properties for interfacial charge transfer reactions (Kharkats and Volkov, 1985). The thermodynamic analysis of redox and mixed potentials at the liquid–liquid interface was discussed in Part B of this book. It is possible to shift the redox potential scale in a desired direction by selecting appropriate solvents, thereby permitting reactions to occur that are highly unfavorable in a homogeneous phase. As follows from thermodynamics, if the resolvation energies of substrates and products are very different, the interface between two immiscible liquids may act as a catalyst. The kinetic mechanism underlying the catalytic properties of the liquid–liquid interface was discussed first by Kharkats and Volkov (1985, 1987).

The quantum theory of chemical reactions in polar media (Kharkats and Kuznetsov, 1996) can be used as the basis for the theory of charge transfer at the interface between two dielectric media like oil–water or biomembrane–water. This theory one can express the electron transfer rate in terms of the dielectric properties of the medium and electronic properties of reactants.

The problem of theoretically describing the elementary charge-transfer process across interfaces between two condensed media has a long history. The earliest studies of interfacial phenomena and electrochemistry focused on metal–electrolyte or semiconductor–electrolyte interfaces. Considerable progress has been made recently in extending the theory to liquid–liquid interfaces (Kharkats and Kuznetsov, 1996; Kharkats and Volkov, 1985; Marcus, 1990a,b, 1991, 1995).

Kharkats (1976) and Kharkats and Volkov (1985, 1987) first calculated the energy of activation and solvent reorganization of charge transfer across the interface between two immiscible liquids. The expression for the probability of electron transfer can be written as

$$W = A \exp\left\{-\frac{U_i}{kT} - \frac{[E_s + \Delta G_c + U_f - U_i]^2}{4E_s kT}\right\}, \tag{8.1}$$

where U_i is the work that must be performed upon the system to place the reactants at distances h_1 and h_2 from the interface (Fig. 8.1), U_f is the corresponding work for the reaction products, ΔG_c is the configurational Gibbs free energy, E_s is the solvent reorganization energy, and A is the pre-exponential factor, which is proportional to the transmission coefficient. Theoretical analysis shows that the most effective electron transfer takes place at the closest disposition of reaction centers (Kharkats, 1990; Kharkats and Volkov, 1985, 1987).

The difference between Gibbs free energies for substrates and products can be

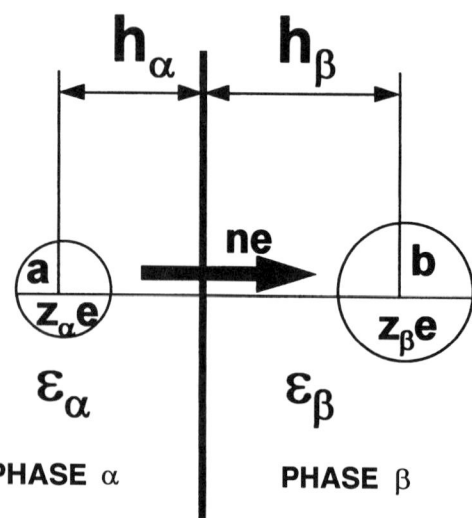

Figure 8.1. Locations of charge donors and acceptors at the interface and vectorial electron transport.

found by using the Born equation (see Chapter 6) corrected for the solvophobic effect

$$\Delta G_c = \Delta G_c(\text{svph}) + \frac{e_0^2}{4\pi\varepsilon_0}\left(\frac{n^2 + 2z_\alpha n}{2\varepsilon_\alpha a} + \frac{n^2 - 2z_\beta n}{2\varepsilon_\beta b}\right), \tag{8.2}$$

where the solvophobic component of Gibbs free energy $\Delta G_c(\text{svp})$ does not depend on the dielectric properties of media α and β. The Born electrostatic contribution to the solvation Gibbs free energy does not yield good agreement with experiment, and a more accurate calculation can be performed using non-local electrostatics (Kornyshev and Volkov, 1984; Markin and Volkov, 1987d; Volkov and Kornyshev, 1985).

The configurational Gibbs free energy ΔG_c in equation (8.1) is the part of the free energy not depending on the transpositional contribution to entropy (Krishtalik and Kuznetsov, 1986). The configurational Gibbs free energy is different from the standard Gibbs free energy ΔG^0:

$$\Delta G_c = \Delta G^0 + RT \ln \frac{\prod_n X_i^0}{\prod_m X_f^0}. \tag{8.3}$$

Here X_i^0 and X_f^0 are the mole fractions of n initial (i) and m final (f) reagents in their standard states. If the reaction is not accompanied by changes in the number of particles, there is no difference between the standard free energy and the configurational free energy. However, if the number of particles change, for instance, by decomposition of a molecule, the expression for the basic reaction involves only the configurational free energy which does not include the entropy related to commutation of the particles. ΔG_c is the free energy of the reaction and it differs from the work of the reaction by the work of mixing reagents depending on the concentration. In the case of interfacial reactions, the quantity which corresponds to ΔG_c will be a configurational interfacial potential (Krishtalik and Kuznetsov, 1986):

$$\Delta_\beta^\alpha \phi_c = -\frac{\Delta G_c}{nF}. \tag{8.4}$$

When interfacial potential equals $\Delta_\beta^\alpha \phi_c$, the free energy of the interfacial reaction

$$\text{Ox} + e^- \iff \text{Red} \tag{8.5}$$

is zero. At interfacial potentials $\Delta_\beta^\alpha \phi$ different from $\Delta_\beta^\alpha \phi_c$, the free energy of the reaction will be $nF(\Delta_\beta^\alpha - \Delta_\beta^\alpha \phi_c)$. Calculations of the configurational Gibbs energy, configurational redox and electrode potentials, and the comparison of their values with standard Gibbs energies and standard redox potentials can be found in the literature (Krishtalik, 1986; Krishtalik and Kuznetsov, 1986).

Marcus (1990a, 1990b, 1991, 1995) estimated the rate constant k_r for an electron transfer reaction between two redox components dissolved in two different liquid phases (Fig. 8.1):

$$-\frac{dN_1}{dt} = -k_r c_1 c_2 S, \tag{8.6}$$

where N_1 is the number of molecules of type 1 in phase α, c_1 is the mean concentration of reactant 1 in phase α, c_2 is the mean concentration of reactant 2 in phase β, and S is the interfacial area. The rate constant k_r can be approximately found as:

$$k_r = \kappa \nu \nu \exp\left(-\frac{E_a}{RT}\right), \tag{8.7}$$

where ν is the characteristic frequency for the molecular motion, and E_a is the activation energy. If the liquid–liquid interface is a sharp boundary and if $h_\alpha \geq a$ and $h_\beta \geq b$, Marcus's expression for ν is:

$$\nu = 2\pi(a + b)(\Delta R)^3, \tag{8.8}$$

where ΔR is the center-to-center distance between reagents (Fig. 8.2). In the system shown in Fig. 8.1, $\Delta R = h_\alpha + h_\beta$. If ions penetrate into second contacting phase but the reactants do not overlap, Marcus (1990b) derived another equation:

$$\nu \approx \pi(a_1 + a_2)^3 \Delta R. \tag{8.9}$$

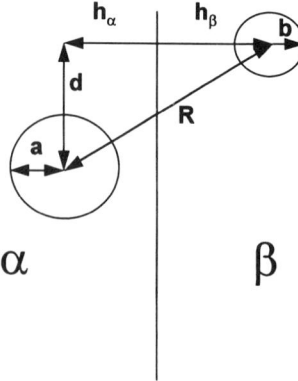

Figure 8.2. Locations of charge donors and acceptors at the interface between two immiscible electrolyte solutions.

8.3. SOLVENT REORGANIZATION ENERGY

The solvent reorganization energy is an important parameter in the quantum theory describing charge transfer in polar media. In the case of homogeneous reactions which take place in one phase it can be estimated by the equation:

$$E_s = \frac{1}{2\varepsilon_0}\left(\frac{1}{\varepsilon_{op}} - \frac{1}{\varepsilon_s}\right)\int_{\infty-V_a-V_b}(\boldsymbol{D}_i - \boldsymbol{D}_f)^2\,dV, \tag{8.10}$$

where ε_{op} and ε_s are the optical and the static dielectric permittivities of the medium, and $\boldsymbol{D}_i$ and $\boldsymbol{D}_f$ are the inductions of the electric fields which are created in the solvent during the initial and final state of charge transfer. Integration in equation (8.10) is carried out over the entire volume of the medium except the reactant volumes.

An approximate calculation of the solvent reorganization energy during charge transfer was first performed by Marcus (1956). Assuming that the distance h_{12} between the reactant centers of the reactant is much larger than their radii a and b, and that the reactants can be described as non-polarizable spheres with charges rigidly and uniformly distributed over the surfaces, the expression for the reorganization energy is:

$$E_s = \frac{e_0^2 n^2}{4\pi\varepsilon_0}\left(\frac{1}{\varepsilon_{op}} - \frac{1}{\varepsilon_s}\right)\left(\frac{1}{2a} + \frac{1}{2b} - \frac{1}{h_{12}}\right). \tag{8.11}$$

If the charge transfer occurs between reactants in two different dielectric media, equation (8.10) can be written as

$$E_s = \frac{1}{2\varepsilon_0}\left[\int_{\infty-V_a-V_b}\frac{1}{\varepsilon_s(r)}(\boldsymbol{D}_f - \boldsymbol{D}_i)^2\,dV\right]_{static}$$
$$- \frac{1}{2\varepsilon_0}\left[\int_{\infty-V_a-V_b}\frac{1}{\varepsilon_{op}(r)}(\boldsymbol{D}_f - \boldsymbol{D}_i)^2\,dV\right]_{optical}. \tag{8.12}$$

Girault (1995) stressed the need for heterogeneous electron transfer reactions to differentiate static from optical integrals. For a sharp interface between two immiscible liquids the solvent reorganization energy can be written as:

$$E_s = \frac{(ne_0)^2}{8\pi\varepsilon_0 a}\left(\frac{1}{\varepsilon_{op\alpha}} - \frac{1}{\varepsilon_\alpha}\right) + \frac{(ne_0)^2}{8\pi\varepsilon_0 b}\left(\frac{1}{\varepsilon_{op\beta}} - \frac{1}{\varepsilon_\beta}\right)$$
$$+ \frac{(ne_0)^2}{16\pi\varepsilon_0 h_\alpha}\left(\frac{\varepsilon_{op\alpha} - \varepsilon_{op\beta}}{\varepsilon_{op\alpha}(\varepsilon_{op\alpha} + \varepsilon_{op\beta})} - \frac{\varepsilon_\alpha - \varepsilon_\beta}{\varepsilon_\alpha(\varepsilon_\alpha + \varepsilon_\beta)}\right)$$
$$- \frac{(ne_0)^2}{16\pi\varepsilon_0 h_\beta}\left(\frac{\varepsilon_{op\alpha} - \varepsilon_{op\beta}}{\varepsilon_{op\beta}(\varepsilon_{op\alpha} + \varepsilon_{op\beta})} - \frac{\varepsilon_\alpha - \varepsilon_\beta}{\varepsilon_\beta(\varepsilon_\alpha + \varepsilon_\beta)}\right)$$

$$- \frac{(ne_0)^2}{2\pi\varepsilon_0(h_\alpha + h_\beta)} \left(\frac{1}{\varepsilon_{op\alpha} + \varepsilon_{op\beta}} - \frac{1}{\varepsilon_\alpha + \varepsilon_\beta} \right), \tag{8.13}$$

where ne_0 is the charge transferred in the reaction, subscripts α and β denote the dielectric permittivities of media α and β, and the reactants are spheres of radii a and b with charges $z_1 e_0$ and $z_2 e_0$ which are located at distances h_1 and h_2 from the interface (Fig. 8.1).

Figure 8.2 shows the boundary where the direction of electron transfer between reactants is not perpendicular to the interface. The expression for solvent reorganization energy in this case is:

$$E_s = \frac{(ne_0)^2}{8\pi\varepsilon_0 a} \left(\frac{1}{\varepsilon_{op\alpha}} - \frac{1}{\varepsilon_\alpha} \right) + \frac{(ne_0)^2}{8\pi\varepsilon_0 b} \left(\frac{1}{\varepsilon_{op\beta}} - \frac{1}{\varepsilon_\beta} \right)$$

$$+ \frac{(ne_0)^2}{16\pi\varepsilon_0 h_\alpha} \left(\frac{\varepsilon_{op\alpha} - \varepsilon_{op\beta}}{\varepsilon_{op\alpha}(\varepsilon_{op\alpha} + \varepsilon_{op\beta})} - \frac{\varepsilon_\alpha - \varepsilon_\beta}{\varepsilon_\alpha(\varepsilon_\alpha + \varepsilon_\beta)} \right)$$

$$+ \frac{(ne_0)^2}{16\pi\varepsilon_0 h_\beta} \left(\frac{\varepsilon_{op\alpha} - \varepsilon_{op\beta}}{\varepsilon_{op\beta}(\varepsilon_{op\alpha} + \varepsilon_{op\beta})} - \frac{\varepsilon_\alpha - \varepsilon_\beta}{\varepsilon_\beta(\varepsilon_\alpha + \varepsilon_\beta)} \right)$$

$$- \frac{(ne_0)^2}{2\pi\varepsilon_0(l)} \left(\frac{1}{\varepsilon_{op\alpha} + \varepsilon_{op\beta}} - \frac{1}{\varepsilon_\alpha + \varepsilon_\beta} \right). \tag{8.14}$$

Similarly, U_i and U_f can be expressed in terms of integrals of inductions $\mathbf{D}_i$ and $\mathbf{D}_f$:

$$U_i = \frac{1}{32\pi^2\varepsilon_0\varepsilon_\alpha} \int_I D_i^2 \, dV + \frac{1}{32\pi^2\varepsilon_0\varepsilon_\beta} \int_{II} D_i^2 \, dV - \frac{z_1^2 e_0^2}{8\pi\varepsilon_0 a\varepsilon_\alpha} - \frac{z_2^2 e_0^2}{8\pi\varepsilon_0 b\varepsilon_\beta}, \tag{8.15}$$

$$U_f = \frac{1}{32\pi^2\varepsilon_0\varepsilon_1} \int_I D_f^2 \, dV + \frac{1}{32\pi^2\varepsilon_0\varepsilon_2} \int_{II} D_f^2 \, dV - \frac{(z_1 + n)^2 e_0^2}{8\pi\varepsilon_0 a\varepsilon_\alpha} - \frac{(z_2 - n)^2 e_0^2}{8\pi\varepsilon_0 b\varepsilon_\beta}, \tag{8.16}$$

where the integration ranges I and II represent the two half-spaces of media α and β excluding the volume of reactants.

Calculation of the integrals in (8.15) and (8.16) is conveniently carried out by changing to surface integrals. For instance, Kharkats (1976) calculated for $\int_I D_i^2 \, dV$ with an accuracy of $(a/h)^3$, $(b/h)^3$:

$$\int_I D_i^2 \, dV = \frac{4\pi e_0^2 z_1^2}{a} - \frac{2\pi e_0^2 z_1^2}{h_1} + \pi e_0^2 \left(\frac{2\varepsilon_\alpha}{\varepsilon_\alpha + \varepsilon_\beta} \right)^2 \left(\frac{z_1^2}{h_1} + \frac{z_2^2}{h_2} + \frac{4z_1^2 z_2^2}{h_1 + h_2} \right). \tag{8.17}$$

The expression for $\int_{II} D_i^2 \, dV$ is obtained from equation (8.15) by making the substitutions $a \to b$, $\varepsilon_\alpha \to \varepsilon_\beta$, $h_1 \to h_2$ and $z_1 \to z_2$. The corresponding expression for $\int_{1,2} D_f^2 \, dV$ can be obtained by making substitutions $z_1 \to (z_1 + n)$ and $z_2 \to (z_2 - n)$.

The simplest expressions for E_s, U_i and U_f are obtained when the reactions take place at equal distances from the interface, $h_1 = h_2 = h$:

$$E_s = \frac{(ne_0)^2}{8\pi\varepsilon_0 a}\left(\frac{1}{\varepsilon_{op\alpha}} - \frac{1}{\varepsilon_\alpha}\right) + \frac{(ne_0)^2}{8\pi\varepsilon_0 b}\left(\frac{1}{\varepsilon_{op\beta}} - \frac{1}{\varepsilon_\beta}\right)$$

$$- \frac{(ne_0)^2}{4\pi\varepsilon_0 h}\left(\frac{1}{\varepsilon_{op\alpha} + \varepsilon_{op\beta}} - \frac{1}{\varepsilon_\alpha + \varepsilon_\beta}\right), \tag{8.18}$$

$$U_i = \frac{z_\alpha z_\beta e_0^2}{4\pi\varepsilon_0(\varepsilon_\alpha + \varepsilon_\beta)h} + \frac{z_\alpha^2 e_0^2(\varepsilon_\alpha - \varepsilon_\beta)}{16\pi\varepsilon_0\varepsilon_\alpha(\varepsilon_\alpha + \varepsilon_\beta)h} - \frac{z_\beta^2 e_0^2(\varepsilon_\alpha - \varepsilon_\beta)}{16\pi\varepsilon_0\varepsilon_\beta(\varepsilon_\alpha + \varepsilon_\beta)h}, \tag{8.19}$$

$$U_f = \frac{(z_\alpha + n)(z_\beta - n)e_0^2}{4\pi\varepsilon_0(\varepsilon_\alpha + \varepsilon_\beta)h} + \frac{(z_\alpha + n)^2 e_0^2(\varepsilon_\alpha - \varepsilon_\beta)}{16\pi\varepsilon_0\varepsilon_\alpha(\varepsilon_\alpha + \varepsilon_\beta)h} - \frac{(z_\beta - n)^2 e_0^2(\varepsilon_\alpha - \varepsilon_\beta)}{16\pi\varepsilon_0\varepsilon_\beta(\varepsilon_\alpha + \varepsilon_\beta)h}, \tag{8.20}$$

$$E_a = U_i + \frac{(E_s + \Delta G_c + U_f - U_i)^2}{4E_s}. \tag{8.21}$$

In the case of homogeneous electron transfer in a dielectric medium, the work required to bring the reactants or reaction products together approaches zero when one of the reactants or products is electrically neutral, whereas in the process discussed here, U_i values are never zero because of the interactions with image charges.

8.4. SELECTIVE CATALYTIC PROPERTIES OF LIQUID INTERFACES

The activation energy of electron transfer depends on the charges of the reactants and dielectric permittivity of the non-aqueous phase. This can be useful when choosing a pair of immiscible solvents to decrease the activation energy of the reaction in question or to inhibit an undesired process (Volkov, 1984a, 1984b, 1985a, 1985b). For example, suppose that an electron is transferred from a donor in aqueous phase α to an acceptor in organic phase β (Fig. 8.1). Assuming that ΔG_c is negligible compared to E_s, the activation energy, E_a, depends on the dielectric permittivity of non-aqueous phase, ε_β:

$$E_a \cong U_i + \frac{(E_s + U_f - U_i)^2}{4E_s}. \tag{8.22}$$

The reorganization energy increases with ε_β to a maximum asymptotic value at $\varepsilon_\beta \gg \varepsilon_{op\beta}$:

$$E_s = \frac{(ne_0)^2}{8\pi\varepsilon_0 a}\left(\frac{1}{\varepsilon_{op\alpha}} - \frac{1}{\varepsilon_\alpha}\right) + \frac{(ne_0)^2}{8\pi\varepsilon_0 b}\left(\frac{1}{\varepsilon_{op\beta}}\right) - \frac{(ne_0)^2}{2\pi\varepsilon_0 h}\left(\frac{1}{\varepsilon_{op\alpha} + \varepsilon_{op\beta}} - \frac{1}{\varepsilon_\alpha + \varepsilon_\beta}\right), \tag{8.23}$$

and at $\varepsilon_\beta = \varepsilon_{op\beta}$ reorganization energy is minimal and equal to

$$E_s = \frac{(ne_0)^2}{8\pi\varepsilon_0 a}\left(\frac{1}{\varepsilon_{op\alpha}} - \frac{1}{\varepsilon_\alpha}\right) - \frac{(ne_0)^2}{2\pi\varepsilon_0 h}\left(\frac{1}{\varepsilon_{op\alpha} + \varepsilon_{op\beta}} - \frac{1}{\varepsilon_\alpha + \varepsilon_\beta}\right). \tag{8.24}$$

Examples of the dependencies of E_s, E_a and U_i for different sets of parameters z_1, z_2, and h/a are plotted in Figs. 8.3–8.5.

Figure 8.3 shows how the dielectric constant of the organic phase and the distance from the interface or between reagents affect the medium reorganization energy. A

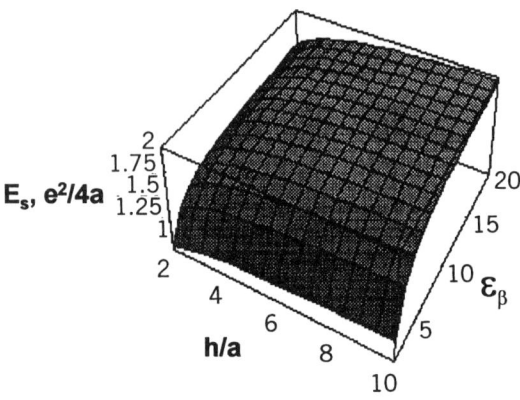

Figure 8.3. Dependence of solvent reorganization energy E_s on dielectric permittivity of non-aqueous phase ε_β and distance $h = h_1 = h_2$ from the interface with parameters: $\varepsilon_{op\alpha} = \varepsilon_{op\beta} = 1.8$, $\varepsilon_\alpha = 78$, $a = b$, $n = 1$.

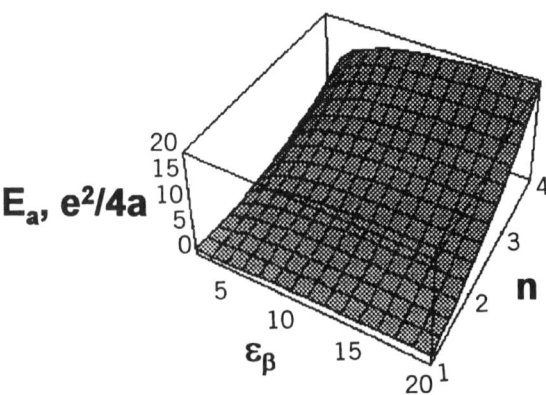

Figure 8.4. Dependence of activation energy E_a on dielectric permittivity of non-aqueous phase ε_β and number of transferred electrons: $\varepsilon_{op\alpha} = \varepsilon_{op\beta} = 1.8$, $\varepsilon_\alpha = 78$, $a = b = h/2$, $z_\alpha = -1$, $z_\beta = +1$, $\Delta G_c = e^2/2a$.

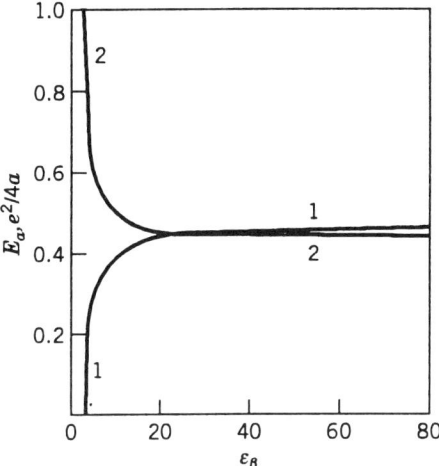

Figure 8.5. Dependence of the activation energy, E_a, on dielectric permittivity of a non-aqueous phase ε_β, calculated by equation (8.21) with parameters: $a = b = h$, $n = 1$, $\Delta G_c = 0.01 e^2/a$ and (1) $z_\alpha = 0$, $z_\beta = +1$; (2) $z_\alpha = 0$, $z_\beta = -1$.

decrease of h or ε_β dramatically decreases E_s (Fig. 8.3), which reaches a minimum value as ε_β approaches 2.

Equation (8.22) is plotted in Figs. 8.4 and 8.5 which show that the activation energy of the process decreases (or increases) greatly at small ε_β and, accordingly, the rate constant of charge transfer across the interface increases (or decreases) sharply at relatively small ε_β. Figure 8.5 shows that the liquid–liquid interface has selective properties and can catalyze or inhibit interfacial charge transfer reactions due to electrostatic effects.

To summarize, the kinetic parameters of interfacial charge transfer depend on the charge being transferred, the charges of reactants, their location in relation to the interface, as well as the dielectric properties of the media forming the liquid–liquid interface. Charge transfer processes in simple models are described by equations that can in turn be extended to more complicated processes, including phase transfer, micellar catalysis and bioenergetic processes taking place in biomembranes.

8.5. CHARGE TRANSFER REACTIONS AT OIL–WATER INTERFACES

Redox reactions at the interface between immiscible liquids fall into two classes. The first class includes spontaneous processes that occur in the absence of external electromagnetic fields. This type of redox transformation has been investigated in bioenergetics (Yaguzhinsky *et al.*, 1975a, 1975b, 1976, 1977), model membrane systems (Volkov *et al.*, 1975), and at oil–water interfaces (Gugeshashvili *et al.*, 1983a, 1983b, 1991, 1992, 1995). Redox reactions in the second class occur at the interface between immiscible electrolytes when external electrical fields are applied to the interface, and under these conditions interfacial charge transfer reactions take place at controlled interfacial potentials (Cunnane *et al.*, 1988a,b, 1995; Cunnane and Murtomaki, 1996; Samec, 1996). Such electrochemical interfacial reactions are usually multi-stage processes (Boguslavsky and Volkov, 1987; Menger, 1972) that have the following steps:

- diffusion of reactant to the interface;
- adsorption of reactants onto the interface;
- chemical reaction at the interface;
- desorption of products from the interface;
- diffusion of products from the interface.

8.6. EXAMPLES OF CHEMICAL REACTIONS AT LIQUID INTERFACES

Long before interfacial charge transfer reactions began to be studied systematically, Bell (1928) observed a multielectron transfer reaction across a benzene–water interface. This involved permanganate oxidation of benzoyl-*o*-toluidine to benzoyl-anthranilic acid:

$$\text{(structure with NHCOPh and CH}_3\text{)} \xrightarrow{\text{KMnO}_4} \text{(structure with NHCOPh and COOH)} \qquad (8.25)$$

This reaction occurred at the water–benzene interface and its kinetics did not depend on the rate of stirring of each of the contacting phases.

Other examples of interfacial reactions are summarized in Fig. 8.6.

An entirely novel type of heterogeneous process, called *"phase transfer catalysis"* or PTC (Starks *et al.* 1994), has been widely used by chemists for preparative purposes. The essence of the method is to create a two-phase system (usually with an organic and an aqueous phase), in which non-polar and ionic reactants are present in different phases, and to use catalysts as sources of lipophilic cations. The role of the catalysts is to form lipophilic ion pairs between the cation of the catalyst and the reacting anion which are then capable of migrating within the organic phase. PTC is one of the simplest and most economical methods of intensifying the production of a wide range of organic materials. The main advantage of the PTC method is that it is general, mild, and catalytic. The limitation of this method is due to the difficulty of separating the catalyst from the reaction medium, the tendency of the system to form stable emulsions, and the impossibility of performing repeated or continuous processes. These disadvantages can be eliminated by three-phase catalysis (TPC) in which a catalyst is immobilized on a polymeric support. The insoluble catalysts are easily separated from the reaction medium by simple filtration and can be used repeatedly.

Phase transfer catalysis requires a phase transfer agent in catalytic amounts which transfers substrates from one phase to a second phase where they can react with other

1. Alkyllation of benzylketones (Makosza, 1977):

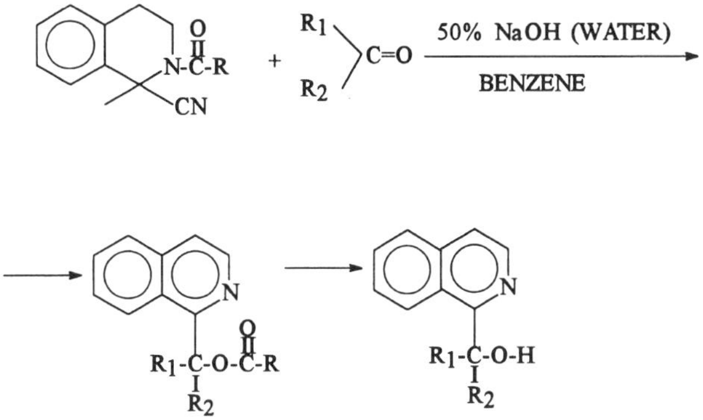

$$+ \quad \begin{array}{c} R_1 \\ R_2 \end{array}\!\!\!\!> C=O \quad \xrightarrow[\text{BENZENE}]{\text{50\% NaOH (WATER)}}$$

$$\longrightarrow \quad R_1-\overset{R_2}{\underset{R_2}{C}}-O-\overset{O}{\overset{\|}{C}}-R \quad \longrightarrow \quad R_1-\overset{}{\underset{R_2}{C}}-O-H \tag{8.26}$$

2. Synthesis of azipiridines at the benzene/water interface in the presence of 50% NaOH in the aqueous phase was discovered by Greibrokk (1972):

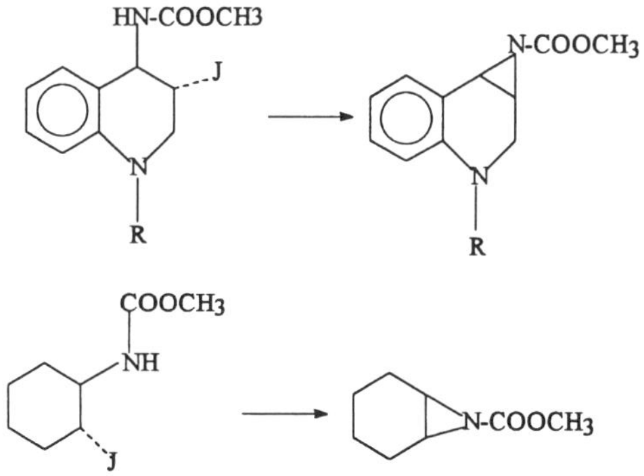

$$\text{HN-COOCH3} \quad \longrightarrow \quad \text{N-COOCH3}$$

$$\text{COOCH3} \quad \longrightarrow \quad \text{N-COOCH3} \tag{8.27}$$

3. Interfacial competitive reactions (Makosza, 1977):

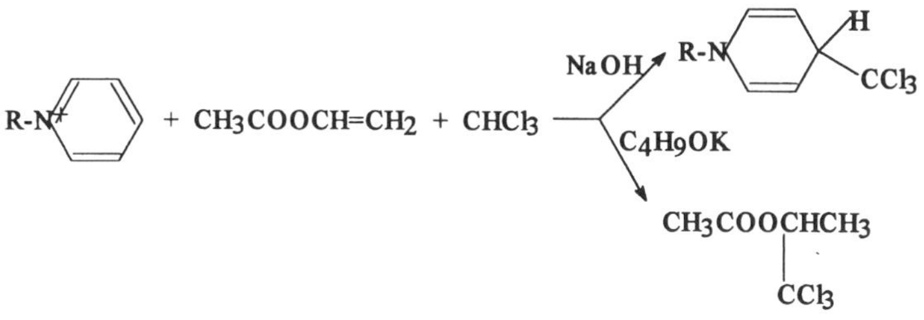

$$R-N^+ \quad + \quad CH_3COOCH=CH_2 \quad + \quad CHCl_3 \quad \xrightarrow[C_4H_9OK]{NaOH}$$

$$\begin{array}{c} R-N \cdots \cdots CCl_3, H \\[4pt] CH_3COOCHCH_3 \\ | \\ CCl_3 \end{array} \tag{8.28}$$

Figure 8.6. Chemical reactions at liquid–liquid interfaces.

4. Oxidation of borneol or iso-borneol to camphora by an aqueous solution of chromic acid:

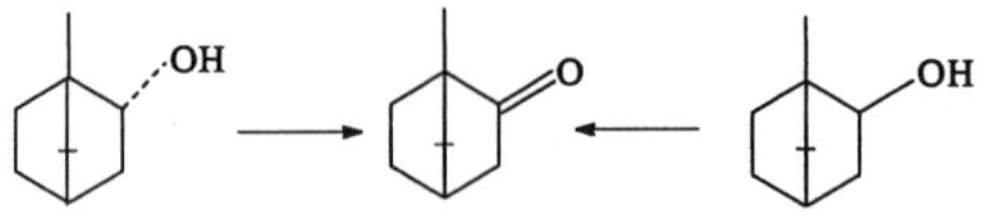

$$(8.29)$$

5. Hydrolysis of the *n*-nitrophenyl ester of lauric acid at the heptane–water interface (Menger, 1972):

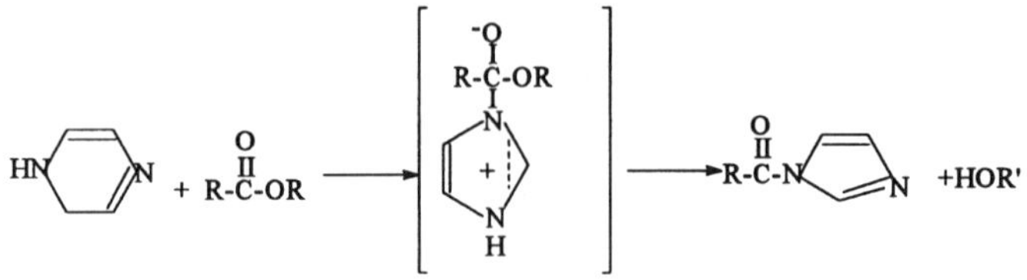

$$(8.30)$$

6. Hydrolysis of *p*-methylbenzyl chloride and of methyl nicotinate in water–CCl$_4$ and water-isopropyl myristate systems (Albery *et al.*, 1974):

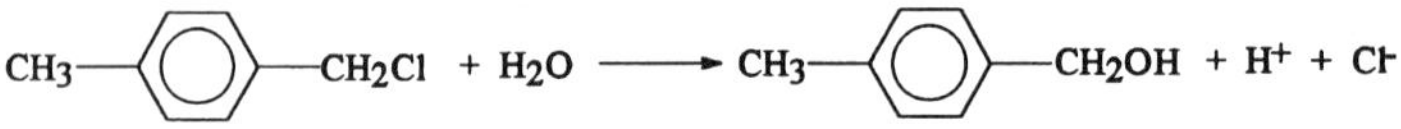

$$(8.31)$$

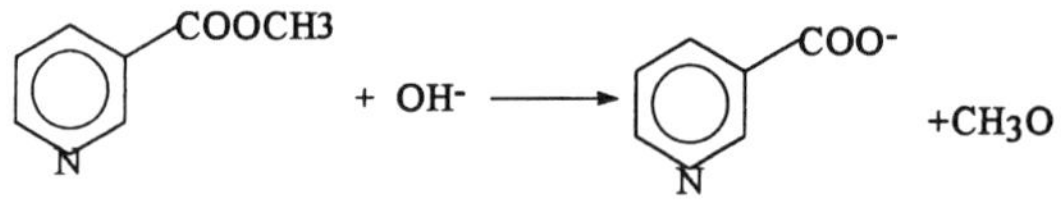

$$(8.32)$$

7. The electrochemical metallization at the water–1,2-dichloroethane and water–dichloromethane interfaces was studied by Guainazzi *et al.* (1975):

$$[V(CO)_6]^-_{\{in\ oil\}} + 2\ Cu^{2+} \rightarrow V^{3+}_{(aq)} + 6CO + 2\ Cu \qquad (8.33)$$

$$[V(CO)_6]^-_{\{in\ oil\}} + 4Ag^+ \rightarrow V^{3+}_{(aq)} + 6CO + 2\ Ag \qquad (8.34)$$

8. Continuous peptide synthesis at an oil–water interface with an immobilized enzyme therolysin (Nakanishi and Matsuno, 1986, 1990):

N-(benzyloxycarbonyl)-L-aspartic acid (Z-L-Asp) + L-phenylalanine methyl ester (L-PheOMe) ↔ *N*-

(benzyloxycarbonyl)-L-aspartyl-L-phenylalanine methyl ester (Z-L-Asp-L-PheOMe) + H$_2$O.   (8.35)

Figure 8.6. (continued)

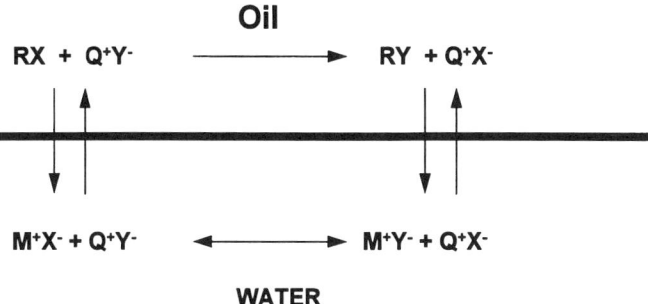

Figure 8.7. Schematic representation of phase transfer catalysis.

reagents. A typical scheme of phase transfer catalysis is shown in Fig. 8.7. Details of PTC catalysis can be found in numerous books (Dehmlow and Dehmlow, 1993; Starks *et al.*, 1994; Yufit, 1984).

If a catalytic reaction of electron or ion transfer takes place at the oil–water interface between reagents located in two different contacting phases, we are dealing with an *"interfacial catalysis"* as discovered by Volkov and Kharkats (1985, 1986, 1987). The interface itself can serve as a catalyst for such heterogeneous charge transfer reactions (Volkov, 1984a,b). If the interfacial catalysis requires an electrical field, the reaction can take place at the interface between two immiscible electrolyte solutions having a fixed interfacial potential, a process called *electrocatalysis* (Cunnane and Murtomaki, 1996).

Redox reactions without an illumination at the bilayer lipid membrane (BLM)–water interface were studied by many authors (Boguslavsky *et al.*, 1975b; Tien, 1986; Volkov *et al.*, 1975a). Usually the BLM, doped by ubiquinone, tetracyano-*p*-quinodimethane (TCNQ), or by ferrocene, has the properties of bipolar electrode (Tien, 1980, 1986, 1989). A variety of redox couples in aqueous phases have been used, including ascorbic acid/dehydroascorbic acid, KI/I_2, $[PtCl_4]^{2-}$, Sn^{4+}/Sn^{2+}, Cd^{2+}, vitamin K_3, benzoquinone/hydrobenzoquinone, HADH, and O_2 (Boguslavsky *et al.*, 1975b; Tien, 1986; Volkov *et al.*, 1975a). Tien (1986) used traditional electrochemical techniques such as cyclic voltammetry for detecting the redox reaction products and electrometric titration of Fe^{2+} with $KMnO_4$ using a pigmented BLM as the indicator electrode. According to Tien, a modified BLM can be considered either as a reversible electrode, or as a polarizable electrode, analogous to a platinum electrode that merely serves as a conductor for facilitating electron transfer (Tien, 1986). Tien introduced the term *electrostenolysis* which means that reduction occurs on one side of the BLM while the oxidation reaction takes place on another side of a membrane.

A number of other reactions have been demonstrated in which an oxidative transformation occurs at an interface, but the factors affecting the rates have not been examined (Yufit, 1984). Some progress in this direction was made in a simple redox reaction between hydrophobic porphyrin dissolved only in an octane and a hydrophilic donor—sodium dithionite dissolved only in water (Volkov *et al.*, 1976b, 1982a,b, 1983a,b).

8.7. CHLOROPHYLL AS A CATALYST OF ELECTRON TRANSFER REACTIONS IN BILAYERS AND AT THE LIQUID HYDROCARBON–WATER INTERFACE

The photosynthetic pigment chlorophyll mediates the primary act of photosynthesis in higher plants. However, chlorophyll can also catalyze electron transfer reactions at oil–water interfaces and on bilayer lipid membranes without illumination (Boguslavsky and Volkov, 1987; Volkov *et al.*, 1975a). The reversible chemical reduction of chlorophyll by zinc was first studied by Timiriazeff (1885, 1886), and the role of chlorophyll in oxidation–reduction reactions has been the subject of numerous later investigations (Goedher, 1966; Kutyurin *et al.*, 1973; Seely, 1977; Volkov *et al.*, 1975a).

Rabinowitch and Weiss (1937) first reported a simple redox reaction of chlorophyll (Chl) with ferric chloride:

$$Chl + Fe^{3+} \iff oChl^+ + Fe^{2+}, \tag{8.36}$$

where $oChl$ is oxychlorophyll, a yellow oxidized form of the pigment. If ferric salts are added to chlorophyll dissolved in CH_3OH, the solution changes color from green to yellow, and spectroscopic analysis of the products shows a nearly complete disappearance of the chlorophyll adsorption spectrum. The oxychlorophyll can be returned to chlorophyll by quickly adding ferrous chloride or other reducing agents, but with time the oxychlorophyll forms allomerized pigments in an irreversible reaction (Watson, 1953). The standard redox potentials and polarographic half-wave potentials of different chlorophylls are described in the review by Seely (1977).

Chlorophyll *a* adsorbed at the oil–water interface catalyzes the electron transport between donors and acceptors of electrons located in different phases (Volkov *et al.*, 1975a). The chemical structure of chlorophyll (Fig. 8.8) accounts for its high catalytic activity at the oil–water interface. The asymmetric amphiphilic chlorophyll molecule consists of a hydrophilic "head" formed by four pyrrole rings surrounding magnesium and a long "tail"—the hydrophobic chain of phytol. The hydrophilic head faces the aqueous phase, while the hydrophobic tail attaches chlorophyll to the non-polar phase of the thylakoid membrane.

Electron transfer reactions have been studied by Volkov *et al.* (1975a) using the vibrating-plate method in the electrochemical circuit:

$$\left. Au \right| Air \left| \begin{array}{c} Octane \\ acceptor \\ catalyst \end{array} \right| \begin{array}{c} Water \\ substrate \\ tris\text{-}HCl \end{array} \left| \begin{array}{c} Water \\ saturated \\ KCl \end{array} \right| Hg_2Cl_2, Hg, \tag{8.I}$$

where chlorophyll was used as a catalyst. If a hydrophobic acceptor of electrons such

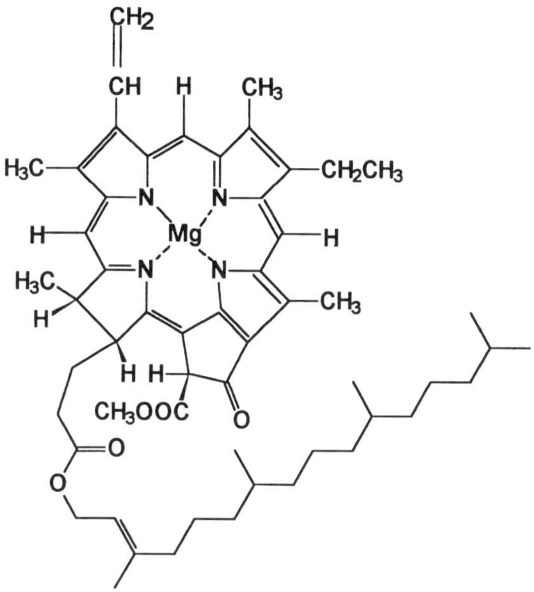

Figure 8.8. Chlorophyll *a*.

as 2*N*-methylamino-1,4-naphtoquinone or vitamin K_3 is added to the octane and a donor of electrons (NADH, NADPH, potassium ascorbate) to the aqueous phase, the Volta-potential measured in the chain (8.I) shifts in the negative direction (Fig. 8.9) due to the electron-exchange reaction at the oil–water interface catalyzed by chlorophyll:

$$\text{NADH}_{aq} + Q_{oil} \xrightarrow{\text{chlorophyll}} \text{NAD}^+_{aq} + H^+ + Q^{2-}_{oil}, \tag{8.37}$$

where Q is the 2N-methylamino-1,4-naphthoquinone. The magnitude of the potential does not depend on how the redox reaction is initiated, that is by addition of substrate, chlorophyll, or a charge acceptor. No change in the Volta potential was observed in the absence of any component of the reaction. The interfacial transfer of electrons catalyzed by chlorophyll has an optimum at pH 6.5–8.1 and is inhibited in the acid and alkaline regions (Fig. 8.10). Inhibition of the electron transfer reaction in the acidic region is probably due to conversion of chlorophyll to pheophytin. Inhibition in the alkaline region may be due to the fact that at $pH \geq 10$ the carboxylic ring breaks and salts of chlorophilins are obtained since the complex ester bonds are simultaneously hydrolyzed (Volkov *et al.*, 1975a).

The occurrence of oxidation–reduction reactions in BLMs containing chlorophyll leads to the generation of transmembrane potentials (Boguslavsky *et al.*, 1975g; Tien, 1986; Volkov *et al.*, 1975a). The transmembrane potential is caused by charge injection from the aqueous electrolyte solution into the BLM when an oxidation–

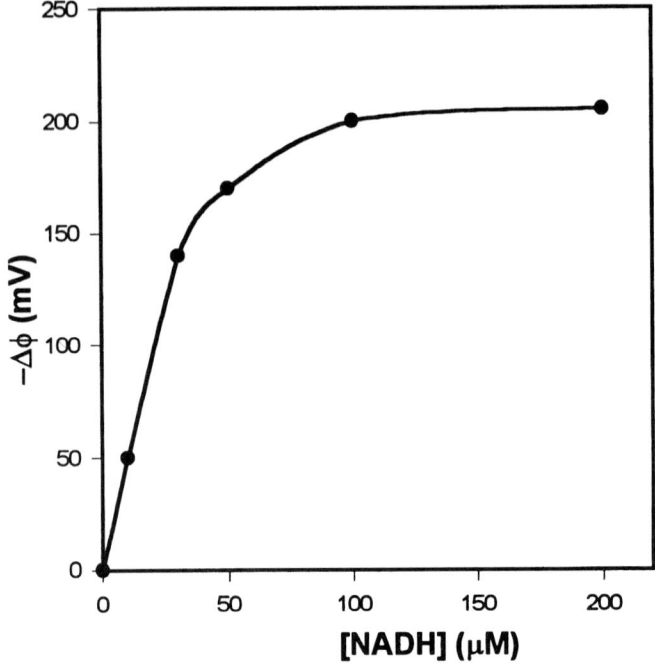

Figure 8.9. Volta potential, measured in the chain (I), as a function of NADH concentration. Medium: 10^{-5} M 2-N-methylamino-1,4-naphtoquinone + 10^{-2} M tris-HCl + 5 µg ml^{-1} chlorophyll.

reduction reaction takes place at the boundary between the membrane and the aqueous solution. To test this hypothesis, decane–water and octane–water interfaces containing chlorophyll adsorbed at the interface and oxidation–reduction systems in both phases were investigated (Boguslavsky *et al.*, 1976d,e,f; Volkov *et al.*, 1975a). The results were compared with data obtained in an investigation of transmembrane potentials on BLM-containing chlorophyll in the presence of NADH and oxygen or ferricyanide as the oxidant in the aqueous solution.

During a redox reaction on BLM-containing chlorophyll, a layer adjacent to the membrane is formed with a proton concentration different from that in the bulk phase (Boguslavsky *et al.*, 1975g; Volkov *et al.*, 1975a). Boguslavsky *et al.* (1976d,e,f) has demonstrated to record the formation of boundary layers with different properties in the octane–water system. It follows from equation (8.27), that a charge transfer reaction must involve proton ejection into the aqueous phase. Figure 8.11 shows the dependence of the Volta potential on the incubation time of the reaction mixture at different tris-HCl buffer concentrations. The aqueous phase pH was chosen such that acidification of the boundary layer inactivated the chlorophyll due to its pheophytinization. As shown in Fig. 8.11, at low Tris-HCl concentrations $\Delta\phi$ decreases at pH <6.5, while in the buffered solution this effect is absent.

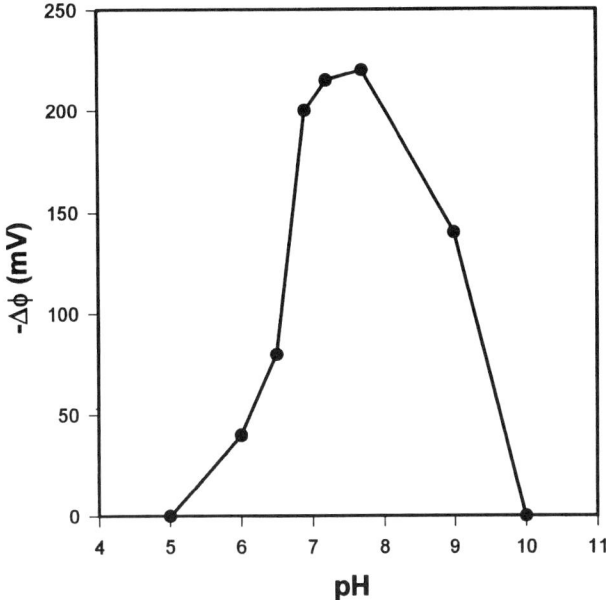

Figure 8.10. Volta-potential as a function of pH of an aqueous solution. Composition of reaction mixture: 20 mM tris-HCl + 1 mM NADH + 10^{-5} M 2-N-methylamino-1,4-naphto-quinone + 5 μg ml^{-1} chlorophyll.

Volkov *et al.* (1975a) investigated the following systems in a BLM system:

$$NADH, O_2 \mid BLM, chlorophyll \mid O_2, \tag{8.II}$$

$$NADH, O_2 \mid BLM, chlorophyll \mid K_3Fe(CN)_6, O_2. \tag{8.III}$$

Plots of the transmembrane potential against NADH concentration are shown in Fig. 8.12 (curves 1, 2). The potential of the half-cell containing NADH was positive relative to the other half-cell, which corresponded to transfer of a anions across membrane. As Fig. 8.12 indicates, the electromotive force (EMF) in systems (8.II) and (8.III) depends on the concentration of the reductant, and higher potentials are attained with ferricyanide as an oxidant. Since redox reactions in BLM can involve protons, it was important to test whether unstirred layers were present at the membrane boundary. The protonophore tetrachlorotrifluorobenzimidazole (TTFB), was therefore added to the membrane to make it proton selective. If the reaction at the interface is accompanied by the formation of a membrane boundary layer with a proton concentration differing from the bulk concentration, then the BLM will acquire a potential which is determined by the proton-concentration gradient across the membrane. If there is no proton-concentration gradient, the protonophore will simply shunt the potential previously present. From Fig. 8.12 (curve 1) it is clear that the proton gradient and transfer of negative charges from NADH go in opposite directions in system (8.II). Since the effect

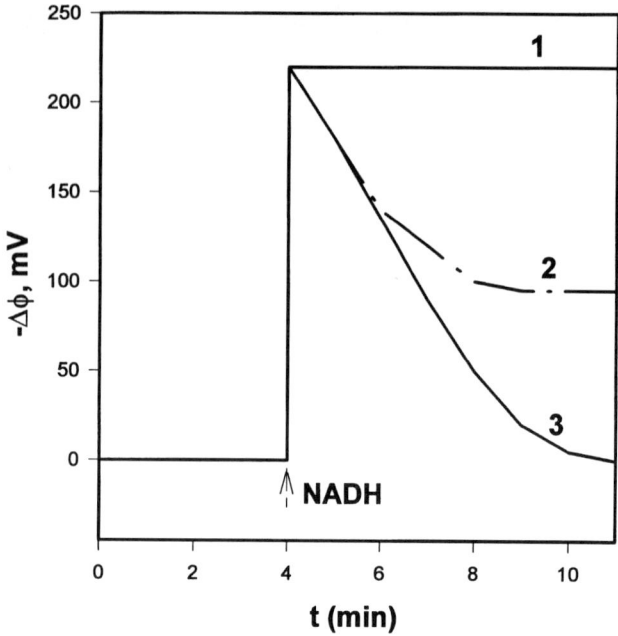

Figure 8.11. Volta potential in the chain (I) as a function of incubation time of reaction mixture. Medium: 20 μM NADH + 10 μM vitamin K_3 and tris-HCl: (1) 10 mM, (2) 1 mM, (3) 0 mM, pH 6.5.

of TTFB disappeared when a 0.01 M buffer was present on the oxidant side, the membrane boundary layer must be acidified on the side of the oxidant, which in this case is oxygen. In the presence of potassium ferricyanide as an oxidant (curve 2, Fig. 8.12) the addition of TTFB acts as a proton shunt and discharges the potential due to the transmembrane electron transfer.

From these results, we conclude that chlorophyll adsorbed at an interface can catalyze redox reactions between solutes in two liquid phases, and these reactions are accompanied by injection of negative charges into the low dielectric membrane interior. Metallo-complexes of porphyrins also have similar catalytic properties at the oil–water interface without illumination (Gugeshashvili *et al.*, 1983a,b; Volkov *et al.*, 1976b, 1982a,b, 1983a; Volkov and Boguslavsky, 1976). An advantage of porphyrins as catalysts in such experiments is their chemical stability in different media.

8.8. PORPHYRINS AS INTERFACIAL CATALYSTS

We will now consider adsorption and catalytic properties of iron coproporphyrin III (CP) at the octane–water interface (Volkov *et al.*, 1982a,b). Figure 8.13 shows the adsorption isotherms of CP at the octane–water interface. The shape of the curve of

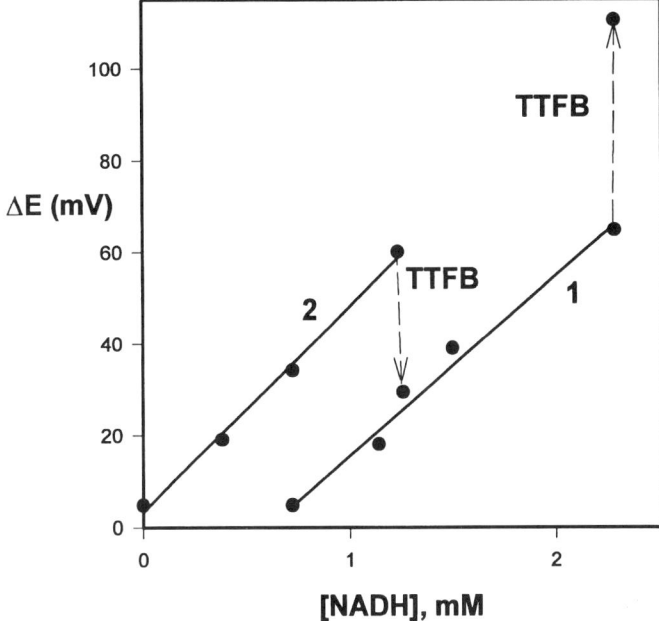

Figure 8.12. Transmembrane potential as a function of NADH concentration in systems: (1) O_2, NADH/BLM, chlorophyll/O_2; (2) O_2, NADH/BLM, chlorophyll/$K_3Fe(CN)_6$. The electrolyte was 1 mM KCl and the TTFB concentration was 2 μM.

the Gibbs adsorption of CP as a function of concentration shows attraction interactions between adsorbed molecules of CP. The free energy of adsorption is $\Delta G_{ads} = -38.9$ kJ mol^{-1}. If the number of CP particles present in 1 cm^2 of the interface at $\Gamma = \Gamma_{max}$ is known, the limiting area occupied by a CP molecule can be calculated. The comparison of the limiting area with the area of a CP molecule of 82 Å^2 permits evaluation of the angle between the CP molecule plane and the octane–water interface. When the interface is completely covered by CP molecules this angle is equal to 65°, a value close to those obtained by optical techniques (Tien, 1974) for other porphyrins at a membrane–water interface and in monolayers at the water–air interface.

Figure 8.14 shows the dependence of the Volta potential measured in chain (8.I) on the CP concentration at the interface (calculated on the basis of experimental data shown in Fig. 8.14). The deviation from a linear dependence when the projection of a molecule area is less than 1 nm^2 reflects interaction between molecules of the adsorbed particles, which decrease the effective molecular dipole moment. This deviation is 2.4 times larger than the expected value due to the change of the angle between the porphyrin ring and the interface.

In the presence of CP and 2-N-methylamino-1,4-naphthoquinone in octane, addition of a reducing agent such as NADH or ascorbate into the aqueous phase leads

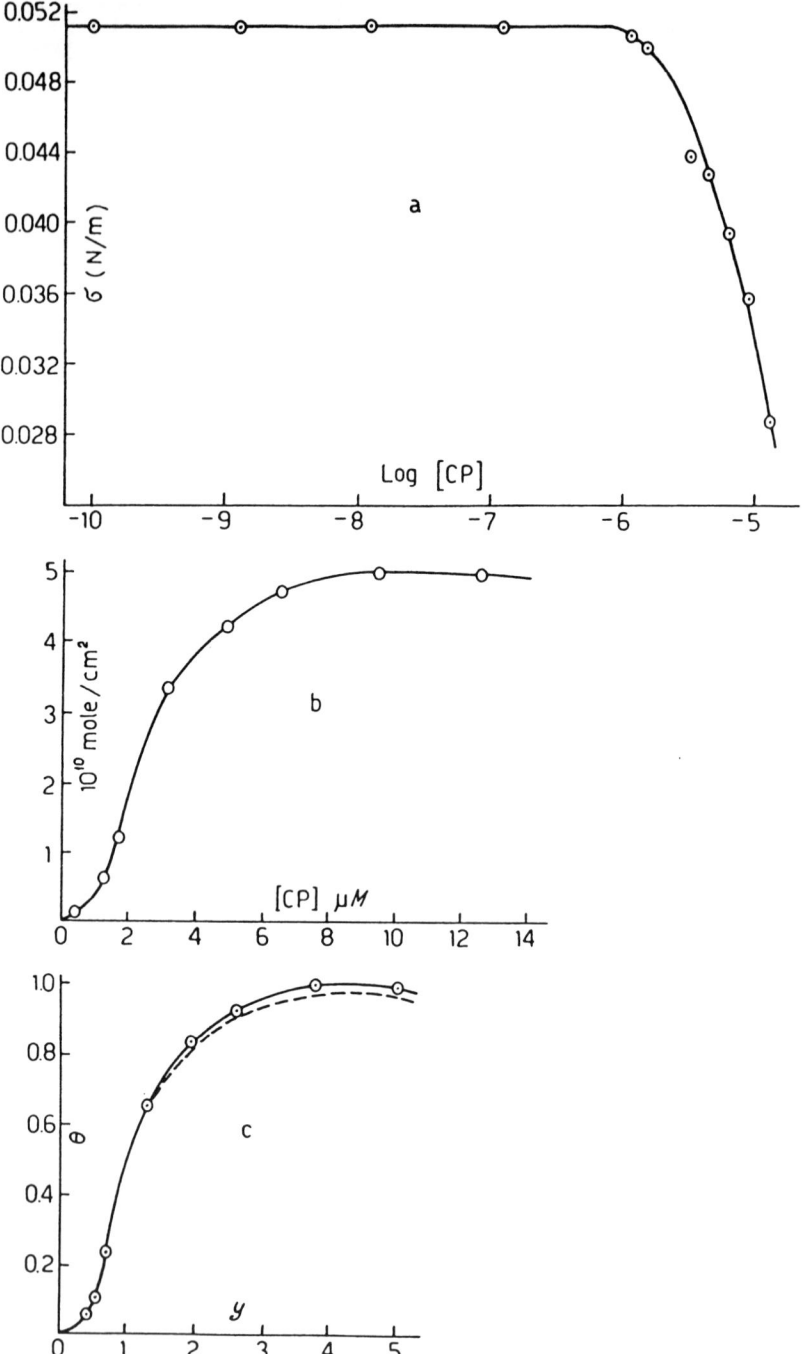

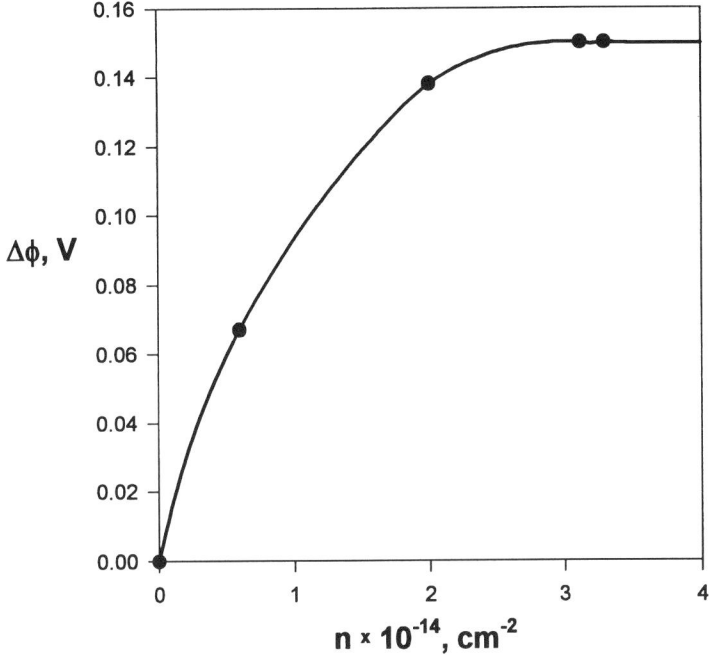

Figure 8.14. Dependence of the adsorption potential on the density of iron coproporphyrin III particles on the surface.

to a shift of potential in a negative direction (Fig. 8.15). This change depends on the concentrations of all reaction components and can be interpreted as a reduction of acceptor molecules during substrate oxidation in an electron exchange reaction catalyzed by CP.

Using the phenomenological theory of a catalytic charge-transfer process through the interface of two immiscible liquids and extrapolating the experimental curve shown in Fig. 8.15, the Michaelis constant can be determined (Volkov *et al.*, 1983a). At pH 7.3 K_m is equal to 3×10^{-4} M. Secondly, Fig. 8.16 shows that the interfacial potential due to the electron exchange reaction is proportional to the number of catalyst molecules located at the interface, which agrees with equation (2.101). The

Figure 8.13. Adsorption isotherms of iron coproporphyrin III at the octane–water interface: (a) dependence of interfacial tension on CP concentration; (b) dependence of adsorption on iron coproporphyrin III concentration; (c) dependence of the extent of the surface coverage on iron coproporphyrin III concentration in reduced coordinates $y = c/c_{\vartheta=0.5}$. The solid line is plotted using the Gibbs equation on the basis of experimental data from interfacial tension, and the dashed line is plotted using the amphiphilic isotherm.

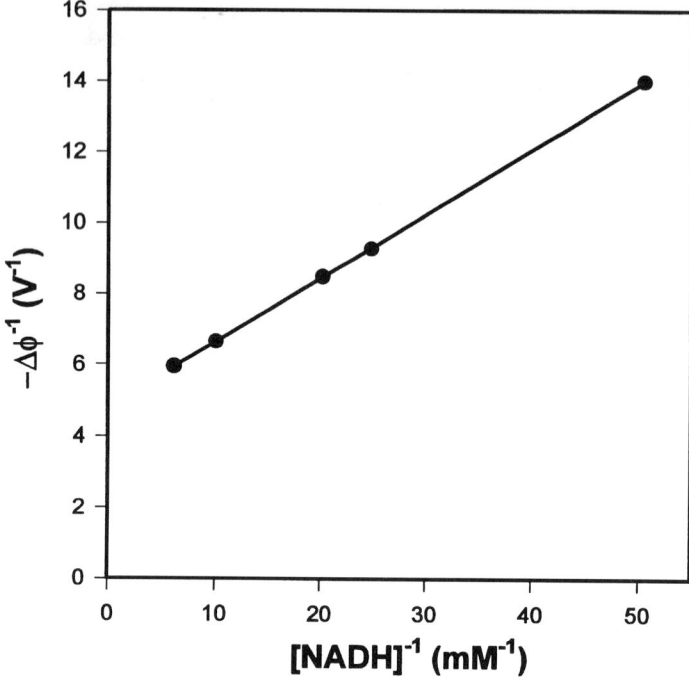

Figure 8.15. Dependence of Volta potential on the NADH concentration in reverse coordinates. Medium: 10^{-4} M 2-*N*-methylamino-1,4-naphthoquinone, 7×10^{-2} M tris-HCl, 7×10^{-6} M on iron coproporphyrin III, pH 7.3. The potential is measured against an aqueous solution free of NADH.

linear dependence shown in Fig. 8.16 indicates that the orientation of the catalyst in all substrate concentrations remains unchanged.

Another electron-transfer reaction was investigated by Volkov *et al.* (1976b).

$$(\text{NADH})_{\text{water}} + (\text{MNQ})_{\text{octane}} = (\text{NAD}^+)_{\text{water}} + (\text{MNQH}^-)_{\text{octane}}. \qquad (8.38)$$

This reaction occurs in the dark, and light (70 mW cm^{-2}) does not affect the potential change. In the absence of one of the reaction components, i.e., CP, octane, NADH, or naphthoquinone, no change of the Volta potential was observed. The Volta potential measured in chain (8.I) is also independent of the buffer capacity of the solvent (tris-HCl concentration range from 10 to 100 mM). This indicates that the change in the potential jump is caused by the electron-transfer reaction. Thus, physical characteristics of the catalyst molecules at the interface (orientation, interaction between the molecules, adsorption energy, Michaelis constant) may be studied using adsorption isotherms and measurements of the Volta potential in chain (8.I). The results of these studies (Volkov *et al.*, 1983a) are in a good agreement with the previously proposed model (Kharkats *et al.*, 1975, 1976) of the catalytic charge transfer through the interface of two immiscible liquids.

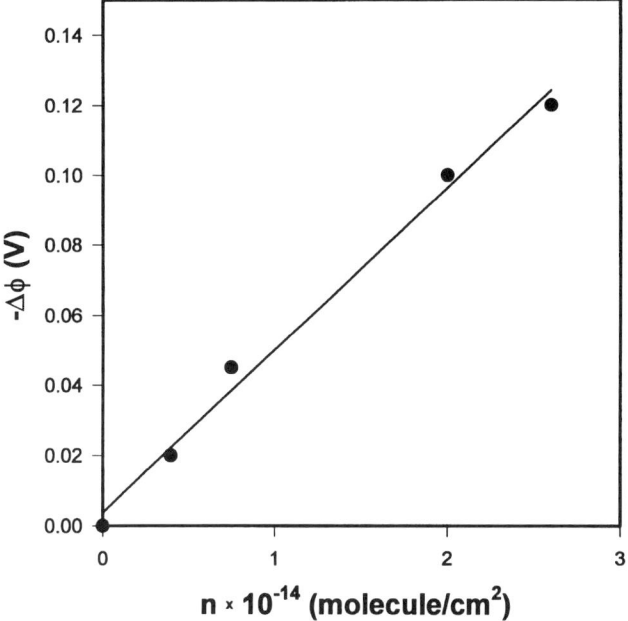

Figure 8.16. Dependence Volta potential on the number of iron coproporphyrin III molecules, adsorbed at the octane–water interface. Medium: 10^{-3} M NADH, 10^{-4} M 2-*N*-methylamino-1,4-naphthoquinone, 10^{-2} M tris-HCl, pH 7.3.

8.9. REDUCTION OF PORPHYRIN AT THE OCTANE–WATER INTERFACE CONTROLLED BY SPECIFIC ADSORPTION

The redox reaction between the ferric complex of hydrophobic porphyrin FeP (ferric complex of 2,7,12,17-tetramethy1-3,13-octadecy1-8,18-bis(carbomethoxyethyl)-porphyrin) and sodium dithionite in two different phases occurring at the interface between two immiscible liquids has been investigated by Volta-potential measurements and UV-Vis spectroscopy (Volkov *et al.*, 1982a,b). The reduction of the ferric complex of hydrophobic porphyrin adsorbed at the interface was found to be accompanied by a potential shift in the negative direction and to depend significantly on the nature of the anion and the ionic strength of the supporting electrolyte. Specifically adsorbed halogen anions inhibited the redox reaction in the series Cl^-, Br^-, I^-. Depending on pH, the ferric complex of hydrophobic porphyrin exists in the uncharged FeP-O-FeP form or in the cation FeP^+ form.

The Volta-potential measurements in the chain (8.I) showed a positive change when plotted with respect to the water phase and increased smoothly up to 0.46 V with increasing FeP concentration up to 10^{-6} M. Ferric complexes of porphyrins are known to be salts in which the metal complex is a cation. When dissolved in dry hydrocarbon these salts are undissociated, but at the aqueous interface or in octane

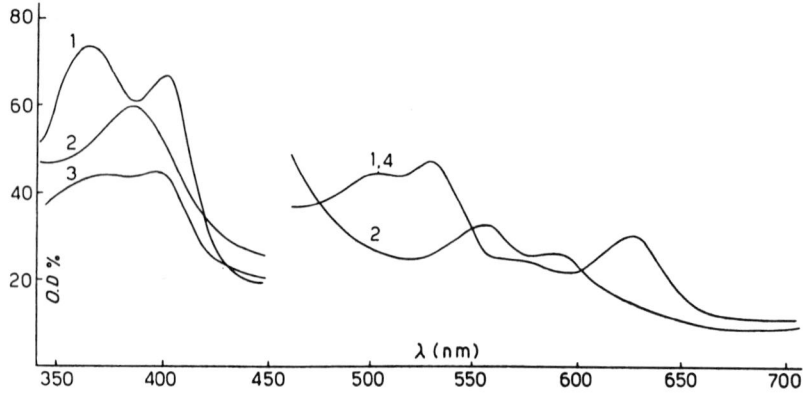

Figure 8.17. Adsorption spectra of 10^{-6} M FeP in the octane phase. (1) FeP in *dry* octane; (2) FeP in *wet* octane; (3) FeP in octane (with 1 mM sodium dithionite in the aqueous phase); (4) FeP in *wet* octane in the presence of 100 mM HCl.

with water present as a microemulsion, the following hydrolysis reaction takes place:

$$FeP^+Cl^- + H_2O \iff FeP^+OH^- + H^+ + Cl^-. \tag{8.39}$$

The product of the reaction can undergo dimerization to form a μ-complex:

$$2\,FeP^+OH^- \iff FeP\text{-}O\text{-}FeP + H_2O. \tag{8.40}$$

Comparison of the FeP absorption spectra in dry and in wet octane (Fig. 8.17) showed that hydrolysis in fact transforms the FeP^+Cl^- spectrum into the FeP^+OH^- or FeP-O-FeP spectra. Acidifying the aqueous solution with hydrochloric acid shifts the equilibrium of reaction (8.29) to the left, and the initial FeP^+Cl^- form appears again. It should be noted that addition of 1 M KCl is not sufficient for the equilibrium of reaction (8.29) to shift to the left.

A second method was used to determine the form of porphyrin at the octane–water interface. It is known that the surface potential χ_s consists of two components

$$\chi_s = \chi_D + \chi_G. \tag{8.41}$$

The first term of the sum represents the potential caused by the dipole moment of the

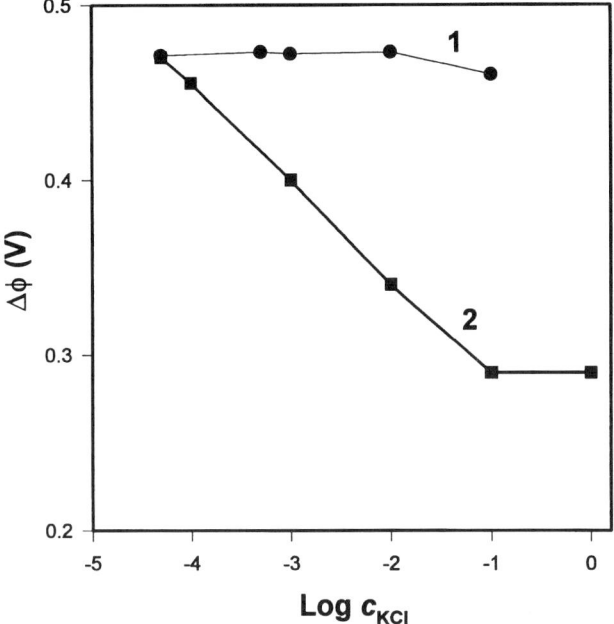

Figure 8.18. Dependence of the chain (I) Volta-potential change on KCl concentration in the system containing 10^{-6} M FeP in octane. pH = 8.5 (curve 1) and 6.5 (curve 2).

adsorbed particles which set up a potential determined by the Helmholtz equation

$$\chi_D = \frac{n_S \mu_{eff}}{\varepsilon\varepsilon_0}, \qquad (8.42)$$

where μ_{eff} is the normal component of the effective dipole moment, n_S is the surface concentration and ε is the static permittivity of the medium surrounding the dipole. The second term of equation (8.41) is due to the presence of the diffuse ion layer and is determined by the well-known Gouy–Chapman equation. For charged monolayers

$$\frac{d\chi_s}{d\log c} = \frac{d\chi_G}{d\log c} = 58\ mV. \qquad (8.43)$$

A study was made of the influence of the ionic strength and pH of the aqueous phase on the interfacial potential change, using the assumption that the area per molecule and the normal component of the effective dipole moment remain unchanged. In the alkaline pH range the reaction is shifted toward µ-complex formation, and this potential change should not depend on the supporting electrolyte concentration unless a significant reorientation of adsorbed molecules takes place. Figure 8.18 shows the dependence of the Volta potential on KCl concentration at pH = 8.5 (curve 1) and pH = 6.5 (curve 2). As expected, the interfacial potential remains constant when the electrolyte content increases at alkaline pH values. This means that the main contribution to the surface potential is made by χ_D. In a more acid region at pH = 6.5 the electric double layer consists of FeP complexes in the octane phase and anions in water. The gradient $d\chi_s/d \log c_{KCl}$ is 55–60 mV per decade of salt concentration between 10^{-4}–10^{-1} M KCl. To determine whether aqueous phase anions make any contribution to the potential change, the Volta potentials of the system under study were measured in KCl, KBr, and KI solutions. If no specific halogen adsorption at the octane–water interface takes place, χ_s should not depend on the nature of the anion. However, as follows from the experimental values of χ_s in the octane–water system, when 10^{-3} M KCl, KBr, and KI solutions are added to the water χ_s measured against the gold reference electrode at pH = 7 equals 580, 570, and 530 mV, respectively. The results obtained may be due to the fact that halogens are specifically adsorbed at the interface and shift $\Delta\phi$ toward more negative values in the sequence Cl^-, Br^-, I^-.

Upon addition of sodium dithionite, FeP undergoes reduction (Fig. 8.17, curve 3). This process may also be detected from the Volta-potential shift in the negative direction. Higher sodium dithionite concentrations increase this shift, which attains a maximum value of -110 mV. A more careful study of the FeP reduction showed the reaction to proceed only at low electrolyte concentrations. The spectrum of FeP treated with dithionite in the presence of 10^{-1} M KCl coincides with that of the oxidized form of FeP, which indicates that no reduction takes place under these conditions.

Two explanations are possible for the observed inhibition of FeP reduction. The first assumes that the dithionite concentration is defined by the Gouy–Chapman diffuse layer theory. As the potential at the interface decreases with increasing KCl concentration (Fig. 8.17), the dithionite anion concentration at the interface must also decrease, which results in a decrease in the final product concentration (reduced FeP). The second explanation assumes that a dithionite anion is displaced from the interface by specifically adsorbed halides. This explanation, contrary to the first model, suggests that the observed effect of the inhibition of porphyrin reduction is based on the forces of specific adsorption. According to the first model, it does not matter what salt is used to increase the ionic strength of the solution. In the second model the concept of specific adsorption is used, and, therefore, the effect should depend on the nature of the anion. To choose between two models, the ionic strength of the aqueous phase was increased by adding Cl^-, Br^-, and I^-.

Figure 8.19 shows that inhibition of porphyrin reduction takes place at different concentrations of halide salts, and therefore depends on the nature of the anion. The I^- ions, which are known to have the highest adsorbability cause inhibition of the reaction at concentrations lower than that of Br^- anions, which, in their turn, are more effective than Cl^- anions. For the dithionite anion to transfer an electron to adsorbed porphyrin, it must be near the porphyrin molecule and occupy a position in the aqueous phase near the interface. The inhibition of porphyrin reduction as halide salt

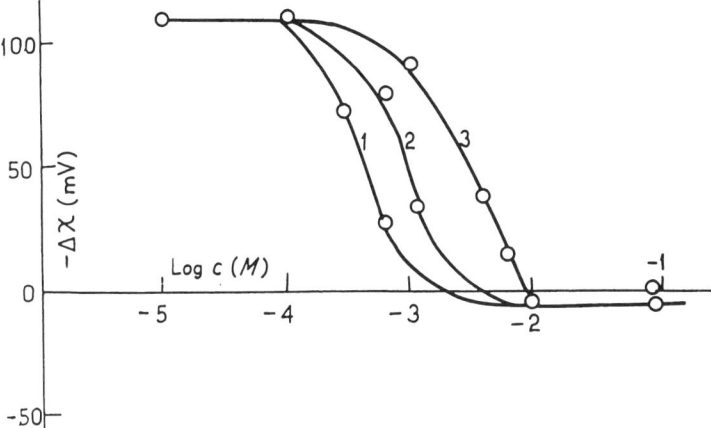

Figure 8.19. Dependence of the Volta potential on the supporting electrolyte concentration with different halide anions. The system contains 10^{-6} M FeP and 10^{-3} M sodium dithionite. (1) KI; (2) KBr; (3) KCl. The Volta potential is plotted against the potential of the dithionite-free solution.

concentration increases means that halides displace dithionite anions from the adsorption layer.

In summary, hydrophobic porphyrin adsorbed at the octane–water interface can be reduced with a soluble agent exclusively in the aqueous phase. This process can be recorded by the measurement of the Volta-potential change in the chain (8.I). The influence of specifically adsorbed anions on the reduction process suggests that the Gouy–Chapman model may not provide a complete description of the electric double layer at the interface between two immiscible liquids or at the membrane–electrolyte interface.

8.10. COUPLING OF TWO REDOX REACTIONS AT THE OCTANE–WATER INTERFACE

Adsorption of FeEP at the octane–water interface is accompanied by a positive potential shift of 340 mV which depends neither on the ionic strength of the KCl solution, whose concentration varied from 10^{-4} M to 10^{-1} M, nor on the pH value within the range 5–8. This indicates that the potential drop is caused by adsorption of dipoles in the form of undissociated FeEP hydroxide or the μ-complex FeEP-O-FeEP which is formed by hydrolysis of FeEP (chloride of the ferri-complex of ethioporphyrin II).

No reduction of porphyrin is observed if NADH is added to the octane–water system containing only FeEP. However, the absorption maximum in the Sauret region shifts toward the long-wave region by 3 nm, which indicates that interaction between the molecules of FeEP and NADH takes place. In the octane–water system containing FeEP, TMQ, and NADH, oxidation of NADH proceeds at the interface and is catalyzed by the ferri-complex of ethioporphyrin. The consumption of NADH, followed spectroscopically at the 340 nm wavelength, is proportional to the reaction mixture incubation time. If the octane–water system in an inert atmosphere does not contain one of the components, NADH is not consumed to any significant extent. In the presence of an oxygen atmosphere, the consumption of NADH is noticeable even in the system without TMQ. In the complete system, however, consumption of NADH in the presence of oxygen is considerably higher than in a control system without one of the ingredients or in an inert atmosphere. Comparison of the consumption of NADH in air and in argon clearly illustrates that in the presence of

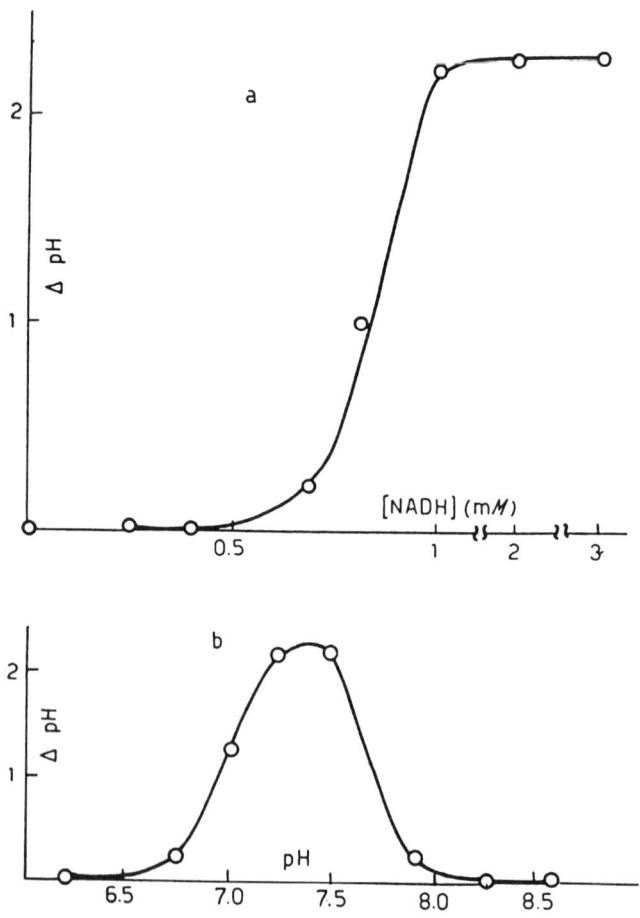

Figure 8.20. Dependence of pH change in the aqueous fraction on (a) NADH concentration, (b) the initial pH value; in experiment (b) 2×10^{-3} M NADH was added.

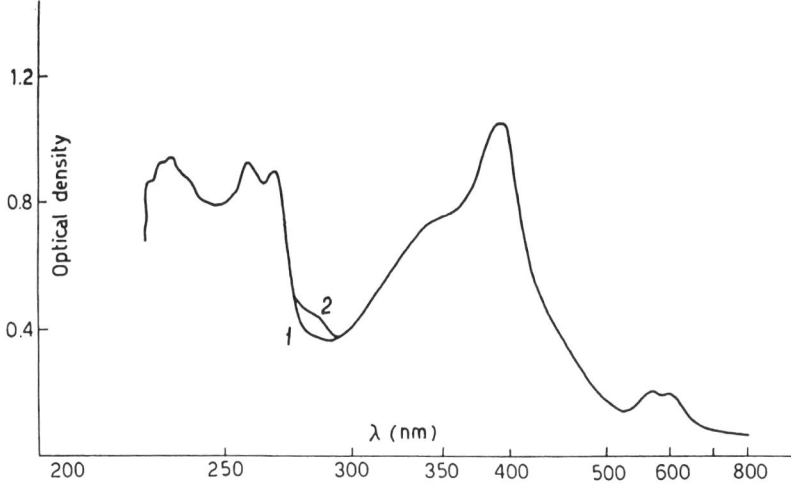

Figure 8.21. Absorption spectrum of the octane fraction after a five-hour incubation in argon atmosphere. Medium: 10^{-6} M FeEP, 2×10^{-4} M NADH, 20 mM tris-HCl, pH = 7.5; (1) prior to incubation with NADH; (2) after incubation with NADH.

oxygen a reaction takes place at the interface which may be regarded as oxidation of NADH by oxygen:

$$\text{NADH} + \text{O}_2 + \text{H}^+ \xrightarrow{\text{FeEP} \cdot \text{TMQ}} \text{NAD}^+ + \text{H}_2\text{O}. \tag{8.44}$$

It is known that oxygen reduction is accompanied by an increase in pH of the solution in the reaction zone. Measurement of pH of the aqueous solution in the complete system in air showed that the pH does increase to an extent which depends on the NADH concentration and on the initial pH of the system (Fig. 8.20). It should be noted that when vitamin K_3 or 1,4-naphthoquinone is substituted for TMQ no pH change is observed in a weakly buffered solution (5 mM tris-HCl). This may be due to a different structure of the reaction complex at the interface. Tetramethylquinone absorbs in the ultraviolet region (260, 265 nm). The peak of the reduced form is at 285 nm. $TMQH_2$ can be again oxidized under vigorous stirring with freshly prepared silver oxide.

Spectroscopic investigation of the system containing all the reaction participants (Fig. 8.21) showed that the reduced form of the acceptor, $TMQH_2$, is formed during the reaction in an inert gas atmosphere (argon or nitrogen). It is important to emphasize that within the NADH concentration range used in the present experiments no noticeable direct reduction of TMQ takes place in the absence of FeEP in the system. The following experimental fact is important for the interpretation of the results. The interface is large when argon is bubbled; after the supply of gas is stopped it is markedly reduced. Under such experimental conditions it remains unclear

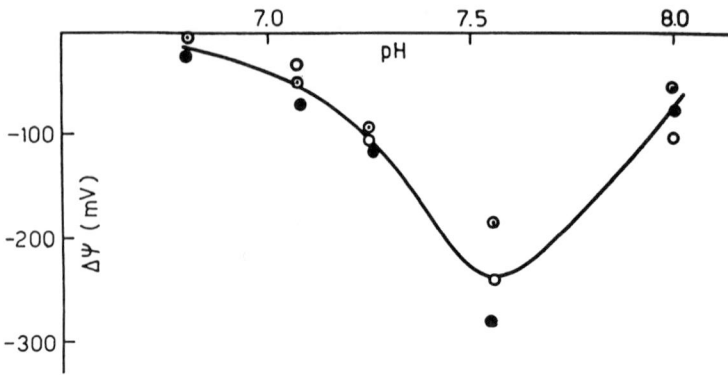

Figure 8.22. Dependence of the potential measured in chain (I), on the solution pH at the beginning of the reaction. Medium: 10^{-6} M FeEP, 10^{-5} M TMQ and NADH: (⊙) 10^{-4} M; (O) 2×10^{-4} M; (●) 5×10^{-5} M.

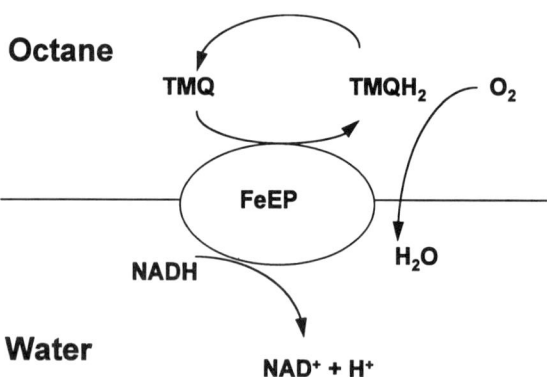

Figure 8.23. A schematic representation of a coupling of two redox reactions at the oil–water interface.

whether desorption of $TMQH_2$ occurs spontaneously during the reaction or whether the presence of $TMQH_2$ in octane is exclusively the result of a significant reduction in the area of the interface. Therefore, we are only able to state that $TMQH_2$ is formed in the octane–water system at the interface. Reduction of TMQ in the whole system was also demonstrated by chromatography. For the purpose of control, TMQ and $TMQH_2$ were applied to a chromatographic system. When the reaction starts, a potential shift occurs at the interface which depends on the concentration of all ingredients and on the pH of the aqueous fraction. According to the spectral data, the reduced form of tetramethylquinone can be obtained in the pH region from 7 to 8. A study of the Volta-potential dependence on pH in the complete system shown in Fig. 8.22 indicates that the maximum negative potential drop, as well as maximum

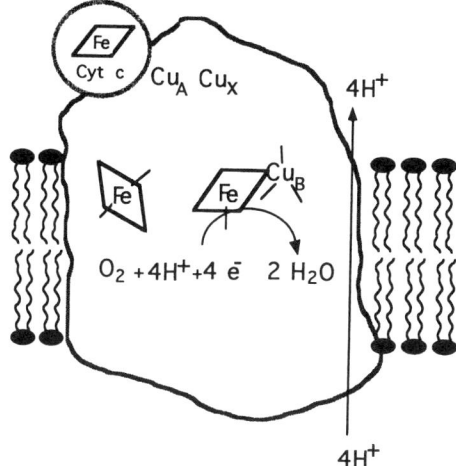

Figure 8.24. Scheme of cytochrome oxidase functioning in the mitochondrial membrane.

increase of pH of the solution, is attained in the same pH region. This means that NADH oxidation and oxygen reduction are probably coupled through TMQH. The reactions may be represented schematically as shown in Fig. 8.23.

This combination of spectroscopic and chromatographic analyses as well as the Volta-potential measurement methods enabled Gugeshashvili *et al.* (1983a,b) to demonstrate NADH oxidation by oxygen at the octane–water interface, the reaction being catalyzed by ethioporphyrin and tetramethylquinone. In this process, redox transformations of tetramethylquinone take place.

8.11. ENZYME COMPLEXES OF THE MITOCHONDRIAL RESPIRATORY CHAIN

The function of the enzymes of the mitochondrial respiratory chain is to transform the energy of redox reactions into an electrochemical proton gradient across the hydrophobic barrier of a coupling membrane. Isolated oligoenzyme complexes of the respiratory chain of mitochondria cytochrome *c* oxidase, succinate-cytochrome *c* reductase, and NADH-CoQ reductase are able to catalyze charge transfer of charges between water and octane, which can be followed by change in the potential at the octane–water interface (Boguslavsky *et al.*, 1975a,c,d,e, 1976c,g,h,i; Boguslavsky and Volkov, 1987; Goodall and Sachs, 1977). A necessary condition for this measurement is the presence of the enzymes and substrates in the aqueous phase and a charge acceptor in the octane phase.

Cytochrome oxidase (EC 1.8.3.1.) is the terminal electron acceptor of the

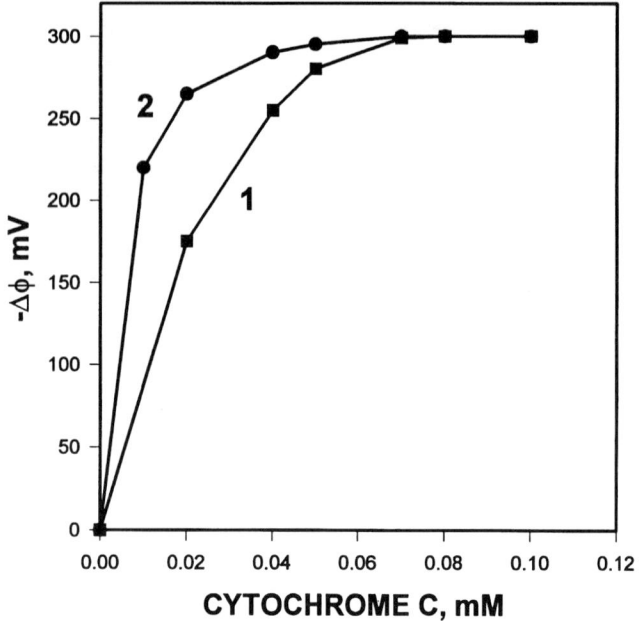

Figure 8.25. Dependence of the Volta potential at the octane–water interface on the concentration of cytochrome c. Incubation medium: (1) 0.05 M tris-HCl (pH 7.4), 1 mM ascorbate, 20 μg of cytochrome oxidase protein per 1 ml, and 10^{-5} M 2-methylamino-1,4-naphtoquinone; (2) the same with 0.3 mM tetramethyl-p-phenylenediamine.

mitochondrial respiratory chain. Its main function is to catalyze the reaction of dioxygen reduction to water using electrons from ferrocytochrome c:

$$4\,H^+ + O_2 + 4\,e^- \underset{\leftarrow \text{photosynthesis}}{\overset{\text{respiration} \rightarrow}{}} 2\,H_2O. \qquad (8.45)$$

Reaction (8.45) is exothermic, and the energy can be used to transport protons across the mitochondrial membrane. The enzyme is composed of several protein subunits (Fig. 8.24), two hemes A, cytochromes a and a_3 and two organo-copper complexes Cu_a and Cu_b coordinated to the protein and has a molecular weight of about 180 000–200 000 kDa for the most active form. Cytochrome oxidases can transport up to eight protons across the membrane per four electrons. Four of the protons bind to the reaction complex during dioxygen reduction to water and up to four other protons are transported across the membrane. The resulting chemiosmotic proton gradient is used in ATP synthesis (Mitchell, 1961, 1976).

Aerobic oxidation of cytochrome c is accomplished by cytochrome c oxidase and represents the terminal step of the respiratory chain in mitochondria. Cytochrome c oxidase can be considered as being the acceptor and donor of four electrons. Experiments with cytochrome c oxidase in the octane–water system showed that,

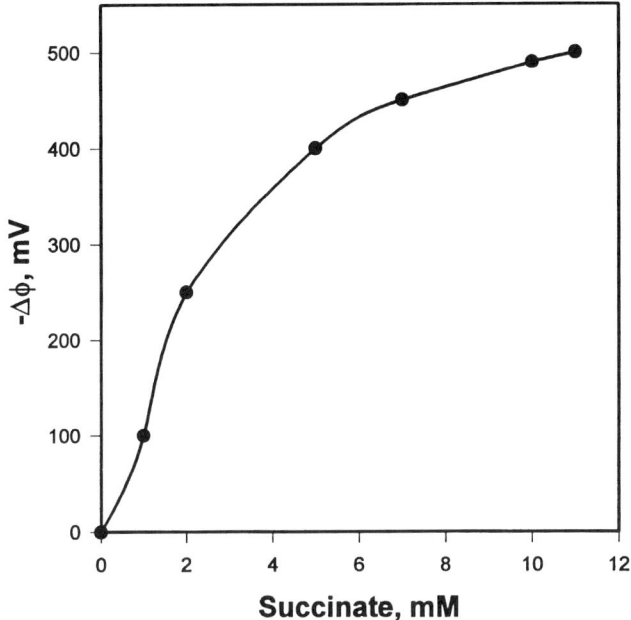

Figure 8.26. The succinate-cytochrome c-reductase-mediated Volta-potential changes at the octane–water interface as a function of the succinate concentration. Incubation medium: 0.05 M tris-HCl (pH 7.4), 0.5 mM cytochrome c, 0.1 mM 2-methylamino-1,4-naphtoquinone and 0.4 mg of succinate-cytochrome c reductase protein per ml.

when an enzyme suspension is added to the octane–water system, some of the enzyme molecules adhere to the interface. As a result, when cytochrome c and naphthaquinone are present, a negative Volta-potential shift occurs upon addition of adsorbate. The dependence of $\Delta\phi$ on the cytochrome c concentration is shown in Fig. 8.25. The half-maximal effect is achieved at a cytochrome c concentration of 16 μM, or 5 μM when the lipophilic electron carrier tetramethyl-N-phenylenediamine is substituted for naphthaquinone.

The negative potential shift arising in the full system decreased to zero when inhibitors of cytochrome c oxidase such as cyanide and polylysine were added. No change of Volta potential was observed if a small amount of the cationic detergent cetyltrimethylammonium bromide was present initially. This agent covered the interface, thereby inhibiting the subsequent interaction of enzyme complexes. The detergent was ineffective if it was added after the enzyme had bound to the interface.

Figure 8.26 shows similar experiments performed with a succinate cytochrome c reductase complex isolated from mitochondria. As the cytochrome c reductase reacts with succinate in the aqueous phase a potential appears at the interface that is negative in the octane phase. Half-saturation is achieved at 2 mM succinate. As in the experiments with cytochrome oxidase, the transfer of negative charges into the octane phase by cytochrome c reductase occurs only when 2-methylamino-1,4-naphto-quinone (MNQ) and cytochrome c are present in the system.

The half-saturation of the system with cytochrome c was achieved at a concentration of about 10 μM. In all cases, the effect was reversed by antimycin, an inhibitor of succinate-cytochrome c reductase activity, but was not inhibited by cyanide.

To conclude this series of experiments the isolated NADH-CoQ reductase complex of mitochondria was investigated (Boguslavsky et al., 1975a,c). The presence of both NADH and MNQ in the system proved to be necessary for the effect to be achieved. Half-saturation was reached at an enzyme concentration of 1.5 μg ml^{-1} and 25 μM NADH. The reduction of MNQ was not inhibited by rotenone, but the electrogenic function of NADH-MNQ reductase in the octane–water system was sensitive to 2 μM rotenone.

In the case of cytochrome c oxidase, the sensitivity to cyanide showed that MNQ takes up electrons either from cytochrome a_3 which is localized in the octane phase, or from ionized oxygen produced by the cytochrome oxidase. In the succinate-cytochrome c reductase complex MNQ probably "removes" electrons from the cytochrome c localized at the phase boundary or in the octane phase. All attempts to transfer charge from an aqueous phase into octane in the absence of enzymes were unsuccessful.

The transfer of charges into octane catalyzed by the NADH-CoQ reductase complex requires special consideration. When a lipophilic electron acceptor (MNQ) is present in the octane, this system catalyzes the transfer of an electron into the octane by a route sensitive to rotenone. In addition, transfer of protons into the octane occurs, as indicated by a suppression of the negative charge of the octane phase in the presence of the lipophilic proton acceptor 2,4-dinitrophenol (DNP). The latter process becomes the only possible one when MNQ is used as the electron acceptor in the aqueous phase and DNP as the proton acceptor in the octane phase.

The increased positive charge in the octane phase cannot be explained simply by a shift in the pH value of the aqueous phase during the enzymatic reaction, since the magnitude and the kinetic characteristics of the effect did not change when the tris-buffer in the aqueous phase was increased from 3 to 50 mM. The transfer of protons into octane by NADH-CoQ reductase models the function of the first energy coupling site, since NADH oxidation in mitochondria generates a trans-membrane proton gradient.

The succinate-cytochrome c reductase segment of the respiratory chain, which includes the second energy coupling site, contains at least two portions interacting with the non-polar octane phase. The first probably consists of cytochrome c associated with the cytochrome c reductase complex and is immersed in the octane phase so that electrons can be transferred to the acceptor in the octane. In addition, catalysis by the succinate-cytochrome c reductase reaction in the aqueous phase (final acceptor ferricyanide) is accompanied by the transfer of protons to a lipophilic proton acceptor (DNP) in the octane phase. This functional response may represent the proton pump of the second energy coupling site which catalyzes the transfer of hydrogen ions through the mitochondrial membrane. It should be mentioned that the reduction of ferricyanide by this enzyme complex is insensitive to antimycin. The suppression by antimycin of the electrogenic function of the succinate-cytochrome c reductase complex in the octane–water system means that the transfer both of negative and of positive charges into the octane requires the transport of electrons through all the components of the complex.

8.12. ATPase

Two types of reactions supplying the cell with adenosine triphosphate (ATP) exist: (i) heterogeneous membrane phosphorylations localized in the inner membrane of the mitochondria, in the membranes of thylakoids of chloroplasts, and in bacterial membranes; (ii) homogeneous ATP formation during glycolysis, fermentation and the substrate-level phosphorylation in Krebs' cycle as a result of the transfer of a high-energy phosphoryl group from the product of the substrate oxidation to ADP.

The native mitochondrial H^+-ATPase complex can be divided into two independent systems, one of which hydrolyzes ATP (coupling factor F_1, or soluble mitochondrial ATPase), while the other consists of several hydrophobic proteins and remains in the mitochondrial membrane after the removal of F_1. This complex of hydrophobic proteins is responsible for the oligomycin-sensitive proton conductance of submitochondrial particles depleted of F_1. The hydrophobic proteins contained in the ATPase complex facilitate the transfer of H^+ from F_1 to the opposite (external) side of the mitochondrial membrane. In this case F_1 will ensure the ATP-dependent transfer of H^+ from the matrix to the hydrophobic part of the membrane, where the oligomycin-sensitive components of the ATPase complex are located.

Volta potential can also be used to study the mechanism of spatial separation of charges at the interface with the participation of soluble ATPases from mitochondria (Boguslavsky *et al.*, 1975a,b,d, 1976g; Goodall and Sachs, 1977), bacteria (Boguslavsky *et al.*, 1975e, 1976h), and chloroplasts (Boguslavsky *et al.*, 1976i). In the presence of all the components of the ATPase reaction (Mg^{2+}, ATP, and factor F_1) and of a lipophilic proton acceptor, for which 2,4-dinitrophenol has been used in the majority of experiments, the transfer of protons from water into octane is accompanied by a change in the interfacial potential (Fig. 8.27). The direction of the electric field generated by ATPase in biomembranes is the same as that at the oil–water interface, so that the aqueous phase toward which the ATPase faces becomes negatively charged. The data indicate that F_1 can catalyze proton transfer from water to octane, coupled with hydrolysis of ATP.

The presence of a proton acceptor was an essential condition for the change in Volta potential at the octane–water interface. The proton acceptor used in most of the experiments was 2,4-dinitrophenol, but other compounds capable of protonation such as trifluoromethoxycarbonylcyanide phenylhydrazone (FCCP) or the methyl ester of phenylalanine were also effective. The magnitude of the effect depended on the nature of the acceptor and its concentration.

The effect remained constant only in the initial period when the ATPase reaction

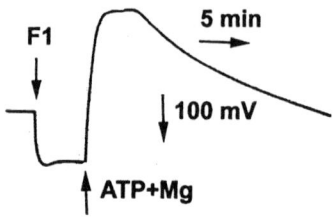

Figure 8.27. The ATP-induced, F_1-mediated changes in the Volta potential at the octane–water interface. Incubation mixture: 1×10^{-3} M DNP, 5×10^{-2} M Tris HCl (pH 7.4). Additions: 2.5×10^{-9} M factor F_1, 1×10^{-2} M Mg ATP (Kozlov and Skulachev, 1977). Reproduced by permission of Elsevier Science – NL.

rate was constant. The change in Volta potential over time as the ATP was used up was similar to the ATPase reaction rate measured in the same conditions (Boguslavsky *et al.*, 1975d; Kozlov and Skulachev, 1977; Skulachev and Kozlov, 1977).

The observed potential changes when the enzymatic reaction is initiated are not due to a change in pH of the aqueous phase, because the potential difference was not altered by a wide range of tris-HCl concentrations (from 3 mM to 100 mM). Furthermore, the measured potential could be observed at pH 6.2, where ATP hydrolysis does not involve a significant change in pH.

Therefore, the most plausible explanation of the potential difference at the octane–water interface is that mitochondrial ATPase can transfer protons from the water to the octane phase. The octane–water system has also been used to study the H^+-transporting function of ATPase from the membrane of *Micrococcus lysodeikticus*, bacteriorhodopsin, and some enzymes of the respiratory chain (Boguslavsky *et al.*, 1976c,g,h; Gugeshashvili *et al.*, 1991).

9

LIGHT ENERGY CONVERSION AT LIQUID–LIQUID INTERFACES: ARTIFICIAL PHOTOSYNTHETIC SYSTEMS

The annual consumption of energy by mankind is currently about 4×10^{17} kJ, rising rapidly and doubling every 20 years. The known reserves of fossil fuels are limited to an estimated energy equivalent of 5×10^{19} kJ, so new sources of energy are of fundamental importance. One obvious possibility is solar energy. The amount of solar energy incident on the Earth is about 5×10^{21} kJ per year, of which 3×10^{18} kJ is converted into chemical energy by photosynthesis in plants and micro-organisms. This light energy is harvested by photosynthetic pigment systems in which the electronic structure of excited-state chlorophyll donates an electron to a primary acceptor pheophytin, the first component of an electron transport chain. The electron carries with it the energy of the original photon of light that was absorbed, and in the process of electron transport the energy is captured in two ways. The first involves coupling a proton pump mechanism to the sequential redox reactions in one part of the electron transport chain, so that a proton gradient is established across the thylakoid membrane. The electrochemical energy of the proton gradient is then used to drive ATP synthesis by the ATP synthase enzymes embedded in the membrane. The second energy capture occurs when an acceptor molecule such as NADP is reduced to NADPH, which in turn is used to reduce carbon dioxide in the Calvin cycle. Systems modeling photosynthesis should have the capability of carrying out relatively simple versions of these fundamental reactions.

The interface between two immiscible liquids with immobilized photosynthetic pigments can serve as a convenient model for investigating photoprocesses that are accompanied by spatial separation of charges. The efficiency of charge separation defines the quantum yield of any photochemical reaction. Heterogeneous systems will be most effective in this regard, where the oxidants and the reductants are either in different phases or sterically separated. Different solubilities of the substrates and reaction products in the two phases of heterogeneous systems can alter the redox potential of reactants, making it possible to carry out reactions that cannot be performed in a homogeneous phase.

The present chapter focuses on electrochemical mechanisms of photocatalytic systems at the oil–water interface. The term photocatalyst is used with regards to a substance that can adsorb light energy and thereby catalyse a reaction by forming activated intermediates. In order to provide a biological perspective, we will first describe the photocatalytic processes that occur during photosynthesis.

9.1. STRUCTURE AND COMPOSITION OF THE OXYGEN-EVOLVING COMPLEX *IN VIVO*

The redox map of photosynthesis in green plants can be described in terms of the well-known Z-scheme proposed by Hill and Bendall (1960). The main advantage of the currently accepted Z-scheme depicted in Fig. 9.1 lies in the specific mechanism of charge transfer at the stage of water oxidation, which is a multi-electron reaction mechanism involving no unknown intermediates.

The molecular organization of a thylakoid membrane is shown in Fig. 9.2. The spectral characteristics of photosystem II indicate that the primary electron donor is the dimer of chlorophyll P680 with absorption maxima near 680 and 430 nm. Water can be oxidized by an oxygen-evolving center (OEC) composed of several chlorophyll molecules, two molecules of pheophytin, plastoquinol, several plastoquinone molecules and a manganese-protein complex containing four manganese ions. The oxygen-evolving complex is a highly ordered structure in which a number of polypeptides interact to provide the appropriate environment for co-factors such as manganese, chloride, and calcium, as well as for electron transfer within the complex.

Manganese binding centers were first revealed in thylakoid membranes by electron paramagnetic resonance (EPR) methods, and it is now understood that four manganese ions are necessary for oxygen evolution during water photooxidation. Plastoquinone (PQ) acts as a transmembrane carrier of electrons and protons between reaction centers of two photosystems in the case of non-cyclic electron transfer and may also serve as a molecular "tumbler" that switches between one-electron reactions and two-electron reactions. Pheophytin is an intermediate acceptor in photosystem II. Direct formation of P680 pheophytin ion radical pairs was revealed by experiments on magnetic interactions between pheo- and PQ- reflected in the EPR spectra.

9.2. THERMODYNAMICS OF WATER OXIDATION

The photocatalytic oxidation of two molecules of water to dioxygen cannot be a single-quantum process since the total energy expenditure of a catalytic cycle cannot

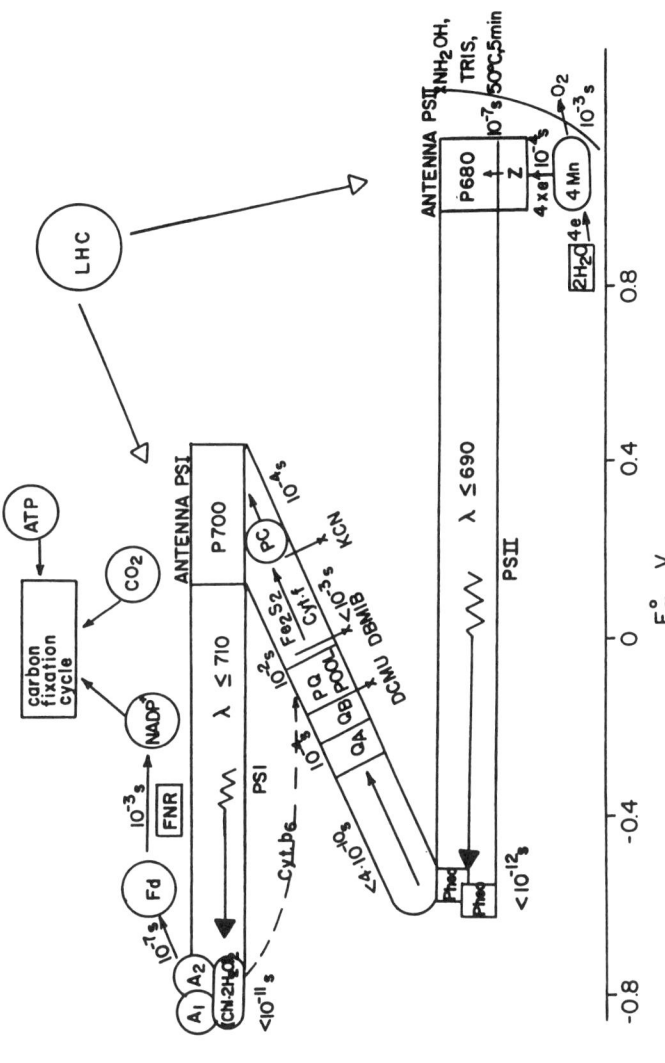

Figure 9.1. Electron transfer during photosynthesis in higher plants. The abscissa shows the midpoint redox potential at pH 7.0. Light quanta (*hv*) are absorbed in two sets of antenna chlorophyll molecules, and the excitation energy is transferred to the reaction center chlorophyll *a* molecules of photosystem II (P680) and photosystem I (P700) forming (P680)* and (P700)*. The latter two initiate electron transport. Abbreviations: LHC—light harvesting complex; Z and D are tyrosine residues; Cyt b_{559} is cytochrome b_{559} of unknown function; Pheo is pheophytin; Q_A, Q_B and PQ are plastoquinone molecules; Fe_2S_2 represents the Rieske iron-sulfur center; Cyt f stands for cytochrome f, PC is plastocyanin; A_0 is suggested to be a chlorophyll molecule, A_1 is possibly vitamin K; FNR is ferredoxin NADP oxidoreductase; x—inhibitors (DBMIB: dibromo-3-methyl-6-isopropyl-p-benzo-quinone; DCMU: 3-(3′,4′-dichlorophenyl)-1,1-dimethylurea) (Volkov et al., 1995a). Reproduced by permission of Elsevier Science, Ltd.

363

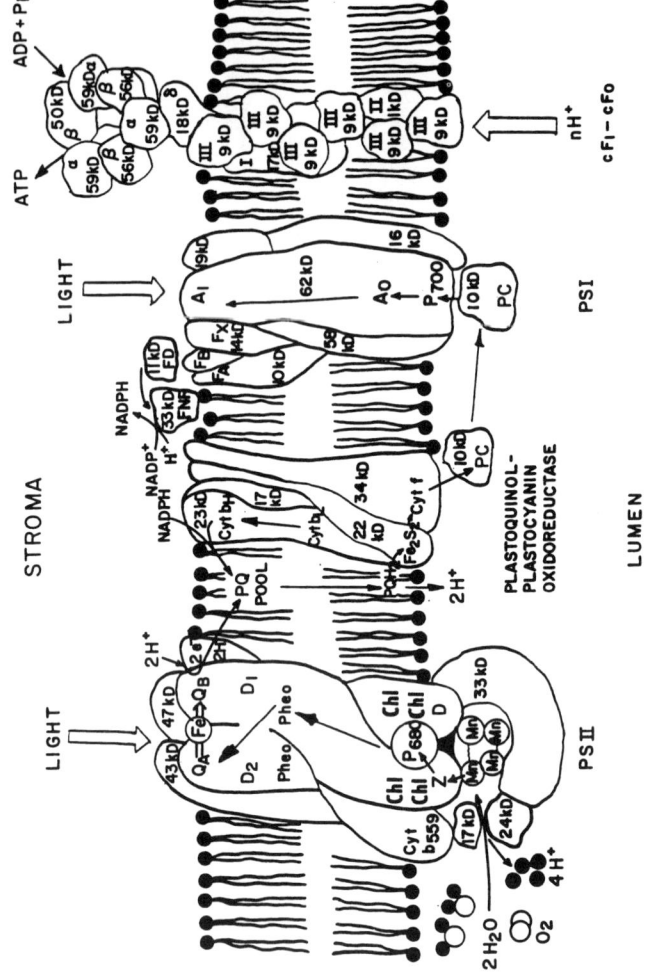

Figure 9.2. A schematic model of the electron transport chain with most of the light-harvesting pigment-protein complexes omitted (Volkov *et al.*, 1995a). Reproduced by permission of Elsevier Science, Ltd.

be less than $476 \, kJ/mol^{-1}$. However, there is no fundamental reason why one quantum should not induce the transfer of several electrons. For instance, a two-quantum process would require light with a wavelength less than 504 nm while a four-quantum process would involve a sequential mechanism in which each light quantum is used to transfer one electron from photocatalyst to an electron acceptor. The threshold wavelength for the oxidation of water in this case is 1008 nm. The eight-quantum scheme which is actually used in photosynthesis can be explained by the need to compensate for energy losses in a long electron-transfer chain of redox reactions.

Water oxidation to molecular oxygen is a multi-electron process that proceeds with a surprisingly high quantum efficiency (Fig. 9.3). The water oxidation reaction can proceed upon illumination at 680 nm, a wavelength of light that excludes one-electron mechanisms using hydroxyl and oxygen radicals. For a three-electron reaction a stronger oxidant than the cation-radical $P680^{+}$ is needed. A synchronous two by two $2:2$-electron pathway of the reaction is thermodynamically possible if the standard free energy of binding of the two-electron intermediate is about $-40 \, kJ \, mol^{-1}$. This value corresponds to the energy of two hydrogen bonds forming between H_2O_2 and the catalytic center. For this case a molecular mechanism can be proposed (Fig. 9.4) and will be discussed below. Synchronous four-electron oxidation of water to molecular oxygen (Fig. 9.5) is also thermodynamically possible.

One-electron mechanisms of water oxidation are likely to be operative in some model systems with a low quantum efficiency, but two- or four-electron reactions cannot occur due to kinetic limitations. The intermediates formed in these systems would be highly reactive and could enter into side reactions of hydroxylation, oxidation, and destruction of chlorophyll and other components of the reaction center.

9.3. KINETIC ASPECTS OF MULTI-ELECTRON REACTIONS

An important parameter in the quantum theory of charge transfer in polar media is the medium reorganization energy E_s that determines activation energy. The energy of medium reorganization in systems with complicated charge distribution was calculated by Kharkats (Kharkats, 1978). Reagents and products can be represented by a set of N spherical centers arbitrarily distributed in a polar medium. The charges of each of the reaction centers in the initial and final state are z^i and z^f, respectively. Taking R_k to represent coordinates of the centers and ε_i for dielectric constants of the reagents it follows that:

$$E_s = 0.8 \left(\frac{1}{\varepsilon_{opt}} - \frac{1}{\varepsilon_{st}} \right)$$
$$\times \left\{ \sum_{p=1}^{N} \left[\frac{(\delta z_k)^2}{2a_p} + \sum_{\substack{k=1 \\ k \neq p}}^{N} \frac{(\delta z_p)(\delta z_k)}{2R_{pk}} + \sum_{\substack{k=1 \\ k \neq p}}^{N} \sum_{\substack{l=1 \\ l \neq p}}^{N} \frac{(\delta z_p)(\delta z_l) a_p^3 (\mathbf{R}_{pk} \mathbf{R}_{pl})}{R_{pk}^3 R_{pl}^3} \left(\frac{3\varepsilon_{st}^2}{(2\varepsilon_{st} + \varepsilon_i)^2} - \frac{1}{2} \right) \right] \right\}$$

$$(9.1)$$

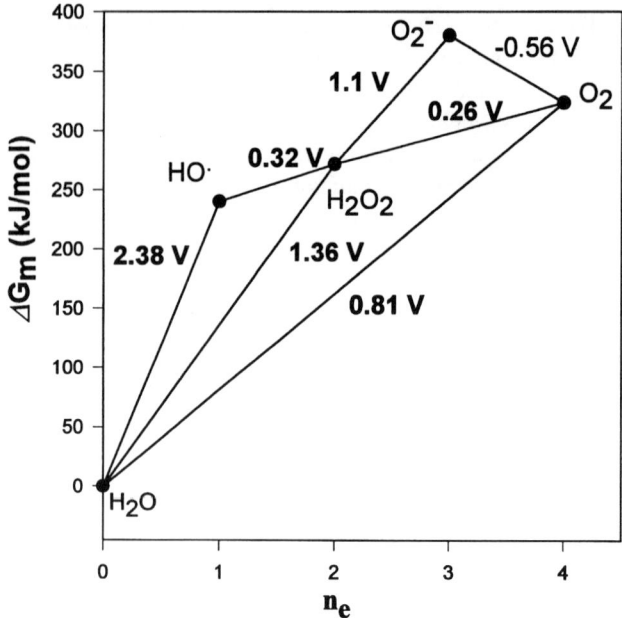

Figure 9.3. Energy scheme for water oxidation to molecular oxygen.

where $\delta z_k = z_k^f - z_k^i$, $R_{pk} = R_p - R_k$, z_k^f, z_k^i are charge numbers of particle k in the final and initial states, respectively, a_p is the radius of particle p, R_k is the coordinate of k-particle center, and ε_i is the dielectric constant of the reactant. Reactions with synchronous transfer of several charges present a particular case of equation (9.1).

It follows from equation (9.1) that E_s is proportional to the square of the number of charges transferred. This factor makes multi-electron processes impossible in the majority of homogeneous redox reactions due to the high activation energy resulting from a sharp rise in the energy of solvent reorganization. For multi-electron reactions the exchange currents of n-electron processes are small compared to those of one-electron multiple steps processes, which makes the stage-by-stage reaction mechanism more advantageous. Therefore, multi-electron processes can proceed only if the formation of an intermediate is energetically disadvantageous. However, conditions can be chosen which reduce E_s during transfer of several charges to the level of the reorganization energy of ordinary one-electron reactions. These conditions require systems with a low dielectric constant and large reagent radii. Furthermore, the substrate must be included in the coordination sphere of the charge acceptor, with several charge donors or acceptors bound into a multicenter complex. Theoretical analysis of the kinetics of heterogeneous multi-electron reactions at water–oil interfaces which proved to be capable of catalyzing multi-electron reactions and sharply reducing the activation energy was given by Kharkats and Volkov (1985) and Volkov et al. (1995a).

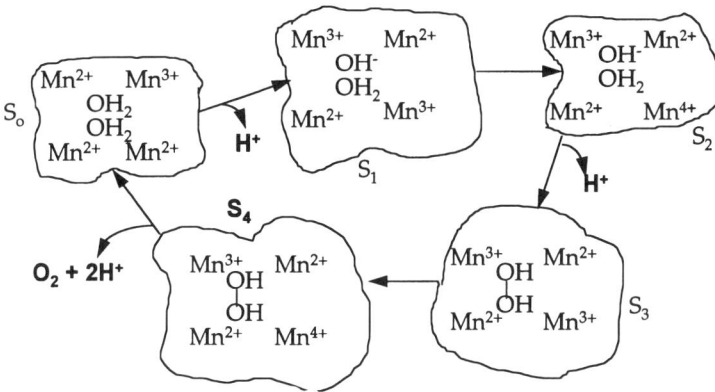

Figure 9.4. Proposed 2:2 electron mechanism of water photo-oxidation (after Volkov, 1986a).

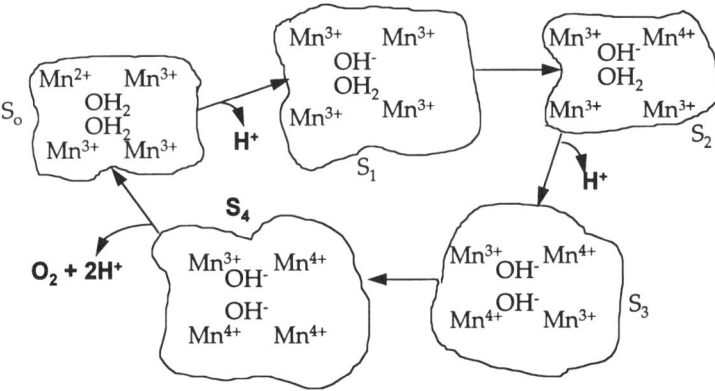

Figure 9.5. Proposed four-electron mechanism of water photo-oxidation (after Volkov, 1986a).

The most effective coupling of ATP synthesis to electron transport can be obtained if the activation energy of the coupled process is lower than that of the charge transfer in the electron transport chain. It is obvious from equation (9.1) that this requires a simultaneous transfer of opposite charges, so that the second and the third terms of equation (9.1) are negative. An optimal geometry between the centers of charges of donors and acceptors must also be chosen.

To illustrate this point, we can consider two instances of multi-electron reactions: simultaneous transfer of n charges from one donor to an acceptor and simultaneous transfer of several charges (one from each of the centers) to m acceptors ($m \le n$). In the former case E_s is proportional to n^2, while in the latter it may be significantly lower (depending on the sign of the charge being transferred and the positions of reagents). A multicenter process with $E_s \sim n$ is also possible. The concerted multicenter mechanism of multi-electron reactions markedly reduces E_s, and hence the activation energy, compared to a two-center multi-electron process. Electrostatic interactions between reactants also reduce the activation energy of multi-electron processes at the interface. Such electrostatic energies in a heterogeneous process can never be equal to zero due to interactions with image charges. However, by appropriate arrangement of the reactants the electron transfer activation energy in a heterogeneous multi-electron reaction can be much lower than the energy of medium reorganization.

9.4. MOLECULAR MECHANISM OF OXYGEN EVOLUTION *IN VIVO*

Membrane-bound P680 becomes excited upon illumination. In dimers and other aggregated forms of chlorophyll the quantum efficiency of triplet states is low, and it is the singlet excited states that undergo photochemical transformations. In several picoseconds, an electron is first transferred to pheophytin, then to plastoquinone Q_A, and from plastoquinone Q_A to another polypeptide-bound plastoquinone Q_B in a thylakoid membrane (Fig. 9.1) resulting in an oxidized pigment and a reduced acceptor. The cation radical P680$^+$ successively oxidizes four manganese ions, which in turn drives the production of molecular oxygen. Formation of a cation radical of chlorophyll or oxidation of manganese ions is accompanied by dissociation of water bound to the reaction center and ejection of protons. A synchronous multi-electron process that describes all four oxidizing states of the oxygen-evolving complex was proposed earlier (Volkov, 1989). The transfer of electrons in a $1:1:1:1$ series from a manganese cluster to the electron-transport chain is accompanied by the ejection of $1:0:1:2$ protons and the evolution of molecular oxygen.

Protons are released from reaction centers either by regulators of proton distribution or by hydrogen bond transfer (analogous to a Grotthus mechanism) through the hydration shell of manganese ions. The hydration sphere of manganese is known to contain water molecules that rapidly exchange protons with bulk water. The presence of divalent cations at the interface between two immiscible electrolyte solutions facilitates strong adsorption of water molecules belonging to the second hydration shell of ions. Thus, a portion of coordinatively bound water enters the compact part of the electric double layer, which changes its differential capacity at the interface. In the case of multivalent ions with small radii, the electric field near a cation is large. This can disturb the microstructure of the adjacent intrathylakoid space and bring about dielectric saturation effects.

Manganese ions play a particularly important role in the evolution of dioxygen during photosynthesis. Although there are several manganese pools in chloroplasts, only one is involved in water oxidation. The manganese ions associated with chloroplast oxygen evolving complex (OECs) can perform a number of functions:

- Mn-polypeptide complex is a redox intermediate that protects the reaction center from redox and radical destruction;

- Mn-clusters are redox buffers facilitating accumulation of four holes in the reaction center of photosystem II, which are needed to ensure water photooxidation;
- hydrated multivalent manganese cations bring water to the reaction center so that rapid proton exchange and transport through the hydration shell of manganese ions in the zone of water oxidation can take place;
- multivalent manganese ions induce dielectric saturation effects in the polar region of the reaction center of photosystem II, which reduces the reorganization energy of the medium during charge transfer.

9.5. PHOTOINDUCED CHARGE TRANSFER ACROSS AN OIL–WATER INTERFACE

Three types of artificial photosynthetic reactions at the liquid–liquid interface can occur:

1. Photoredox reactions at the interface between two immiscible liquids.
2. Phototransfer of ions or electrons.
3. Photochemical reactions in one phase following extraction of products in two different phases.

Many photochemical reactions of artificial photosynthesis studied in homogeneous systems are usually characterized by low quantum yield and low efficiency. Mathai and Rabinowitch (1962), and Frankowiak and Rabinowitch (1966) suggested that only heterogeneous systems with liquid–liquid interfaces can increase yields due to separation of photoproducts in different phases. A classic example is the photochemical reaction at the ether–water interface:

$$T + nFe^{2+} \underset{\text{DARK}}{\overset{\text{LIGHT}}{\longleftrightarrow}} (S + L) + nFe^{3+}, \tag{9.2}$$

where T is the purple-colored cation of thionine (or methylene blue), and S and L its reduced forms semithionine (or semiquinone) and leukothionine (leucomethylene blue), respectively. Potential differences ranging from 234 to 358 mV could be measured in the cell, depending on the concentration of the reaction products.

From thermodynamic and kinetic principles the interface between two immiscible liquids can have catalytic properties for interfacial charge transfer reactions. For example, at the oil–water interface the following redox reaction can occur:

$$Red \leftrightarrow Ox + ne^- + mH^+. \tag{9.3}$$

The electrons which are the products of reaction (9.3) can be accepted at the interface by another substance if it is dissolved in one of these phases. It is possible to shift the redox potential scale in a desired direction by selecting appropriate solvents, thereby permitting reactions to occur that are highly unfavorable in a homogeneous phase. If the resolution energies of substrates and products are very different, the interface between two immiscible liquids may act as a catalyst.

9.6. ARTIFICIAL PHOTOSYNTHESIS AT THE OIL–WATER INTERFACE IN THE PRESENCE OF CHLOROPHYLL

The photoelectrochemical properties of chlorophyll *a* in monolayers, multilayers, and thin films at oil–water, gas–water, and solid–water interfaces have been studied extensively to obtain a thorough understanding of its role in the primary events in photosynthesis: light harvesting, energy transfer, and charge separation in the photosynthetic unit. The state of chlorophyll molecules in the photosynthetic apparatus is still under active investigation.

Several forms of chlorophyll are involved in the primary events of photosynthesis. Absorbance peaks at longer wavelengths than those characteristic of chlorophyll in solution have been observed *in vivo* and are attributed to varying degrees of chlorophyll aggregation. Chlorophyll *a* dissolved in anhydrous liquid hydrocarbons (hexane, octane, decane) gives the characteristic absorption spectrum shown in Fig.

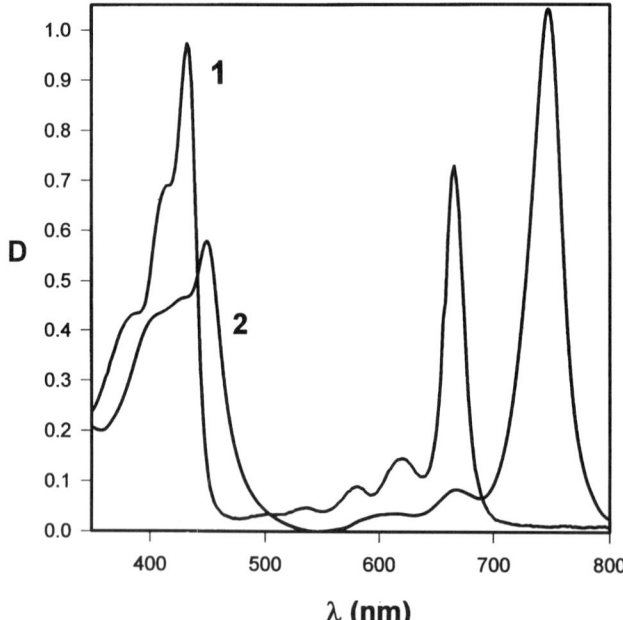

Figure 9.6. Electronic absorption spectra of dry (1) and hydrated (2) chlorophyll *a* in *n*-hexane.

9.6 (curve 1). This spectrum does not change if the solution is allowed to contact water for a short time (5 min). However, when chlorophyll a is dissolved in water-saturated hydrocarbon an additional band appears in the absorption spectrum with a maximum between 740 and 745 nm, which reflects the hydrated oligomer of chlorophyll a (Fig. 9.6, curve 2). The electronic adsorption spectrum of anhydrous chlorophyll a solution shows two bands at 428 and 660 nm similar to the absorption spectra of chlorophyll a in non-polar solvents. The hydrated oligomer of chlorophyll a is characterized by shifts of its absorption spectrum to lower energy wavelengths with maxima at 448 and 745 nm. It is worth noting that the intensity ratio of the blue to red adsorption bands is 1.3 for the dry sample and 0.6 after hydration. The differences in the absorption spectra of dry and wet solutions, coupled with marked decreases in energy of the lowest electronic transition of hydrated oligomer of chlorophyll a ($\Delta v = 1700 \, cm^{-1}$) show that in water-saturated hydrocarbon chlorophyll a exists as organized aggregates. Such self-organized molecular assemblies of hydrated chlorophyll a molecules are characterized by lower energy electronic transitions with an absorption band between 740 and 745 nm and a fluorescence band at around 755 nm. The lifetime of the emission is 0.1 to 0.2 ns.

Chlorophyll a dissolved in anhydrous hydrocarbon solvents at relatively high concentrations (1–10 mM) also undergoes aggregation to form oligomers. For instance, in hexane at higher concentration ranges, chlorophyll a is present as tetramers and dimers, provided its concentration is sufficiently high. In the range of chlorophyll a concentrations examined in water-saturated hexane ($<10^{-5}$ M), it is present mostly as a mixture of its hydrated oligomers and monomers, as can be inferred from the fluorescence and circular dichroism spectra.

Circular dichroism spectra of chlorophyll a in water-saturated hexane (Fig. 9.7) have negative bands at 765 nm in the region of Q_y transition. This suggests that the oligomer consists of at least four or six chlorophyll molecules. The aggregate exhibits a relatively strong optical rotation, with a Kuhn anisotropy factor of 0.05 at 748 nm, 500 times that of monomeric chlorophyll.

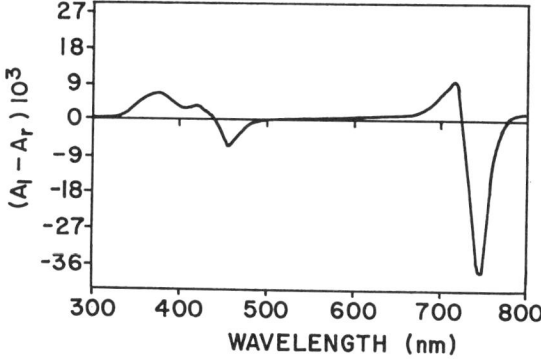

Figure 9.7. Circular dichroism spectrum of chlorophyll a in water-saturated octane. Chlorophyll a concentration in octane is 10^{-5} M, optical density $A_{742} = 0.27$ (Kandelaki *et al.*, 1983). Reproduced by permission of Elsevier Science S.A.

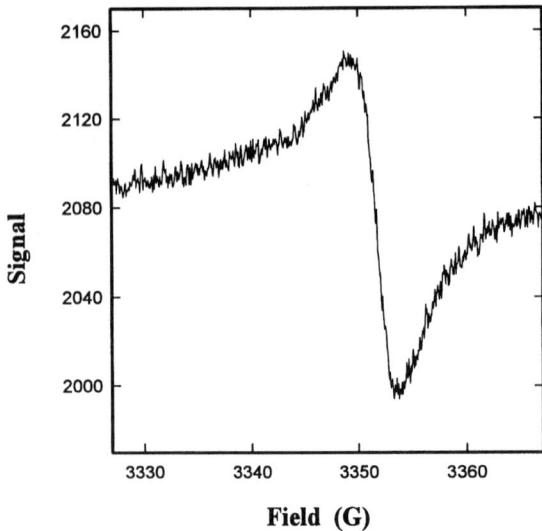

Figure 9.8. X-band ESR spectrum of wet chlorophyll *a* radicals at 298 K. Microwave power—20.7 mW; gain—10^6, time constant—200 ms; modulation amplitude—1 G peak-to-peak. The spectrum is an average of 25 scans.

Hydrated oligomers of chlorophyll *a* in *n*-octane have an intense narrow dark electron spin resonance (ESR) signal (Fig. 9.8) with a line width of 4.7 G and a *g*-value of 2.0025 ± 0.0003. Hydrated oligomers of chlorophyll *a* in the form of solid colloidal particles have been investigated as well and it was found that colloidal particles of dry and hydrated chlorophyll *a* have different ESR signals. The line width for dry chlorophyll *a* was about 12 G (Anderson and Calvin, 1964), while the line width for colloidal particles of hydrated chlorophyll *a* was found as 1.6 G, 6 G, 2–3 G, or 1–14 G (see review Volkov *et al.*, 1995) due to different aggregation states of hydrated chlorophyll in solid films or colloidal suspension. The number of chlorophyll molecules (N) in the cluster of hydrated oligomers of chlorophyll *a* can be found from the equation:

$$N = (\Delta H_m/\Delta H_n)^2, \tag{9.4}$$

where ΔH_m is the linewidth of chlorophyll *a* monomer and ΔH_n is the linewidth when the unpaired spin is delocalized over N molecules. It follows from equation (9.4), that a hydrated oligomer of chlorophyll *a* in hexane solution consists of six molecules of chlorophyll *a*. CD-spectra and surface tension measurements show the same number of molecules in a cluster of hydrated chlorophyll.

Chlorophyll molecules in monolayers, multilayers and thin films also form aggregates due to the strong attraction between the pigment molecules in the densely packed monolayer. Results of nuclear magnetic resonance (NMR) and infrared (IR) spectroscopic studies revealed formation of coordination linkage between the carbonyl group of a chlorophyll molecule and the magnesium atom of another molecule. In densely packed structures such as monolayers or films, thermal vibrations lead to attenuation of luminescence due to overlap of the porphyrin ring electron clouds during molecular collisions.

Hydration of chlorophyll oligomers inhibits the expected coordination interaction between carbonyl oxygen and magnesium in a porphyrin-like ring referred to as a "chlorine" ring:

$$C{=}O\cdots Mg$$

Instead, water molecules are coordinated by the central magnesium atom of chlorophyll:

$$={=}C{=}O\cdots H{-}O{-}H\cdots O{=}C{=}$$
$$\vdots$$
$$Mg$$

with an adsorption maximum at 720 nm or:

$$={=}C{=}O\cdots H{-}O{-}H$$
$$\vdots$$
$$Mg$$

with an adsorption maximum at 740–745 nm.

Chlorophyll molecules readily form monolayers which considerably reduce interfacial tension during adsorption at the interface. This property can be used to determine the Gibbs surface excess of chlorophyll. Figure 9.9 shows the dependence of interfacial tension γ on the chlorophyll a concentration in dry (curve 1) and water-saturated (curve 2) octane. A gradual reduction of γ was observed in chlorophyll a concentration ranges from 10^{-6} to 5×10^{-6} M for wet chlorophyll and from 10^{-6} to 1.5×10^{-5} M for anhydrous chlorophyll.

Using this interfacial tension data and the Gibbs adsorption equation,

$$\Gamma = -\frac{1}{RT}\frac{d\gamma}{d\ln c} = -\frac{c}{RT}\frac{d\gamma}{dc}, \tag{9.5}$$

the adsorption chlorophyll a isotherms at the octane–water interface can be plotted (Fig. 9.10). It is clear that adsorption isotherms of dry and wet chlorophyll a are quite different. The packed monolayer areas of the monolayer at the n-octane–water interface correspond to surface areas of 29 Å² per molecule for anhydrous chlorophyll a and 101 Å² for an adsorbed cluster of hydrated chlorophyll oligomer.

Figure 9.11 shows the dependence of surface pressure of anhydrous and hydrated chlorophyll a on the area per molecule at the octane–water interface. Using the isotherm of adsorption of amphiphilic compounds at the liquid–liquid interface, it is

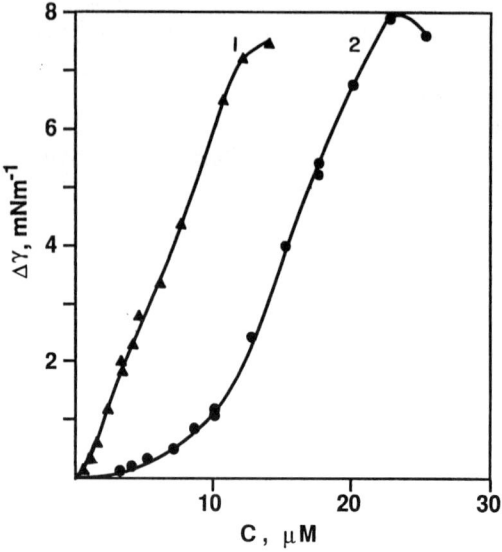

Figure 9.9. The dependence of surface tension at the octane–water interface on concentration of chlorophyll *a*: (1) dry monomer of chlorophyll *a*, (2) hydrated oligomer of chlorophyll *a*; pH = 7.4.

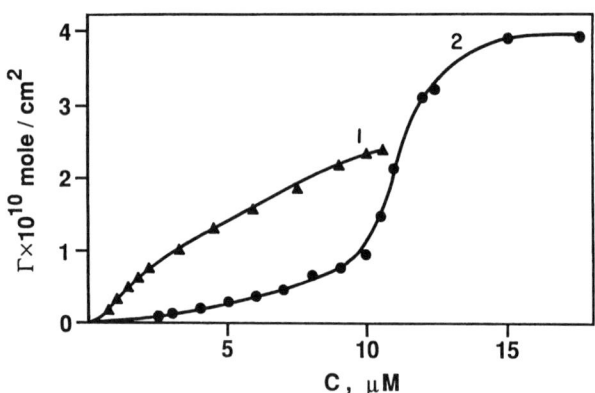

Figure 9.10. Concentration dependence of adsorption of dry chlorophyll *a* (1) and a hydrated oligomer of chlorophyll (2) at the octane–water interface.

possible to calculate the parameters of adsorption. The standard Gibbs free energy of adsorption at the n-octane–water interface is equal to $-27.4 \text{ kJ mol}^{-1}$ for dry chlorophyll *a* and $-35.2 \text{ kJ mol}^{-1}$ for a hydrated oligomer of chlorophyll. The positive sign of the attraction constant $a = 0.59$ for dry chlorophyll *a* shows attractive forces between adsorbed molecules of chlorophyll. The negative constant

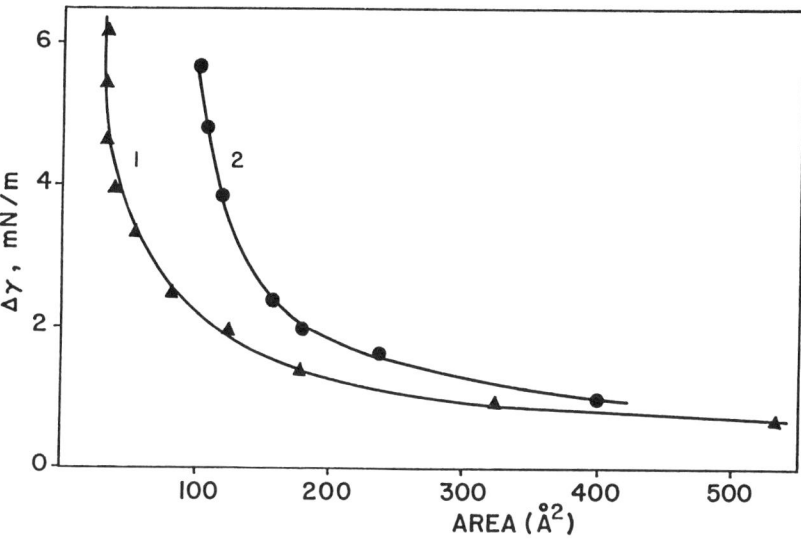

Figure 9.11. Concentration dependence of interfacial pressure at the octane–water interface for dry chlorophyll *a* (1) and a hydrated oligomer of chlorophyll (2).

$a = -0.215$ shows repulsion between clusters of the hydrated chlorophyll *a* (Chl *a*) oligomer.

Singlet electronic excitation of dry Chl *a* in dilute solution (λ_{exc} = 428 nm) leads to the appearance of fluorescence emission with a strong band at 665 nm and a low intensity band at 722 nm (see Fig. 9.12(*a*)). Moreover, fluorescence excitation spectrum (see Fig. 9.13(*a*)) measured at maximum emission corresponds well to the Soret absorption band of dry Chl *a* in *n*-hexane. This indicates that Chl *a* in dry diluted solution exists in monomeric form.

Fluorescence emission of a wet chlorophyll *a* solution when excitation is at 448 nm (Fig. 9.12(*b*)) exhibits two emission bands at 666 and 755 nm. However, when excitation is set at 428 nm, the fluorescence emission spectrum shows one strong band at 666 nm and two low-intensity bands at 720 nm and 755 nm. These emission spectra along with the fluorescence excitation spectra of wet Chl *a* in *n*-hexane when emission is at 755 nm or at 666 nm (Fig. 9.13(*b*)) indicate that the pigment exists not only as a hydrated oligomer but also as hydrated monomers and dimers. Such conclusions can also be supported by the fluorescence lifetime values presented in Table 9.1 and in Fig. 9.14. The decay of the emission at 755 nm is biexponential with the lifetime value of $\tau_1 = 5.54 \pm 0.05$ ns (30%) and $\tau_2 = 0.17 \pm 0.06$ ns (70%). The short-lived component belongs to well-aggregated forms of Chl *a*. Monoexponential decay has been measured for the 666 nm emission band with $\tau_1 = 5.68 \pm 0.02$ ns. For diluted dry Chl *a* solution, the monoexponential decay of the emission at 665 nm yields $\tau_1 = 5.75 \pm 0.02$ ns (Table 9.1); this lifetime value corresponds to emission of *a* Chl *a* monomer. The difference between dry

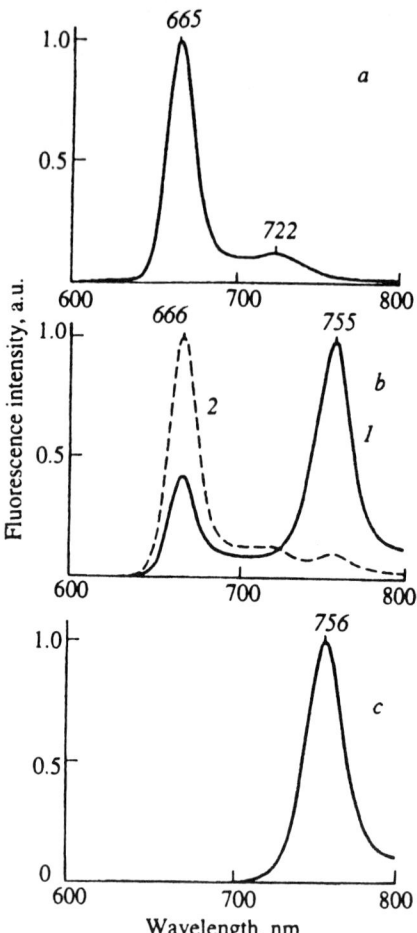

Figure 9.12. Corrected fluorescence spectra of: (*a*) dry Chl *a* in *n*-hexane ($c = 3 \times 10^{-7}$ M, l = 1.0 cm), excitation at 428 nm; (*b*) wet Chl *a* in *n*-hexane solution, excitation at 448 nm (1) and at 428 nm (2); (*c*) wet Chl *a* in a thin film on a quartz slide, excitation at 448 nm (after Zeleut *et al.*, 1993).

($\tau_1 = 5.75 \pm 0.02$ ns) and wet ($\tau_1 = 5.68 \pm 0.02$ ns) Chl *a* in *n*-hexane seems to be due to Chl *a*–water interactions.

Electronic absorption spectrum of thin films of wet chlorophyll *a* (Fig. 9.15(*b*)) shows a Soret band at 448 nm and a low-energy band at 746 nm, similar to the spectrum of wet Chl *a* in *n*-hexane and in LB film. Nevertheless, the emission spectrum of the thin film, when excitation is at 448 nm exhibits only one band at 756 nm (Fig. 9.12(*c*)). Moreover, fluorescence excitation spectrum, when emission is at 756 nm (Fig. 9.13(*c*)) corresponds to the absorption spectrum of the hydrated

Table 9.1. Fluorescence excitation and emission maxima and fluorescence lifetime of dry and wet chlorophyll (Chl) *a* in *n*-hexane solutions and wet Chl *a* in monolayers, multilayers, and thin films on a quartz slide (Zeleut *et al.*, 1993).

Pigment conditions	Excitation (nm)	Emission (nm)	τ_1 (ns)	τ_2 (ns)
Dry Chl *a* in *n*-hexane	428	665	5.75 ± 0.02 (100%)	—
Wet Chl *a* in *n*-hexane	428	666	5.68 ± 0.02 (100%)	—
	448	755	5.54 ± 0.05 (30%)	0.17 ± 0.06 (70%)
Wet Chl *a* monolayer	448	755	3.31 ± 0.43 (8%)	0.12 ± 0.05 (92%)
Wet Chl *a* multilayers	448	755	3.25 ± 0.33 (5%)	0.15 ± 0.05 (95%)
Wet Chl *a* thin films	448	756	3.22 ± 0.74 (3%)	0.12 ± 0.03 (97%)

Chl *a* thin film. The observed lifetime of the emission at 756 nm, $\tau_2 = 0.12 \pm 0.03$ ns is calculated to be 97% (Table 9.1). This is also consistent with the presented properties of wet Chl *a* in thin films. The fluorescence lifetime components of wet Chl *a* in monolayer and multilayer LB films, previously reported, well correspond to the lifetime values of wet Chl *a* in thin films (see Table 9.1). This clearly indicates that the aggregation states of the described molecules are of similar type in those model systems.

The low-energy emission transition has also been measured for a multilayer of dry Chl *a* in a water-saturated nitrogen atmosphere. It was observed that fluorescence maximum of Chl *a* in a wet atmosphere gradually shifts from 740 to 760 nm when temperature decreases from 298 K to 85 K. Moreover, Kooyman *et al.* (1977) have studied Chl *a*–water complexes in *n*-octane, using fluorescence-detected magnetic resonance and fluorimetry. From the concentration dependence of the pigment fluorescence at 4.2 K, they detected four emission bands at 669 nm, 687 nm, 725 nm, and 750 nm. These bands have been assigned to the emission of (Chl $a \cdot H_2O$), (Chl $a \cdot 2H_2O$), (Chl $a \cdot H_2O)_2$, and (Chl $a \cdot H_2O)_n$ species, respectively.

It is well known that Chl *a* molecules form self-aggregates *via* the ring V C_9 keto $C=O \cdots Mg$ interactions and Chl *a*-nucleophile ligand adducts *via* electron donor–acceptor interactions. In aliphatic or cycloaliphatic hydrocarbon solvents, Chl *a* can form Chl *a*–*Chl a* aggregates even at the most diluted solutions. However, in the presence of water, which is a polar molecule, the dimers or oligomers of Chl *a* can be easily disrupted to form hydrated monomers and oligomers as (Chl $a \cdot H_2O)_n$ or (Chl $a \cdot 2H_2O)_n$.

FTIR (Fourier Transform Infrared) spectroscopy has been used to study the

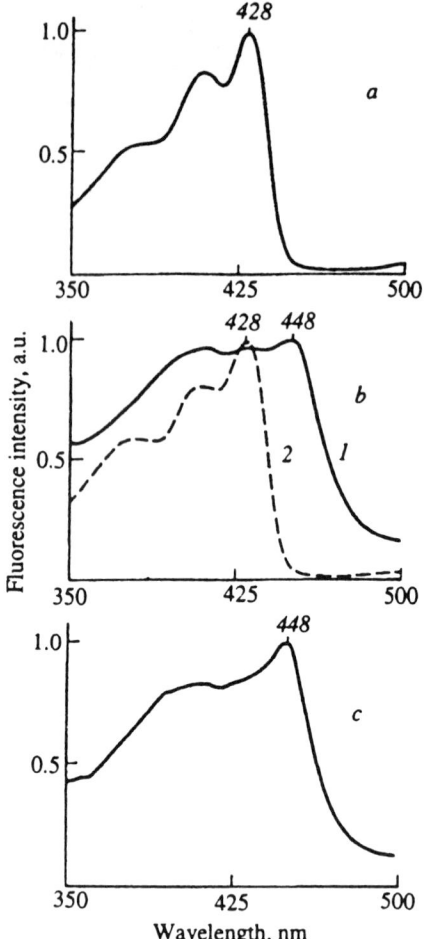

Figure 9.13. Fluorescence excitation spectra of (*a*) dry Chl *a* in *n*-hexane (*c* = 3 × 10⁻⁷ M, l = 1.0 cm), emission at 665 nm, (*b*) wet Chl *a* in *n*-hexane solution, emission at 755 nm (1) and at 666 nm (2); (*c*) wet Chl *a* in a thin film on a quartz slide, emission at 756 nm (after Zeleut *et al.*, 1993).

different aggregation states of Chl *a*. The band at 1733–1735 cm⁻¹ is the vibration frequency of the C=O ester group positioned at the C_{10} in the porphyrin ring. Another important band in that region is situated at 1688–1692 cm⁻¹ which corresponds to the vibration of the free (uncoordinated) C_9 keto C=O group. The 1659–1661 cm⁻¹ region is related to the frequency of the coordinated C_9 keto C=O group, which is bound to the magnesium of another Chl *a* molecule by the oxygen of the carbonyl functional group in dry dimers or oligomers. The fourth important band in the C=O vibration region appears at 1633–1647 cm⁻¹. This may represent carbonyl functional group in the C_9 keto C=O···H—O(H)···Mg hydrated aggregation state, which is

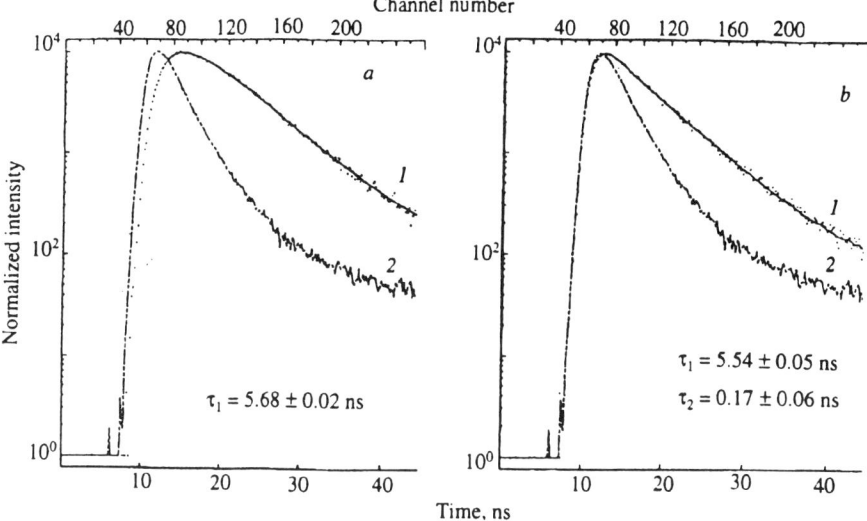

Figure 9.14. Time-resolved fluorescence decay curves (dotted) associated with the lamp profiles (dashed) showing (a) the single exponential decay ($\lambda_{em} = 666$ nm) and (b) the biexponential decay ($\lambda_{em} = 755$ nm) of the wet Chl a in n-hexane. Excitation wavelength, 440 ± 50 nm (after Zeleut et al., 1993).

observed in hydrated oligomers, or it may be the bending mode of the water molecule that is trapped between the two Chl a molecules in the same aggregation state as indicated before.

In the case of amorphous thin films of hydrated oligomers, the absorption spectrum shows strong maxima at 746 nm (Fig. 9.15, curve (a)1) which indicates a high oligomer concentration. The first IR spectrum (curve (b)1) supports the presence of a strong water band at 1639 cm^{-1}, completely overlapping the aggregated C_9 keto C=O$\cdots$Mg band (which is possible to observe at around 1660 cm^{-1} in samples less concentrated in hydrated oligomers). Drying of the sample for 18 hours under vacuum (curves (a)2 and (b)2) produced a small decrease in the water band in the IR spectrum due to removal of excess water that was trapped in the sample during preparation. The water band indicates that water molecules present in the sample are strongly coordinated to the Chl a in the hydrated oligomeric form.

9.7. WATER PHOTO-OXIDATION

In the case of photocatalysis in organized molecular assemblies, the reaction of most current interest is water photo-oxidation. The design of systems for this reaction is often based on mimicking the function and structure of natural photosynthesis.

Photocatalytic splitting of water is usually a sequence of three steps. The first is charge separation under the action of solar radiation in the presence of a photocatalyst.

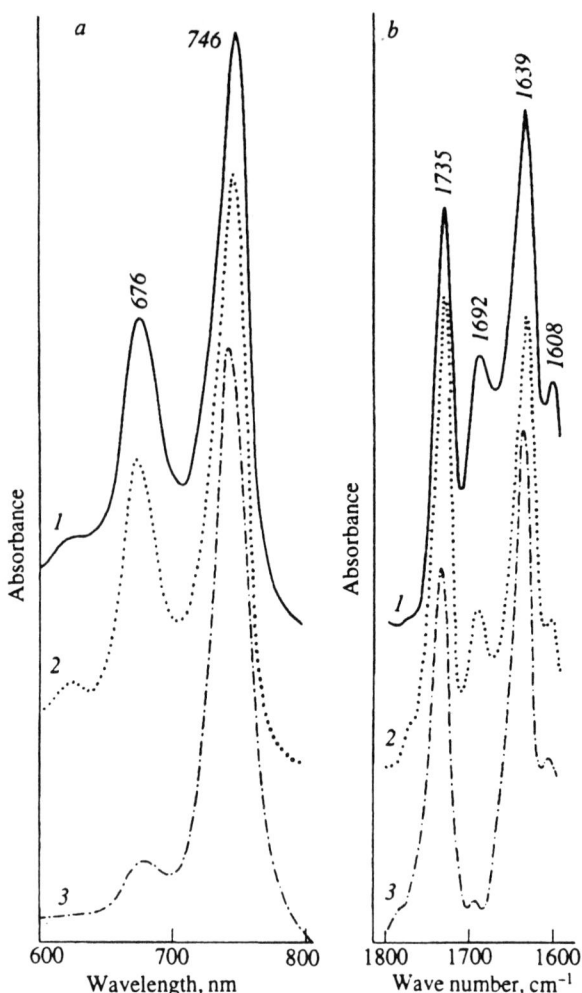

Figure 9.15. Electronic absorption (*a*) and FTIR spectra (*b*) of thin films of hydrated Chl *a* oligomers: (1) on cleaned CaF_2 crystal; (2) after 18 hours of vacuum on cleaned CaF_2; (3) on Cd arachidate-covered CaF_2 (after Gugeshashvili *et al.*, 1995).

A strong oxidant and reductant are formed, followed by the subsequent catalytic reaction of oxygen evolution. The major problem in this approach is suppressing the recombination of the strong oxidant and reductant, a simple biomolecular process that is very rapid. The best way to overcome this problem might be to use heterogeneous liquid–liquid photocatalytic systems.

The ability of chlorophyll molecules to be involved in reversible redox reactions underlies the primary process by which chloroplasts convert solar energy into electrochemical energy. A few years ago, hydrated oligomers of chlorophyll immobilized at the interface between two immiscible liquids were found to catalyze oxygen evolution during photo-oxidation of water (Boguslavsky *et al.*, 1976f, 1977b, 1978; Boguslavsky and Volkov, 1987; Kandelaki *et al.*, 1983a,b, 1984, 1987, 1988; Kandelaki and Volkov, 1991, 1993; Volkov, 1984a,b, 1985a,b, 1988, 1989; Volkov *et al.*, 1985, 1986). The quantum efficiency of oxygen production was 10–20%. Comparative measurements with the Clark oxygen electrode and mass spectrometry proved conclusively that the oxygen resulting from the reaction was not due to thermal effects of illumination on the Clark electrode. Visible light absorption spectra (Fig. 9.6, curve 2) indicated that hydrated oligomers of Chl *a* were present. If Chl *a* was replaced by Chl *b*, no oxygen evolution occurred, presumably due to different supramolecular structures of Chl *a* and *b* aggregates.

Other interfaces with immobilized photosynthetic pigments can serve as models of photoprocesses accompanied by spatial separation of charges. For instance, oxygen evolution occurs upon illumination of the octane–water interface in the presence of a hydrated oligomer of Chl *a* (P745), a proton acceptor in the octane, and electron acceptors such as nicotinamide adenine dinucleotide (NAD^+), nicotinamide adenine dinucleotide phosphate ($NADP^+$), or $K_3Fe(CN)_6$) (Kandelaki and Volkov, 1991, 1993). Oxygen production was proportional to the area of the interface and did not depend on the volume of the oil or water phases.

Illumination also induces an interfacial photopotential (Boguslavsky *et al.*, 1976f, 1977b, 1978; Boguslavsky and Volkov, 1987) if Chl *a* and $K_3Fe(CN)_6$ are added to an equilibrated octane–water system containing 2,4-dinitrophenol (Fig. 9.16). The magnitude of this potential depends on the wavelength and intensity of the incident light and the reagent concentrations. Similar effects were observed in other experiments in which chlorophyll was used to catalyze redox reactions. The electrochemical chain for such experiments is shown below:

Au	Air	Octane, chlorophyll, proton acceptor	Water, electron acceptor	Water KCl-saturated	Hg_2Cl_2, Hg.
					(9.I)

The photopotential results from the following reaction taking place at the interface:

$$2H_2O + 4A + 4\ PCP^- \Rightarrow O_2 + 4\ PCPH + 4A^-. \tag{9.6}$$

The reaction is accompanied by the capture of protons released by proton acceptors. We also found that dinitrophenol (DNP) or pentachlorophenol (PCP^-) are adsorbed at the interface between two immiscible liquids such as octane and water. Such compounds dissociate to produce a net negative charge on the octane phase relative to water. The magnitude of the measured potential was -0.2 V, following addition of DNP to the octane–water system. When the photoreaction (9.6) was initiated, the potential shifted abruptly in the positive direction by 0.1 V. The stoichiometry of reaction (8) was previously verified polarographically.

Figure 9.17 shows an absorption spectrum of a hydrated chlorophyll oligomer in hydrated octane and an action spectrum of the water photo-oxidation reaction. The action spectrum in Fig. 9.17 suggests that hydrated chlorophyll oligomers P745 and

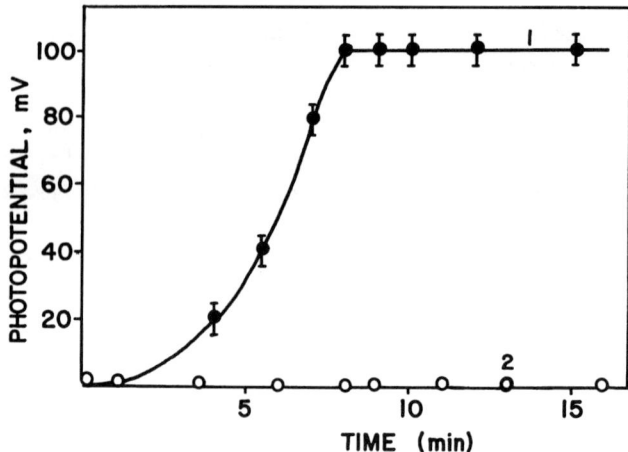

Figure 9.16. Dependence of photopotential on the time of illumination with different contact times of aqueous phase with dry *n*-octane. Medium: 1 mM dinitrophenol, 10 mM K₃Fe(CN)6, 10 mM Tris-HCl (pH 7.7). Chlorophyll *a* in concentration 10 mM was added after: (1) 10 min; (2) 0 min (Kandelaki *et al.*, 1983). Reproduced by permission of Elsevier Science S.A.

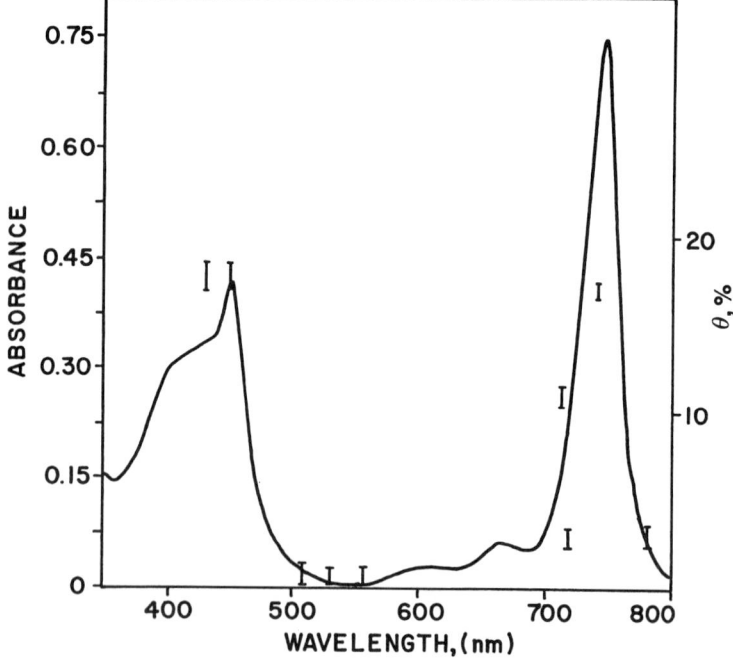

Figure 9.17. Action spectrum of water photo-oxidation in the presence of chlorophyll (I) (Kandelaki *et al.*, 1987). Medium: 1 mM pentachlorophenol, 10 mM tris-HCl, pH 7.7, 10^{-5} M chlorophyll *a*, 1 mM NAD⁺, octane–water. For comparison, absorption spectrum of hydrated oligomer of chlorophyll solution in octane equilibrated with water is presented (solid line). Reproduced by permission of Elsevier Science S.A.

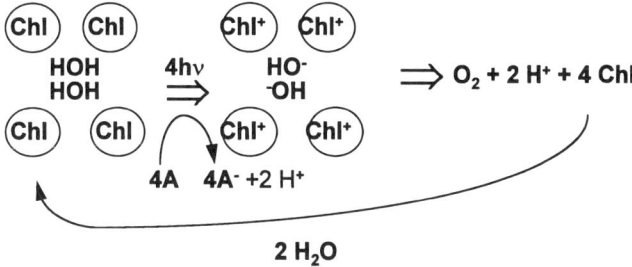

Figure 9.18. Proposed mechanism of water photo-oxidation by a hydrated chlorophyll oligomer at the octane–water interface.

chlorophyll monomers are involved in the reaction. Monomeric forms of Chl *a* may serve as sensitizers or antennae in this system.

The hydrated chlorophyll oligomer, $Chl_6(H_2O)_n$ ($n > 2$), is adsorbed on the surface and closely packed so that overlapping porphyrin rings enter an excited state upon illumination which leads to pigment oxidation and acceptor reduction (Fig. 9.18). Water is coordinated to the magnesium atom of one chlorophyll molecule and linked by hydrogen bonds to the carbonyl group of a second chlorophyll molecule and a pentachlorophenol anion, which is also adsorbed at the interface and forms part of the catalytic reaction center. The PCP^- anion is required for water binding at the reaction center and protection of chlorophyll against pheophytinization during the course of the reaction. The redox potential of the hydrated chlorophyll oligomer is about 0.92 V, so only one possible pathway is available for water oxidation, in which a direct four-electron oxidation produces molecular oxygen. It should be mentioned that one- and two-electron reactions shown in Fig. 9.3 are formally possible if their intermediates are absorbed at the interface and their binding energy is high enough. However, the radicals formed in the course of multistage processes would inevitably oxidize and destroy the catalytic complex. The quantum yield of fluorescence of the aggregate is 10^3 less than the quantum yield of monomer fluorescence. This indicates that the molecular ensemble responsible for water oxidation to molecular oxygen involves other catalytic sites such as chlorophyll dimers, trimers, and monomers.

In the reaction described above, P745 acts both as a photosensitizing agent and a catalyst. It is possible to uncouple these functions by adding a dye (β-carotene) that can also sensitize the photo-oxidation of water. A sensitizer is a substance that induces reactions by energy transfer to the reactants rather than by forming chemical intermediates. Carotene not only protects chlorophyll against photodynamic destruction, but is also capable of absorbing light and imparting the excitation energy to chlorophyll *a*. Adsorption spectrum of β-carotene is shown in Fig. 9.19. The van der Waals interaction known to exist between the tetrapyrolle ring of chlorophyll and β-carotene apparently stabilizes a weak complex, which can then participate in water photo-oxidation at wavelengths where β-carotene rather than chlorophyll absorbs light (Fig. 9.20).

Figure 9.20 illustrates the relationships between evolved oxygen, the length of the

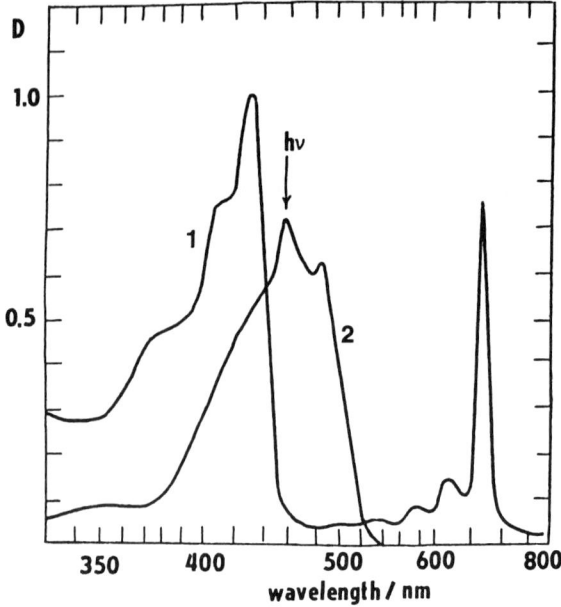

Figure 9.19. Absorption spectra of dry chlorophyll *a* (1) and β-carotene (2) in dry octane.

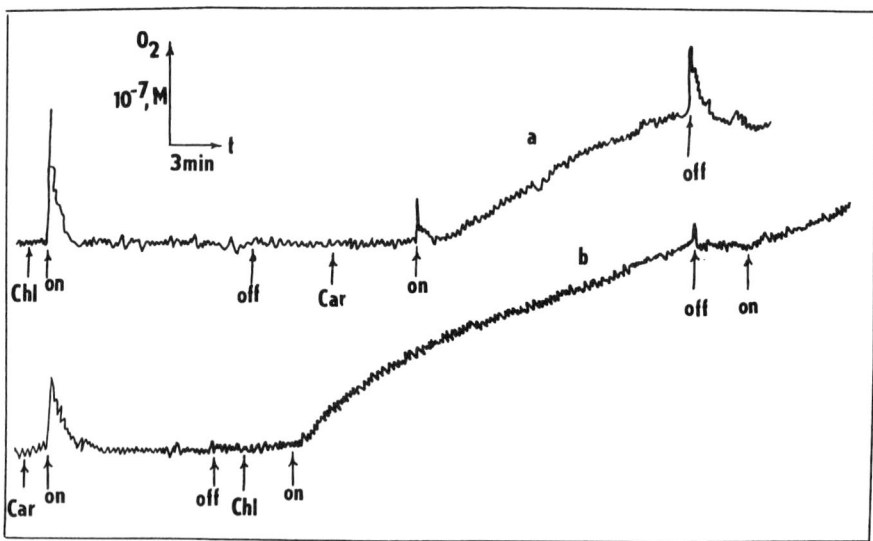

Figure 9.20. Dependence of the amount of oxygen evolved on the incubation time and the sequence of addition of reagents. Composition: water–octane, 1 mM NAD$^+$, 20 mM tris-HCl, pH 7.4, 1 mM pentachorophenol. Chlorophyll and β-carotene were added to the octane fraction in concentrations of 4 μM and 10 μM, respectively. The interface was illuminated with monochromatic light at a wavelength of 462 nm.

illumination period, and different sequences of reagent addition to the octane–water system. Addition of a proton acceptor (PCP), an electron acceptor (NAD^+) and chlorophyll results in light-induced evolution of oxygen. No oxygen was evolved during illumination at 462 nm (Fig. 9.20(a)) but addition of β-carotene caused oxygen evolution at this wavelength. The process stops completely in the dark and returns to the initial rate upon reillumination. In the presence of β-carotene the quantum yield at 462 nm is approximately 0.5% (calculated on the basis of incident light quanta). Figure 9.20 shows that oxygen evolution in the complete system is independent of the order of chlorophyll and β-carotene addition. However, no oxygen was produced in the absence of chlorophyll (Fig. 9.20(b)).

Artificial Photosynthesis in Monolayers and Langmuir–Blodgett Films.

Chemical models of photosynthetic processes can be used to investigate two types of reactions. In photosynthesis, the standard Gibbs free energy of the reaction is positive, and solar energy is utilized to perform chemical work. In photocatalytic processes, the Gibbs free energy is negative and solar energy is used to overcome the activation barrier. A detailed description of some chemical models of photosynthesis in heterogeneous systems can be found in numerous reviews (Deamer and Volkov, 1996; Gratzel, 1989; Volkov, 1989; Zamaraev and Parmon, 1983).

Chlorophyll *a* in photosystem II serves as the electron donor, P680, while pheophytin *a* is the electron acceptor of P680. These two photosynthetic pigments are directly involved in the conversion of light energy to chemical energy in the primary process of charge separation. Pheophytin *a* is a derivative of the Chl *a* molecule in which magnesium has been replaced by two hydrogens in the center of the porphyrin ring. Spectroscopic and electrochemical properties of pheophytin *a* are different from those of Chl *a*. For instance, pheophytin *a* absorbs more weakly in the red region and more strongly at 505 and 535 nm than Chl *a*, and the midpoint redox potential of pheophytin *a* (-0.84 V against saturated calomel electrode) is less negative than that of Chl *a*. Moreover, it is likely that pheophytin *a* is monomeric in the reaction center.

Optically transparent electrodes (OTE), modified with monolayers of photosynthetic pigments using the Langmuir–Blodgett technique, are suitable for the investigation of chemical models of photosynthesis. During photocatalytic oxidation of hydroquinone by Chl *a* or pheophytin *a* in monolayers, the standard Gibbs free energy is negative, so that light energy can be used to overcome the activation barrier. In the absence of hydroquinone, it is possible to use water as an electron donor and $NADP^+$ or methylviologen as an electron acceptor. Photoelectrochemistry has been widely used to investigate charge separation phenomena following the absorption of light on a modified electrode.

The optically transparent SnO_2 (tin oxide) electrode doped with antimony is a quasimetallic conductor. The tin oxide electrode is transparent in the visible region of the optical spectrum (Fig. 9.21, curve 1) and has a high reflectivity for infrared radiation. This type of electrode is often used as a working electrode in photoelectrochemistry due to its large potential window and its optical transparency in the visible region.

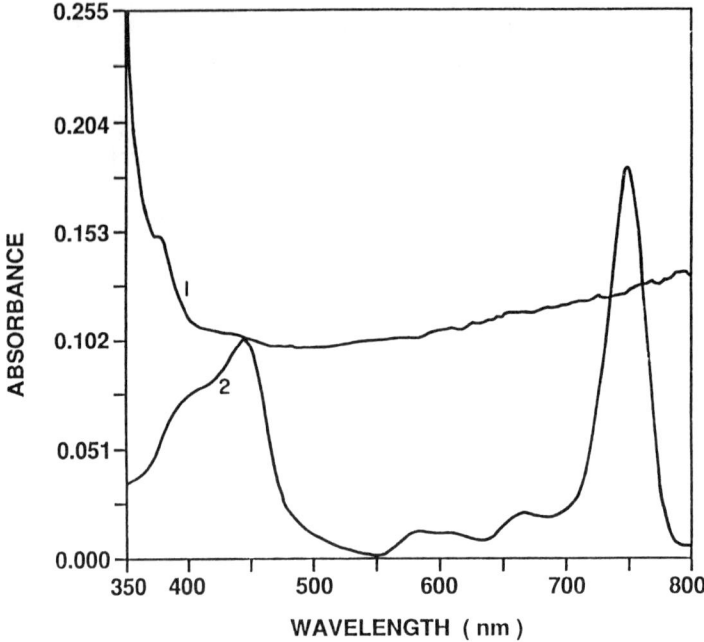

Figure 9.21. Electronic absorption spectra of: (1) SnO$_2$–OTE; (2) a hydrated chlorophyll *a* monolayer on SnO$_2$–OTE.

Illumination of Chl *a* or pheophytin *a* monolayers at SnO$_2$-OTE in the aqueous solution, in the absence of artificial donors of electrons, gave rise to a small anodic photocurrent at potentials higher than +20 mV (less than 2 nA cm^{-2}), or to a small cathodic photocurrent in the acidic pH range under negative polarization up to −100 mV (less than 4 nA cm^{-2}). If hydroquinone QH$_2$ is added to the aqueous solution, the anodic photocurrent increases by a factor of from ten to as much as a hundred times. Cathodic photocurrent can be explained by the reduction of impurities of oxygen on the order of 10^{-12} mol s^{-1}).

Figure 9.22 illustrates the dependence of photocurrent on the wavelength of light in the presence of QH$_2$. The action spectra correlated with absorption spectra of photosynthetic pigments at the SnO$_2$-OTE. The photoeffect is a function of time of illumination, the electrode potential, the pH of the solution, and the concentration of hydroquinone. If NADH is used as an electron donor rather than QH$_2$, the photocurrent is smaller than in the presence of QH$_2$, in spite of the fact that the redox potential of NADH is more negative. This phenomenon probably reflects a kinetic barrier to the formation of a one-electron intermediate during the oxidation of NADH, or catalytic oxidation of NADH on the Pt electrode.

The quantum efficiency Φ_λ for photocurrent generation at the wavelength λ was calculated from the equation

$$\Phi_\lambda = (I_\lambda/q)/n(1 - 10^{-A_\lambda}), \tag{9.7}$$

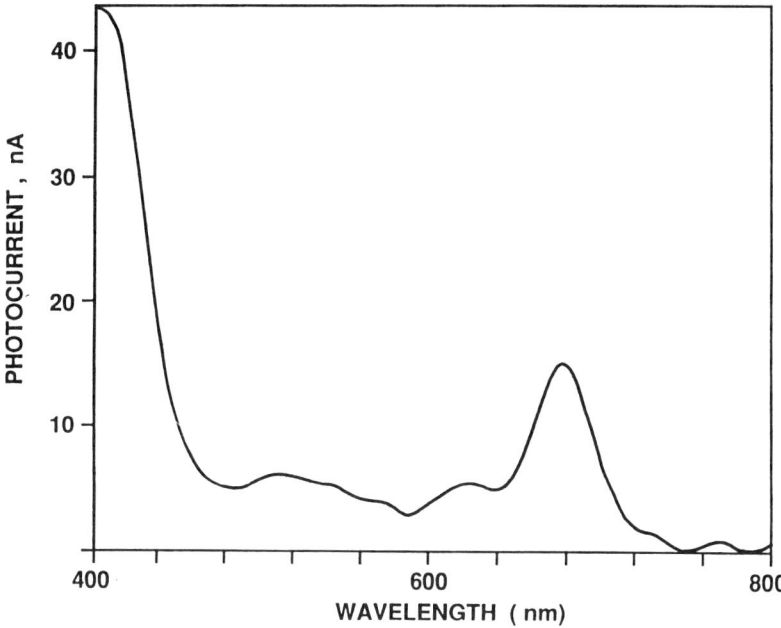

Figure 9.22. Action spectrum of the chlorophyll *a* monolayer at SnO₂–OTE. Medium: 0.1 M KCl, 0.05 M QH₂, 0.025 M KH₂PO₄, pH 3.6. Electrode potential: 100 mV.

where I_λ is the photocurrent density at wavelength λ, q is the elementary charge, n is the number of incident photons, and A_λ is the absorbance of the Chl *a* or pheophytin *a* monolayer in contact with the aqueous solution. Quantum yields for pheophytin *a* and Chl *a* monolayers in the presence of QH₂ at pH 6.9 were 0.7% and 0.4% for the red absorption peak, and 0.6% and 0.3% for the blue peak.

Pheophytin *a* in *n*-hexane possesses two well-defined electronic transitions which in the absorption spectrum are manifested by two bands in the blue region at $\lambda_{max} = 409$ nm and in the red region at $\lambda_{max} = 670$ nm (Fig. 9.23, curve 2). The comparison of the absorption spectra of pheophytin *a* in *n*-hexane and in benzene indicates that the long-wavelength band is not affected by solvent polarity. However, the short-wavelength band, going from benzene to *n*-hexane, is slightly shifted to higher energy ($\Delta\nu = 300$ cm^{-1}) due to the decrease of the static dielectric constant of the solvent from 2.28 to 1.88, correspondingly. Absorption spectra of pheophytin *a* visibly change going from a diluted *n*-hexane solution to a Langmuir–Blodgett monolayer on a quartz slide. This is shown in Fig. 9.23 (curve 1). Two broad absorption bands of pheophytin *a* in a Langmuir–Blodgett film in the blue and the red regions possess maxima at 412 and 674 nm, respectively. A more dramatic change is observed in the fluorescence spectra of pheophytin *a* in the same samples (see Fig. 9.24). A typical emission spectrum of pheophytin *a* in diluted *n*-hexane solution possesses a maximum at $\lambda_{em} = 674$ nm and a low-intensity band at $\lambda_{em} = 723$ nm.

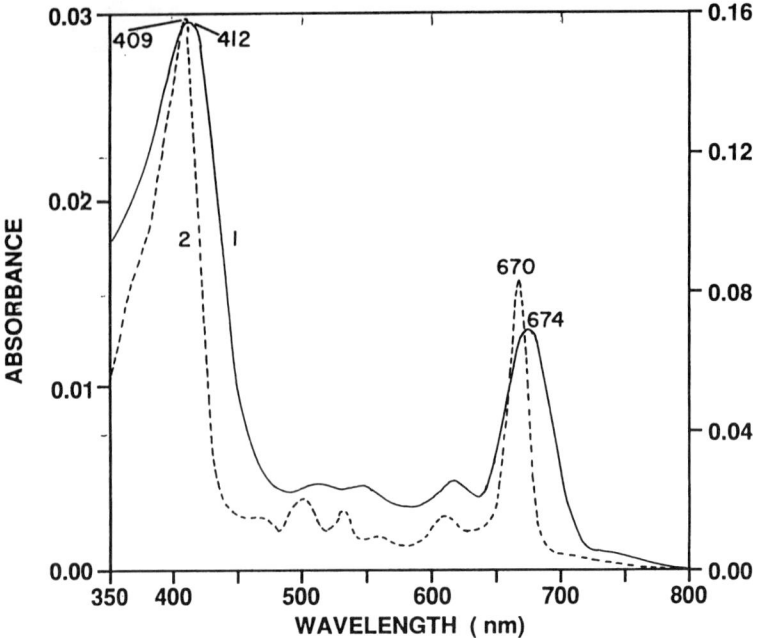

Figure 9.23. Absorption spectra of pheophytin *a* in: (1) *n*-hexane (10^{-5} M); (2) a monolayer on a quartz slide.

However, the emission spectrum of pheophytin *a* in a Langmuir–Blodgett film is shifted to lower energy displaying a maximum at λ_{em} = 695 nm and a large shoulder at λ_{em} = 750 nm. The fluorescence excitation spectra of pheophytin *a* (see Fig. 9.25) measured at the maxima of the emission spectra at 674 nm and 695 nm correspond well with the absorption spectra of pheophytin *a* in *n*-hexane and in the Langmuir–Blodgett film (see Fig. 9.23), respectively. The observed changes in the absorption and fluorescence spectra indicate that pheophytin *a* in a Langmuir–Blodgett film forms molecular aggregates, a conclusion supported by the fluorescence lifetime values presented in Table 9.2. Monoexponential decay has been measured for a 674 nm emission band of pheophytin *a* in a dilute *n*-hexane solution with τ_1 = 5.79 ± 0.02 ns. Such a lifetime value indicates a monomer emission of pheophytin *a*. Similar lifetime values have also been reported for monomer emission of Chl *a* in solution. However, the decay of the emission of pheophytin *a* in a Langmuir–Blodgett film is biexponential, with lifetimes at 695 nm of τ_1 = 3.03 ± 0.27 ns (17%) and τ_2 = 0.46 ± 0.06 ns (83%). These values can be assigned to dimeric and oligomeric forms of pheophytin *a*, respectively. Only small changes in τ_1 and τ_2 are observed when measurements were carried out at 750 nm (Table 9.2). This is consistent with the fact that the fluorescence excitation spectra of pheophytin *a* in a Langmuir–Blodgett film measured at λ_{em} = 695 nm and at λ_{em} = 750 nm are the same (Fig. 9.25).

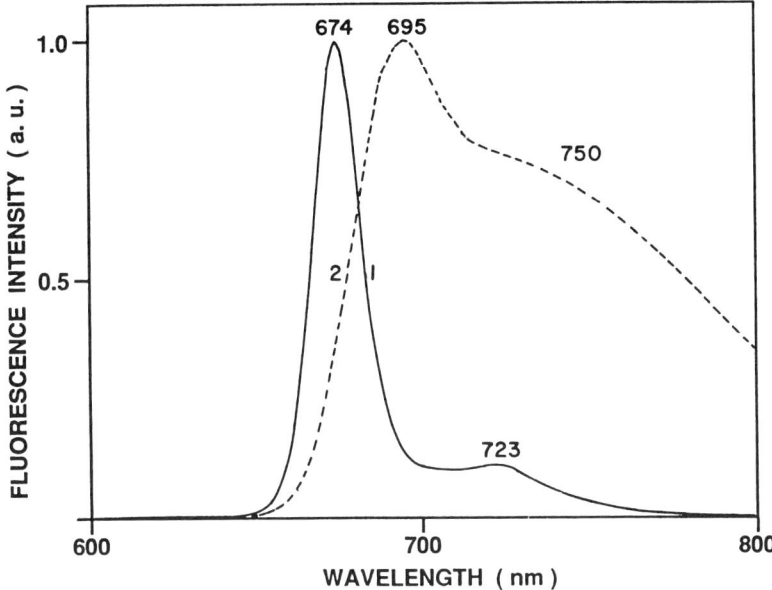

Figure 9.24. Corrected fluorescence spectra of pheophytin *a* in: (1) *n*-hexane ($c = 3 \times 10^{-7}$ M, $l = 1.0$ cm); (2) a monolayer on the quartz slide; $\lambda_{exc} = 410$ nm.

It is known from monolayer studies of pheophytin *a* at the air–water interface that at pressure above 15 mN m^{-1}, pheophytin *a* forms aggregates with coplanar stacking of the porphyrin rings. We can expect a similar arrangement of pheophytin *a* molecules in Langmuir–Blodgett films, since the deposition of a pheophytin *a* monolayer on the hydrophilic surface of a quartz slide was performed at 16 mN m^{-1}.

In summary, Chl *a* and pheophytin *a* in monolayers effectively transform light energy into electrical energy. The quantum yields of the process in the presence of hydroquinone are $0.7 \pm 0.1\%$ for pheophytin *a* and $0.4 \pm 0.1\%$ for Chl *a*. The photoeffect depends on the electrode potential, concentration of the redox components and pH of the aqueous solution.

Emulsion Photobioelectrochemistry. The mechanism of active ion transport by biological membranes is a central problem of bioelectrochemistry, and it can be investigated in model systems of microemulsions. We note that although emulsions are attractive experimental systems, earlier investigators sometimes have not defined their properties, which can lead to conflicting results. We will therefore briefly outline the relevant parameters.

A system of two immiscible liquids can form six types of structures, as shown in Fig. 9.26:

(a)   two individual immiscible liquids;

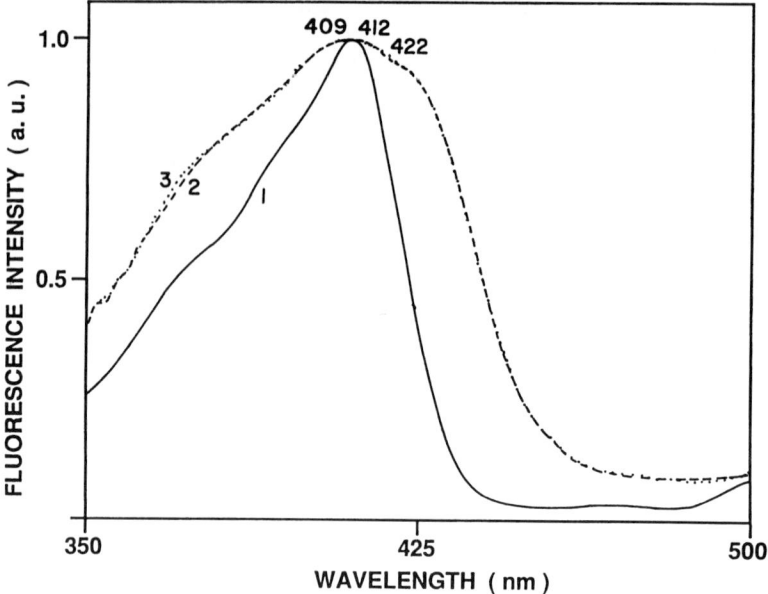

Figure 9.25. Fluorescence excitation spectra of pheophytin *a* in: (1) *n*-hexane ($c = 3 \times 10^{-7}$ M, $l = 1.0$ cm) at $\lambda_{em} = 674$ nm; (2) a monolayer on quartz slide at $\lambda_{em} = 695$ nm; (3) as for (2) at $\lambda_{em} = 750$ nm.

    (b)   water-in-oil emulsion/aqueous phase;

    (c)   oil/oil-in-water emulsion;

    (d)   two immiscible liquids and emulsion upon the interface;

    (e)   oil-in-water emulsion or water-in-oil emulsion;

    (f)   water-in-oil emulsion/oil-in-water emulsion.

When studying reactions at the oil–water interface, one of the above structures should be specified. It should also be noted that transitions from one structure to another can occur if concentrations of surface-active compounds or electrolyte are varied, or if components of the system are altered by chemical reactions.

Bacteriorhodopsin in Emulsions. Bacteriorhodopsin functions in the photo-electrochemical transformation of light energy, and is one of the most extensively studied ion pumps. Bacteriorhodopsin sheets from *Halobacterium halobium* were among the first membrane proteins to be investigated at the octane–water interface. These studies were carried out in several laboratories, with differing results and interpretation. Post *et al.* (1984) established a method to immobilize bacteriorhodop-sin sheets in octane-in-water emulsions. This considerably simplified measurements, so that proton pumping in aqueous emulsions could be carried out using conventional pH meters. The emulsion system also facilitates quantitative measurements of other

Table 9.2. Fluorescence lifetimes of pheophytin _a_ in _n_-hexane, thin films and LB films.

Conditions	Excitation $\lambda_{max}(\pm 1$ nm$)$	Emission $\lambda_{max}(\pm 1$ nm$)$	τ_1 (ns)	τ_2 (ns)
_n_-hexane $(3 \times 10^{-7}$ M$)$	409	674	5.79 ± 0.02 (100%)	—
Thin film on quartz slide adsorbed from	410	680	5.61 ± 0.02 (100%)	—
dry benzene		726	5.82 ± 0.10 (87%)	1.32 ± 0.35 (13%)
Thin film on quartz slide adsorbed from	410	675	5.16 ± 0.14 (74%)	2.40 ± 0.32 (26%)
wet benzene		722	5.16 ± 0.28 (56%)	2.45 ± 0.27 (44%)
LB films: one monolayer on quartz	412 (422)	695	3.03 ± 0.27 (17%)	0.46 ± 0.06 (83%)
slide		750	3.10 ± 0.51 (12%)	0.72 ± 0.06 (88%)

ion transport processes through the water–lipid interface, using ion selective electrodes. The large interfacial area makes it possible to obtain the products of heterogeneous reactions in macroscopic quantities, and the low dielectric constant of the non-aqueous phase can decrease the activation energy of transport reactions. Emulsion enzymology thus allows study of naturally immobilized membrane enzymes in conditions close to native states.

Earlier papers showed that bacteriorhodopsin immobilized at the water–lipid-in-octane interface is capable of phototransfer of protons from water into octane (Boguslavsky _et al._, 1975a; Boguslavsky and Volkov, 1977, 1987). The photoeffect was not inhibited by uncouplers of oxidative phosphorylation and depended upon the ionic strength of the solution. However, the molecular mechanism by which immobilized bacteriorhodopsin carried out the pumping remained unclear. Some authors claimed that bacteriorhodopsin is denatured at the octane–water interface and, in the presence of phospholipids, forms closed structures with a "third water" at the octane–water interface. Upon illumination the bacteriorhodopsin was thought to pump protons from the aqueous phase into the "third water" rather than transferring protons directly from water into octane. In order to measure the photoeffect Volta-potential measurements were used. This method yielded information on the variation of the potential difference and the charge upon the interface but did not specify whether transfer of charges from one phase into another occurred, since both the orientation of dipoles at the interface and transfer of charges could affect the value of the Volta potential.

The capacity of bacteriorhodopsin to phototransfer protons through the water–lipid interface in octane was investigated by Gugeshashvili _et al._ (1991), who analyzed the mechanism of fusion of purple membranes with monolayers of lipids

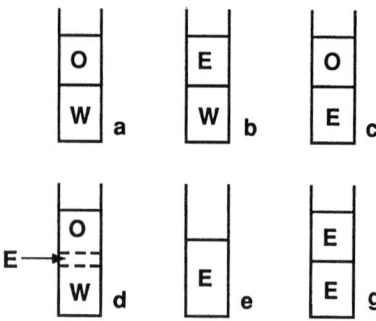

Figure 9.26. Various types of structure arising in the oil–water system.

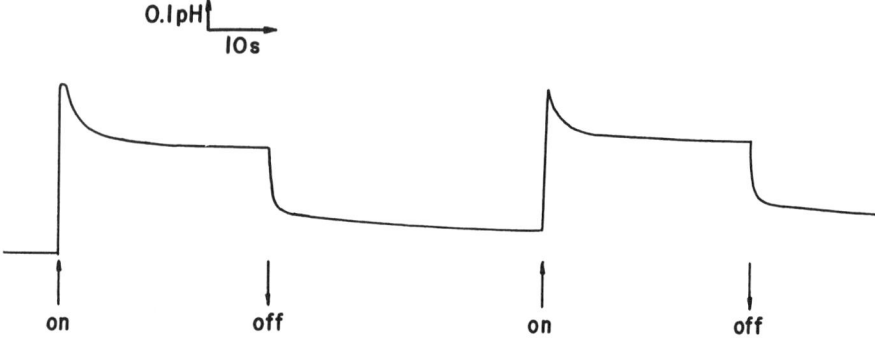

Figure 9.27. The reversible light-dependent pH shift in an octane-in-water emulsion containing bacteriorhodopsin sheets; the pH of the emulsion in dark condition is 6.0. The aqueous phase contained 2 M NaNO₃.

at the octane–water interface and proposed a theoretical model of the photoeffect in the emulsion. The measurements were carried out in octane-in-water emulsions, the weight ratio being 1 (octane):8 (water). 2 M NaNO$_3$ or 2 M NaCl was initially dissolved in the water phase, and sonication resulted in octane-in-water emulsions covering the layer of aqueous electrolyte solution.

Upon illumination of the emulsions with white light, the pH increased in the aqueous phase of the emulsion (Fig. 9.27). If one of the components (either bacteriorhodopsin or inorganic salt) was excluded from the emulsion, the pH shift was not observed. Upon illumination the photoresponse rose very quickly and attained a steady state in approximately 2–3 s. In the dark, the pH returned to the initial level more slowly. The photoeffect was reversible and could be repeated many times. Introducing pentachlorophenol (PCP) or tetrachloro-2-trifluoromethylben-zimi-dazole (TTFB) into the emulsion did not inhibit proton transport.

Sonicating the system without asolectin leads to denaturation of bacteriorhodop-sin. Substitution of a more polar solvent—1,2-dichlorethane—for octane also caused

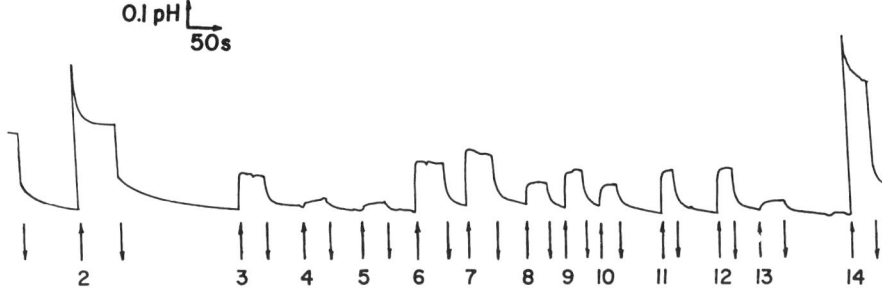

Figure 9.28. The reversible light-dependent alkalization of an aqueous phase containing 2 M NaNO$_3$ in the presence of an octane-in-water emulsion comprising bacteriorhodopsin sheets. The wavelengths of incident light are: 1, 2, 14—white light; 3—593 nm; 4—429 nm; 5—591 nm; 6—563 nm; 7—572 nm; 8—511 nm; 9—542 nm; 10—480 nm; 11—583 nm; 12—593 nm; 13—660 nm.

partial denaturation. However, dichlorethane is the most extensively used non-aqueous solvent in electrochemical studies of two immiscible electrolyte solutions. Therefore classic electrochemical methods such as cyclic voltametry could be used to quantitatively estimate the phototransfer of protons through the dichloroethane interface.

Figure 9.28 shows the reversible light-induced pH shift in the aqueous phase during illumination with light of different wavelengths. The magnitude of the photoeffect clearly depends upon the wavelength of the incident light. The absorption spectra of bacteriorhodopsin sheets in water and the action spectrum are compared in Fig. 9.29, which shows that the action spectrum corresponds to the absorption spectrum. A mathematical model for the photoeffect in emulsions was developed by Markin and Portnov (Gugeshashvili *et al.*, 1991). The reversible light-induced pH shift in emulsions depends on the activity of the proton pump of bacteriorhodopsin (br) in transferring protons through the octane–water interface. However, the standard free energy of proton resolvation is so high that the concentration of protons in octane is negligibly small. If one assumes that in octane protons are present only as protonated acid molecules HA (A$^-$ = anion, for example, nitrate), the simplest diagram of proton transport through the water–octane interface (Fig. 9.30) undergoes the following stages:

1. active transfer of proton through the water–octane interface by bacteriorhodopsin;
2. formation of an HA complex in octane from the entering proton and resident anion A$^-$;
3. diffusion of HA within the octane droplet;
4. diffusion of HA to the aqueous phase and rapid dissociation into H$^+$ and A$^-$.

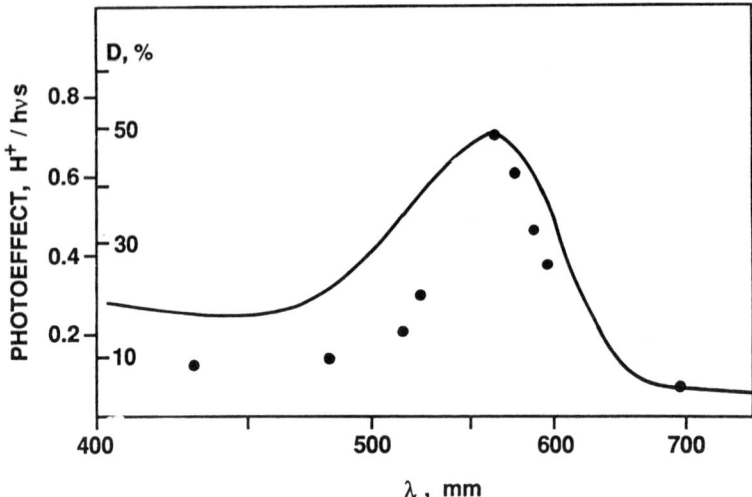

Figure 9.29. The absorption spectrum of bacteriorhodopsin sheets in water and the action spectrum of the reaction. The photoeffect was calculated with respect to the energy of incident quanta. The conditions are the same as those given in Fig. 9.27.

The emulsion in Fig. 9.30 consists of octane (volume V_o) and water (volume V_w). We can now introduce the concentrations c_H^w, c_{OH}^w, c_A^w, c_C^w of ions H^+, OH^-, A^-, C^+, respectively (C^+ being the salt cation dissolved in water). The values of concentrations are close to c_A^w, $c_A^w \sim 1$ M, c_H^w, $c_{OH}^w < 1$ μM.

We will also denote the concentrations of neutral complexes HA and CA in octane as c_{HA}^o and c_{CA}^o and concentrations of ions A^- and C^+ in octane as c_A^o and c_C^o. The latter may be assumed to be evenly distributed over the droplet volume since the thickness of the Debye layer (δ) in octane is

$$\delta \approx \sqrt{\frac{\varepsilon\varepsilon_0 kT}{2ec_{A,C}^o}} \approx 1.0\text{–}2.0 \text{ nm} \ll R, \tag{9.8}$$

where ε is the relative dielectric constant of octane, ε_0 the electric constant equal to 8.85×10^{-11} F m^{-1}, k the Boltzmann constant, and e the elementary charge.

Creation of the emulsion does not significantly affect the pH of the solution. Because the phase volumes are approximately equal, we may write

$$c_{HA}^{o(dark)} \ll c_H^w, \tag{9.9}$$

otherwise the concentration of protons in water would have been decreased, resulting in a pH increase. The indices "dark" and "light" will designate parameters differing in the dark and light conditions. Illumination shifts the pH by a small amount, no more than 0.5–0.7 pH units, assuming the $c^{o(light)}$ and c_H^w are of the same order of magnitude.

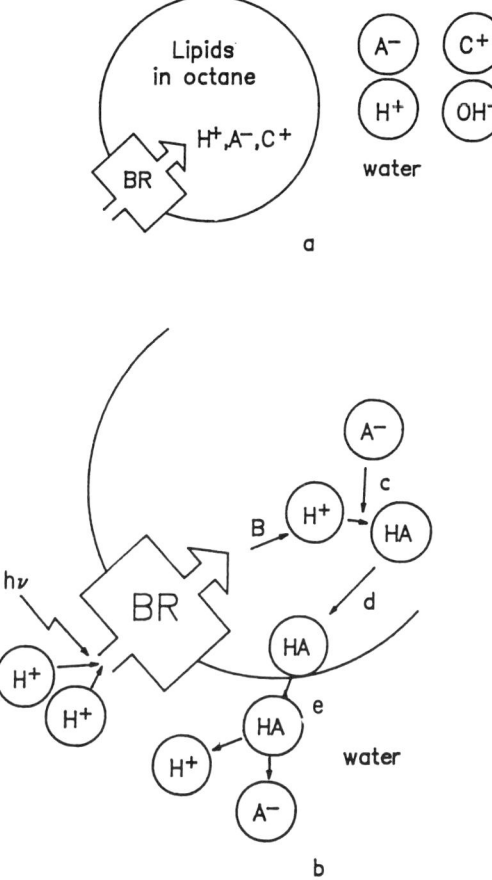

Figure 9.30. A schematic diagram of the proton transport in the water–octane system: (*a*) octane droplet with adsorbed bacteriorhodopsin in water; (*b*) the active transport of proton through bacteriorhodopsin; (*c*) creation of HA complex in octane; (*d*) diffusion of HA inside the octane droplet; (*e*) emission of HA into the aqueous solution and decomposition into H^+ and A^-.

If one neglects the pH dependence of the charge of the lipid adsorbed at the interface, the equilibrium solute concentrations in the system and potential differences depend mainly upon the ions C^+ and A^-, due to the absence of protons in octane at neutral pH ranges. The concentrations of ions in water and in octane are described by the following relationships

$$c_A^o = c_A^w \exp\left[\frac{F}{RT}(\Delta_w^o \phi - \Delta_w^o \phi_A^o)\right], \qquad c_C^o = c_C^w \exp\left[\frac{F}{RT}(\Delta_w^o \phi - \Delta_w^o \phi_A^o)\right], \quad (9.10)$$

where $\Delta_w^o \phi = \phi^o - \phi^w$ is the Galvani potential difference, and $\Delta_w^o \phi_A^o$ and $\Delta_w^o \phi_C^o$ are the standard distribution potentials of the ions between octane and water (Markin and Volkov, 1989f).

The electroneutrality condition can be written as follows:

$$c_A^w = c_C^w = c^w, \qquad c_A^o = c_C^o = c^o. \tag{9.11}$$

It is known that:

$$\Delta_w^o \phi = \frac{\Delta_w^o \phi_A^0 + \Delta_w^o \phi_C^0}{2} \tag{9.12}$$

and

$$c^o = c^w \exp\left[\frac{F}{RT}(\Delta_w^o \phi_C^0 - \Delta_w^o \phi_A^0)\right]. \tag{9.13}$$

It should be noted that $\Delta_w^o \phi_C^0 < 0$ and $\Delta_w^o \phi_A^0 > 0$, since the equation for the resolution energy expressed in the standard potential of distribution of ions $\Delta_w^o \phi_C^0$ is

$$\Delta_w^o G_i^o = -\frac{ze\Delta_w^o \phi_i^0}{kT}. \tag{9.14}$$

The resolution energy of ions decreases in absolute value as ionic radius increases. Therefore one can expect that an increase of cation radius would increase the standard distribution potential, whereas an increase of anion radius would decrease $\Delta_w^o \phi_i^0$.

The transfer of protons from water to octane occurs in two stages: active transport through bR, followed by association of the proton with anion A^-. Until the proton associates with A^-, the bacteriorhodopsin cannot transfer another proton. Therefore, the active transport caused by illumination must take place in bacteriorhodopsin molecules that have not previously transferred protons. The flux can be described by the relationship

$$j_{bR} = \eta I N_{bR}^{free} c_H^w, \tag{9.15}$$

where I is the light intensity, N_{bR}^{free} the number of vacant bacteriorhodopsin molecules and η the proportionality coefficient. The activity of the proton pump is controlled by interfacial potential, and is a monotonically decreasing function of $\Delta_w^o \phi$. Because the dependence of η on $\Delta_w^o \phi$ is unknown, one can initially assume a linear voltametric characteristic for bacteriorhodopsin.

Proton flux due to HA formation can be described by the following expression

$$j_{HA} = q(N_{bR} - N_{bR}^{free})c_A^o - r_{HA}^o, \tag{9.16}$$

where $N_{bR} - N_{bR}^{free}$ is the number of bacteriorhodopsin molecules retaining a proton at the interface with octane; N_{bR} is the total number of bacteriorhodopsin molecules and q and r are the reaction rate constants. The first term in equation (9.16) gives the rate of proton binding with anion A^- upon bacteriorhodopsin and the second term corresponds to the dissociation of HA and the transport of H^+ back to bacteriorhodopsin.

If we assume that the protons in octane are mainly present as HA, rather than free in solution, we can write

$$j_{bR} = j_{HA}.$$ (9.17)

Proton flux results in a highly nonequilibrium situation at the interface so that the second term in equation (9.16) can be neglected:

$$r_{HA}^o = 0.$$ (9.18)

The proton flux through the interface free from bacteriorhodopsin is equal to

$$j_{free} = \xi_H (K_{HA} c_H^w c_A^w - c_{HA}),$$ (9.19)

where K_{HA} is the equilibrium distribution coefficient between octane and water and ξ_H the coefficient proportional to the rate constant and to the free interfacial area. As we noted earlier, the pH does not vary in the dark emulsion, so we will assume that

$$K_{HA} = 0.$$ (9.20)

An increase in the number of protons in octane depends upon the entire flux:

$$j_H = j_{bR} + j_{free} = j_{HA} + j_{free}.$$ (9.21)

Solving equation (9.21), and taking into account the relationships (9.18) and (9.19), one obtains

$$j_H = N_{bR} \bigg/ \left(\frac{1}{qc^o} + \frac{1}{I\eta c_H^w} \right) - \xi c_{HA}^o.$$ (9.22)

The number of protons present as HA in the octane phase can be determined assuming the concentration to be evenly distributed:

$$N_H^o = V_o c_{HA}^o.$$ (9.23)

The variation of the proton number in octane is controlled by the flux j:

$$\frac{dN_H^o}{dT} = j_H.$$ (9.24)

The same quantity of protons leaves the aqueous phase. In this process some water molecules dissociate to keep the ionic product constant:

$$c_H^w c_{OH}^w = K_w = 10^{-14} \, M^2 \tag{9.25}$$

or

$$N_H^w N_{OH}^w = V_w^2 K_w. \tag{9.26}$$

Thus,

$$\frac{dN_H^w}{dt} - \frac{dN_{OH}^w}{dt} = -j_H. \tag{9.27}$$

Summing equations (9.24) and (9.17) we get the proton balance equation

$$N_H^o + N_H^w - N_{OH}^w = \text{constant} = N_H. \tag{9.28}$$

The system of equations (9.26), (9.27), (9.28), and (9.22) completely describes the behavior of the system. Equation (9.27) yields:

$$N_H^o = N_H - N_H^w + N_{OH}^w,$$

$$c_{HA}^o = \frac{N_H - N_H^w + V_w^2 K_w / N_H^w}{V_o}. \tag{9.29}$$

Substituting equation (9.28) into equation (9.24), one obtains an equation for variation of proton concentration in the aqueous phase:

$$\left(1 + \frac{K_w V_w^2}{(N_H^w)^2}\right) \frac{dN_H^w}{dt} = \frac{\xi_H}{V_o}\left(N_H + \frac{K_w V_w^2}{N_H^w} - N_H^w\right) - \frac{N_{bR}}{\dfrac{1}{qc^o} + \dfrac{V_w}{I\eta N_H^w}}. \tag{9.30}$$

Although equation (9.30) is not very complex, it does contain a number of physical parameters that should be defined. The factor $1 + [K_w V_w^2/(N_H^w)^2]$ reflects the buffer properties of the system, since transport of protons into the aqueous phase only partially increases the number of free protons. The first term on the right-hand side of equation (9.30) is the rate at which HA leaves an octane droplet. The second term on the right-hand side of equation (9.30) describes proton influx into the droplet through the bacteriorhodopsin. The term $V_w/I\eta N_H^w$ in the denominator describes the inhibition of the transport at the stage of active transfer, whereas $1/qc^o$ describes the inhibition at the stage of formation of the HA complex.

A Steady State. When net proton flux approaches zero we obtain

$$\frac{\xi_H}{V_o}\left(N_H + \frac{K_w V_w^2}{N_H^w} - N_H^w\right) = \frac{N_{bR}}{\dfrac{1}{qc^o} + \dfrac{V_w}{I\eta N_H^w}} \tag{9.31}$$

yielding the dimensionless concentration of protons (x):

$$x = \frac{N_H^w}{V_w\sqrt{K_w}}, \qquad \log x = 7 - \text{pH}. \tag{9.32}$$

The equation for the stationary state in dimensionless variables can then be written:

$$\frac{1}{\bar{c}^w} = \frac{1}{x_0 - \dfrac{1}{x} - \left(x - \dfrac{1}{x}\right)} - \frac{1}{jx}. \tag{9.33}$$

In this equation,

$$\bar{c}^w = \frac{V_o}{V_w}\frac{N_{bR}}{\xi_H}\frac{q}{\sqrt{K_w}}c^w \exp\left[\frac{F}{2RT}(\Delta_w^o\phi_C - \Delta_w^o\phi_A)\right], \tag{9.34}$$

$$j = \frac{I\eta N_{bR}}{\xi}\frac{V_o}{V_w}, \tag{9.35}$$

$$\log x_0 = 7 - \text{pH}_0, \tag{9.36}$$

where pH_0 is the equilibrium value of pH in the dark.

Let us consider two limiting cases:

(a) A high dimensionless concentration of salt

$$\bar{c}^w \gg 1. \tag{9.37}$$

This inequality corresponds to the non-limiting stage of creation of the HA complex. Solving equation (9.33) one obtains the expression for ΔpH:

$$\Delta\text{pH} = \log\frac{2(j+1)x_0}{x_0 - \dfrac{1}{x_0} + \sqrt{\left(x_0 - \dfrac{1}{x_0}\right)^2 + 4(j+1)}}. \tag{9.38}$$

Equation (9.38) shows that the dependence of ΔpH on the light intensity is non-linear (Fig. 9.31). The origin of the non-linearity is simply that the concentrations of protons and hydroxide ions vary non-linearly.

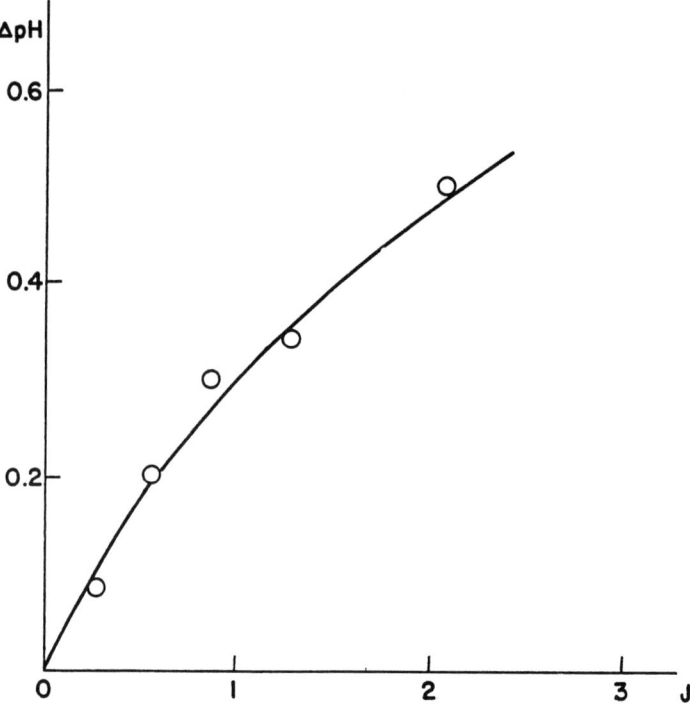

Figure 9.31. Theoretical dependence of the light-induced pH shift as a function of light intensity in the approximation of a high concentration of salt (solid line). The dark value of pH is 6.1. The coefficient $J/I = V_0 V_{bR} \eta / V_w \xi_H$, determining the relationship between the light intensity I and the dimensionless intensity J is equal to $0.14 \text{ cm}^2 \text{ mW}^{-1}$. The points correspond to experimental data of the dependence of light-induced pH increase on the intensity of the incident light. The medium is the same as that described in Fig. 9.27.

Figure 9.31 depicts the dependence of $\Delta pH(I)$ in approximation (9.33) at $pH = 6.1$. Comparison with experimental points yields an estimation of the parameters (see Table 9.3):

$$\frac{\eta N_{bR}}{\xi_H} \frac{V_0}{V_w} \quad \text{and} \quad \frac{V_0}{V_w} \frac{N_{bR}}{\xi_H} \frac{q}{\sqrt{K_w}} \exp\left[\frac{F}{2RT} (\Delta_w^o \phi_{Na} - \Delta_w^o \phi_{NO_3}) \right]. \quad (9.39)$$

(b) A high light intensity corresponds to the rate-limiting step at which HA complexes are formed and the limiting pH shift is attained:

$$\Delta pH_{max} = \log\left(\frac{\left(\dfrac{1}{x_0} - x_0 + \overline{c}^w + \sqrt{\left(\dfrac{1}{x_0} - x_0 + \overline{c}^w\right)^2 + 4}\right)}{2} \right). \quad (9.40)$$

Table 9.3. Parameters of the system calculated for various salts at $I = 70$ mW/cm^{-2} (Gugeshashvili *et al.*, 1991).

Parameter	NaNO$_3$	NaCl	KNO$_3$
$\dfrac{\eta N_{bR}}{\xi_H}\dfrac{V_o}{V_w}I$	10	4	0.95
$\dfrac{V_o}{V_w}\dfrac{N_{bR}}{\xi_H}\dfrac{q}{\sqrt{K_w}}\exp\left[\dfrac{F}{2RT}(\Delta^o_w\phi_{Na} - \Delta^o_w\phi_{NO_3})\right]$	0.34	0.14	3.4

At higher concentration the dependence of $\Delta pH(\bar{c}^w)$ is purely logarithmic

$$\Delta pH = \log \bar{c}^w. \tag{9.41}$$

The dependence of ΔpH on the salt concentration at fixed intensity is a transition from the limiting case at low concentrations to the limiting case at high concentrations. Figure 9.32 shows the experimental and theoretical curves of $\Delta pH(c^w)$, and optimal values of parameters are given in Table 9.3. It is clear that the pH shifts are a function of ionic radius, with large anions and small cations producing the greatest effect.

Non-Stationary State. The approximate analytical solution of equation (9.30) can be obtained at small pH values (see equation (9.40)) and high concentrations of salts. In this case equation (9.30) can be rewritten as follows:

$$\frac{dN^w_H}{dt} = \frac{\xi_H}{V_o}N^w_{H(dark)} - \left(\frac{\xi_H}{V_o} + \frac{I\eta N_{bR}}{V_w}\right)N^w_H, \tag{9.42}$$

yielding the pH shift upon illumination:

$$pH = pH_{dark} - \log\left(\frac{1 + \dfrac{V_o}{V_w}\dfrac{I\eta N_{bR}}{\xi_H}\exp\left[-\left(\dfrac{\xi_H}{V_o} + \dfrac{I\eta N_{bR}}{V_w}\right)(t - t_0)\right]}{1 + \dfrac{V_o}{V_w}\dfrac{I\eta N_{bR}}{\xi_H}}\right). \tag{9.43}$$

In the dark

$$pH = pH_{dark} - \log\left(1 - \frac{V_o}{V_w}\frac{I\eta N_{bR}}{\xi_H}\frac{\exp\left[-\dfrac{\xi_H}{V_o}(t - t_0)\right]}{1 + \dfrac{V_o}{V_w}\dfrac{I\eta N_{bR}}{\xi_H}}\right), \tag{9.44}$$

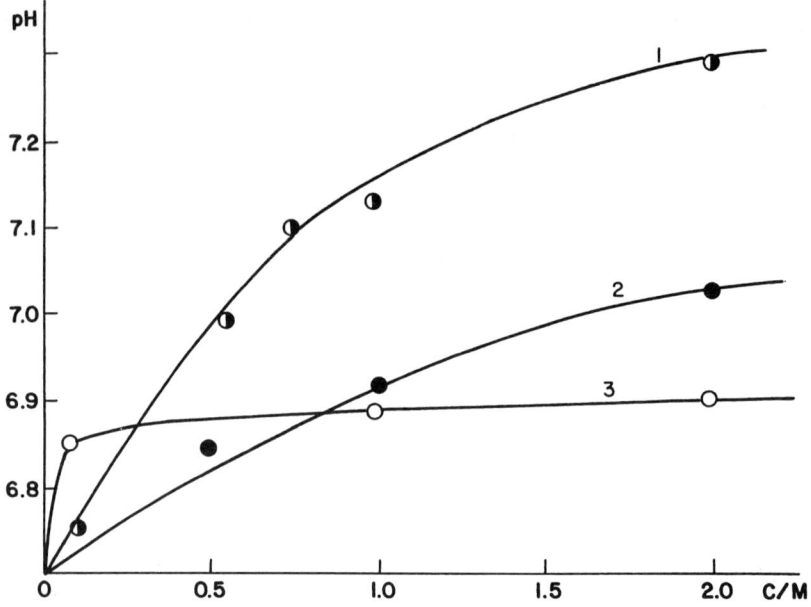

Figure 9.32. The light-induced pH increase plotted as a function of the salt concentration at constant intensity $I = 70$ mW cm^{-2}: (1) sodium nitrate; (2) sodium chloride; (3) potassium nitrate. The curves are plotted using equation (9.40). Experimental points are taken from (Post *et al.*, 1984).

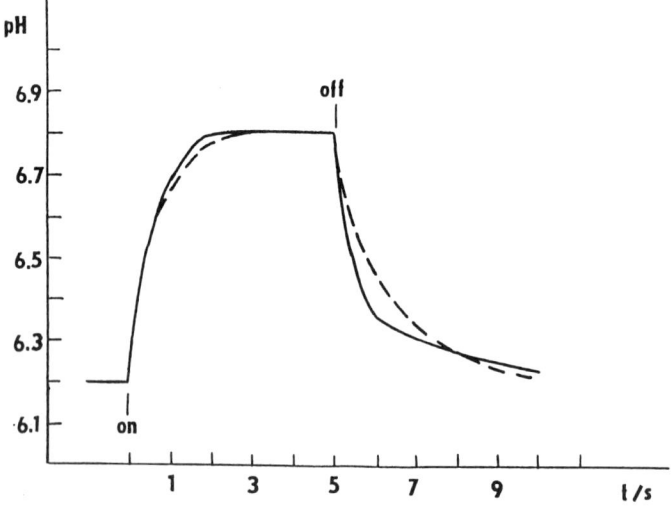

Figure 9.33. Theoretical (dashed line) and experimental (solid line) dependence of pH on time according to equations (9.43) and (9.44).

where pH_{dark} is the initial value of pH in the aqueous phase of the emulsion prior to illumination. These dependences are given in Fig. 9.33 as a broken curve. The time needed to attain the stationary state upon illumination is τ_l, and that in dark is τ_d. It is significant that these characteristic times are different. They can be expressed through the system's parameters as follows

$$\tau_l = \frac{V_o}{\xi_H}\left(1 + \frac{I\eta N_{bR}V_o}{V_w \xi_H}\right)^{-1}, \tag{9.45}$$

$$\tau_d = \frac{V_o}{\xi_H}. \tag{9.46}$$

The theoretical and experimental curves in Fig. 9.33 are close enough. The small difference in τ_l and τ_d can be explained if an additional proton transport channel is opened upon illumination.

E

MEMBRANES

10

MEMBRANE THERMODYNAMICS AND ELECTROSTATICS

10.1. STRUCTURE AND PROPERTIES OF BIOLOGICAL MEMBRANES

Cell membranes serve several important roles: 1) they separate individual cells from the external environment; 2) within eukaryotic cells they define the compartments of intracellular organelles such as mitochondria, the nucleus, Golgi, endoplasmic reticulum. lyposomes, and vacuoles; 3) they allow concentration and potential gradients to be maintained by the plasma membrane and intracellular compartments; and 4) cell membranes incorporate specialized proteins that are used to transport ions and molecules through the cell membrane and serve as recognition sites (receptors) for signaling molecules.

A fundamental constituent of the cell membrane is the lipid bilayer which composes about half of the mass of typical membranes. All of the lipids in a bilayer are amphiphilic, having both hydrophobic and hydrophilic moieties. A common example of a membrane lipid is a phospholipid, which has a phosphate-containing hydrophilic head group and two hydrophobic tails. In an aqueous medium phospholipids self-assemble into bilayers with their hydrophobic tails oriented to the center and hydrophilic head groups forming a charged layer on both surfaces. The lipid composition can vary greatly since over a hundred lipid species can be incorporated into a cell membrane. The distribution of lipids within each monolayer of the bilayer can be asymmetric, which is necessary for the functional roles that different membranes serve in the cell.

The lipid bilayer is a two-dimensional fluid and the fluidity of a given bilayer depends upon the composition of the bilayer and the temperature. Because of the fluid nature of the bilayer, individual lipid molecules can laterally change places with other lipid molecules up to 10^7 times a second. In contrast, there is much less mobility across the bilayer (flip-flop) which typically has half-times of hours to days. Individual lipid molecules can also rotate along their long axis.

The lipid bilayer of plasma membranes is an effective barrier to the diffusion of ions and polar molecules. This property is very important in that it allows a cell to maintain concentration gradients across membranes that define intracellular and extracellular compartments. In order to transport molecules across membranes specialized processes have evolved which are mediated by membrane proteins. Such transport proteins fall into three major classes; carrier proteins (permeases, carriers, or transporters), channel proteins and active transport enzymes. Carrier proteins bind specific molecules and through conformational change transport them across the plasma membrane. These carrier proteins act like enzymes, except that they do not covalently modify the solute. They have a maximal transport rate and a characteristic solute binding constant. Carrier proteins can transport one species of solute molecule vectorially (uniport) or couple it to transport of a second species (coupled transport).

Ions in solution are hydrophilic and strongly attract polar water molecules. Ions therefore are surrounded by electrostatically bound water, called the water of hydration. In order for the ion to move through the non-polar region of a lipid bilayer it must lose its hydration shell which requires considerable energy. Cell membranes therefore have channels incorporated into the lipid bilayer to facilitate the diffusion of ions through hydrophilic pores. Ionic solutes permeate the pore by passive diffusion, the direction being determined by the transmembrane concentration gradient of the solute. In the case of ions, the transmembrane electrical potential is also a factor and net transfer of solute is determined by both its concentration and electrical gradient (electrochemical gradient).

These channels have evolved to be fairly ion selective, which is thought to be achieved by the narrow region of the ion channel pore acting as a molecular sieve. The channel pore size is such that a fully hydrated ion cannot pass through the pore. However, if the ion loses its waters of hydration the effective radius will be reduced and it will be able to move through the pore down an electrochemical gradient. Apparently the ion loses its waters of hydration when it is energetically favorable to form a weak chemical bond with ionic amino acids in the wall of the channel. It follows that channels regulate their ion selectivity by placing the appropriate amino acid within their pore which will interact in an energetically favorable way with a particular ion. Certain membrane-bound enzymes are also capable of transporting ions and molecules against an electrochemical gradient by using metabolic energy, a process called active transport. Such transport processes make it possible for cells to store potential energy in the form of electrochemical gradients that can later be used to drive other forms of transport, synthesize ATP, and produce membrane potentials in excitable cells.

Because the lipid bilayer acts as a diffusion barrier to ions, a cell can maintain a separation of charge across its membrane. This charge separation results in an electrical potential difference across the cell (resting membrane potential), which in a typical excitable cell ranges from 60 to 70 mV with the inside being negative. A reduction in charge separation is called depolarization and an increase is called hyperpolarization. The resting membrane potential is determined by the distribution and concentration of ions across the membrane and by the permeability of the resting cell membrane to the various ion species. In electrically excitable cells, voltage-gated ion channels exist which open sequentially and result in the propagation of current along the neuronal cell membrane.

Planar Bilayer Lipid Membranes (BLMs). In order to understand the biophysics of the cell membrane it has been necessary to restrict many basic studies to a planar bilayer lipid membrane (BLM) model system. The BLM offers the advantages of having a known lipid composition as well as good reproducibility for experimental purposes. In addition, functional membrane proteins can be selectively integrated into a BLM and studied. In the following section we will examine such model membranes in detail.

The first successful attempts to prepare BLMs were achieved when Mueller *et al.* (1962) found that a suspension of oxidized ox brain phospholipids in aqueous solution spontaneously formed a bimolecular film. The details of the experimental protocol proved amazingly simple. A certain amount of phospholipid was dissolved in a liquid hydrocarbon such as *n*-decane. A Teflon diaphragm having a small aperture was placed in a vessel filled with water and a drop of lipid solution was painted across the aperture. The drop first converted into a thick film showing interference patterns, and then thinned and became "black" when it no longer reflected light. One of the first experiments was to determine the thickness of the film. The simplest and most widespread method consisted of measuring its electrical capacitance. If one knows the dielectric constant of the lipid (which lies in the range of 2 to 3), one can calculate the thickness directly from the formula for a planar capacitor. While varying slightly as a function of the lipid composition, the thickness of a BLM is approximately 50 Å, about twice the length of the lipid molecule hydrocarbon chains. Optical measurements give comparable values.

We can represent a lipid bilayer as shown in Fig. 10.1. What is the state and arrangement of the solvent molecules (i.e., *n*-decane) in the membrane? In principle, they can lie either between the two lipid layers, or within each layer between the lipid molecules. Investigations of BLMs made of phosphatidylcholine showed that one lipid molecule in the film takes up an area of about 75 Å, whereas the minimum possible area it can occupy is 58 Å. It follows that a BLM is a thin film of hydrocarbons stabilized in the aqueous phase by the lipid molecules, which themselves contribute substantially to the volume of this film. Thermodynamic analysis of these films suggests that molecules forming such structures must possess a high energy of adsorption, both on oil and on water (Haydon and Taylor, 1963). These conditions are satisfied by lipids which contain long hydrocarbon chains and short, strongly polar groups. Another requirement is that the polar head cross-section of lipids should not differ greatly from their hydrocarbon chain cross-section. A unique feature of BLMs is that they have molecular dimensions in one direction, but macroscopic dimensions in another. Usually these distances differ by a factor of 10^6.

If suitable functional groups are integrated into BLMs, it is possible to ascribe certain functions to this membrane. For instance, by using alamethicin together with a surface-active protein, Mueller and Rudin (1968) were the first to reconstruct the

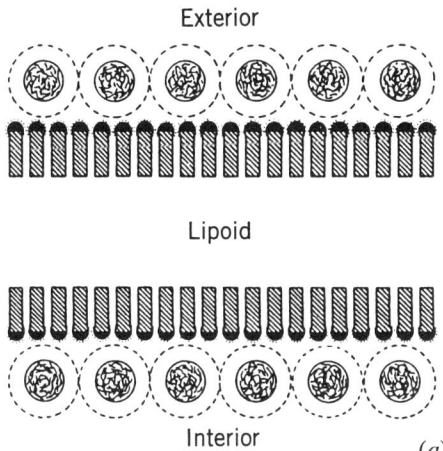

Exterior

Lipoid

Interior

(a)

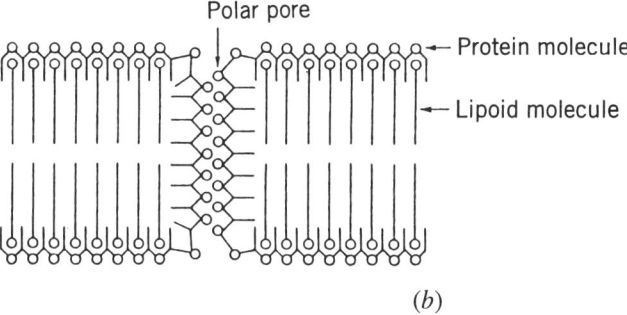

Polar pore

Protein molecule

Lipoid molecule

(b)

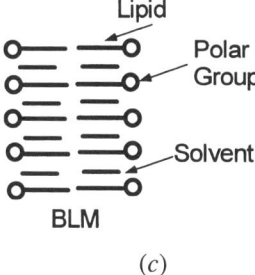

Lipid

Polar
Group

Solvent

BLM

(c)

Figure 10.1. Schematic structures of a membrane: (a) The original proposal of Danielli and Davson (1935) showing a lipoid sheet of undefined thickness; (b) Danielli's modification (1954) showing a lipid bilayer and hydrophilic pores; (c) BLM.

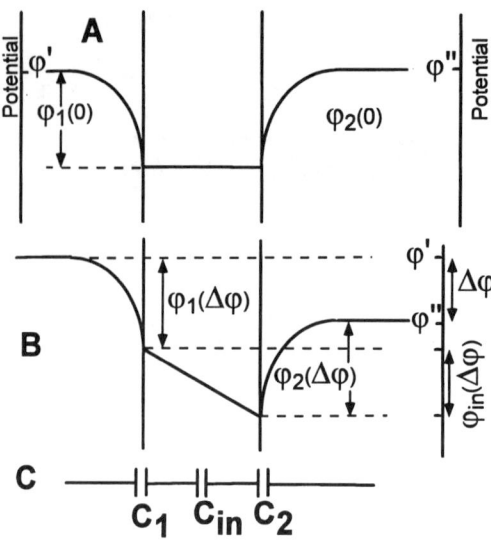

Figure 10.2. The potential distribution in a symmetrical membrane: (*a*) no applied transmembrane potential; (*b*) an external transmembrane potential $\Delta\varphi = \varphi' - \varphi''$ is applied to the membrane; (*c*) a three-capacitor equivalent membrane circuit.

phenomenon of electric excitability in an artificial lipid membrane. Subsequently, other investigators have been able to duplicate many biological membrane phenomena in BLM model systems.

10.2. MEMBRANE ELECTROSTATICS

Potential distribution. Electrostatics plays a crucial role in the behavior of biological membranes: it determines the distribution and transport properties of ions within or near membranes, influences the conformation and function of membrane proteins, and governs interactions between molecules in membranes. Electrical potential in membranes is generated predominantly by electrical charges distributed near membrane surfaces, since the concentration of charges is extremely low within bilayers. Biological membranes carry a surface charge of the order of $0.05\,\mathrm{C\,m^{-2}}$ while model lipid membranes can have a charge up to $0.4\,\mathrm{C\,m^{-2}}$. The sign of this surface charge is usually negative. In addition to surface charge, lipid head groups carry electrical dipoles which contribute to the potential drop between the bulk solution and the membrane interior.

Figure 10.2(*a*) illustrates the distribution of electrical potential in a membrane with symmetrical surfaces in the absence of external potential. The electrical potential of the left bulk solution φ' is equal to the electrical potential of the right bulk solution φ'', and therefore $\Delta\varphi = \varphi' - \varphi'' = 0$. The average potential profile of the electrical potential inside a membrane is generally linear and in this case it is horizontal as well. If membrane surfaces are charged and carry electrical dipoles the potential drop φ_1 between the left bulk solution and the membrane interior is comprised of the electrical double layer potential $\varphi_{1,DL}$ (usually described by the Gouy–Chapman theory), and the dipole potential $\varphi_{1,DP}$:

$$\varphi_1 = \varphi_{1,DL} + \varphi_{1,DP}. \tag{10.1}$$

There is a similar contribution to the potential drop φ_2 at the right side of the membrane. When discussing membrane potential it is important to define the sign of the potential drop. The convention we will use is to define the electrical potential difference as the potential drop between two points, from the left to the right. In this convention, the electrical potential drop φ_1 in Fig. 10.2(a) is positive, while φ_2 is negative.

If an external potential $\Delta\varphi = \varphi' - \varphi''$ is applied to the membrane (Fig. 10.2(b)), then the initial potential drops $\varphi_1(0)$ and $\varphi_2(0)$ change to $\varphi_1(\Delta\varphi)$ and $\varphi_2(\Delta\varphi)$. These changes can be described by the three-capacitor equivalent circuit presented in Fig. 10.2(c) where C_1 is the capacitance of the electrical double layer at the left surface, C_2 is the capacitance at the right surface and C_{in} is the interior capacitance. The total membrane capacitance is

$$C = \left(\frac{1}{C_1} + \frac{1}{C_{in}} + \frac{1}{C_2}\right)^{-1}. \tag{10.2}$$

Usually the capacitance of the membrane's interior is much smaller than the capacitance of electrical double layers in solutions:

$$C_{in} \ll C_1, C_2. \tag{10.3}$$

The total external transmembrane potential $\Delta\varphi$ is divided between these three capacitors in three parts $\Delta\varphi_1$, $\Delta\varphi_{in}$, $\Delta\varphi_2$ such that:

$$\Delta\varphi_1 = \frac{C}{C_1}\Delta\varphi, \quad \Delta\varphi_{in} = \frac{C}{C_{in}}\Delta\varphi \quad \text{and} \quad \Delta\varphi_2 = \frac{C}{C_2}\Delta\varphi. \tag{10.4}$$

Therefore, the total potential drops are:

$$\varphi_1(\Delta\varphi) = \varphi_1(0) + \Delta\varphi_1 = \varphi_1(0) + \frac{C}{C_1}\Delta\varphi, \tag{10.5}$$

$$\varphi_{in}(\Delta\varphi) = \varphi_{in}(0) + \Delta\varphi_{in} = \varphi_{in}(0) + \frac{C}{C_{in}}\Delta\varphi, \tag{10.6}$$

$$\varphi_2(\Delta\varphi) = \varphi_2(0) + \Delta\varphi_2 = \varphi_2(0) + \frac{C}{C_2}\Delta\varphi. \tag{10.7}$$

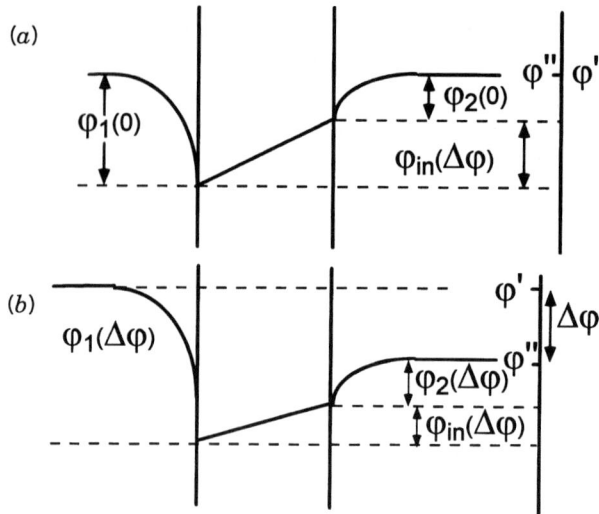

Figure 10.3. The potential distribution in an asymmetrical membrane: (a) no applied transmembrane potential; (b) an external transmembrane potential $\Delta\varphi = \varphi' - \varphi''$ is applied to the membrane.

This quasi-equilibrium analysis is possible because electrical current across a membrane is usually low and the membrane equilibrium is perturbed very little.

A membrane with asymmetrical surfaces is shown in Fig. 10.3 and the same considerations are valid for this case.

Charges in a Membrane. Membranes have very low concentrations of free charge which results in cellular and planar lipid membranes having low conductance. Very often the analysis of electrical properties of membranes is based on considering them as a homogeneous phase. The extremely low dielectric constant of lipids (ranging from 2 to 3) is very unfavorable for the incorporation of charged particles into membranes. We can estimate the partition coefficient of a charged particle between a lipid and an aqueous phase:

$$K = \frac{c_m}{c_w} = \exp\left(-\frac{\Delta G_{tr}}{kT}\right). \qquad (10.8)$$

ΔG_{tr} is the energy of the charged particle in the lipid with respect to water. In a simplified form (see also Chapter 6) it is composed of the electrostatic energy and the energy of hydrophobic interaction:

$$\Delta G_{tr} = \Delta G_{el} + \Delta G_n. \qquad (10.9)$$

For spherical particles of radius a, ΔG_{el} takes the form

$$\Delta G_{el} = \frac{q_0}{a}\left(\frac{1}{\varepsilon_m} - \frac{1}{\varepsilon_w}\right), \qquad \text{where } q_0 = \frac{z^2 e_0^2}{8\pi\varepsilon_0 kT}. \qquad (10.10)$$

At a temperature of 25 °C and a valence of one, $q_0 = 28.2$ nm. If we assume the radius of the ion to be 0.2 nm, and the dielectric constant of the membrane $\varepsilon_m = 3$, then the partition coefficient K is equal to 10^{-20}. If this value is used to calculate membrane conductivity the result is much lower than actual measurement would show. The estimate for K can be improved by taking into consideration the hydrophobic interactions which can increase the partition coefficient (Chapter 6). In addition there are several membrane specific factors that can diminish the energy of a charged particle inside a membrane.

Parsegian (1969) analyzed four factors that diminish the energy of an ion in a membrane:

1. finite membrane thickness;
2. ion-pair formation inside the membrane;
3. membrane pores of high dielectric permeability through which the ions can pass; and
4. ion–carrier complex formation in which ions are wrapped in a neutral molecule of high polarizability that solvates it (increases its effective radius a), and thus facilitates its partitioning in the membrane phase.

Calculation of each of these effects yields the following results.

(a)   Due to the finite thickness of membrane (l) image forces arise at the boundary between the membrane and the aqueous phase. The electrostatic energy ΔG_{el} of the ion in the membrane is diminished, and it takes the form of the curve shown in Fig. 10.4(a) (Neumke and Laüger, 1969). At the center of the membrane, the image forces decrease the energy by the following amount:

$$\Delta G_{im} = \frac{e_0^2}{4\pi\varepsilon_0\varepsilon_m l}\ln\frac{2\varepsilon_w}{\varepsilon_w + \varepsilon_m}. \tag{10.11}$$

When $\varepsilon_w = 80$ and $\varepsilon_m = 2$, the relative decrease of the Born energy $e_0^2/(4\pi\varepsilon_0\varepsilon_m l)$ of the ion due to image forces is equal to $1.4a/l$. This value is negligible unless the ion is very large and/or the membrane is very thin. This example makes it clear that the hydrocarbon part of the membrane constitutes a substantial barrier to the passage of ions. The height of the barrier amounts to about a hundred kJ mol^{-1}.

(b)   Formation of ion pairs also does not yield an appreciable gain. The electrostatic energy of two particles of radii a_+ and a_-, separated by a distance d [see Fig. 10.4(b)] is

$$\Delta G_{el} = \frac{e^2}{8\pi\varepsilon_0\varepsilon_m a_+} + \frac{e^2}{8\pi\varepsilon_0\varepsilon_m a_-} - \frac{e^2}{4\pi\varepsilon_0\varepsilon_m d}. \tag{10.12}$$

From this we see that the maximum reduction in energy will be no more than twofold. Only a covalent bond between the charged particles would substantially diminish the electric field around them, but this would now imply creating a neutral molecule.

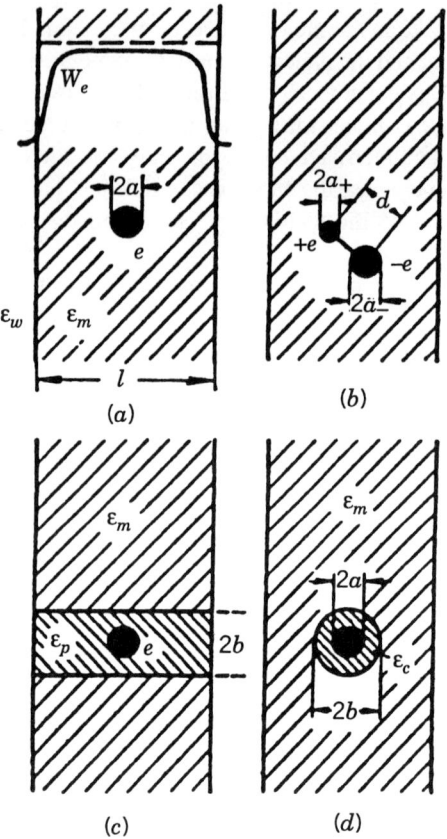

Figure 10.4. Four factors decreasing the energy of an ion in a membrane: (a) image force effects; (b) ion-pair formation; (c) presence of hydrophilic pores; (d) carrier-ion complex formation.

(c)   Pores having high polarizability can substantially diminish the energy of a charge in a membrane (see Fig. 10.4(c)). If the radius of pore b is much smaller than the membrane thickness, $b \ll l$, the energy of a particle at the axis of the pore is

$$\Delta G_p = \frac{e^2}{8\pi\varepsilon_0\varepsilon_p a} + \frac{e^2}{4\pi\varepsilon_0\varepsilon_m b} P\left(\frac{\varepsilon_m}{\varepsilon_p}\right). \tag{10.13}$$

The second term in this formula arises from the image forces in the walls of the pore which are inversely proportional to the pore radius. The function $P(x)$ has been calculated numerically (Parsegian, 1969). Its maximum value does not exceed 0.25. If, for example, the dielectric permittivity of the pore, ε_p, is comparable with the dielectric permittivity of water, then the height of the barrier for an ion passing through the membrane is determined mainly by the second term. When $\varepsilon_m = 2$, this energy is

$$\Delta G_p \approx \frac{e^2}{4\pi\varepsilon_0 \varepsilon_m b} P\left(\frac{1}{40}\right) = \frac{11.8}{b} \text{ kJ mol}^{-1}. \tag{10.14}$$

Here b is expressed in nanometers.

(d)   Finally let us examine the possibility of complex formation between a carrier and an ion. Let a neutral molecule of high polarizability form a spherical complex with the ion. If the outer radius of the complex is b, then its energy in the medium is equal to

$$\Delta G_c = \frac{e^2}{8\pi\varepsilon_0 \varepsilon_m b} + \frac{e^2}{8\pi\varepsilon_0 \varepsilon_c}\left(\frac{1}{a} - \frac{1}{b}\right). \tag{10.15}$$

If the molecules of the carrier have high polarizability and we neglect the second term, the energy of the complex formed is still significant in comparison with the thermal energy, though considerably smaller than the energy of a "bare" ion. For example, if b is in the range of 0.5–1 nm, then at 25 °C we have: $\Delta G_c = 69.3 - 34.4 \text{ kJ mol}^{-1}$ or $9.8 - 4.9 \text{ kT ion}^{-1}$.

Of course, these estimates are very crude. Yet, they clearly indicate that the barrier for passage of an ion through a membrane can be considerably reduced by the formation of a carrier complex or by special pores.

Gating Charge. Passage of electrical current through cellular membranes involves conformational changes in membrane proteins including displacement of certain charged groups. When an external electric field varies, this manifests itself in an additional component of a displacement current that has been given the descriptive name of the "gating current."

By integrating the gating current over the time, one can obtain the size of the equivalent charge that is transported from one side of the membrane to the other. Figure 10.5 shows the relationship of the displaced charge to the membrane potential (Keynes and Rojas, 1974). Curve 2 was obtained semi-empirically under the assumption that the gate particles have one stable state near each side of the membrane, and that they have a charge of $1.3e$. This curve is described by the formula

$$\frac{Q(\varphi)}{Q_{\max}} = \frac{1}{1 + \exp\left(-\dfrac{zF\varphi}{RT}\right)}, \qquad (z = 1.3), \tag{10.16}$$

where $Q_{\max}$ is the maximum displaced charge.

In the Hodgkin–Huxley model of nerve excitation one can consider the gating currents to result from jumping of the m-particles during the activation of a membrane

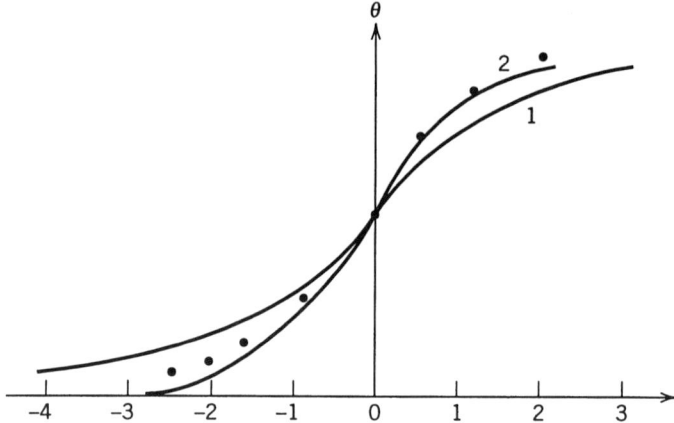

Figure 10.5. Fraction of transported charge as a function of membrane potential. Curve 1 is plotted according to equation (10.17) with a linear approximation of u' (θ). Curve 2 is plotted according to equation (10.16) and the experimental points are from Keynes and Rojas (1974). Reproduced by permission of Professor R. Keynes, Cambridge University.

across the whole membrane thickness. From a physical standpoint, the movement of several charged particles through the entire membrane seems hardly possible. It would be more plausible to ascribe the gating current to a change in the total dipole moment of certain proteins in the membrane. If the lipoprotein complex comprising a membrane channel is rigid enough, then reorientation of the elementary dipoles in it will be cooperative. This is essential for explaining the steep dependence of the displaced charge on the membrane potential that is observed experimentally (Fig. 10.5). If we assume that every elementary dipole has only two possible orientations, then we can describe the fraction θ of dipoles that have changed orientation by the relationship

$$\frac{\theta}{1-\theta}e^{-w\theta} = k\exp\left(-\frac{F\varphi d}{RT\delta}\right). \qquad (10.17)$$

Here w is a cooperativity parameter, d is the actual displacement of the charge, and δ is the thickness of the membrane. Figure 10.2 shows the $\theta(\varphi)$ relationship calculated for $w = 3.5$ and $d/\delta = 0.2$. The gating current, which is proportional to $d\theta/dt$, and the voltage dependence of its relaxation time are in qualitative agreement with experimental results.

Change in ion permeability during the reorientation of the dipoles can arise from various factors. If the reorientation is associated with a significant conformational change in the system, then the "opening" or "closing" of the channel can literally amount to a rearrangement of the molecular geometry of the channel. Alternatively, dipole reorientation does not change the geometry of the channel, but affects the electrostatic component of the energy of ions in the channel; thereby, both their concentration in the membrane and their effective mobility can be altered. Simple estimates show that the electrostatic mechanism of regulation can in principle give rise to the observed potential dependence of the conductivity.

By finding out whether the conductivity of an individual channel varies discretely or continuously one can determine which of the two mechanisms is occurring. It is obvious that the "all-or-none" rule (discrete step) will favor the conformational hypothesis, while continuous variation of the conductivity should implicate an electrostatic regulation principle. The Hodgkin–Huxley model postulated that an ion channel, e.g., the potassium channel, can exist in one of five conformational states, only one of them being conductive. Of course, this picture of the channel is not the only possible one. For example, one can assume that all conformational states are conductive, while one seeks the conductivity distribution function from the condition which best matches the experimental data. Further details on channel gating can be found in standard neurophysiology textbooks.

10.3. CONSECUTIVE STAGES OF MEMBRANE ION TRANSPORT

The passage of ions through a membrane from one bulk solution to another occurs in consecutive stages: first, diffusion occurs in the electrolyte solution; second, there is a heterogeneous reaction at the outer membrane surface; third, transfer occurs through the membrane interior; fourth, another reaction takes place at the inner membrane surface; and finally the ion diffuses into the interior solution. The processes involved at the first and the last two stages are fundamentally the same and will be considered as such. The overall electrical conductance of the membrane characterizes the ease of passage of ions from one compartment to another.

The electrical conductance of a cellular membrane is rather low, and for black lipid membranes is even lower. While the electrical conductivity of cell membranes is about 10^{-3} S cm^{-2}, the electrical conductivity of BLMs varies widely over a range from 10^{-10} to 10^{-5} S cm^{-2}, depending on experimental conditions. One can safely select 10^{-8} S cm^{-2} as an average value for black lipid membrane conductance. These values are low compared with the conductivity of the surrounding electrolyte solution. If this solution contains 0.01 M KCl, the conductivity of an aqueous layer having the same thickness as the BLM is equal to 10^4 S cm^{-2}. Thus, the difference in conductance amounts to a factor of 10^{12}.

The concentration of ions in the bulk solution is kept constant due to thermal convection and the conductance is governed by Ohm's law. However, near the membrane convection is hindered, resulting in the appearance of an unstirred or diffusion layer with a linear concentration gradient. The steady-state flux of particles across this diffusion layer of thickness h is given by the equation

$$j = D\frac{c_0 - c_s}{h},\tag{10.18}$$

where c_0 and c_s are the concentrations at the ends of the unstirred layer and D is the diffusion coefficient.

The thickness of an unstirred layer h varies depending on hydrodynamic conditions. Very often it is close to 100 μm. The relative importance of these unstirred layers for membrane transport depends on a given membrane's permeability. The properties of an unstirred layer can be characterized by the permeability coefficient $P = D/h$. For many substances the diffusion coefficient D is on the order of 1000 $\mu m^2 s^{-1}$ and hence the permeability coefficient $2P$ is of the order on 10 $\mu m\ s^{-1}$. This value of P for an unstirred layer is not very different from the permeability of biological membranes and hence unstirred layers can significantly contribute to the overall membrane resistance.

Suppose that a membrane has permeability P_m. Then, the total permeability P_t of this membrane along with its unstirred layers is

$$P_t = \left(\frac{1}{P_m} + \frac{2h}{D} \right)^{-1}. \tag{10.19}$$

Very often a membrane is characterized by its conductance g rather than permeability P. The relationship between these parameters can be defined by the following equation:

$$g = \frac{z^2 F^2 cP}{RT}, \tag{10.20}$$

where c is the concentration of the permeating ion.

The conductance of an unmodified black lipid membrane ranges from 10^{-10} to $10^{-7}\ S\ cm^{-2}$ and corresponds to a permeability $P = 10^{-8}$–$10^{-5}\ \mu m\ s^{-1}$. These numbers are much lower than the permeability of the unstirred layers estimated above and in this case the unstirred layers do not contribute significantly to the total membrane permeability.

High Membrane Permeability Conditions. If the membrane permeability is very high, then the permeability of the total system is determined by the unstirred layers only. If one designates the concentration of the permeating ion in the left bulk solution to be c_1 and in the right solution c_2, then if the circuit is open, the membrane will achieve the Nernst potential

$$\varphi = \frac{RT}{zF} \ln \frac{c_2}{c_1}. \tag{10.21}$$

If a current is passed through this membrane, a linear profile of concentration will develop in the unstirred layers with concentrations at the membrane surfaces being equal to c_{s1} and c_{s2}. The fluxes in the two layers should be equal to each other:

$$D\frac{c_1 - c_{s1}}{h} = D\frac{c_{s2} - c_2}{h}. \tag{10.22}$$

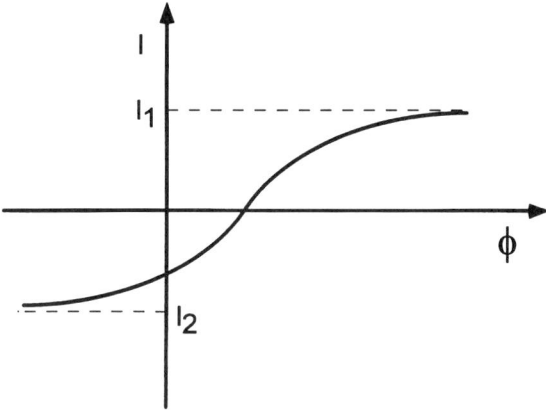

Figure 10.6. Current–voltage characteristics of unstirred layers. The case of a highly permeable membrane.

Because of the high intrinsic permeability of the membrane it will be in a state of quasi-equilibrium such that the potential difference φ is equal to:

$$\varphi = \frac{RT}{zF} \ln \frac{c_{s2}}{c_{s1}}. \tag{10.23}$$

By solving equations (10.22) and (10.23) one can find the current–voltage characteristics:

$$I = \frac{zFD}{h} \ln \frac{c_1 \exp\left(\dfrac{zF\varphi}{RT}\right) - c_2}{\exp\left(\dfrac{zF\varphi}{RT}\right) + 1}. \tag{10.24}$$

This function is presented in Fig. 10.6. This curve saturates at both high positive and negative voltages. The limiting currents are equal to

$$I_1 = \frac{zFDc_1}{h} \quad \text{and} \quad I_2 = \frac{zFDc_2}{h}. \tag{10.25}$$

If the membrane surfaces and bulk solutions are the same at both sides of the membrane the current–voltage characteristics also become symmetrical:

$$I_1 = \frac{zFDc}{h} \tanh \frac{zF\varphi}{2RT}. \tag{10.26}$$

At small potentials the conductance of this system is

$$g_0 = \frac{z^2 F^2 Dc}{2hRT}.$$

(10.27)

In this case the conductance g_0 is entirely determined by the unstirred layers.

Electrodiffusion in Membranes. There are two general approaches to ion transport: discrete and continuous. In the discrete approach, which is based on the Eyring theory of absolute reaction rates, one assumes that a particle gets through a membrane by making several discrete jumps over activation barriers. The continuous approach is based on the concept of free diffusion and migration of particles in the membrane, which is treated as a continuous, homogeneous phase. The ion current is given by the expression

$$I_k = z_k u_k RT \left(\frac{z_k F c_k E}{RT} - \frac{dc_k}{dx} \right).$$

(10.28)

Here z_k is the charge number of a type k ion, u_k is the mobility, which is related to the diffusion coefficient D_k by the Einstein relationship $u_k = FD_k/RT$, and E is the electric field intensity, which satisfies the Poisson equation:

$$\frac{dE}{dx} = \frac{F}{4\pi\varepsilon_0\varepsilon} \sum_k z_k c,$$

(10.29)

where ε is the dielectric constant of the membrane and c_k is the concentration of type k ions. At the steady state, $I_k =$ constant, i.e., equation (10.28) is a non-linear first-order differential equation that contains the unknown functions c_k and E, and the unknown constant I_k. Total current (I) is a sum of all partial currents:

$$I = \sum_k I_k.$$

(10.30)

At non-steady state, the partial current and concentration are related by the continuity equation

$$\frac{\partial I_k}{\partial x} + Fz_k \frac{\partial c_k}{\partial t} = 0.$$

(10.31)

The total current I now includes a displacement current and an ionic current:

$$I = 4\pi\varepsilon_0\varepsilon \frac{\partial E}{\partial t} + \sum_k I_k.$$

(10.32)

In the steady-state case, one can derive a useful integral relationship from equation (10.28):

$$I_k = g_k(\varphi - \varphi_k). \tag{10.33}$$

Equation (10.33) is useful for calculating membrane conductivity in the presence of small electrical fields and also when the external field is changing rapidly. In the presence of a rapidly changing external field the membrane concentration distribution does not have time to adjust to the new field and therefore the current–voltage relationship is linear. However, under steady-state conditions the current–voltage relationship is non-linear.

The electrodiffusion problem is considerably simplified when the space charge in the membrane is small and hence $E = $ constant. This is the so-called Goldman's approximation of a constant field. In this approximation, the partial current–voltage characteristics take the form

$$I_k = \frac{z_k^2 F u_k \varphi}{\delta} \frac{c_k(0) - c_k(\delta) \exp(z_k F\varphi/RT)}{1 - \exp(z_k F\varphi/RT)}. \tag{10.34}$$

Here $c(0)$ and $c(\delta)$ are the ion concentrations in the membrane phase, which are related to the corresponding ion concentrations in the solutions by each partition coefficient K.

This mechanism based on the concept of diffusion and migration of ions in a continuous, homogeneous phase was used to describe the resting state of biological membranes. One can easily calculate from equation (10.34) the resting membrane potential which is one of the most fundamental characteristics of the nerve cell. The membrane of a nerve cell is permeable to sodium, potassium and chloride ions. Hence, the resting state is a steady state, rather than a thermodynamic equilibrium. For an open circuit, the sum of all partial ionic currents should be zero:

$$\varphi = \frac{RT}{F} \ln \frac{P_K c_K^{out} + P_{Na} c_{Na}^{out} + P_{Cl} c_{Cl}^{in}}{P_K c_K^{in} + P_{Na} c_{Na}^{in} + P_{Cl} c_{Cl}^{out}}. \tag{10.35}$$

Relationship (10.35) describes the experimental data fairly well, provided that the concentration range is not too broad. However, one should not put too much weight on this agreement since the permeability parameters (P) are determined from the same set of experiments. All things considered, treating a membrane as a continuous homogeneous phase as in the electrodiffusion theory is not accurate; however, we use relationships (10.34) and (10.35) for their simplicity and because of our lack of reliable information on the true mechanisms of ion transport in a membrane at rest. The majority of what is known about ion transport through membranes, is known during the process of excitation, when the conductivity of the system is greatly elevated. It has been firmly established that during excitation of membranes ion transport occurs through specialized lipoprotein structures that are called ion channels.

10.4. MECHANISMS OF PASSIVE PERMEATION OF IONS AND DIPOLES THROUGH MEMBRANES

We demonstrated in the previous section that membrane conductance is very low because the concentration of charged species in a membrane is very low. It has been a general assumption that the low permeability reflects the Born energy for dissolving (partitioning) an ion or dipole in the low-dielectric hydrocarbon interior, an energy term so large that it represents a nearly insurmountable barrier (Parsegian, 1969). But this simple approach predicts an extremely low concentration of ions and therefore too low a conductance. Earlier we discussed a number of obvious corrections for Born theory that could improve the agreement between theory and experiment. Nevertheless, certain discrepancies remain that should be considered in more detail.

Probably the most dramatic inconsistency with Born theory is seen with proton gradients. A remarkably high proton permeability was first reported by Nichols and Deamer (1978, 1980) and confirmed by other laboratories (Elamrani and Blume, 1983; Gutknecht, 1984; Perkins and Cafiso, 1986; Rossignol et al., 1982). There is no reason to expect proton permeation to differ from that of other cations, yet calculated fluxes of protons down a concentration gradient of 10^{-6} to 10^{-7} M (pH 6 to 7) are in the same range as the flux of potassium ions down a gradient of 10^{-1} to 10^{-2} M. That is, proton and potassium fluxes down tenfold gradients are similar even though the potassium concentration at 0.1 M is 10^5 times that of protons at pH 6. Other investigators have noticed certain shortcomings of the Born theory with respect to experimental results. Among the first was Bell (1931) who noted that measured partitioning of ions between polar and non-polar media was only in qualitative agreement with expectations from Born calculations.

We will discuss two possibilities that in a sense represent alternative hypotheses. The first hypothesis is that ion or dipole permeation can be understood in terms of energy related to the partitioning of ions into the non-polar phase of the lipid bilayer, and that electrostatic considerations, including Born energy, are primary concerns. The second hypothesis is that high-dielectric defects in the form of transient pores occur in the bilayer and allow permeating ions or dipoles to bypass the partitioning energy barriers. The two hypotheses were tested by experimental measurements of proton and potassium flux across lipid bilayers of varying thickness (Paula et al., 1996).

Partition Model. We will first discuss the partition model for calculating the Gibbs free energy of ion or dipole transfer through bilayers (Fig. 10.7). In this model, an ion with a bare radius a and hydrated radius b is transferred from the aqueous phase to the center of a bilayer with thickness d of the non-polar phase.

The relationship between an ionic radius and the electrostatic (Born) component of the Gibbs energy of ion transfer from medium w into the medium m is calculated from the expression

$$\Delta_w^m G(\text{Born}) = \frac{q^2}{8\pi\varepsilon_0 a}\left(\frac{1}{\varepsilon_m} - \frac{1}{\varepsilon_w}\right), \qquad (10.36)$$

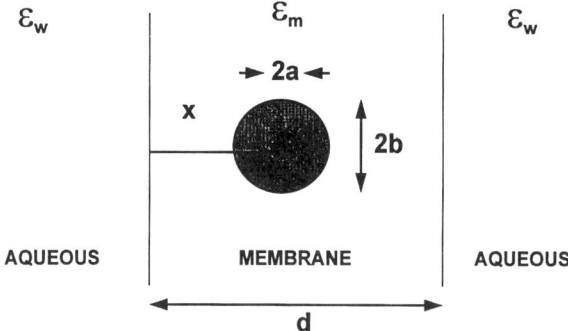

Figure 10.7. The partition model for ion permeation of a lipid bilayer. ε_w, ε_m, and ε_s are the dielectric constants of the aqueous phase, membrane phase, and the ion solvation shell; a is the bare ionic radius, b is the hydrated ionic radius, x is the distance the ion penetrates into the bilayer, q is the ionic electrical charge, and d is the thickness of the hydrophobic part of the bilayer.

where ε_w and ε_m are dielectric permittivities of phases w and m. The dependence of the Born energy of ion transfer between two immiscible liquid phases such as water and hexadecane is shown in Fig. 10.8, curve 1.

The solvophobic contribution to the Gibbs free energy of ion or dipole resolvation can be calculated using the solvophobic equation (6.36). It has been surprisingly useful in calculations of partition coefficients in systems consisting of two immiscible liquids (Abramson, 1981). Hydrophobic effects are a function of the interfacial tension at the oil–water interface (equation (6.36) and Table 2.1). Although the interfacial tension at the liquid hydrocarbon–water interface is about 50–52 mN m^{-1} for saturated alkanes (Table 2.1), it is much lower in unsaturated hydrocarbons, which in turn should be reflected in the related hydrophobic effect. For example, Fig. 10.8 shows the dependence of the hydrophobic effect on the radius of transferred particles (ions, molecules) for the water–alkane and water–alkene interface. It is clear that the presence of double bonds should decrease diffusional transport across a membrane so long as the dielectric permittivity of the hydrocarbon (and Born component of the resolvation energy) is unchanged. Experimental measurements, of course, show just the opposite, that double bonds generally increase permeability of a bilayer to ions and polar molecules. This is presumably due to the fact that although alkanes have dielectric moment equal to zero, the *cis* double bond in alkenes induces a dielectric moment of about half a Debye unit that can cause specific ion–dipole and dipole–dipole interactions among solvent molecules. This in turn produces a higher dielectric permittivity and increased partition coefficient, with the end result that unsaturated lipid bilayers are more permeable than saturated lipid bilayers.

A significant point is that the free energy of the solvophobic effect is opposite in a sign to the electrostatic effect (Fig. 10.8). As a result, the sum of electrostatic and hydrophobic components of Gibbs free energy decreases with ionic size, so that $\Delta G(\text{tr}) > 0$ only for small ions. For ions with radii larger than 0.45 nm $\Delta G(\text{tr}) < 0$. This prediction is consistent with experimental data (Markin and Volkov, 1989b).

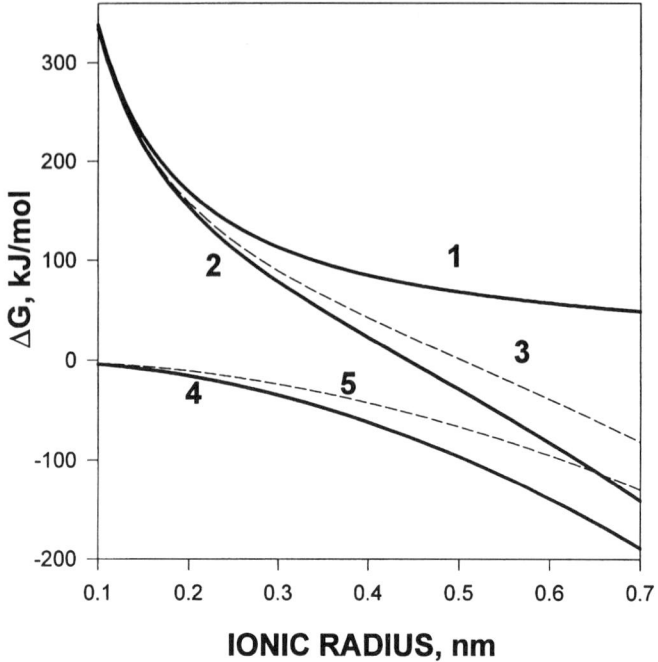

Figure 10.8. Gibbs free energy of ion transfer from water to liquid hydrocarbon calculated according to: the Born equation (1); Born equation, image energy, and solvophobic effect together (2, 3); the solvophobic effect (4, 5) according to equation (6.36); $T = 25\,°C$, $\varepsilon_w = 78.5$, $\varepsilon_m = \varepsilon_s = 2$, $\gamma_{wm} = 51$ mN m^{-1} (2, 4), $\gamma_{wm} = 35$ mN m^{-1} (3, 5), $d = 6$ nm.

In the water–hydrocarbon system the force of attraction (or repulsion) of charge in the oil phase with its image in the aqueous phase is given by

$$F(x) = -\frac{\varepsilon_w - \varepsilon_{hc}}{\varepsilon_w + \varepsilon_{hc}} \frac{q^2}{16\pi\varepsilon_0\varepsilon_{hc}x^2},$$ (10.37)

where x is the distance from the interface. If $\varepsilon_w > \varepsilon_{hc}$, the charge in the non-polar phase is attracted to its image, but if $\varepsilon_w < \varepsilon_{hc}$ there is repulsion between the charge in a hydrocarbon phase and its image. From equation (10.37) it follows that charge in the hydrocarbon phase is attracted to the water–hydrocarbon interface. In the case of a thin membrane placed in a medium with high dielectric permittivity (Fig. 10.7), the sum from an infinite number of images must be calculated. Neumke and Laüger (1969) estimated the combined Born and image energy of an ion in a thin membrane as follows:

$$\Delta G(\text{el}) = \Delta G(\text{Born}) - \frac{q^2}{16\pi\varepsilon_0\varepsilon_w}\left\{\frac{1}{x} + \frac{1}{d}\sum_{n=1}^{\infty}\left[\frac{\theta^{2n}}{n + x/d} + \frac{\theta^{2n-2}}{n - x/d} - \frac{\theta^{2n}}{n + a/d} - \frac{\theta^{2n-2}}{n - a/d}\right]\right\},$$ (10.38)

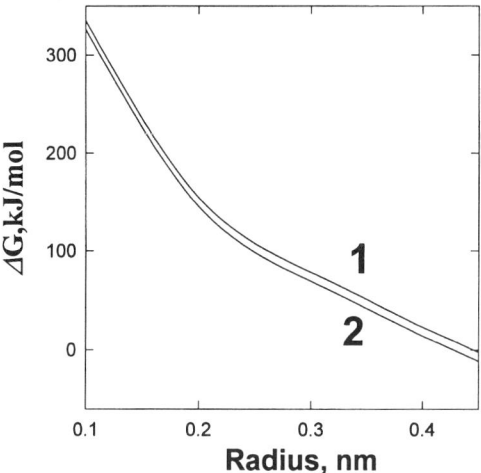

Figure 10.9. Line 1—dependencies of the Gibbs free energy of ion transfer on ion size calculated according to the Born equation and solvophobic effect together. Line 2 is the same as line 1 but corrected for image forces ($d = 6$ nm, $T = 25\,°C$).

where

$$\theta = \frac{\varepsilon_{hc} - \varepsilon_w}{\varepsilon_{hc} + \varepsilon_w}. \tag{10.39}$$

The effect of image forces is small compared with solvophobic and Born electrostatic effects and depends on the ion size (Fig. 10.9) and membrane thickness (Volkov *et al.*, 1997).

One can calculate the Gibbs free energy of ion or dipole transfer ΔG(tr) from an aqueous phase to a hydrocarbon phase according to the relation:

$$\Delta G(\text{tr}) = \Delta G(\text{el}) + \Delta G(\text{solv}) + \Delta G(\text{si}), \tag{10.40}$$

where ΔG(el) is the electrostatic contribution, ΔG(solv) is the hydrophobic effect and ΔG(si) is caused by specific interactions of the transferred particle (ion, dipole) with solvent molecules, such as hydrogen bond formation, donor–acceptor and ion–dipole interactions. Calculating the Gibbs free energy of ion transfer from water to a non-polar phase according to equation (10.40) requires interpolation, and this apparently is the reason for its success, despite the simplifying assumptions used. For small ions, the electrostatic contribution is predominant in ΔG(tr), so that the discrepancies of the solvophobic relation, which becomes evident with small ions are negligible. At large ionic radii the solvophobic term predominates, so that calculations according to the solvophobic equations are consistent with experimental results (Markin and Volkov, 1989b).

Bilayers differ from water–liquid hydrocarbon membranes by the presence of a two-dimensional layer of adsorbed water dipoles and lipid heads. Even if lipids are neutral there is a dipolar potential within the membrane surface layer, which should be taken into account when describing specific interaction of ions with the membrane surface. For instance, the dipole–dipole specific interaction creates different concentrations of cations and anions at water–membrane interfaces. Rusanov *et al.* (1984) calculated this energy as $-4\pi z e\Delta P$, where P is a characteristic of the surface excess polarization. We can estimate the specific energy of the ion–dipolar layer interaction as a function of a dipole potential $\Delta G(si) = -zF\phi_s$, where z is the charge number, F is the Faraday potential and ϕ_s is the dipolar membrane surface potential. Usually ϕ_s is estimated for bilayers to be between -100 and $-200\,\text{mV}$. At nitrobenzene–water and 1,2-dichloroethane–water interfaces dipolar potential was measured directly at the point of zero free charge and dipolar potential is between -20 and $-50\,\text{mV}$ (Volkov, 1996) in the absence of phospholipids and between -100 and $-200\,\text{mV}$ in the presence of phospholipid monolayer (Kakiuchi, 1996b).

It is possible to calculate the ion permeability of bilayers for the partition model of ion transfer using equations (6.36), (10.36), (10.38) and (10.40). The permeability coefficient (P_i) for an ion crossing a bilayer is related to the partition coefficient by the expression:

$$P_i = D_i K_i / d, \qquad (10.41)$$

where D_i is the ion diffusion coefficient and d is the membrane thickness. Substituting K_i from equation (10.8) into equation (10.41) allows P_i to be calculated directly, and the results are plotted in Fig. 10.10. From this, it is clear that the partition coefficient and the calculated permeability coefficient have an extreme parameter sensitivity with respect to an ionic radius, varying over 65 orders of magnitude as the radius increases from 1 to 5 Å. Substituting radii of bare (non-hydrated) ions (Table 6.1) such as sodium and potassium in this calculation gives much too small values for bilayer permeability when compared with experimental data, and the radius of a bare proton leads to absurd values. On the other hand, substituting hydrated radii taken from Table 6.1 gives permeability values surprisingly close to those measured experimentally. We will use the same method of calculation of the Gibbs energy of transfer for ideal dipoles. The Gibbs energy of a dipole transfer between two phases is presented by equation (6.39). The free Gibbs energy of image forces for a dipole in a membrane was calculated by Arakelyan and Arakelyan (1983b) and Arakelyan *et al.* (1985, 1986):

$$\Delta G_{I(dip)} = -\frac{2d^2}{12\pi\varepsilon_0\varepsilon_m l^3} \sum_{i=1}^{\infty} \left(\frac{\theta^{2i-1}}{(2i-1)^3} + \frac{\theta^{2i-1}}{(2i-1)^3} \right). \qquad (10.42)$$

The dependence of the electrostatic image energy of the water dipole on the thickness of a liquid membrane is shown in Fig. 10.11. The image energy strongly depends on the membrane thickness and decreases with increasing thickness.

The water concentration in a hydrophobic region of a membrane can be calculated from the estimated free Gibbs energy of water molecule transfer from water to the

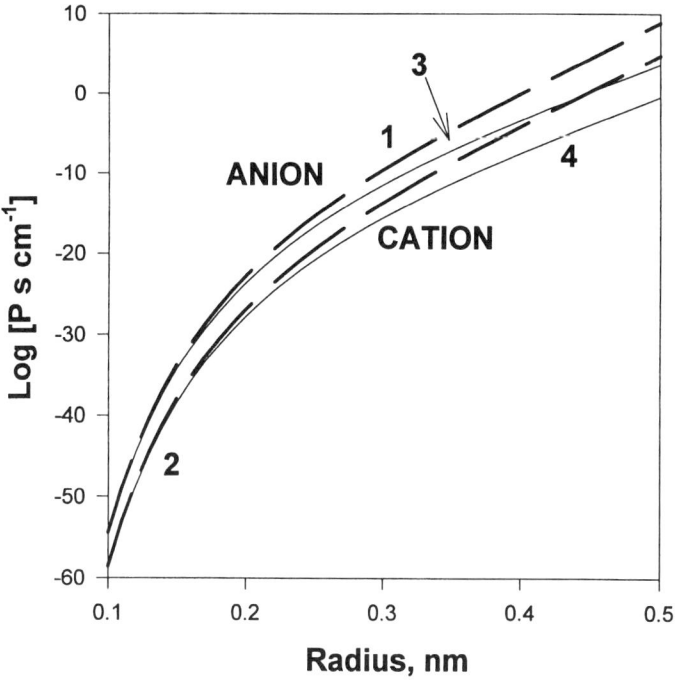

Figure 10.10. Dependence of the ionic permeability of a bilayer ($d = 3$ nm) on the radius of a permeating ion at 25 °C ($\varepsilon_w = 78.5$, $\varepsilon_m = \varepsilon_s = 2$, $\gamma_{wm} = 51$ mN m^{-1} (1, 2), $\gamma_{wm} = 35$ mN m^{-1} (3, 4), $\Delta G(si) = 12z$ kJ mol$^-$, $z = +1$ for cations and -1 for anions).

liquid hydrocarbon using equation (10.12). The dependence of water concentration on the liquid membrane thickness is shown in Fig. 10.12.

Using equations (6.36), (6.37), (10.41) and (10.42) it is possible to calculate the permeability of dipolar molecules across the liquid membranes (Fig. 10.13). $\Delta G(si)$ is approximated by 33 kJ/mol^{-1} for the breakage of a hydrogen bond between water molecules and by 75 kJ/mol^{-1} for hydrogen bonds between three water molecules and a urea or a glycerol molecule (Pimentel and McClellan, 1960).

Despite the interesting results arising from considerations of both electrostatic and solvophobic contributions to ionic permeation, we must be careful in applying them uncritically to ionic permeation events in bilayers, and particularly to proton permeation. The main difficulty is that lipid bilayers are not ideal slabs of a hydrocarbon with uniform dielectric properties, but instead are highly dynamic, with considerable fluctuation around the means of thickness and intermolecular distance within the bilayer (Volkov et al., 1997). Secondly, given the steepness of the slope relating ionic permeability to an ionic radius (Fig. 10.10) in which a change of 0.05 nm radius can produce more than ten orders of magnitude variation in permeability, it is a concern that estimates of hydrated ionic radii are highly dependent on the technique used to make the measurement. Last, the hydronium ion is not a hard

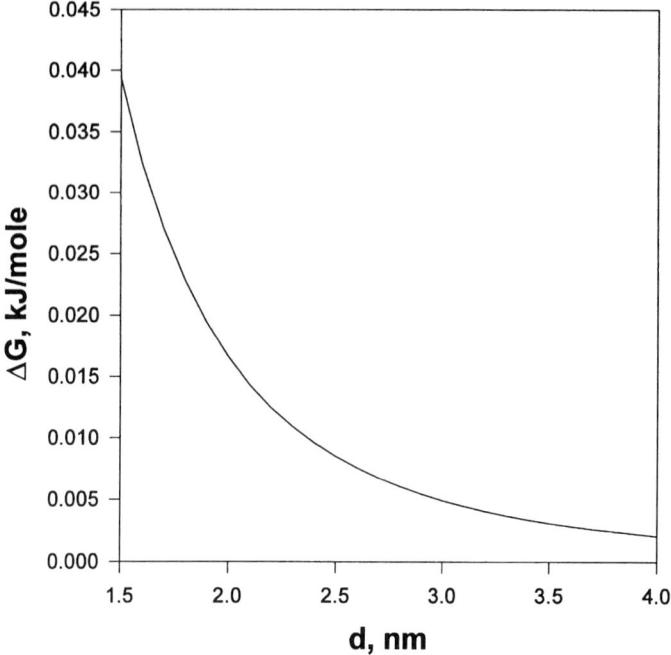

Figure 10.11. Dependence of the image energy for a water dipole on liquid membrane thickness. Parameters: $d = 1.85$ D, $l = 1.38$ Å, $\varepsilon_w = 78$, $\varepsilon_m = 2$.

shell of four water molecules bound tightly to a central proton. Instead, proton–water interactions are dynamic, with the proton rapidly entering and leaving clusters of water of various sizes. In the section to follow, we will examine an alternative mechanism for the permeation of ions across lipid bilayers, then compare experimental results with theoretical predictions from both models.

Transient Pore Mechanism. Although the partition model can describe the barrier property of lipid bilayers under specified conditions, certain assumptions regarding the hydrated radius of the permeating species are required. We will now consider an alternative hypothesis, in which fluctuations in bilayer structure produce rare transient defects that allow solutes to bypass the electrostatic and solvophobic energy barriers.

Two types of pore structures in the lipid bilayer are possible, which can be roughly classified as hydrophobic and hydrophilic defects. During formation of a transient hydrophobic defect lipid molecules are moved apart by thermal fluctuations so that the membrane hydrophobic core makes contact with and is penetrated by the aqueous bulk phase. Hydrophilic defects are formed if the lipid molecules are tilted into the transient defect so that it is lined with lipid polar head groups. In both hydrophobic

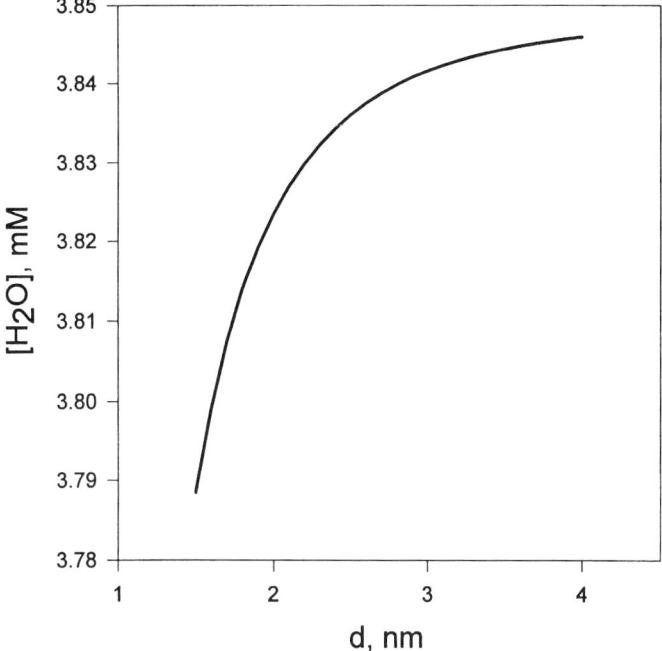

Figure 10.12. Dependence of the water concentration in a liquid membrane on the thickness of the membrane. Parameters: $d = 1.85$ D, $l = 1.38$ Å, $\varepsilon_m = 2$, $\varepsilon_w = 78$, $r_{H_2O} = 1.2$ Å, $\gamma_{w,m} = 51$ mN m^{-1}, $\Delta G(\text{si})(\text{water}) = 33$ kJ mol^{-1}.

and hydrophilic defects, pore formation results from dynamic properties of the lipid bilayer, and the equilibrium pore distribution is relatively constant over time.

Hamilton and Kaler (1990a,b) assumed that thermal fluctuations in bilayers produce transient pores lined with lipid head groups. Ion flux was then calculated from the collision frequency of the ions with the bilayer surface, and the total area of pores in the bilayer that are large and deep enough to allow an ion to cross the bilayer. Given these assumptions the ionic permeability P_i is:

$$P_i = \frac{D_i \sigma n_0 RT}{R_{avg} A_{mem} k_1} \left[s_i + \frac{RT}{k_1} \right] \exp[-k_1 s_i / RT] \exp[-k_2 d / RT]. \qquad (10.43)$$

Here D_i is the ion diffusion coefficient, n_0 is the pore formation frequency factor, $\exp(-k_1 s / RT)$ is the probability of forming a pore of area s, $\exp(-k_2 s / RT)$ is the probability of forming a pore of depth d, σ is the concentration enhancement at the vesicle surface due to the electric double layer effects, R_{avg} is the vesicle radius, A_{mem} is the membrane area, R is the gas constant, and T the absolute temperature.

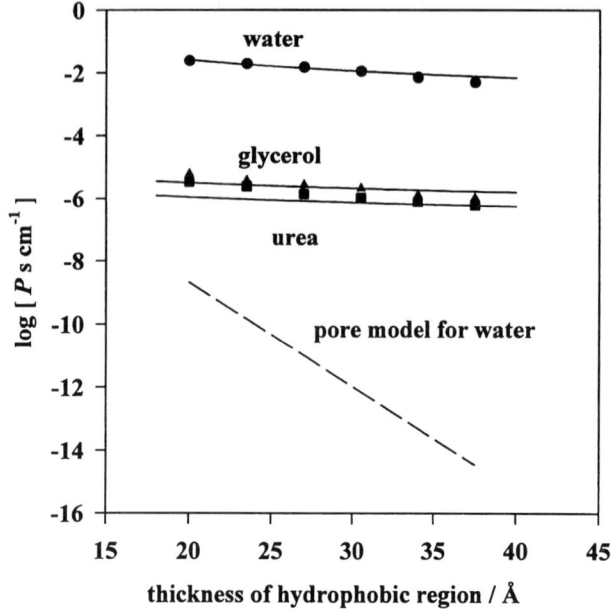

Figure 10.13. Dependencies of water, urea, and glycerol permeability on membrane thickness d. Points are experimental data, solid lines are calculated according to equation (10.41) for the solubility-diffusion model. Parameters: $r_{water} = 1.2$ Å, $r_{urea} = 2.8$ Å, $r_{glycerol} = 2.9$ Å, $\gamma = 51$ mN m^{-1}, $p_{water} = 1.85$ D, $a_{water} = 1.38$ Å, $\Delta G(si)_{(water)} = 33$ kJ mol^{-1}, $\Delta G(si)_{(urea/glycerol)} = 3.25$ kJ mol^{-1}, $D_{water} = 4.59 \times 10^{-5}$ cm^2 s^{-1}, $D_{urea} = 1.38 \times 10^{-5}$ cm^2 s^{-1}, $D_{glycerol} = 1.06 \times 10^{-5}$ cm^2 s^{-1}.

Tests for Partitioning and Transient Pores. In testing the two models described above, an important variable under experimental control is bilayer thickness, which can be changed by choosing lipids with longer or shorter hydrocarbon chains.

In the results to be discussed in this section, one approach is to measure proton and potassium flux across liposomes composed of phospholipids with fatty acid chain lengths varying from 14 to 24 carbons (Paula *et al.*, 1996). One can guess that, in lipids with short-chain fatty acids, non-polar forces would not be sufficient to stabilize bilayer structures. However, as chain length increased, at some point stable bilayer vesicles will form that can provide a permeability barrier to ionic flux. If only partition energy is involved, ionic flux would be reduced to a minimum when stable bilayers formed and would not change significantly as longer-chain lipids were used. However, if transient defects were the primary factor regulating ionic flux, shorter-chain lipids would have many more such defects, leading to a much higher ionic permeability than longer-chain lipids.

Results expressed as proton and potassium permeability coefficients are shown in Fig. 10.14, and compared with the theoretical lines calculated from equation (10.43)

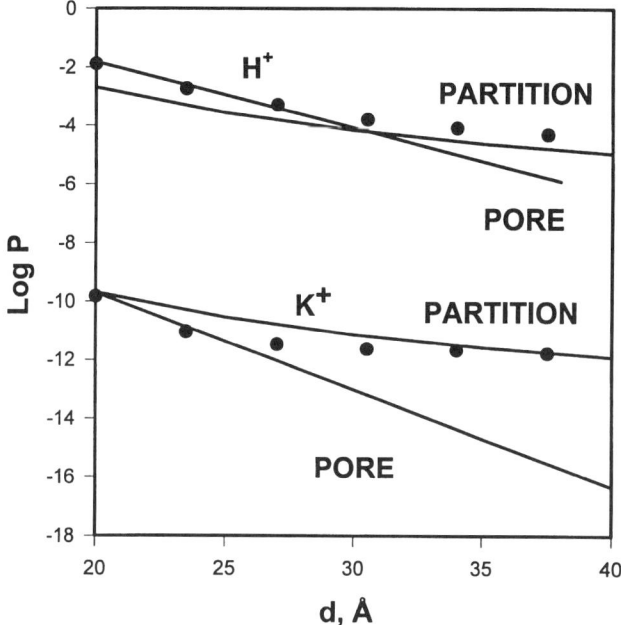

Figure 10.14. Lower section: dependence of K^+ permeability on d calculated from equations (10.41) and (10.43). Points (•)—experimental data. Parameters: $D_i = 2 \times 10^{-5}\,cm^2\,s^{-1}$, $s = 10$, $R_{avg} = 1.05 \times 10^{-6}\,cm$, $A_{mem} = 26\,000\,cm^{-1}$, $T = 300\,K$, $k_1 = 2.2 \times 10^{15}\,kJ\,mol^{-1}\,cm^{-2}$, $k_2 = 1.9 \times 10^8\,kJ\,mol^{-1}\,cm^5$, $n_0 = 1.04 \times 10^{33}\,cm^5$, $\phi_d = -125\,mV$.
Upper section: dependence of the H^+ permeability on d calculated from equations (10.41) and (10.43). Points (•)—experimental data. Parameters: $D = 1.4 \times 10^{-4}\,cm^2\,s^{-1}$, $s = 10$, $R_{avg} = 1.05 \times 10^{-6}\,cm$, $A_{mem} = 26\,000\,cm^{-1}$, $T = 300\,K$, $k_1 = 2.2 \times 10^{15}\,kJ\,mol^{-1}\,cm^{-2}$, $k_2 = 1.2 \times 10^8\,kJ\,mol^{-1}\,cm^5$, $n_0 = 1.04 \times 10^{33}\,cm^5$, $\phi_d = -125\,mV$. Ionic radii are taken from Table 6.2.

(transient pore mechanism) and equation (10.40) (partitioning model). Permeability decreased logarithmically as bilayer thickness increased in the shorter-chain lipids, following the slope of the line predicted by the transient pore mechanism. However, permeability tended to level off in the thicker bilayers (C16 and C18 lipids) approaching the lines predicted by the partition model. Although more data are needed to reach a conclusion, the results suggest that the mechanism of ionic permeation may depend on bilayer thickness: thinner bilayers have many transient defects that allow rapid permeation of small ionic species, while in thicker bilayers defects become so rare that partitioning mechanisms dominate ionic flux.

The permeability coefficients predicted by the solubility–diffusion mechanism display only a moderate dependence on membrane thickness. For uncharged permeating particles without a significant dipolar moment, a semi-logarithmic plot of P against d should ideally give a straight line because in this case, neither D nor K depends on d. Only in the case of ions and small, strong dipoles and very thin membranes will K be affected by the membrane thickness and a small deviation from a straight line is expected. Nonetheless, the overall influence of d on P remains rather small.

The assignment of a permeation mechanism based on the dependence of P on d is easiest for the neutral molecules such as water, urea, and glycerol. Here, the moderate dependence of P on d is consistent with a solubility–diffusion mechanism for these molecules. Using the calculated partition coefficients, the diffusion coefficient of water in a hydrocarbon phase and the diffusion coefficients of urea and glycerol in water, reasonable fits to the data points can be achieved (see Fig. 10.13). Although slightly lower than the experimental points, the calculated permeabilities are of the right order of magnitude. In addition, the slopes of calculated P versus d plots are similar to the experimental slopes, although for urea and glycerol, they are slightly smaller. Analogous calculations for neutral permeates based on the pore mechanism show two inconsistencies (Fig. 10.13 illustrates this for water). First, the computed permeabilities are too low by orders of magnitude. Second, the dependence of P on bilayer thickness is not properly described. These findings suggest that permeation of uncharged polar molecules is more appropriately characterized by the solubility–diffusion than by the pore formation. The contribution of pores to the net permeability seems to be negligible.

Additional support for a solubility–diffusion mechanism for water was provided by measurements of the diffusional and osmotic water permeability (P_d and P_f) of planar lipid bilayers (Finkelstein, 1987). The observation that these two quantities were equal discounted any involvement of a single-file pore mechanism in water permeation. These findings were confirmed by measurements of P_d and P_f simultaneously in phospholipid liposomes by optical methods, which gave a P_f/P_d ratio of unity (Ye and Verkman, 1989).

Using calculations based on literature values, Markin and Kozlov (1985) ruled out the involvement of pores in the permeation of uncharged particles across a membrane. Permeabilities obtained from pore statistics gave values that were markedly lower than experimental observations, in agreement with the data presented in Fig. 10.13. Overall, the solubility–diffusion mechanism seems to impose a significantly lower permeation barrier on neutral molecules than the pore mechanism and is therefore the more likely permeation mechanism for these molecules.

In contrast to the uncharged permeants, the data indicate that a combination of both mechanisms is responsible for ion permeation. For potassium ion permeation, the solubility–diffusion mechanism can properly describe P only for the longer lipids, provided that the permeating species is a hydrated potassium ion of radius 0.331 nm. If the number of carbon atoms in the lipid acyl chain is less than 18, the solubility–diffusion model fails to describe the experimental data accurately, since the computed values and the slope of a P versus d graph are both too small. In this region, permeation through pores better describes the experimental evidence. Equation (10.43) and the k_2 parameter provided by Hamilton and Kaler (1990a,b) give reasonable fits to the data points, as depicted in Fig. 10.14. Apparently, there is a transition from one permeation mechanism to the other in the chain length region between 16 and 18 carbons. It follows that the net permeability of a bilayer can be considered to be the sum of two individual permeabilities, in which the dominating permeation mechanism will account for the majority of the actual value of P and will also dictate the slope of a curve obtained from plotting permeability against bilayer thickness.

Proton Permeation Mechanisms. Earlier work has shown that the permeability of lipid bilayers to protons is five–six orders of magnitude greater than to other monovalent cations (Nichols and Deamer, 1980). This result is consistent with the partitioning model only if every proton carries at least four waters of hydration into the bilayer phase. The alternative is that transient hydrated defects occur in the bilayer. If water in the defects has significant hydrogen bonding, as it does in the bulk phase, hydrogen bond exchange could account for the vastly greater proton permeation rates without any assumptions regarding water of hydration. That is, protons could cross the membrane by "hopping" along hydrogen-bonded chains of water in the defect. This concept was first proposed by Nagle and Morowitz (1978) who coined the term "proton wire" for hydrogen-bonded networks of amino acid side chains. Nagle and Tristram-Nagle (1983) extended this concept to more general proton transport mechanisms, including water, and developed the concept of "transient hydrogen bonded chains" (tHBC) to describe proton transport in clusters of water molecules associated through hydrogen bonds.

Thermal fluctuations in shorter-chain phospholipids produce large numbers of transient defects that cause such bilayers to be relatively permeable to ions, including protons. Because the slopes of the curves for proton and potassium flux are similar, with a reasonable fit to the theoretical expectation of the pore model, both protons and potassium ions presumably have access to the same transient defects. If protons are able to cross the bilayer by hydrogen-bond exchange along water chains in the transient defects, the relatively high permeability of protons can be understood. These results are consistent with the transient pore model described earlier.

As chain length increases, the transient fluctuations become rarer and it is possible that partitioning is the dominant factor limiting permeation of ions such as potassium, and perhaps protons if they cross as $H_9O_4^+$. As a matter of fact both models—partition and transient aqueous pores—give contribution to the membrane permeability.

10.5. FACILITATED TRANSPORT—MOBILE CARRIERS AND CHANNELS

Mobile Carriers. Apart from passive diffusion, all other membrane transport is mediated by specialized protein molecules. This kind of transport is called facilitated transport and proceeds at much higher rates than diffusion alone. Stein (1967, 1986) defined a mobile carrier in the following way. A mobile carrier is a hypothetical particle present in a membrane having the following properties: 1) the carrier alone can be either mobile or immobile in the membrane; 2) it can combine with an ion or a molecule from the surrounding solution to form a relatively stable complex; 3) this complex can move within the membrane; 4) the rate of substrate penetration through a membrane without the carrier is negligible.

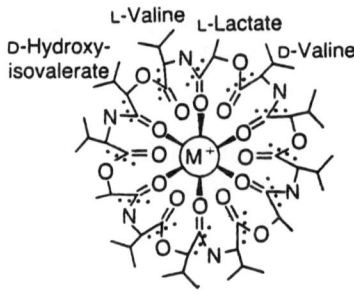

L-Valine L-Lactate
D-Hydroxy-isovalerate D-Valine

Figure 10.15. Structure of valinomycin.

Much of our understanding of the mechanisms of facilitated membrane transport has been gained by studying valinomycin (Fig. 10.15), a naturally occurring polypeptide antibiotic soluble in organic solvents and biological membranes but not in water. Valinomycin can form a complex with inorganic cations rendering it soluble in membranes. This reaction is very selective as demonstrated by the fact that valinomycin preferentially interacts with potassium versus sodium by a ratio of 1000:1 because its internal pocket, lined with negatively charged groups, is just large enough to accommodate K^+ but not larger ions such as sodium. Valinomycin easily extracts potassium from water into organic phases, thus functioning as a mobile carrier.

This type of transport mechanism is presented schematically in Fig. 10.16. A transporter, or a carrier molecule (T), can shuttle between two sides of a membrane and we will designate its state at the left side of the membrane in Fig. 10.16 to be state 1. At this surface the carrier can combine with a substrate A (an ion or a molecule) from the left solution to form a complex AT (state 2) with association and dissociation rates k_{12} and k_{21}, respectively. This complex can cross the membrane (state 3), release the substrate into the right solution (state 4), and then the empty carrier can return to the left membrane surface (state 1).

The principle of detail equilibrium (Markin and Chizmadzhev, 1974) demands that the product of all rate constants in the clockwise direction should be equal to the same product in the counter-clockwise direction:

$$k_{14}k_{43}k_{32}k_{21} = k_{12}k_{23}k_{34}k_{41}. \tag{10.44}$$

To calculate the rate of facilitated membrane transport, one needs to write down a set of differential equations that represent the probability of each transport state P_1, P_2, P_3, and P_4, and solve them. This set is:

$$\frac{dP_1}{dt} = k_{41}P_4 + k_{21}P_2 - k_{14}P_1 - k_{12}[A']P_1 \tag{10.45}$$

$$\frac{dP_2}{dt} = k_{12}[A']P_1 + k_{32}P_3 - k_{21}P_2 - k_{23}P_2 \tag{10.46}$$

$$\frac{dP_3}{dt} = k_{23}P_2 + k_{43}[A'']P_4 - k_{32}P_3 - k_{34}P_3. \tag{10.47}$$

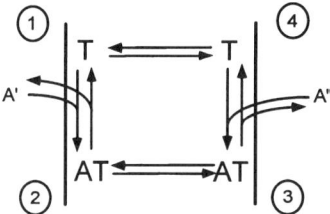

Figure 10.16. Four-state kinetic model of a mobile carrier.

The equation for the last state (P_4) is obtained as a result of solving the first three, taking into consideration that the sum of all probabilities is equal to one:

$$P_1 + P_2 + P_3 + P_4 = 1. \tag{10.48}$$

If one assumes steady-state flux through the membrane, then all time derivatives are zero. Solving this set of equations is simple although the final expressions are rather cumbersome and can be found in several different books (Markin and Chizmadzhev, 1974; Stein, 1986). If these equations are solved and the probabilities of all the states are found, then the flux J_A of particles A through the membrane per carrier can be found using the following equation:

$$J_A = k_{23}P_2 - k_{32}P_3. \tag{10.49}$$

To obtain a simple and convenient expression for J_A, a few assumptions are made, such as: the surface reaction is instantaneous; the membrane is symmetrical in the sense that the rate constants at both sides are equal; and that all membrane transition rates are equal:

$$k_{14} = k_{41} = k_{23} = k_{32}. \tag{10.50}$$

Then the membrane flux is represented by

$$J_A = \frac{1}{2}k_{14}\left(\frac{A'}{K_{des} + A'} - \frac{A''}{K_{des} + A''}\right). \tag{10.51}$$

This flux is driven by the concentration gradient of species A. It becomes zero when the concentrations A' and A'' are equal and it saturates as either concentration increases. Many biological processes follow these principles.

10.6. COUPLED TRANSPORT AND MEMBRANE EQUILIBRIUM

Coupled Transport. In biological membranes transport of one substrate is often coupled to transport of another. There are two types of coupled transport called

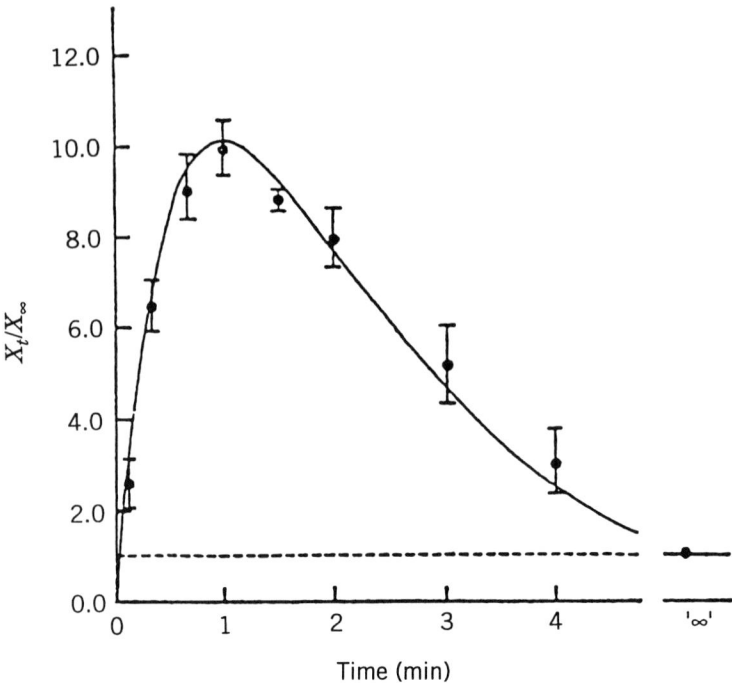

Figure 10.17. Counter-transport of two isotopes of glucose in human red blood cells (Baker and Widdas, 1973; Stein, 1986). The concentration of unlabelled glucose was 76 mM inside and 4 mM outside. The ^{14}C labeled glucose initially was outside only. The ordinate presents the ratio of the concentration of label inside the cell to that outside. The solid curve was computed in the model of mobile carrier and the dashed line shows a concentration ratio of unity driving the uphill flow of the [^{14}C]glucose, such that the gradient of labeled glucose builds up. However, this is a transitive phenomenon and eventually this gradient dissipates (Stein, 1986). Reproduced by permission of Academic Press, Inc.

co-transport (simport) and counter-transport (antiport). As reflected by their names, in co-transport two substrates cross the membrane in the same direction, while in counter-transport they move in opposite directions. In both cases, one of the substrates flows down its concentration gradient generating the electrochemical energy necessary to transport the second substrate against its concentration gradient.

The phenomenon of counter-transport was theoretically predicted by Widdas (1954) and observed experimentally by Rosenberg and Wilbrandt (1957). An example of this type of transport is presented in Fig. 10.17 where the movement of two isotopes of glucose in human red blood cells was studied (Baker and Widdas, 1973; Stein, 1986). Initially, the gradient of unlabelled glucose is able to drive the flow of the [^{14}C]glucose against its concentration gradient. However, this is a transitory phenomenon and eventually this gradient dissipates. The conventional understanding of the mechanism of counter-transport is that both substrates use the same carrier (Fig. 10.18) and that one substrate moves predominantly in one direction

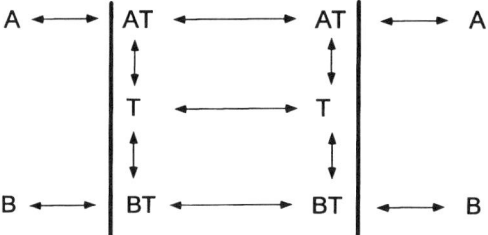

Figure 10.18. The carrier mechanism of counter-transport. The carrier T in the membrane can competitively bind either substrate A or B.

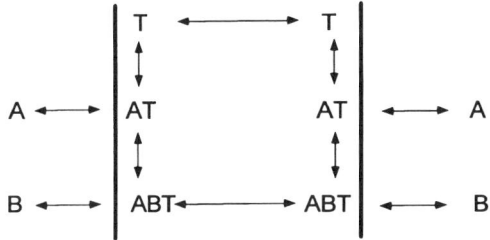

Figure 10.19. The carrier mechanism of co-transport. The carrier T in the membrane can simultaneously bind substrate A or B.

(down its electrochemical gradient) while the other is driven in the opposite direction, against its electrochemical gradient.

In co-transport, the carrier can bind both substrates simultaneously (Fig. 10.19) and in this case the gradient of one species (A) will drive another species (B) in the same direction, even against its own concentration gradient. This transport mechanism commonly occurs in cellular membranes and is demonstrated in Fig 10.20 for the process of methyl glycoside accumulation by isolated intestinal epithelial cells which is driven by the sodium concentration gradient (Kimmich, 1975). In this figure, one can see that the larger the sodium gradient, the greater the accumulation of methyl glycoside.

Coupled transport has an important influence on interfacial equilibrium. Usually interfacial equilibrium is described by the Nernst–Donnan theory which plays a fundamental role in the description of electrical phenomena in membranes selectively permeable for one or several ion species. If only ions of species A penetrate through a membrane from one solution into another they will do so until their electrochemical potential in each solution becomes equal. This leads to a Donnan equilibrium state where a membrane potential ϕ_A is produced across the membrane in accordance with the Nernst formula:

$$\phi_A = \frac{RT}{z_A F} \ln \frac{a''}{a'}. \tag{10.52}$$

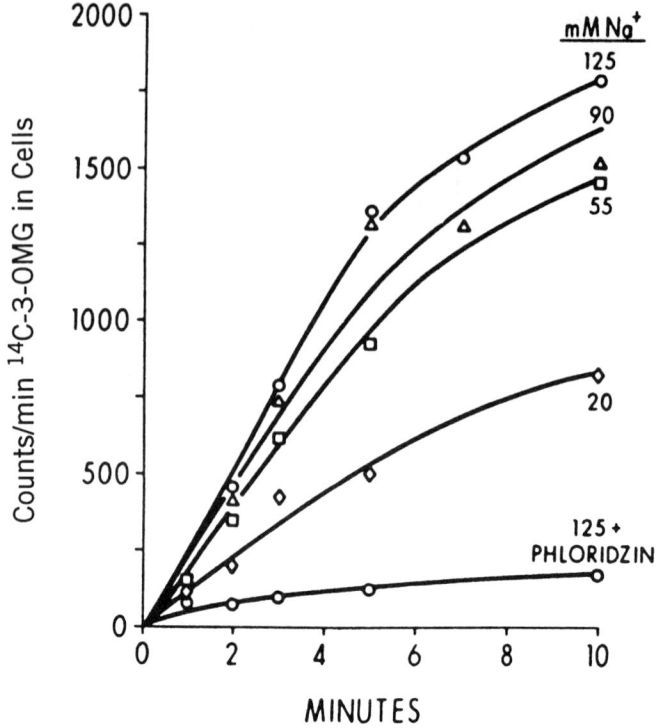

Figure 10.20. Accumulation of methyl glycosides by isolated intestinal epithelial cells driven by the gradient of sodium. Concentration of sodium is indicated at the curves (Kimmich, 1975).

In this equation $\phi_A = \phi' - \phi''$ is the potential difference between one solution (indicated by $'$) and the other ($''$), and a' and a'' are the activities of A-type ions in the respective solutions. The sign of this potential is determined as follows: as equilibrium is being established in the system, ions pass from the more concentrated solution to the less concentrated one and transfer their charge to the latter solution; thus, the sign of the potential on the side where the concentration is lower coincides with the sign of the ion charge.

In reality, Donnan equilibrium does not occur and the open-circuit membrane potential is determined not by the equilibrium state, but during steady-state conditions in which the electrical currents of different ion species balance one another. The steady-state potential is then described under the assumption that different ion species move in the membrane independently. In this case, although the Nernst–Donnan concept does not give the exact magnitude of the membrane potential, it predicts an interval within which the membrane potential may change. From a physical point of view, the situation can be easily understood if the membrane potential is considered to result from multiple-ion Nernst-potential generators connected in parallel. In this case, the potential across the membrane will not exceed the limits specified by individual generators. This is demonstrated in the following case.

Let us consider a membrane permeable for two ion species, A and B. In the case of independent ion transport the partial currents, as a function of the membrane potential ϕ_d, can be represented as

$$i_A = g_A(\phi_d - \phi_A), \qquad i_B = g_B(\phi_d - \phi_B), \qquad (10.53)$$

where the partial conductances g_A and g_B depend on the potential and concentration differences, but must be positive. The stationary diffusion potential ϕ_d is found when the condition $i_A + i_B = 0$ and is equal to:

$$\phi_d = \frac{g_A \phi_A + g_B \phi_B}{g_A + g_B}. \qquad (10.54)$$

If we assume that $\phi_B < \phi_A$, it follows from relation (10.54) that

$$\phi_B \leq \phi_d \leq \phi_A. \qquad (10.55)$$

In other words, in the case of independent ion transport the zero current potential falls between the equilibrium potentials of the individual ion species.

Later we shall be interested in the particular case where a concentration gradient (across the membrane) exists only for one type of ion (for example, A ions), while the activities of other permeable ions (B) are equal on both sides. In this case $\phi_B = 0$, and it follows from relation (10.54) that the potential cannot exceed the equilibrium potential ϕ_A, nor can it have an opposite sign. If the membrane is independently permeable to type B ions, it is clear that this leakage shunts the generator of ϕ_A thereby reducing the magnitude of the potential. However, it cannot, in principle, increase the potential or change its sign.

The Nernst–Donnan theory has been confirmed by experiments with biological membranes and lipid bilayer membranes whose permeability was increased by the addition of different ionophores (Markin and Chizmadzhev, 1974). However, in experiments with nigericin, which generates potassium–hydrogen ion exchange in membranes under certain conditions (Markin et al., 1975, 1977b, Sokolov and Markin, 1984), a quite unexpected result is obtained: the membrane acquires an anomalous potential beyond the limits defined by the Nernst theory. In the case of a pH gradient across a nigericin-containing lipid bilayer membrane, the potential has a sign opposite to the equilibrium potential of a proton-selective membrane (Fig. 10.21). Experiments conducted with a potassium-ion concentration gradient result in a potential whose magnitude is greater than that of the potassium equilibrium potential (Fig. 10.22). These results imply the existence of more complicated mechanisms of potential generation than described by the Nernst–Donnan concept.

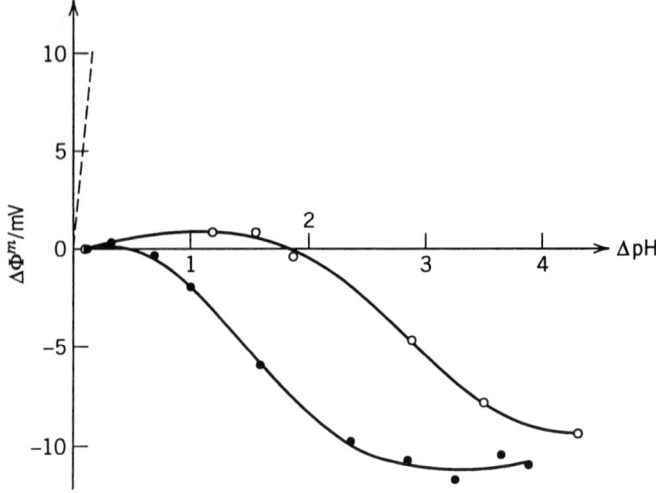

Figure 10.21. Potential difference across a bilayer lipid membrane in the presence of 5 μM nigericin versus the pH gradient. The initial pH value of the buffer solution was 5.8. The concentrations of KCl were: 0.05 M (○) and 0.5 M (●). The dashed line is plotted using the Nernst equation for a proton-selective membrane (Sokolov and Markin, 1984). Reproduced by permission of Elsevier Science S.A.

Membrane Equilibrium. The generalized concept of membrane equilibrium relies upon the condition that ion transport across the membrane ceases. It does not mean complete equilibrium with respect to penetrating ion species. If ions penetrate across the membrane independently, their transport ceases when corresponding electrochemical potentials in the aqueous phases become equal. Therefore, it was assumed that the two conditions—the cessation of ion transport and the equality of the electrochemical potential—are equivalent. However, in more complicated types of transport—coupled transport of several ion species—the transfer of these ions may cease even though their electrochemical potentials on each side of the membrane are not equal. In this case, the membrane potential no longer satisfies the Nernst relation and may have an anomalous magnitude and sign.

Coupled transport of ions across a membrane can be described in terms of the Onsager theory which implies that steady-state flow of ions is linearly related to the generalized forces and that the matrix of kinetic coefficients is symmetric (in this case the generalized force is the difference in electrochemical potential of the dissolved ions). However, since the equilibrium condition, not the steady-state, is of primary interest, a simpler approach can be used to determine the potential (Sokolov and Markin, 1986). Let us consider a membrane where coupled transport of two ion species, A and B, occurs. The activities of dissolved ions on one side of the membrane are a'_A and a'_B and on the other side a''_A and a''_B. The difference in electrical potential between the solutions is $\Delta\phi_m = \phi' - \phi''$. Transport of ions can be represented as a cycle where transfer across the membrane of n_A ions of type A and n_B ions of type

B returns the transport mechanism to its initial state. The quantities n_A and n_B are assumed to be positive if ions are transported from solution (') to solution (") and negative if transport proceeds in the opposite direction. The ratio n_A/n_B is determined by the stoichiometry of the coupled transport system. This ratio can also be defined as the ratio of fluxes of ions A and B because the quantities n_A and n_B divided by the cycle period are proportional to the ion fluxes.

The charge transferred across the membrane per cycle is

$$q = e_0(z_A n_A + z_B n_B). \qquad (10.56)$$

We will define the coupling coefficients θ_A and θ_B as the ratio of charges transferred by A^-- or B^--type ions to the total charge:

$$\theta_A = \frac{z_A n_A}{z_A n_A + z_B n_B}, \qquad \theta_B = \frac{z_B n_B}{z_A n_A + z_B n_B}. \qquad (10.57)$$

The sum of these coefficients is equal to unity:

$$\theta_A + \theta_B = 1. \qquad (10.58)$$

The coupling coefficients show the relative contributions of ions of a given type to the total electric current. Notice that the coefficients θ_A and θ_B resemble the transport numbers usually employed in electrochemistry (Frumkin *et al.*, 1952). It is known that in a solution of constant concentration the transport numbers also give the relative contribution of ions of a given type to the total electric current. However, there is an essential difference between these parameters: in solutions or membranes where ions move independently, the contribution of the coupling coefficients or transport numbers to the total electric current, in the absence of concentration gradients, is always positive. Therefore, while transport numbers range from 0 to 1, coupling coefficients can be less than 0 or greater than 1, provided ions of different types carry current in opposite directions.

Coupled transport can proceed spontaneously provided this process results in a decrease in Gibbs free energy. The decrease in Gibbs free energy per transport cycle $\Delta \bar{G}$ is equal to

$$\Delta \bar{G} = n_A \Delta \bar{\mu}_A + n_B \Delta \bar{\mu}_B \qquad (10.59)$$

when we introduce the differences in the electrochemical potentials of transported ions:

$$\Delta \bar{\mu}_A = z_A e_0 \Delta \phi_m + kT \ln(a'_A/a''_A),$$
$$\Delta \bar{\mu}_B = z_B e_0 \Delta \phi_m + kT \ln(a'_B/a''_B). \qquad (10.60)$$

The Gibbs energy decrease $\Delta \bar{G}$ may be considered as a driving force for coupled transport and as such permits one to determine the direction in which transport can spontaneously proceed and also the values of the parameters at which transport ceases. The potential $\phi_{A,B}$ at which coupled transport ceases can be found when $\Delta \bar{G} = 0$:

$$\phi_{A,B} = \frac{kT}{e_0} \left(\frac{n_A}{z_A n_A + z_B n_B} \ln \frac{a_A''}{a_A'} + \frac{n_B}{z_A n_A + z_B n_B} \ln \frac{a_B''}{a_B'} \right). \qquad (10.61)$$

This equation can be rewritten in a simpler form:

$$\phi_{A,B} = \theta_A \phi_A + \theta_B \phi_B. \qquad (10.62)$$

It follows from this relation that the potential $\phi_{A,B}$ is uniquely determined by the activities of dissolved ions appearing on the right-hand side of equation (10.62) and by the stoichiometry of the coupled transport system which is characterized by the coupling coefficients θ_A and θ_B.

Apparently, under open-circuit conditions the membrane potential $\Delta \phi_m$ can change spontaneously due to coupled electrogenic transport until it reaches the value $\phi_{A,B}$ at which transport ceases. Thus, membrane equilibrium is established in the system. This equilibrium is not equivalent to the Nernst–Donnan equilibrium because the electrochemical potentials of ions on each side of the membrane may not be equal. At the same time, it is easy to demonstrate that the Nernst–Donnan equilibrium is a particular case of the general equilibrium condition considered above. To convert back to selective transport of one ion species, it is sufficient to set one of the coupling coefficients equal to zero: $\theta_B = 0$. In this case, according to equation (10.58), the other coupling coefficient is equal to unity ($\theta_A = 1$) and we revert to the particular case of Donnan equilibrium for a membrane permeable only to type A ions (equation (10.52)).

Unusual Properties of Membrane Potential. Equation (10.62) has rather unusual properties. It can be demonstrated for the case when the activities of type B ions in each solution are equal, $a_B' = a_B''$, and those of type A ions are different $a_A' - a_A''$. In this case $\phi_B = 0$ and we obtain

$$\Delta \phi_m = \theta_A \phi_A. \qquad (10.63)$$

We have to keep in mind two circumstances: 1) no restrictions, either in magnitude or in sign, are imposed on the coupling coefficients; 2) their sum, according to identity (10.58), is always equal to unity. As can be seen from relation (10.63), if the coupling coefficient of ions A is negative, the sign of the membrane potential is opposite to that of the equilibrium Nernst potential ϕ_A. On the other side, if this coefficient exceeds unity, the membrane potential has the same sign as ϕ_A, but is greater in magnitude. Furthermore, it follows from identity (10.58) that while for some ions participating in electrogenic transport the coupling coefficient is negative, for other ions it is positive. It follows then that whereas the concentration gradient of some ions

(A) gives rise to a potential with a sign opposite to that of the equilibrium potential ϕ_A the concentration gradient of the other ions (B) should give rise to a potential which is greater in magnitude than the equilibrium potential (ϕ_B).

Now let us consider an example of coupled transport where there is a 2:1 stoichiometry in the exchange of monovalent cations. Two potassium ions can be exchanged for one hydrogen ion. In such an exchange $n_K = 2$ and $n_H = 1$. Thus, the coupling coefficient of potassium ions is equal to 2, and that of protons is -1. The membrane potential is given by

$$\Delta\phi_m = 2\phi_K - \phi_H. \tag{10.64}$$

If a hydrogen-ion concentration gradient exists, the sign of the potential is always opposite to that of the equilibrium proton potential, and if the concentration of potassium ions exhibits a gradient, the potential is twice the equilibrium potassium potential.

Another example is the electrogenic transport of Na^+ and Ca^{2+} ions which takes place in the sarcolemma. This exchange has been shown to occur with a stoichiometry of three Na^+ ions per one Ca^{2+} ion (Reeves and Hale, 1984). Thus, the coupling coefficient of sodium ions is 3 and that of calcium ions, -2, so that the potential is

$$\Delta\phi_m = 3\phi_{Na} - 2\phi_{Ca}. \tag{10.65}$$

This potential is threefold more sensitive to a change in the concentration of sodium ions than would be the equilibrium potential of the membrane selective to these ions. If, on the other hand, there exists a concentration gradient of calcium ions only, the potential will have the opposite sign and will be twice as high as the equilibrium calcium potential.

Coupled Transport of Ions and Neutral Particles. The potential arising from the coupled transport of ions and neutral particles cannot be determined by equation (10.62) because no equilibrium Nernst potential exists for a neutral particle. Therefore, we shall use equation (10.61) assuming one of the electrical valencies to be equal to zero: $z_B = 0$. Then:

$$\phi_{A,B} = \frac{kT}{z_A e}\left(\ln\frac{a''_A}{a'_A} + \frac{n_B}{n_A}\ln\frac{a''_B}{a'_B}\right) = \phi_A + \frac{kT}{z_A e}\frac{n_B}{n_A}\ln\frac{a''_B}{a'_B}. \tag{10.66}$$

It can be seen that if there is a concentration gradient of type A ions, the potential coincides with the equilibrium Nernst potential ϕ_A. However, when there is a concentration gradient of neutral particles (B), a potential arises whose magnitude and sign depend on the stoichiometric ratio n_B/n_A.

Potentials arising from the concentration gradient of a solute have been observed on industrial liquid membranes (Greyson, 1967). This fact was interpreted as an interaction between ions and water molecules, but there must be a coupling between ion and water fluxes. This coupling, compared with the one considered above, is non-ideal because the coupling coefficients are not constant, instead depending on the difference in electrochemical potential of water and the ions in the solutions. This system is in a steady-state condition rather than in an equilibrium state. The thermodynamic approach below will give a limiting range for the steady-state potential. Because of this uncertainty one can hardly expect quantitative agreement between the equilibrium thermodynamic potential values and the experimental ones measured on industrial membranes. However, qualitative agreement between theoretical and experimental potentials does exist.

Determination of Coupling Coefficients. When coupled transport occurs in a membrane, one can determine the coupling coefficients by measuring the membrane potential. Let us rewrite equation (10.62) in a different form:

$$\phi_{A,B} = -\frac{\theta_A}{z_A e_0}\Delta\mu_A - \frac{\theta_B}{z_B e_0}\Delta\mu_B, \tag{10.67}$$

where $\Delta\mu_A$ and $\Delta\mu_B$ are the differences in the electrochemical potentials of the ions in two solutions

$$\Delta\mu_A = kT\ln\frac{a'_A}{a''_A}, \qquad \Delta\mu_B = kT\ln\frac{a'_B}{a''_B}. \tag{10.68}$$

In the system under consideration, the coefficients θ_A and θ_B do not depend on electrical potential or ion concentrations and can be represented as:

$$\theta_A = -z_A e_0\left(\frac{\partial\phi_{A,B}}{\partial\Delta\mu_A}\right)_{\Delta\mu_B}, \qquad \theta_B = -z_A e_0\left(\frac{\partial\phi_{A,B}}{\partial\Delta\mu_B}\right)_{\Delta\mu_A}. \tag{10.69}$$

If the experimental dependence of membrane potential on chemical potential difference is known, these relations enable one to determine the coupling coefficients. This methodology has led to the determination of the Na^+–Ca^{2+} antiporter stoichiometry (Mullins, 1979; Reeves and Halle, 1984; Toro *et al.*, 1987). However, it should be noted that this problem is only easily solved in ideal systems where coupled transport, with constant stoichiometry, is the only type of transport occurring in the membrane. In real systems, involving several types of transport mechanisms, the problem is much more difficult to solve.

The Case of Several Transport Mechanisms. If other types of transport occur in a membrane along with coupled transport, this can lead to a complicated dependence of membrane potential on ion concentrations such that its unusual properties can be masked. In the general case of an arbitrary concentration distribution, there is no equilibrium in the system, and one can only determine the limits between which the stationary potential can change. As before, the membrane may be considered as having a set of potential generators connected in parallel, which now may include a generator of anomalous potential. The range of stationary potential variation is determined by individual generator potentials, but now the limits are greater, for they include the contribution of the anomalous potential. A definitive value for the potential depends on the kinetic parameters, i.e., on the power of corresponding generators. This can be demonstrated by way of a simple example.

Let independent transport of C-type ions occur in a membrane along with coupled transport of ions A and B. The current carried by type A and B ions can be presented as

$$i_{A,B} = g_{A,B}(\Delta\phi_m - \phi_{A,B}), \qquad (10.70)$$

and the current due to type C ions is given by

$$i_C = g_C(\Delta\phi_m - \phi_C). \qquad (10.71)$$

As in equation (10.53). $g_{A,B}$ and g_C denote the conductances of the respective transport systems. The steady-state potential, determined when the total current is zero, is expressed as

$$\Delta\phi_m = \frac{g_{A,B}\phi_{A,B} + g_C\phi_C}{g_{A,B} + g_C}. \qquad (10.72)$$

The larger the conductance $g_{A,B}$ in comparison to g_C, the closer the potential will be to the value $\phi_{A,B}$. Even in this simple example, determination of the coupling coefficients becomes a difficult task. By using the dependence of the membrane potential on the concentration of type A ions at constant concentrations of ions B and C, one can find the derivative

$$\left(\frac{\partial\Delta\phi_m}{\partial\Delta\mu_A}\right)_{\Delta\mu_B,\Delta\mu_C} = \frac{g_{A,B}}{g_{A,B} + g_C}\left(\frac{\partial\phi_{A,B}}{\partial\Delta\mu_A}\right)_{\Delta\mu_B} = -\frac{g_{A,B}}{g_{A,B} + g_C}\frac{\theta_A}{z_A e_0}. \qquad (10.73)$$

The coupling coefficient is then

$$\theta_A = z_A e_0\left(1 + \frac{g_C}{g_{A,B}}\right)\left(\frac{\partial\Delta\phi_m}{\partial\Delta\mu_A}\right)_{\Delta\mu_B,\Delta\mu_C}, \qquad (10.74)$$

i.e., the result depends on an unknown quantity—the conductance ratio. However, in this case, one can determine the sign of the coupling coefficient because the ratio $g_C/g_{A,B}$ is known to be positive.

One case of interest is that where the same ions are involved in both independent and coupled transport (this means that in the above example type A ions are transported instead of type C ions). In this case type A ions participate in coupled

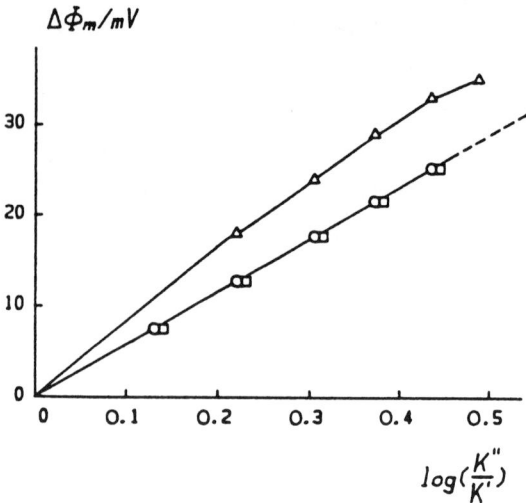

Figure 10.22. Potential difference across a bilayer lipid membrane in the presence of 5 μM nigericin versus the gradient of KCl concentration. The initial KCl concentration in the solution was 0.05 M. The pH values of the buffer solution were: 7.3 (○). 6.0 (□) and 4.2 (△). The dashed line is plotted using the Nernst equation for a membrane selective with respect to K- ions (Sokolov and Markin, 1984). Reproduced by permission of Elsevier Science S.A.

transport together with type B ions as well as being selectively transported alone. The total transport of type A and B ions by both mechanisms can be considered as "partially" coupled. In this type of transport the coupling coefficients are not constant: they depend on the membrane potential and ion concentrations. Nevertheless, all the above-mentioned arguments concerning the steady-state potential remain valid, and we can determine membrane potential by relation (10.72) in which subscript A is substituted for C. This is an important conclusion which implies that one can determine the anomalous potential of a complicated transport system. An example of such a system will be given below.

Nigericin and Grisorixin. Markin *et al.* (1975, 1977b) and Sokolov and Markin (1984) measured the membrane potential during coupled electrogenic transport of potassium and hydrogen ions in black lipid membranes. Membrane transport was induced by the antibiotics nigericin or grisorixin. The magnitude and sign of the membrane potential depended on the experimental conditions. Under certain conditions, membrane potentials were of an anomalous character: they had an opposite sign or a greater magnitude than the equilibrium potentials.

Figure 10.22 shows the membrane potential in the presence of nigericin versus the pH difference of the solutions at both sides of the membrane for several potassium-ion concentrations (equal on both sides of the membrane). The sign of the membrane potential is opposite to that of the equilibrium proton potential (the corresponding dependence for the latter is shown by the dashed line). Figure 10.23 presents the

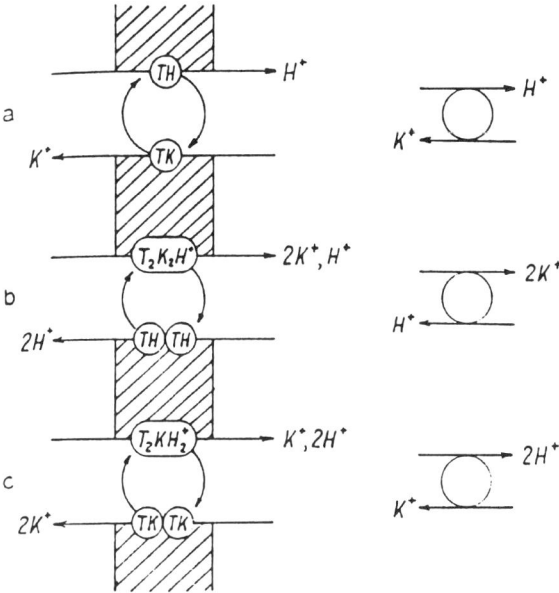

Figure 10.23. Coupled transport of potassium and hydrogen ions realized by different complexes of nigericin in a membrane: (*a*) an electroneutral exchange executed by TH and TK monomers; (*b*) an electrogenic exchange of two potassium ions against one proton executed by $T_2K_2H^+$ dimers and TH monomers; (*c*) an electrogenic exchange of one K^+ ion against two protons executed by $T_2H_2K^+$ dimers and TK monomers. Schematic diagrams of the transport are shown in the left panel and the resulting stoichiometries are presented in the right panel.

membrane potential versus the logarithm of the ratio of potassium-ion concentrations for several pH values. At pH = 6.0 and 7.3 the membrane potential coincides with the equilibrium potassium potential (dashed line), and exceeds the latter at pH = 4.2.

Thus, these experimental observations confirm the existence of potentials of opposite signs or of greater magnitude, as compared with the equilibrium potentials of individual ions. They also confirmed the theoretically predicted relationship between membrane potential and different gradients. With the pH gradient, the sign of the membrane potential is opposite to that of the equilibrium proton potential, and with a potassium-ion concentration gradient, the membrane potential exceeds the equilibrium potential in magnitude. This means that under these conditions the proton coupling coefficient is negative and the potassium coupling coefficient is greater than unity; however, their sum is equal to unity.

Detailed analyses of the mechanisms of nigericin and grisorixin, performed by Amblard *et al.* (1985), Markin *et al.* (1975, 1977b), Sandeaux *et al.* (1978), Sokolov and Markin (1984), and Toro *et al.* (1987) have shown that these antibiotics do not ensure ideal coupling of the transport with constant stoichiometry. The transport involves at least four different complexes of the antibiotic T with potassium and hydrogen ions (Sokolov and Markin, 1984): neutral monomers TK and TH, which can realize electroneutral ion exchange with 1:1 stoichiometry, and positively charged dimers $T_2H_2K^+$ and $T_2K_2H^+$ which make the exchange transport electrogenic (Fig. 10.24). Such a model allows electrogenic exchange transport of potassium and hydrogen ions with 2:1 stoichiometry (Sokolov and Markin, 1986). If transport across a membrane involves two types of complex of the antibiotics with these ions, a neutral monomer TH and a positively charged dimer $T_2K_2H^+$, then the transfer of two potassium ions and one hydrogen ion by $T_2K_2H^+$ in one direction is accompanied by transfer of two hydrogen ions in one direction and two TH monomers in the opposite direction (Fig. 10.24(b)), which ensures the exchange stoichiometry of two potassium ions for one hydrogen ion.

Biological Membranes. Due to the complicated structure and small dimensions of biological membranes it is not possible to measure the membrane potential of coupled transport under ideal conditions. There are a number of factors which hinder the detection of anomalous potentials. Several types of transport always proceed simultaneously in biological membranes; therefore, the potential generated by electrogenic coupled transport can be shunted by other transport mechanisms. Cells control their internal ion concentrations which limit the concentrations at which membrane potential can be measured. In addition one has to keep in mind that anomalous potentials occur at only certain ion concentrations. For example, if a membrane contains a mechanism for coupled transport of type A and B ions and an independent channel for type A ions, there is a possibility of redistribution of concentrations such that Donnan equilibrium will be reached at the equilibrium Nernst potential.

Anomalous potentials across biological membranes have been detected *in vitro* under conditions when ion concentrations were controlled on either side of the membrane. Reeves and Hale (1984) studied a Na^+–Ca^{2+} antiporter in isolated vesicles of sarcoplasmic reticulum. Although no direct measurements were made, the authors succeeded in determining the rate of coupled transport as a function of applied external potential. The coupled transport rate was measured from the change in the ion concentration outside the vesicles, and membrane potential was kept constant due to the gradient of potassium ions in the presence of valinomycin. Transport was induced by a stepwise change in the concentration of sodium ions. It was found that the potential at which coupled transport ceases depends on the concentrations of Na^+ and Ca^{2+} ions both inside and outside the vesicles in accordance with equation (10.65).

Active Coupled Transport. The analysis of coupled transport in previous sections can be extended to active transport generated by ATPases. In this case the membrane potential includes an additional constant equal to the EMF of active transport. If the ATPase exchanges sodium and potassium ions with a stoichiometry of 3 to 2, the membrane potential is given by the following equation:

$$\Delta\phi_m = 3\phi_{Na} - 2\phi_K + E \tag{10.75}$$

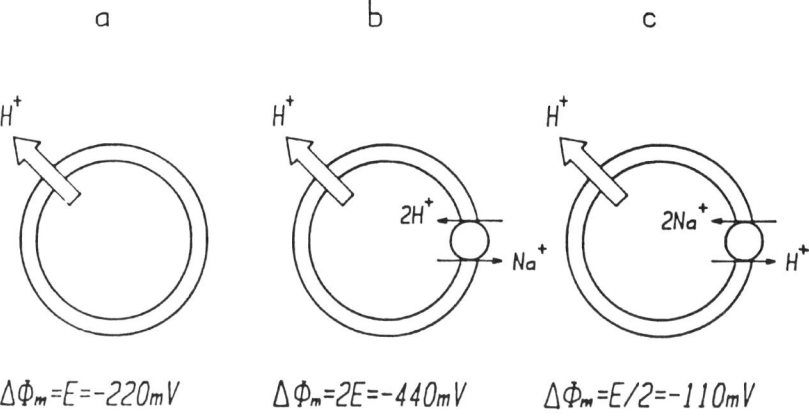

Figure 10.24. A schematic diagram of control of membrane potential by the simultaneous action of active proton transport and electrogenic exchange transport of sodium and hydrogen ions. The membrane contains: (*a*) a proton pump; (*b*) a proton pump and an $Na^+/2H^+$ antiporter; (*c*) a proton pump and a $2Na^+/H^+$ antiporter.

where E is the EMF of active transport. Using equation (10.69), one can determine the relative potential change due to variation of the potassium concentration:

$$\theta_K = -\left(\frac{\partial \Delta \phi_m}{\partial \Delta \mu_K}\right)_{\Delta \mu_{Na}} = -2. \tag{10.76}$$

It is clear that the sign of potential change due to variation of potassium concentration is opposite to the sign of the equilibrium potassium potential. This effect has been observed in membranes of epithelial cells from isolated frog skin by Carrasquer *et al.* (1985, 1986). These authors called the potential they measured, which revealed an extraordinary dependence on the potassium ion concentration, the "anomalous membrane potential of the Na^+-K^+-ATPase."

Membrane Voltage Transformer. As mentioned before, coupled transport can pump certain ions and neutral particles by using the concentration gradients of other ions (see, for example, the review by West, 1980). This kind of transport, which is also called secondary-active transport, does not require an ATPase to transfer each substance. An even more interesting possibility arises when coupled transport can be combined with primary active transport—with an electrogenic pump. This combination can control membrane potential in an unusual way: it can both decrease and increase the membrane potential generated by the ion pump. Let us consider the following example.

Let a vesicle contain an ion pump P(A) (for example, a bacteriorhodopsin proton pump P(H)). If the EMF of the pump giving rise to a flux of type A ions is equal to E, the equilibrium potential at which the pump stops is

$$\phi_{P(A)} = E + \phi_A, \tag{10.77}$$

where $\phi(A)$ is the equilibrium potential of passive transport of type A ions. Let the outer and inner solutions have the same composition, and let a bacteriorhodopsin molecule act as a pump. In this case $\phi_A = \phi_H = 0$ and $\phi_{P(H)} = E$. The value of E, in accordance with Markin *et al.* (1987a), is equal to $-220\,\text{mV}$. The minus sign means that inner volume of the vesicle is negative. In the absence of other transport systems, the pump comes to equilibrium, and the potential across the membrane reaches $-220\,\text{mV}$ (Fig. 10.24(*a*)).

Now, suppose another transport system is incorporated into the membrane: an electrogenic antiporter of type A and B ions with coupling coefficients θ_A and θ_B. This antiporter will operate until the potential $\phi_{A,B}$ given by relation (10.62) is established across the membrane. Equilibrium in the vesicle can be established for both types of transport, provided the potential created by the pump is equal to the antiporter potential

$$\phi_{P(A)} = \phi_{A,B}. \tag{10.78}$$

Solving this equation together with equations (10.62) and (10.76), we obtain an expression for the equilibrium membrane potential

$$\Delta\phi_m = \frac{\theta_A}{\theta_A - 1} E - \frac{\theta_B}{\theta_A - 1} \phi_B. \tag{10.79}$$

If there is an excess of type B ions, operation of the ion antiporter A/B results in a transmembrane gradient of type A ions and negligibly alters the concentration ratio for type B ions (in the initial state their concentration on each side of the membrane is supposed to be equal). Consequently, $\phi_B = 0$ and

$$\Delta\phi_m = \frac{\theta_A}{\theta_A - 1} E. \tag{10.80}$$

Thus, the initial voltage E across the membrane produced by the pump is changed by the coefficient

$$\xi = \frac{\theta_A}{\theta_A - 1}. \tag{10.81}$$

This parameter can be called the voltage transformation coefficient which can be either greater or less than unity, depending on the coupled transport stoichiometry. Hence, the transformer can both raise and lower the voltage generated by the pump. If we assume, as in Fig. 10.24(*b*), the exchange stoichiometry to be $H^+:Na^+ = 2:1$, then $\theta_H = 2$ and $\xi = 2$, i.e., we obtain a step-up transformer with a transformation coefficient of 2. The potential that arises across the vesicle membrane in this case will be equal to -440 mV (Fig. 10.24(b)). For an exchange stoichiometry $H^+:Na^+ = 1:2$, the transformation coefficient is 1/2, i.e., this device transforms -220 mV into -110 mV (Fig. 10.24(c)). Other values for the transformation coefficient can occur depending on the antiporter stoichiometry.

The process of voltage transformation just described is related to the fact that the ion antiporter A/B, together with the ion pump P(A), redistributes ion concentrations. In the first case where the stoichiometry of the antiporter is $H^+:Na^+ = 2:1$ (Fig. 10.24(b)) the situation seems paradoxical: in response to each proton ejected by the pump the antiporter brings two protons back into the vesicle and although the pump ejects protons from the vesicle, their concentration inside the vesicle increases. This is a remarkable feature of an electrogenic antiporter operating in conjunction with an ion pump. Figure 10.24(c) shows the situation corresponding to a stoichiometry where $H^+:Na^+ = 1:2$. In this case the system operates in such a way that, together with each proton ejected by the pump from the vesicle, the antiporter ejects another proton. As a result, the concentration of protons inside the vesicle decreases, as does the membrane potential.

The transformer just described is only hypothetical. However, experimental data exist suggesting the possibility of such a mechanism. Many cells contain ion pumps and electrogenic exchangers. For example, *Halobacterium cells* have not only a proton pump (bacteriorhodopsin), but also an electrogenic H^+/Na^+ antiporter (Murokami and Konishi, 1988). This combination can provide a sophisticated control mechanism for membrane potential.

11

MECHANICS OF INTERFACES

11.1. DEFINITION OF SURFACE TENSION AT A NON-SPHERICAL INTERFACE

The concept of surface tension has existed for centuries, but controversy still surrounds its definition in regard to non-spherical surfaces (Boruvka *et al.*, 1985; Buff, 1956; Gibbs, 1928; Markin *et al.*, 1988; Markin and Kozlov, 1989; Melrose, 1968; Rowlinson, 1983). The problem of capillary phenomena at arbitrarily shaped surfaces was first pursued by Gibbs (1928). Buff (1956) later expanded upon Gibbs approach and analyzed capillarity in more detail. Melrose (1968) subsequently reviewed Buff's work and made several additions. However, Boruvka and Neumann (1977) reconsidered Buff's work and concluded it did not adequately describe the capillarity problem for non-spherical surfaces because of an improper subdivision between extensive and intensive variables leading to an incorrect generalization of the Laplace equation. In order to rectify this drawback, they formulated a fundamentally new thermodynamic equation for a non-spherical surface in terms of new variables and developed a new version of the generalized Laplace equation.

Their paper was later extensively cited by many scientists, none of whom expressed an opinion concerning the argument between Boruvka and Neumann and Buff. However, Rowlinson (1983) expressed doubts about the correctness of Boruvka and Neumann's equations and Boruvka *et al.* (1985) offered a new interpretation of a dividing surface which was later criticized by Markin and Kozlov (1989). The following discussion presents the current description of surface tension for non-spherical surfaces.

Fundamental Equations for an Interface. Gibbs Approach. As discussed previously, in the absence of an external field, the Gibbs equation for the grand thermodynamic potential of a surface, Ω^s, can be presented in the following form:

$$d\Omega^s = -S^s\,dT - \sum_i N_i^s\,d\mu_i + \gamma_G\,dA + AC_{xG}\,dc_x + AC_{yG}\,dc_y \qquad (11.1)$$

where γ_G is Gibbs surface tension, c_x and c_y are the principal curvatures of the Gibbs dividing surface. The products AC_{xG}, and AC_{yG} in front of the differentials of curvature include the area A of the interface. The subscript G stands for Gibbs representation. The surface is assumed uniform with respect to the curvature.

Surface tension is defined as the derivative with respect to area:

$$\gamma_G = \left(\frac{\partial\Omega^s}{\partial A}\right)_{T,\mu_i,c_x,c_y}, \qquad (11.2)$$

and the coefficients:

$$C_{xG} = \frac{1}{A}\left(\frac{\partial\Omega^s}{\partial c_x}\right)_{T,\mu_i,c_y}; \qquad C_{yG} = \frac{1}{A}\left(\frac{\partial\Omega^s}{\partial c_y}\right)_{T,\mu_i,c_x} \qquad (11.3)$$

represent bending moments per unit length of a line in the dividing surface.

Area (A) is the only extensive variable in equation (11.1). The remaining variables are intensive ones. Thus, using Euler's theorem, we obtain the equation:

$$\Omega^s = \gamma_G A. \qquad (11.4)$$

If we write the grand potential for an entire two-phase system, we will have volume contributions from the α and β phases enter the expression:

$$\Omega = -P^\alpha V^\alpha - P^\beta V^\beta + \gamma_G A, \qquad (11.5)$$

where P^α and P^β are the pressures in the bulk phases.

Note that no simplifications have been made up to now and no terms have been neglected, such that all formulae are of a general nature. The next step made by Gibbs consisted of changing from the principal curvatures c_x and c_y to their sum, $c_x + c_y$, and difference, $c_x - c_y$. Gibbs then neglected the difference term, because it is much smaller than the sum, and made the sum term equal to zero by an appropriate choice of dividing surface. Further analysis was carried out for the surface of tension.

The Buff and Melrose Approach. Buff (1956) repeated Gibbs transformation of equation (11.1) to incorporate the sum and difference of the principal curvatures and also neglected the difference term. However, he retained the sum term with the sum of the curvatures in his calculations and obtained a generalized Laplace equation. In addition, Buff presented two hydrostatic interpretations for surface tension and torque. The neglected difference term was later accounted for by Melrose (1968) and included in the generalized Laplace equation. We will present these results in their updated version.

Let us introduce the mean (or first) curvature $J = c_x + c_y$ and the Gaussian (or second) curvature $K = c_x c_y$. Using these variables, equation (11.1) can be written in the following form:

$$d\Omega^s = -S^s\,dT - \sum_i N_i^s\,d\mu_i + \gamma_G\,dA + AC_1\,dJ + AC_2\,dK. \qquad (11.6)$$

The coefficients C_1 and C_2 are now referred to as the first and the second torques.

The surface tension is defined by the following relationship

$$\gamma_G = \left(\frac{\partial\Omega^s}{\partial A}\right)_{T,\mu_i,J,K}. \qquad (11.7)$$

Despite the fact that expressions (11.2) and (11.7) are formally different, they produce the same result, since fixing the principal curvatures c_x and c_y means fixing the mean and Gaussian curvatures J and K as well. Therefore, the definition proposed for surface tension by Buff and Melrose does not differ, in principle, from that of Gibbs. The integral formulae (11.4) and (11.5) are not fundamentally altered by this change of variables at all, because J and K remain as intensive variables, as do the original variables c_x and c_y. As for the torques C_1 and C_2, the relation between them and the torques C_{xG} and C_{yG} can be expressed as follows:

$$C_{xG} = C_1 + C_2 c_y, \qquad C_{yG} = C_1 + C_2 c_x, \qquad (11.8)$$

or

$$C_1 = \frac{C_{xG}c_x - C_{yG}c_y}{c_x - c_y}, \qquad C_2 = \frac{C_{xG} - C_{yG}}{c_x - c_y}. \qquad (11.9)$$

Thermodynamic definitions for the new coefficients are given as follows:

$$
\begin{aligned}
C_1 &= \frac{1}{A}\left(\frac{\partial\Omega^s}{\partial J}\right)_{T,\mu_i,A,K} = \left(\frac{\partial\omega^s}{\partial J}\right)_{T,\mu_i,K},\\[2mm]
C_2 &= \frac{1}{A}\left(\frac{\partial\Omega^s}{\partial K}\right)_{T,\mu_i,A,J} = \left(\frac{\partial\omega^s}{\partial K}\right)_{T,\mu_i,J},
\end{aligned}
\qquad (11.10)
$$

where ω^s denotes the grand potential of the surface per unit area.

The equilibrium condition of the surface is determined from the minimum of the grand potential of the two-phase system, equation (11.6). Minimizing this potential at constant T and μ_i, we obtain the generalized Laplace equation:

$$\gamma_G J - C_1(J^2 - 2K) - C_2 JK = p^\alpha - p^\beta, \qquad (11.11)$$

for which we will consider the curvature positive if its center lies in the α phase. Melrose (1968) presented this equation in a slightly different form and called it the Gibbs–Kelvin equation.

The Approach of Boruvka and Neumann. Boruvka and Neumann (1977) believed that Buff and Melrose made a mistake in choosing the intensive variables J and K. Therefore they introduced another pair of extensive variables

$$\bar{J} = \iint_{(A)} J \, dA, \qquad \bar{K} = \iint_{(A)} K \, dA. \tag{11.12}$$

One can regard a small part of the surface as being uniform and therefore:

$$\bar{J} = JA, \qquad \bar{K} = KA. \tag{11.13}$$

Using these variables, the basic equation (11.6) transforms to:

$$d\Omega^s = - S^s \, dT - \sum_i N_i^s \, d\mu_i + \gamma_{BN} \, dA + C_1 \, d\bar{J} + C_2 \, d\bar{K}. \tag{11.14}$$

In this equation a new symbol is introduced, γ_{BN}, as a coefficient for the area differential. The definition of this quantity is

$$\gamma_{BN} = \left(\frac{\partial \Omega^s}{\partial A} \right)_{T, \mu_i, \bar{J}, \bar{K}}. \tag{11.15}$$

Boruvka and Neumann called the quantity γ_{BN} the surface tension, overlooking the fact that it differs significantly from the Gibbs surface tension, γ_G. It is instructive to find the relationship between these quantities. Substituting formulae (11.13) into (11.14) and comparing the result with equation (11.6), one arrives at the following relation:

$$\gamma_G = \gamma_{BN} + C_1 \, dJ + C_2 \, dK. \tag{11.16}$$

The same result can be obtained using Euler's theorem in equation (11.14):

$$\Omega^s = (\gamma_{BN} + C_1 J + C_2 K) A. \tag{11.17}$$

Comparison with equation (11.4) yields equation (11.16).

Based upon the interface being at equilibrium, Boruvka and Neumann (1977) derived their own version of the generalized Laplace equation

$$\gamma_{BN} J + 2 C_1 K = p^\alpha - p^\beta \tag{11.18}$$

which they regard to be a more correct expression than that of Buff and Melrose.

However, the substitution of γ_{BN} from equation (11.16) into (11.18) makes the equation by Boruvka and Neumann identical to the equation of Buff and Melrose. Therefore, the analysis of the mechanical properties of a curved interface introduced by Buff and Melrose is correct. As for using extensive or intensive variables, it is possible to work with either set, as long as derivatives from different sets are not mixed.

Non-spherical Interfaces Described Hydrostatically. Mechanical properties of plane and spherical interfaces are often described using a hydrostatic approach based on local thermodynamic assumptions (Bakker, 1928; Ono and Kondo, 1960; Rusanov, 1967). Buff (1956) was the first to use this approach to describe non-spherical boundaries. This approach employs Gibbs method of describing phase boundaries using surface excesses (in this case excesses of forces and torques).

Recently a new approach to surface thermodynamics has come into practice which is not based upon the concept of surface excesses and which may be called a continuous description of the system (Goodrich, 1969). It has a number of advantages and we will apply it to the hydrostatic description of non-spherical interfaces. Let us consider a system of two liquid phases with a boundary of arbitrary form. Let the z-axis pass through an arbitrary point normal to the phase boundary, while the x- and y-axes specify the principal curvature planes (Fig. 11.1). Assume that the pressure tensor **P** is a diagonal such that $P_{zz} = P_N$, $P_{xx} = P_{yy} = P_T$, where both P_N and P_T depend on the coordinate z. (This assumption is an approximation, since in general there are two different transverse components, P_{xx} and P_{yy}. The treatment of the general case can be found in Markin *et al.* (1988).)

Consider a liquid body, bounded from above and below by two dividing Gibbs surfaces, and at its sides by cross-sections which are normal to the dividing surfaces and define the principal curvatures of the phase boundary. Points Z_x and Z_y designate the positions of the center of curvature for these sections perpendicular to the x- and y-axes, respectively. For any dividing surface its curvatures will be given by:

$$c_x(Z) = \frac{1}{Z - Z_x}; \qquad c_y(Z) = \frac{1}{Z - Z_y};$$
$$J(Z) = c_x(Z) + c_y(Z); \qquad K(Z) = c_x(Z)c_y(Z). \qquad (11.19)$$

The equilibrium condition will be given in Melrose's (1968) form:

$$\frac{dP_N}{dZ} = (P_T - P_N)J(Z). \qquad (11.20)$$

Integration from an arbitrary Z_α within the bulk of the α phase to an arbitrary Z_β within the bulk of the β phase gives the following relationship between the α and β phase pressures:

$$P^\beta - P^\beta = \int_{Z_\alpha}^{Z_\beta} (P_T - P_N)J(Z)\,dZ. \qquad (11.21)$$

To calculate the work which must be done to deform the chosen liquid body, the strain

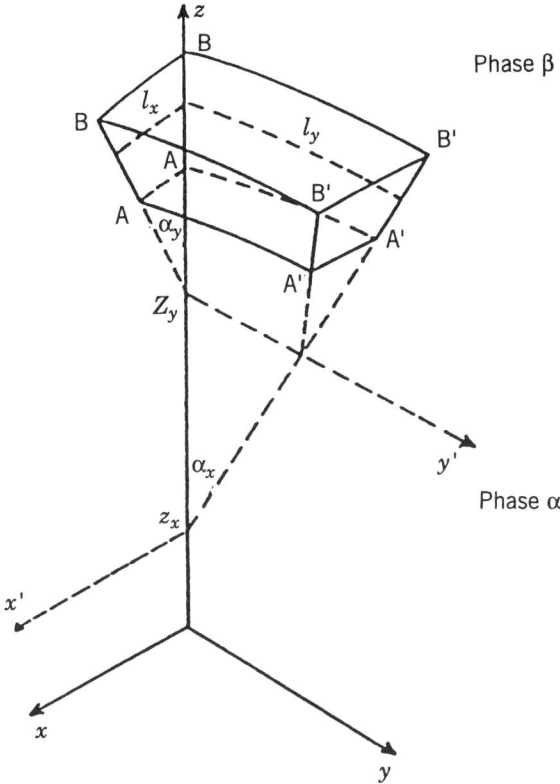

Figure 11.1. The volume selected at an interface.

can be presented as a sum of strains of elementary layers contained between two closely opposed dividing surfaces. Consider such an elementary layer with coordinate Z, thickness h, and area $A(Z)$. As the thickness increases by δh and the area increases by $\delta A(Z)$, the work can be presented as follows:

$$\delta W = -P_N A(Z)\delta h - P_T h\delta A(Z) = -P_N\delta V + (P_N - P_T)\delta A(Z). \quad (11.22)$$

The first term is the work of changing the volume V, while the second is the work of extending the elementary layer. The surface properties we are interested in relate solely to the second term. Therefore, the total work of deformation of the phase boundary at constant volume can be found by summing the last term in each elementary layer over the entire interfacial region. To perform the summation, let us write down a number of simple geometric relationships. The change in the area of the dividing surface is described by the equation:

$$\frac{dA}{dZ} = J(Z)A(Z). \quad (11.23)$$

Integrating this equation one obtains:

$$A(z) = [1 + (z - Z_s) J + (z - Z_s)^2 K]A. \tag{11.24}$$

Here the area of an arbitrary dividing surface is compared with the area of some chosen surface passing through the coordinate Z_s. Let the surface area be denoted by $A(Z_s)$ and curvatures be denoted by $J(Z_s)$ and $K(Z_s)$. To simplify the notation we shall write these quantities simply as A, J and K. It can be shown that

$$\frac{dJ(Z)}{dZ} = 2J^2(Z)K(Z), \qquad \frac{dK(Z)}{dZ} = -J(Z)K(Z), \tag{11.25}$$

and

$$J(Z) = \frac{J + 2(Z - Z_s)K}{1 + (Z - Z_s)J + (Z - Z_s)^2 K}, \tag{11.26}$$

$$K(Z) = \frac{K}{1 + (Z - Z_s)J + (Z - Z_s)^2 K}. \tag{11.27}$$

When the area and curvatures of a chosen dividing surface change by δA, δJ, and δK, respectively, then it follows from the isochoric condition that the surface and coordinate of an arbitrary dividing surface will change by

$$\delta A(Z) = [1 + (Z - Z_s)J + (Z - Z_s)^2 K]\,\delta A + (Z - Z_s)A\,\delta J$$
$$+ (Z - Z_s)^2 A\,\delta K + [J + 2(Z - Z_s)K]A\,\delta Z, \tag{11.28}$$

$$\delta Z = -\frac{(Z - Z_s)[1 + \frac{1}{2}(Z - Z_s)J + \frac{1}{3}(Z - Z_s)^2 K](\delta A/A) + \frac{1}{2}(Z - Z_s)^2\,\delta J + \frac{1}{3}(Z - Z_s)^3\,\delta K}{1 + (Z - Z_s)J + (Z - Z_s)^2 K}. \tag{11.29}$$

Now we can replace h by δZ in the expression for the work of deformation and integrate it over the entire body at constant volume:

$$\delta W = \int_{Z_\alpha}^{Z_\beta} (P_N - P_T)\,\delta A(Z)\,dZ. \tag{11.30}$$

After integration we find

$$\delta W = \gamma_G\,\delta A + AC_1\,\delta J + AC_2\,\delta K, \tag{11.31}$$

where

$$\gamma_G = \int_{Z_\alpha}^{Z_\beta} dZ(P_N - P_T)$$

$$\times \left\{ \frac{[1 + (Z - Z_s)J + (Z - Z_s)^2 K] - }{[(Z - Z_s)J + 2(Z - Z_s)^2 K] \times [1 + \frac{1}{2}(Z - Z_s)J + \frac{1}{3}(Z - Z_s)^2 K]}{1 + (Z - Z_s)J + (Z - Z_s)^2 K} \right\}, \tag{11.32}$$

$$C_1 = \int_{Z_\alpha}^{Z_\beta} dZ(Z - Z_s)(P_N - P_T)\left\{1 - \frac{[(Z - Z_s)J + 2(Z - Z_s)^2 K]}{2[1 + (Z - Z_s)J + (Z - Z_s)^2 K]}\right\}, \tag{11.33}$$

$$C_2 = \int_{Z_\alpha}^{Z_\beta} dZ(Z - Z_s)^2(P_N - P_T)\left\{1 - \frac{[(Z - Z_s)J + 2(Z - Z_s)^2 K]}{3[1 + (Z - Z_s)J + (Z - Z_s)^2 K]}\right\}. \tag{11.34}$$

Eq. (11.31) coincides with the expression for the work of deformation of the phase boundary in equation (11.6) used by Buff and Melrose. Hence, we have arrived at a hydrostatic definition for Gibbs surface tension (11.32), as well as for the first and second moments, equations (11.32) and (11.33). We have not used the Gibbs method, which employs surface excesses. As for the value of the surface tension in the equations of Boruvka and Neumann, one can find it to be

$$\gamma_G = \int_{Z_\alpha}^{Z_\beta} dZ(P_N - P_T)\frac{1 + (Z - Z_s)^2 K}{1 + (Z - Z_s)J + (Z - Z_s)^2 K}. \tag{11.35}$$

Generalized Laplace Equation. The equations that we have obtained make it possible to arrive at the Laplace equation in the most natural way. Let us use the equilibrium condition of the liquid presented in the integral form of equation (11.21). Substituting expression (11.26) into this equation for $J(Z)$, one obtains:

$$P^\alpha + P^\beta = \int_{Z_\alpha}^{Z_\beta} dZ(P_N - P_T)\frac{J + 2(Z - Z_s)K}{1 + (Z - Z_s)J + (Z - Z_s)^2 K}. \tag{11.36}$$

A comparison of the right-hand side of this equation with expressions (11.32)–(11.34) for surface tension and moments immediately yields the generalized Laplace equation in the form of Buff and Melrose (11.11).

Hydrostatic Approach in Gibbs Formalism. For the sake of completeness, and in order to compare these results with well-known hydrostatic formulae, we present the results of similar calculations within the framework of the Gibbs method. These results can be obtained immediately from equations (11.32)–(11.34). Expression (11.32) for surface tension γ_G can be rewritten with equations (11.20) and (11.26) as follows:

$$\gamma_G = \int_{Z_\alpha}^{Z_\beta} dZ(P^{\alpha\beta} - P_T)[1 + (Z - Z_s)J + (Z - Z_s)^2 K] \tag{11.37}$$

where

$$P^{\alpha\beta} = \begin{cases} P^\alpha, & \text{if } Z \leq Z_s, \\ P^\beta, & \text{if } Z > Z_s. \end{cases} \tag{11.38}$$

Similarly, expressions for torques can be presented in the following form:

$$C_1 = \int_{Z_\alpha}^{Z_\beta} dZ(P^{\alpha\beta} - P_T)(Z - Z_s), \tag{11.39}$$

$$C_2 = \int_{Z_\alpha}^{Z_\beta} dZ(P^{\alpha\beta} - P_T)(Z - Z_s)^2. \tag{11.40}$$

Formulae of this type are commonly obtained by using Gibbs method and the appropriate reference system from the very beginning. The original Gibbs method of describing phase boundaries was based on surface excesses of *scalar* characteristics of the system. However, one can work equally well with vector quantities. Consider the pressure acting upon the area AABB (Fig. 11.1) in a real system and in the Gibbs reference system. The density of surface excess of force is:

$$\Gamma_{py} = -\int_{Z_\alpha}^{Z_\beta} dZ(P^{\alpha\beta} - P_T)[1 + (Z - Z_s)c_x]. \tag{11.41}$$

Similarly one may write the surface excess of the force acting along the x-axis:

$$\Gamma_{px} = -\int_{Z_\alpha}^{Z_\beta} dZ(P^{\alpha\beta} - P_T)[1 + (Z - Z_s)c_y]. \tag{11.42}$$

Quantities c_x and c_y in equations (11.41) and (11.42) refer to the chosen surface S and hence are independent of the z-coordinate.

In a similar way we can introduce surface excesses of torques which bend the phase boundary around the lines l_x and l_y:

$$\Gamma_{Cx} = -\int_{Z_\alpha}^{Z_\beta} dZ(P^{\alpha\beta} - P_T)[1 + (Z - Z_s)c_y](Z - Z_s) = C_{xG}, \tag{11.43}$$

$$\Gamma_{Cy} = -\int_{Z_\alpha}^{Z_\beta} dZ(P^{\alpha\beta} - P_T)[1 + (Z - Z_s)c_x](Z - Z_s) = C_{yG}. \tag{11.44}$$

Excesses of tangential pressure and of bending moments appear to be different in different planes, in spite of the fact that the tangential pressure is assumed isotropic. This phenomenon is due to the geometries of the sections (namely, the curvatures of the boundary are different in two sections).

Surface excesses Γ_{C_x} and Γ_{C_y} are essentially the torques in equations (11.1) and (11.4). Surface excesses Γ_{P_x} and Γ_{P_y} resemble, at a first glance, the surface tension taken with the opposite sign. However, as we shall see later, this is not the case. Nevertheless, these excesses are convenient for describing the mechanical equilibrium of a dividing surface. Consideration of forces gives the following analogue of the Laplace equation:

$$P^\alpha - P^\beta = -\Gamma_{P_x} c_y - \Gamma_{P_y} c_x. \tag{11.45}$$

We are now able to find the torques C_1 and C_2 by substituting the expressions for C_{xG} and C_{yG} into equation (11.9), and arrive immediately at equations (11.39) and (11.40).

Surface Tension. Gibbs definition of the surface tension γ_G, treats this quantity as the specific work of formation of a new surface with the same curvature. To produce such work, one must not only extend this surface but also bend it. Consider the work of formation of the surface contained in the volume shown in Fig. 11.1. Let this surface be formed by the motion of the segment l_x along the segment l_y. The first part of the work is the work against the force $-\Gamma_{P_y} l_x$ through the distance l_y, i.e., it is equal to $-\Gamma_{P_y} l_x l_y$. The second part is the work against the torque $C_{xG} l_x$ produced when the surface is bent by an angle $\alpha_x = l_y c_x$. Therefore, the second part of the work is equal to $C_{xG} c_x l_x l_y$. The total work of forming the new surface is as follows:

$$(-\Gamma_{P_y} + C_{xG} c_x) l_x l_y = l_x l_y \int_{Z_\alpha}^{Z_\beta} dZ (P^{\alpha\beta} - P_T)[1 + (Z - Z_s)J + (Z - Z_s)^2 K]. \tag{11.46}$$

Several important conclusions flow from equation (11.46). First, this equation is symmetrical with respect to x and y. This means that the result is independent of the way the new surface is formed, i.e., the movement of the segment l_x or l_y. Secondly, for the Gibbs surface tension we obtain equation (11.37), in agreement with the corresponding result of Melrose. We can rewrite it in a different form:

$$\gamma_G = \int_{Z_\alpha}^{Z_\beta} dZ (P^{\alpha\beta} - P_T) + C_1 J + C_2 K. \tag{11.47}$$

Comparing it with equation (11.1.16) we find the surface tension of Boruvka and Neumann:

$$\gamma_{BN} = \int_{Z_\alpha}^{Z_\beta} dZ (P^{\alpha\beta} - P_T). \tag{11.48}$$

The difference between these two definitions is clearly seen. The physical meaning of the quantity introduced by Boruvka and Neumann is not easy to interpret. One may say that their surface tension is that of a hypothetical plane phase boundary, having the same distribution of pressures $P^{\alpha\beta}$ and P_T as in the system considered. However, this statement would be contradictory, because a plane surface will not have any pressure differential.

Surface of Tension. Gibbs considered a surface of tension which was located in such a position that the sum of coefficients $C_{xG} + C_{yG}$ is zero. As noted by Rusanov (1967), in this case the bending of the surface layer does not affect the surface energy. Kondo (1956) studied the dependence of the surface tension on the position of the dividing surface in a spherical case, and showed that this dependence can be described by a curve with a single minimum. The dividing surface corresponding to this minimum is the surface of tension, while the derivative of the surface tension with respect to the coordinate determining the dividing surface position, $(\partial\gamma/\partial Z)^*$, corresponds to an imaginary displacement of the dividing surface and vanishes at this point. Let us apply this approach to the analysis of γ_G and γ_{BN} in a general case.

Kondo's equation can be presented in the following form:

$$\gamma_G J + \left(\frac{\partial\gamma_G}{\partial Z_s}\right)^* = P^\alpha - P^\beta. \tag{11.49}$$

The function $\gamma_G = \gamma_G(Z_s)$ has an extreme at the point where $(\partial\gamma_G/\partial Z_s)^* = 0$ (Fig. 11.2). The dividing surface passing through this point is the surface of tension. It is commonly believed that the extreme value of the function $\gamma_G = \gamma_G(Z_s)$ is a minimum. In the general case, however, it is not necessarily so. If there is a film (membrane) at the phase boundary, which has an elasticity of bending, then the tension of the phase boundary may be both positive and negative, the boundary remaining stable. In this case the extreme value is a maximum (curve 2 in Fig. 11.2). Combining these two cases, one may say that at the surface of tension it is the absolute value of the surface

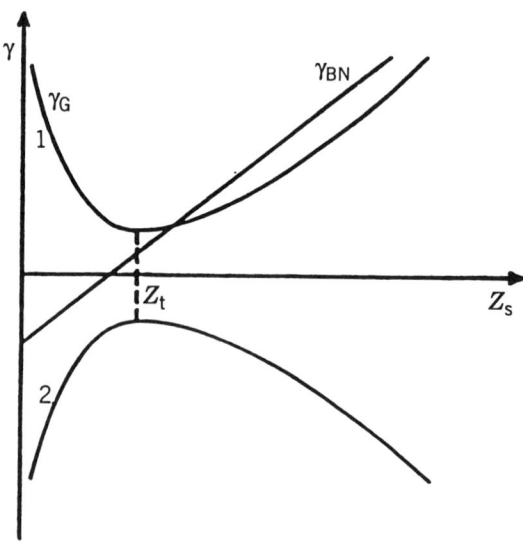

Figure 11.2. Surface tension as a function of position of the dividing surface.

tension which is minimal. Using equation (11.12) we can relate the derivative $(\partial \gamma_G / \partial Z_s)^*$ to the torques:

$$\left(\frac{\partial \gamma_G}{\partial Z_s}\right)^* = -C_1(J^2 - 2K) - C_2 JK. \tag{11.50}$$

It is this expression which becomes zero at the surface of tension. A linear combination of torques C_1 and C_2 (rather than each of them separately) becomes zero at the surface of tension. In Gibbs' variables which contain principal curvatures c_x and c_y, the surface of tension is defined as follows:

$$C_{xG} c_x^2 + C_{xG} c_y^2 = 0. \tag{11.51}$$

On the basis of general expressions (11.50) and (11.51) we can easily consider any special case. The case of a spherical surface of tension is instructive and deserves special attention. In this case equation (11.50) becomes

$$C_1 R + C_2 = 0 \tag{11.52}$$

where R is the radius of the surface of tension, while equation (11.51) transforms into

$$C_{xG} = C_{yG} = 0. \tag{11.53}$$

Gibbs defined the surface of tension by the approximate condition $C_{xG} + C_{yG} = 0$. Comparing formulae (11.8) and (11.50), we see that such an approximation means neglecting the term $C_1(J^2 - 4K)$ in equation (11.50). Note that at a spherical surface of tension this term is zero, and the Gibbs condition becomes exact.

Now let us consider how the surface tension of Boruvka and Neumann depends on the position of a dividing surface Z_s. It follows from equation (11.48) that

$$\left(\frac{\partial \gamma_{BN}}{\partial Z_s}\right)^* = P^\alpha - P^\beta. \tag{11.54}$$

This means that γ_{BN} depends linearly on the z-coordinate. The slope of this dependence is larger than the asymptotic slope of corresponding dependence $\gamma_G(Z_s)$ (Fig. 11.2). Therefore, the definition of surface tension by Boruvka and Neumann does not permit one to choose the surface at which the surface tension has given properties.

Let us now clarify the relationship existing between γ_G and γ_{BN} at the surface of tension determined by equation (11.50). In particular, it is important to know whether they coincide at the surface. By excluding one of the torques at the surface of tension in equation (11.16) we obtain:

$$\gamma_G = \gamma_{BN} + \frac{2C_1 K}{J}. \tag{11.55}$$

As mentioned above, the torque C_1 is generally non-zero at the surface of tension. Therefore γ_G and γ_{BN} do not coincide at this surface, although they may do so at some other surface.

In conclusion, this analysis justifies the existence of two different definitions of surface tension. The definition suggested by Boruvka and Neumann has a simpler hydrostatic representation and leads to a more compact form of the generalized Laplace equation. However, its physical meaning is more complicated and is not as easy to interpret as that of Gibbs definition which overall proves to be much more convenient when searching for the surface of tension.

11.2. ELASTIC PROPERTIES OF INTERFACES AND THE SHAPE OF VESICLES

Interfaces with Low Tension. In the previous section we defined the differential of the grand thermodynamic potential $d\Omega$. A similar equation can be written for the Helmholtz free energy:

$$dF^s = - S^s \, dT - \sum_i \mu_i \, dN_i^s + \gamma_G \, dA + AC_1 \, dJ + AC_2 \, dK. \qquad (11.56)$$

The last three terms define the mechanical work of interface deformation which is important for the analysis of different mechanical processes in biological and lipid bilayer membranes:

$$dW = \gamma_G \, dA + AC_1 \, dJ + AC_2 \, dK. \qquad (11.57)$$

The largest term in this equation is usually the first one; it describes the contribution of surface (membrane) tension to the total work of deformation. However, modern physico-chemical studies often involve interfaces with a small or zero value for surface tension. An example of this type of system is a microemulsion which consists of suspensions of droplets (10 nm diameter) of either water in oil or vice versa. The small value of surface tension at oil–water interfaces, $\gamma_G \sim 1$ mN m^{-1}, is due to the presence of surface-active substances (surfactants or co-surfactants) in the system. In addition to microemulsions, aqueous solutions of lipids at different concentrations are widely used in scientific research and technical applications.

The amphiphilic nature of lipid molecules (the presence of hydrophilic polar heads and hydrophobic hydrocarbon tails) determines the types of structures that are formed by lipids in water. The most common structure is that of a bilayer, which consists of two layers of lipid molecules with their polar heads in contact with water and hydrocarbon tails hidden inside the bilayer. Lipid bilayers can form closed vesicles or plane multilamellar structures. In addition, certain types of lipids in small amounts of water can form non-bilayer structures like hexagonal lattice cylinders, normal and inverted micelles. In the discussion to follow we will present the forces which define the physical properties of lipid bilayers and biological membranes.

Elastic Energy of Membranes. Bilayer lipid membranes have been well described as two-dimensional fluids: they strongly resist changes in their surface area, but readily undergo surface shear deformations. In equilibrium, these membranes are considered free of shear stresses. They also exhibit elastic resistance to changes in their curvature. This means that every conformation of a membrane carries a certain amount of elastic energy and in order to determine the equilibrium shape of cells and vesicles it is useful to calculate the value of this energy.

The expression for membrane elastic energy cannot be derived from general principles of surface thermodynamics and therefore it is based on model assumptions. The simplest approach assumes that the membrane behaves as an ideal elastic system. Total elastic energy E_{elast} per unit area can be divided into the energy of stretching E_A and bending E_B:

$$E_{elast} = E_A + E_B. \tag{11.58}$$

Assume that membrane tension γ is proportional to the area expansion ΔA_{mem},

$$\gamma = \frac{K_A \Delta A_{mem}}{A_0}, \tag{11.59}$$

where A_0 is the area of a non-stretched membrane, or membrane spontaneous area, and K_A is the area compressibility modulus or area expansivity modulus. For lipid bilayers K_A is equal to about $0.2 \, \text{N m}^{-1}$ (Evans and Needham, 1987; Rutkowsky *et al.*, 1991). The area compressibility modulus for a single monolayer is one half of this value, $K_A^{mono} \approx 0.1 \, \text{N m}^{-1}$. When using this approach, the energy of membrane stretching per unit area can be represented as

$$E_A = K_A \left(\frac{\Delta A}{A_{20}} \right)^2. \tag{11.60}$$

Bending Energy. The concept of membrane bending energy was developed quite recently (Canham, 1970; Evans, 1974; Helfrich, 1973, 1974; Israelashvili, 1992; Lipowsky, 1991). The most important contribution was made by Helfrich (1973) who presented the bending energy in the following form:

$$E_B = \tfrac{1}{2}\kappa(c_1 + c_2 - c_0)^2 + \bar{\kappa}c_1 c_2 \tag{11.61}$$

This equation incorporates the mean curvature of the membrane $c_1 + c_2$ and the Gaussian curvature $c_1 c_2$. Helfrich's innovative idea was to introduce the spontaneous membrane curvature which accounts for the membrane's intrinsic tendency to curve. Spontaneous curvature c_0 is equal to the mean membrane curvature at which torque C_1 in the membrane becomes zero. Likewise, the first term in equation (11.61) also becomes zero under this condition. Spontaneous curvature can build up for different reasons; for example, due to the monolayers in a lipid bilayer having different chemical composition or due to the asymmetric surface charge of the membrane (Kozlov *et al.*, 1992). The monolayers can have their own spontaneous curvatures c_0^{in} and c_0^{out} and in some cases they should be considered separately. However, in a symmetrical case they compensate each other so that the total spontaneous curvature of the membrane becomes zero.

The proportionality coefficients κ and $\bar{\kappa}$ in equation (11.60) are called *the bending rigidity* and *the Gaussian bending rigidity*, respectively; they are also called the bending moduli of a membrane. Both κ and $\bar{\kappa}$ characterize *local properties* of membranes, and for this reason they are called *local bending moduli*. The value of κ depends on the chemical composition of the lipid bilayer and varies in different membranes. A good estimate for κ places it in a range from 0.1 to 0.4 aJ. Gaussian bending rigidity $\bar{\kappa}$ is much more difficult to measure and one should rely on theoretical estimates only (Helfrich, 1973).

Equation (11.60) gives the bending energy per unit area of membrane. To find total bending energy, one has to integrate this expression over the entire membrane area:

$$E_B^{tot} = \int [\tfrac{1}{2}\kappa(c_1 + c_2 - c_0)^2 + \bar{\kappa}c_1 c_2] \, dA. \tag{11.62}$$

The goal of this analysis is to determine the shape of closed lipid vesicles and cells. Obviously, this shape should be one that minimizes the elastic bending energy of its membrane within the constraints imposed by the membrane area and enclosed volume.

Because membranes normally form a closed shell, there is another reason for them to accumulate elastic energy. Sheetz and Singer (1974) introduced the bilayer couple hypothesis which states that two monolayers in a membrane are coupled together in a transverse direction but can freely slide along each other. If the area of each monolayer is a constant, then this limits the number of possible conformations of its shell. This is a *global* restriction imposed on the membrane. Using this restriction one can predict possible shapes for the cell or vesicle. This bilayer couple with incompressible monolayers (BCIMs) model has been extensively employed for the analysis of lipid vesicle shapes (Berndl *et al.*, 1990; Markin *et al.*, 1987b; Seifert *et al.*, 1991a,b; Svetina *et al.*, 1982; Svetina and Zeks, 1983, 1989).

Generalized Bilayer Couple Model with Elastic Monolayers and Spontaneous Curvature. In reality, lipid monolayers have finite compressibility and can change their area to a certain extent. In this process they accumulate a certain amount of elastic energy. Taking these properties into account led to the development of the generalized bilayer couple model with elastic monolayers and spontaneous curvature (BCEMSC). It was originally developed by Markin *et al.* (1987b) and later improved upon (Bozic *et al.*, 1992; Heinrich *et al.*, 1993; Markin and Martinac, 1991; Miao *et al.*, 1994; Seifert *et al.*, 1991a,b; Wiese *et al.*, 1992).

The main ideas of the BCEMSC model are as follows. Let us designate the area of the inner and outer monolayers A^{in} and A^{out}, respectively (Fig. 11.3). Due to global restrictions on the bilayer couple, one of the monolayers can be stretched, while

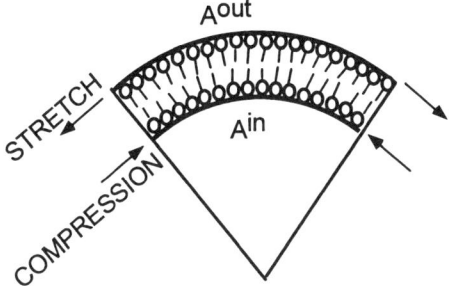

Figure 11.3. Interaction of monolayers in the bilayer couple model.

another is compressed. In the non-stressed state these monolayers will have areas A_0^{in} and A_0^{out}. The corresponding deformations of these monolayers are:

$$\Delta A^{in} = A^{in} - A_0^{in} \quad \text{and} \quad \Delta A^{out} = A^{out} - A_0^{out}. \tag{11.63}$$

The differential area of the two monolayers is:

$$\Delta A = A^{out} - A^{in}, \tag{11.64}$$

while the non-stressed differential area is

$$\Delta A_0 = A_0^{out} - A_0^{in}. \tag{11.65}$$

It is important to define the position within the monolayer where the area is measured, because in a curved monolayer the result will depend on this position. Usually, it is assumed that somewhere within the middle of the monolayer there is a neutral surface that preserves its area in the process of bending. This area is considered as the area of the monolayer. The distance between the neutral surfaces of adjacent monolayers is usually taken to be equal to the thickness of one monolayer h.

The tension in the monolayers is

$$\gamma^m = \frac{K_A^{mono} \Delta A^{in}}{A_0^{in}} \quad \text{and} \quad \gamma^{out} = \frac{K_A^{mono} \Delta A^{out}}{A_0^{out}}. \tag{11.66}$$

For the sake of simplicity we assume that the compressibility modulus K_A^{mono} has the same value for both monolayers. We can also assume to a good approximation that the denominators in (11.66) have the same value as the mean unstressed area $A_0 \approx A = (A^{in} + A^{out})/2$ and use this simplification in the following derivation.

Total membrane tension can be found to be the sum of the tensions in the monolayers:

$$\gamma = \gamma^{in} + \gamma^{out}. \tag{11.67}$$

This sum is determined by the pressure differential ΔP across the membrane according to the Laplace equation:

$$\gamma^{in} + \gamma^{out} = \frac{\Delta P}{c_1 + c_2}. \tag{11.68}$$

This equation was used for the analysis of membrane mechanosensitivity and to describe the role of the two monolayers in this phenomenon (Markin and Martinac, 1991). In the absence of the pressure differential ΔP, the tensions in the two monolayers compensate each other:

$$\gamma^{out} = -\gamma^{in}, \tag{11.69}$$

and

$$\Delta A^{out} = -\Delta A^{in} = \frac{\Delta A - \Delta A_0}{2}. \tag{11.70}$$

The elastic energy accumulated in each monolayer is equal to

$$E_A^{in} = \frac{K_A^{mono}(\Delta A^{in})^2}{A} \quad \text{and} \quad E_A^{out} = \frac{K_A^{mono}(\Delta A^{out})^2}{A}. \tag{11.71}$$

In the absence of a pressure differential, the total membrane elastic energy due to stretch/compression of the monolayers is

$$E_A = \frac{K_A(\Delta A - \Delta A_0)^2}{8A_0}. \tag{11.72}$$

Notice that in this equation we switched to the compressibility modulus of the total bilayer K_A. The area differential ΔA can be expressed as the integral of the mean curvature:

$$\Delta A = h \int (c_1 + c_2) \, dA \tag{11.73}$$

With this expression, one can rewrite equation (11.72) in the following way:

$$E_A = \frac{h^2 K_A}{8A_0} \left[\int \left(c_1 + c_2 - \frac{\Delta A_0}{hA_0} \right) dA \right]^2. \tag{11.74}$$

Now let us consider the local bending energy (11.61). When this expression is integrated as in (11.62), the second term with Gaussian curvature $c_1 c_2$ gives a constant which does not depend on the shape of a closed membrane. This constant does not influence the equilibrium of the membrane and therefore it is usually neglected. The other term with mean curvature is shape-sensitive and we shall calculate its contribution for both monolayers separately. They can have different spontaneous curvature c_0^{in} and c_0^{out}. One can assume that the bending rigidity of a single monolayer is one half of that of the total bilayer. The local bending energy of the bilayer can then be represented as follows:

$$E_B = \tfrac{1}{4}\kappa(c_1 + c_2 - c_0^{in})^2 + \tfrac{1}{4}\kappa(c_1 + c_2 - c_0^{out})^2$$

$$= \tfrac{1}{2}\kappa\left(c_1 + c_2 - \frac{c_0^{in} + c_0^{out}}{2}\right)^2 + \tfrac{1}{8}\kappa(c_0^{in} - c_0^{out})^2. \qquad (11.75)$$

Comparison with equation (11.61) shows that the spontaneous curvature of the bilayer is equal to the mean curvature of two monolayers. The last term in this equation does not depend on the membrane shape and can be neglected.

Therefore, the total expression for the elastic energy of the bilayer in the general BCEMSC model can be presented in any of the two following forms:

$$E_{elast}^{tot} = \tfrac{1}{2}\kappa \int (c_1 + c_2 - c_0)^2 \, dA + \frac{K_A(\Delta A - \Delta A_0)^2}{8A_0}$$

$$= \tfrac{1}{2}\kappa \int (c_1 + c_2 - c_0)^2 \, dA + \frac{h^2 K_A}{8A_0}\left[\int\left(c_1 + c_2 - \frac{\Delta A_0}{hA_0}\right) dA\right]^2. \qquad (11.76)$$

The two terms in this equation account for local and non-local bending energy, correspondingly. In the last line the integrals

$$\int (c_1 + c_2 - c_0)^2 \, dA \qquad \text{and} \qquad \frac{1}{A_0}\left[\int\left(c_1 + c_2 - \frac{\Delta A_0}{hA_0}\right) dA\right]^2 \qquad (11.77)$$

are dimensionless quantities of a similar structure. Their relative importance is determined by the ratio of their coefficients:

$$\alpha = \frac{h^2 K_A}{4\kappa}. \qquad (11.78)$$

Assuming $h = 2$ nm, $K_A = 0.2$ N m^{-1}, and $\kappa = 0.2$ aJ, one can estimate this ratio as

$$\alpha \approx 1. \qquad (11.79)$$

This means that in general both contributions to the elastic energy are equally important unless there is a large difference between the two integrals in equation (11.77).

There are two limiting cases for this BCEMSC model. If $\alpha = 0$, the general model is reduced to Helfrich's spontaneous curvature (SC) model. In the opposite case, when

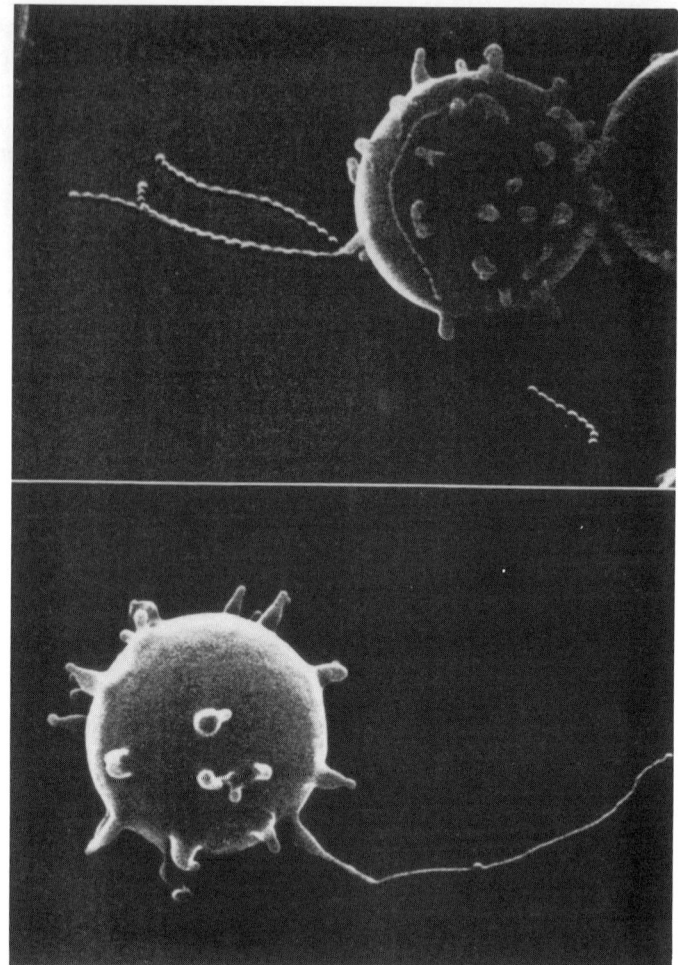

Figure 11.4. Electron micrographs of myelin forms. From Bessis (1973), p. 170. Reproduced by permission of Springer-Verlag New York, Inc.

$\alpha = \infty$, the area difference ΔA cannot deviate from the spontaneous difference ΔA_0 and the BCIM model is recovered. These limited cases provide a simplified analysis of vesicle shape. This is a useful approach to a rather complicated problem, and we will provide a few examples of these types of calculations. However, one has to keep in mind that the generalized version of the BCEMSC model is still required to account for the numerous shapes and shape transitions that can occur in a giant vesicle (Mui *et al.*, 1995; Seifert, 1994).

Myelin Figures. Free-living cells and lipid vesicles have many different and interesting shapes that have intrigued biologists. One notable example is the

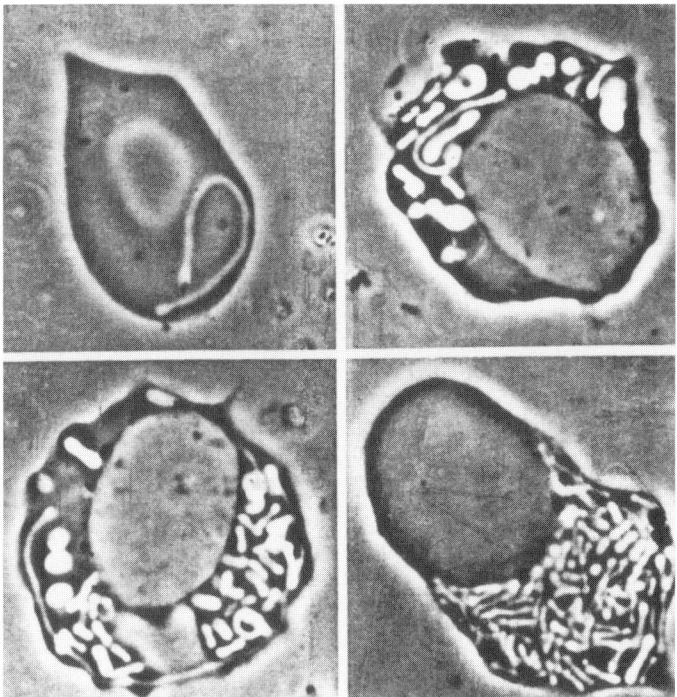

Figure 11.5. Intracellular myelin forms. From Bessis (1973), p. 61. Reproduced by permission of Springer-Verlag New York, Inc.

biconcave disk shape of red blood cells, which can be understood if we assume that the equilibrium shape of a cell or vesicle corresponds to the minimum elastic energy of its membrane provided appropriate constraints are made on the membrane area A and vesicle volume V. Several classes of vesicle shapes were classified by Deuling and Helfrich (1976).

It was noted long ago that aged red blood cells form long thin tubes called myelin forms (Figs. 11.4 and 11.5 from Bessis, 1973). These shapes have been analyzed by Deuling and Helfrich (1977) based on the SC model. Consider a cell which has rotational symmetry. The contour that generates the surface of revolution is presented in Fig. 11.6. The principal curvatures of this surface are

$$c_{\mathrm{p}}(x) = \frac{\sin \psi(x)}{x} \qquad \text{and} \qquad c_{\mathrm{m}}(x) = \cos \psi(x) \frac{\mathrm{d}\psi(x)}{\mathrm{d}x}. \qquad (11.80)$$

The angle $\psi(x)$ can be found as

$$\tan \psi(x) = -\frac{\mathrm{d}z(x)}{\mathrm{d}x}. \qquad (11.81)$$

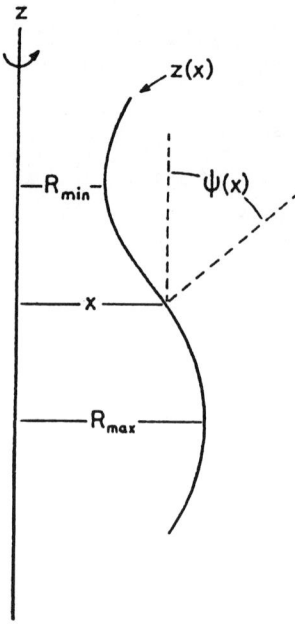

Figure 11.6. Contour of myelin shapes with explanation of parameters of the model.

Deuling and Helfrich assumed that the myelin shapes are infinitely long and exist without any mechanical stress. In this case, the bending energy given by the first term in equation (11.60) should be zero everywhere:

$$c_m(x) + c_p(x) - c_0 = 0. \tag{11.82}$$

Eliminating $c_m(x)$ and $\psi(x)$ from equations (11.79)–(11.81) one obtains

$$\frac{dc_p(x)}{dx} = \frac{c_0 - 2c_p(x)}{x}. \tag{11.83}$$

This equation can be integrated to give

$$c_p(x) = \tfrac{1}{2}c_0 + \frac{x_0}{x^2}, \tag{11.84}$$

where x_0 is a constant of integration. Using the condition that $\psi = \pi/2$ at $x = R_{min}$ and $x = R_{max}$, one can find the unknown constants:

$$c_0 = \frac{2}{R_{max} + R_{min}}, \tag{11.85}$$

$$x_0 = \frac{R_{min} + R_{max}}{4} \left[1 - \frac{(R_{min} - R_{max})^2}{(R_{min} + R_{max})^2} \right]. \tag{11.86}$$

The contour $z(x)$ is found as:

$$z(x) = z(R_{min}) - \int_{R_{min}}^{x} \frac{\xi c_p(\xi)\, d\xi}{\sqrt{1 - \xi^2 c_p^2(\xi)}}. \tag{11.87}$$

The result one obtains following integration is determined by the dimensionless parameter $c_0 R_{max}$, which can vary from 1 to 2. Shape contours are presented for different values of this parameter in Fig. 11.7. These shapes range from a simple cylinder to a string of beads and closely approximate those actually observed in nature (Fig. 11.4).

To model the shorter myelin shapes observed within the red cells Deuling and Helfrich (1976, 1977) introduced the equivalent volume of a sphere V_0 having the same membrane area A as the myelin shape (the radius of this sphere is designated $R_A = (A/4\pi)^{1/2}$). In this case, the result depends on both the dimensionless parameters $c_0 R_{max}$ and the reduced or equivalent volume $v = V/V_0$. The shapes found for $c_0 R_{max} = 0.5$ and for different values of V/V_0 are presented in Fig. 11.8. One can see a close resemblance between these shapes and the photograph in Fig. 11.5.

Vesicles. Investigation of vesicle shapes began with the biconcave disk shape of red blood cells. Later, large unilamellar lipid vesicles (LUV) became the focus of investigations because they can assume a variety of exotic forms (Fig. 11.9) (Mui *et al.*, 1995; Seifert, 1994). Transition between these different forms can be induced by changing the area or volume of the vesicles (Berndl *et al.*, 1990; Käs and Sackmann, 1991). One can observe the transition between stomaticytes, which have no equatorial symmetry, and equatorial symmetric shapes such as ellipses or dumbbells (Seifert *et al.*, 1991a,b).

These forms were successfully described by the generalized BCEMSC model (Mui *et al.*, 1995; Seifert, 1994). Parameters of the model are the vesicle volume V, the membrane area A, the spontaneous curvature c_0, the monolayer equilibrium area difference ΔA_0 and the material parameter moduli ratio α. In practice, the actual number of variables determining the shape of the vesicle is less since the membrane area is usually constant and the volume is controlled by the osmotic effect. In addition, area and volume do not influence the shape of vesicles separately but rather in a combination which is represented by the dimensionless reduced volume v defined above. The spontaneous curvature c_0 and the monolayer equilibrium area difference ΔA_0 also enter the equations as a combination defined by equation (11.75) which is referred to as an effective spontaneous curvature. The material parameter moduli ratio α is a constant for a given lipid composition of a vesicle.

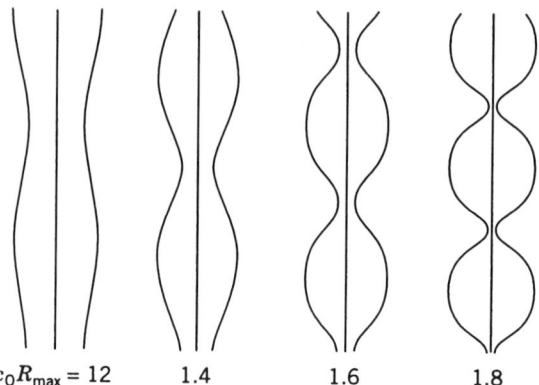

$$c_0 R_{max} = 12 \qquad 1.4 \qquad 1.6 \qquad 1.8$$

Figure 11.7. Myelin forms calculated for different values of spontaneous curvature. From Deuling and Helfrich (1977). Reproduced by permission of Les Edition de Physique.

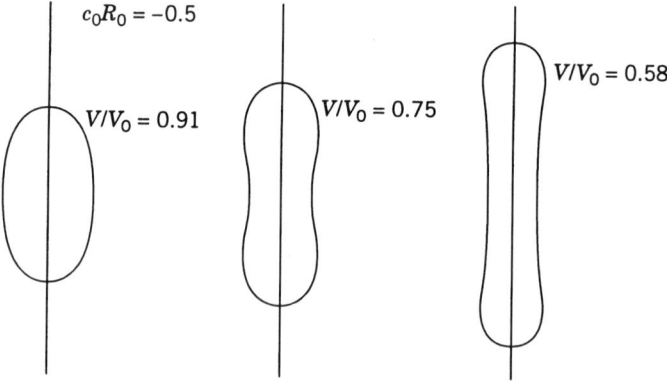

Figure 11.8. Theoretical contours of intracellular myelin shapes. From Deuling and Helfrich (1977). Reproduced by permission of Les Edition de Physique.

Therefore, the shape of vesicles is effectively determined by two parameters: the reduced volume and the effective spontaneous curvature. The observed and calculated vesicles can be placed at a two-dimensional phase diagram with axes representing the reduced volume v and the reduced differential area Δa_0 (Fig. 11.9). These shapes correspond to lowest bending energy. In all of these cases the membrane was assumed to be homogeneous which is not necessarily the case with real cellular membranes. The presence of non-homogeneous membrane regions opens additional interesting possibilities for shape transformations (Bebhardt *et al.*, 1977; Markin, 1981).

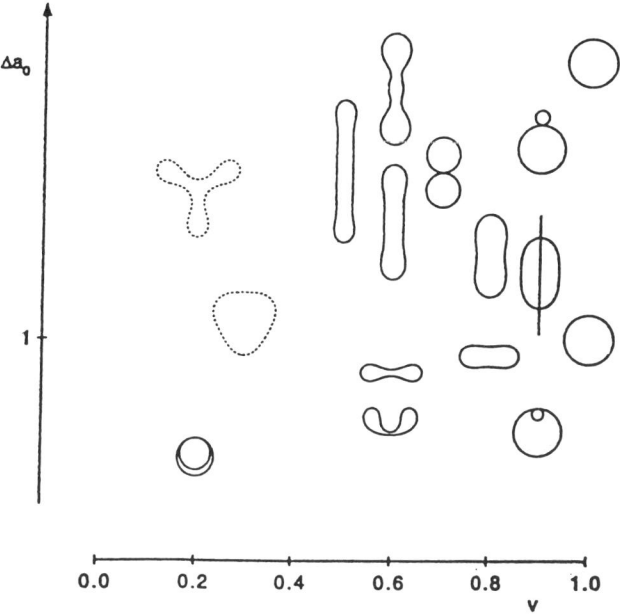

Figure 11.9. A phase diagram of vesicle shapes. These shapes correspond to the lowest bending energy for a given reduced volume v and reduced differential area Δa_0. The vesicles presented by continuous lines have vertical axes of symmetry. The dashed lines depict non-axisymmetric shapes. From Mui *et al.* (1995). Reproduced by permission of Biophysical Society.

11.3. EDGE ENERGY AND PORES IN MEMBRANES

Rupturing a membrane means creation of a pore whose edges require additional energy. In trying to dissipate this energy, the membrane pores are closed and can reappear again, sometimes rupturing the entire membrane. In recent decades there have been many works on the mechanical and electrical stability of biological and artificial membranes (Abidor *et al.*, 1979; Benz *et al.*, 1979; Chizmadzhev and Pastushenko, 1988; Deryaguin and Prokhorov, 1981; Freeman *et al.*, 1994; Glaser *et al.*, 1988; Kinosita and Tsong, 1977; Markin *et al.*, 1990; Neumann and Rosenheck, 1972; Petrov *et al.*, 1980; Taupin *et al.*, 1981; Tsong, 1991; Weaver, 1993; Zimmermann and Neil, 1996). Here we will consider the fundamentals of this phenomenon (Markin *et al.*, 1990).

Thermodynamics of Pore Formation in a Charged Membrane. The bulk of the theoretical studies are associated with modeling approaches, and we will present the general thermodynamics of membrane pore formation based on Hansen's method. Let us consider an insoluble film separating two solutions α and β (Fig. 11.10). Symbols

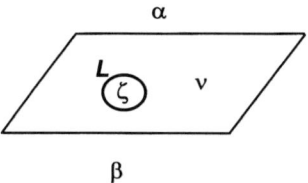

Figure 11.10. The interface between two phases α and β with two two-dimensional phases ζ (inner phase) and ν (outer phase) and dividing line L.

α and β will also denote the solvents proper. The solutions can be different or identical. Use of the Gibbs method to derive the membrane characteristics of an insoluble film is very difficult to implement. In the case of two identical solutions on both sides of a film it is impossible to choose a reference system which has zero solvent excess at a set volume. Sometimes, under these conditions, one is forced to use the cumbersome method of two dividing surfaces (Rusanov, 1983). Hansen's method is more convenient to use for this purpose because it does not preset the volume of the reference system and makes it possible to calculate surface excesses relative to the selected reference substances, for instance, solvents α and β. If α and β are identical, then one of the reference substances can be a salt present in the system. When using this description one has to incorporate Hansen's surface excesses of entropy $S^s_{(\alpha,\beta)}$, components $n^s_{(\alpha,\beta)}$, and volume $V^s_{(\alpha,\beta)}$ relative to reference substances α and β. These values refer to the total surface and not to its unit area. Upon exclusion of the volume parts the fundamental thermodynamic equation for a plane interface will take the form

$$dU^s = T\,dS^s_{(\alpha,\beta)} - P\,dV^2_{(\alpha,\beta)} + \sum_{i \neq \alpha,\beta} \tilde{\mu}_i\,dn^s_{i(\alpha,\beta)} + \gamma\,dA \tag{11.88}$$

where U^s is the interfacial energy, T is the absolute temperature; P the pressure, $\tilde{\mu}_i$ the electrochemical potential of the i-component, γ the interfacial (membrane) tension, and A the area of the dividing surface. The dividing surface is assumed to be a plane coincident with the surface of tension. This permits one to neglect all terms with membrane curvature.

We shall designate the interface as ideally polarizable which allows the term with electrochemical potentials to be transformed such that:

$$\sum_{i \neq \alpha,\beta} \tilde{\mu}_i\,dn^s_{i(\alpha,\beta)} = \sum_{i \neq \alpha,\beta} \mu_i\,dn^s_{i(\alpha,\beta)} + \phi\,dQ, \tag{11.89}$$

where ϕ is the interfacial difference of electrical potential and Q is the total surface charge whose sign is chosen such that ϕ increases with Q. Substances that do not penetrate through the interface but are present on both of its sides should be taken into account twice in the sum (11.88) because formally they are considered to be different. We can now obtain Gibbs absorption equation in Hansen's representation:

$$d\gamma = s^s_{(\alpha,\beta)} \, dT + \tau^s_{(\alpha,\beta)} \, dP + \sum_{i \neq \alpha,\beta} \Gamma_{(\alpha,\beta)} \, d\mu_i - q \, d\phi, \tag{11.90}$$

where q is the surface charge density; in this case it has no indices because at an ideally polarizable interface it does not depend on the choice of reference system.

At constant temperature, pressure and chemical potentials, the membrane tension can be presented in the following integral form:

$$\gamma = \gamma_0 - \int_0^\phi q \, d\phi. \tag{11.91}$$

Suppose that the surface (membrane) consists of two different phases ζ and ν (Fig. 11.10). These can be two different states of lipid molecules or one of these phases can be a pore in a membrane. Let us draw a dividing line L between them so that the total area A is divided into A^ζ and A^ν. The two-dimensional phase ζ shall be referred to as an internal phase and ν as an external phase. Then the fundamental thermodynamic equation will take the form:

$$dU^s = T \, dS^s_{(\alpha,\beta)} - P \, dV^s_{(\alpha,\beta)} + \sum_{i \neq \alpha,\beta} \mu_i \, dn^s_{i(\alpha,\beta)} + \phi \, dQ$$
$$+ \gamma^s \, dA^s + \gamma^\nu \, dA^\nu + f \, dL + LC_L \, dc_L, \tag{11.92}$$

where f is the *linear tension* of the border between two phases; C_L is the two-dimensional bending moment; and c_L is the curvature of the line L. If the line of tension is chosen as a dividing line L, the last term vanishes.

Formation of a pore has the characteristics of classical nucleus formation. Therefore, the following analysis can be conveniently carried out by introducing one of the Gibbs free energy forms, especially designed for the description of a nucleus:

$$\hat{G}^s = U^s - TS^s + PV^s - \phi Q - \gamma^\nu A = \sum_{i \neq \alpha,\beta} \mu_i n^s_{i(\alpha,\beta)} + (\gamma^s - \gamma^\nu)A^s + fL. \tag{11.93}$$

To determine the reversible work, W, required to form the nucleus, we shall consider the ζ phase and a closed two-dimensional system. The initial value of Gibbs energy prior to the appearance of the nucleus is

$$\hat{G}^s_0 = \sum_{i \neq \alpha,\beta} \mu_i n^s_{i(\alpha,\beta)}. \tag{11.94}$$

Assuming that the chemical potentials of the substances do not change in the process of nucleus formation we obtain the following solution for the reversible work:

$$W = \hat{G}^s - \hat{G}^s_0 = (\gamma^s - \gamma^\nu)A^s + fL. \tag{11.95}$$

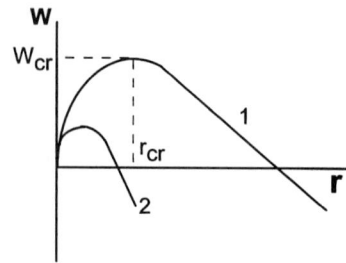

Figure 11.11. Dependence of pore energy W on the pore radius. Curve 2 corresponds to a higher membrane potential than curve 1. Curves are plotted according to equation (11.97) with the additional assumption that linear tension f does not depend on the pore radius.

Suppose that the nucleus has a circular shape with radius r. Then using equation (11.90) one can obtain the work of the nucleus formation:

$$W = \left[(\gamma_0^\xi - \gamma_0^\nu) - \int_0^\phi (q^\xi - q^\nu)\, d\phi \right] \pi r^2 + 2\pi fr. \tag{11.96}$$

Now we assume that the electric capacitance of both surface phases does not depend on the electrical potential and is determined according to the formula of a plane capacitor, and take into account that the surface tension in the pore is zero, $\gamma_0^\xi = 0$. Then we obtain for the work of pore formation:

$$W = - \left[\gamma_0 + \frac{\varepsilon_0(\varepsilon_w - \varepsilon_m)}{2h} \phi^2 \right] \pi r^2 + 2\pi fr, \tag{11.97}$$

where h is membrane thickness, ε_0 is the dielectric constant, ε_w is the relative permittivity of water in the pore, and ε_m is the relative permittivity of the bilayer, γ_0 is the membrane tension without field. If the linear tension f is a constant, then the dependence (11.97) of energy on the pore radius is a bell-shaped function which rises from the origin, reaches a maximum and then decreases (Fig. 11.11). The maximum is reached at a critical radius

$$r_{cr} = \frac{f}{\gamma_0 + \dfrac{\varepsilon_0(\varepsilon_w - \varepsilon_m)}{2h} \phi^2}, \tag{11.98}$$

where the energy is equal to

$$W_{cr} = \frac{\pi f^2}{\gamma_0 + \dfrac{\varepsilon_0(\varepsilon_w - \varepsilon_m)}{2h} \phi^2}. \tag{11.99}$$

This is the activation barrier that prevents the membrane from rupture. If the pore radius is smaller than r_{cr}, the natural tendency for this pore is to decrease the energy and to close down, but if due to a certain fluctuation the pore exceeds the critical radius, it should increase indefinitely.

As usual in the theory of nucleation, the curves in Fig. 11.11 represent nonequilibrium states with only two exceptions: the very beginning where there is no pore in the membrane, and at the top of the barrier, where equilibrium is unstable. Nevertheless, this is the equilibrium state and equilibrium relationships are applied here. For example, in a general case of a 2-d nucleation the relationship between the surface tension of the equilibrium (although unstable) nucleus ζ and the rest of the membrane v can be presented by the generalized Laplace equation:

$$fc_L - C_L c_L^2 = \gamma^v - \gamma^s. \tag{11.100}$$

On the circular line of tension with radius r it reduces to

$$\gamma^v - \gamma^s = \frac{f}{r}, \tag{11.101}$$

or, if it is a pore,

$$\gamma_0 + \frac{\varepsilon_0(\varepsilon_w - \varepsilon_m)}{2h} \phi^2 = \frac{f}{r_{cr}}, \tag{11.102}$$

which is just another form of equation (11.98).

The activation barrier described by equations (11.97) and (11.99) changes with membrane potential. As potential increases, the curves in Fig. 11.11 bend down more quickly and the barrier decreases. This corresponds to the transition from curve 1 to curve 2. Therefore, membrane potential destabilizes membranes and lipid bilayers.

Linear Tension. Estimation of the linear tension of a pore is based on a specific model (Markin and Kozlov, 1985). In large pores the hydrophobic tails of lipid molecules should not be exposed to water, and hence the pore surface should be lined up with polar heads of lipid molecules (Fig. 11.12). Therefore, the lipid monolayer is bent and its elastic energy per unit area of its neutral surface can be found from Helfrich's formula (11.61). Neglecting the Gaussian term, one can reduce it to $\frac{1}{2}\kappa(c_1 + c_2 - c_0)^2$. Before the pore was formed, the monolayer was in a planar shape. If it had no spontaneous curvature, it would not carry any elastic energy. However, if the monolayer had spontaneous curvature c_0, then in the planar state it would already have an elastic energy of $\frac{1}{2}\kappa c_0^2$ per unit area. To find the elastic energy of the pore, one has to integrate the difference between these two expressions over the pore edge:

$$W_{edge} = \frac{1}{2}\kappa \int_{edge} [(c_1 + c_2 - c_0)^2 - c_0^2] \, dA. \tag{11.103}$$

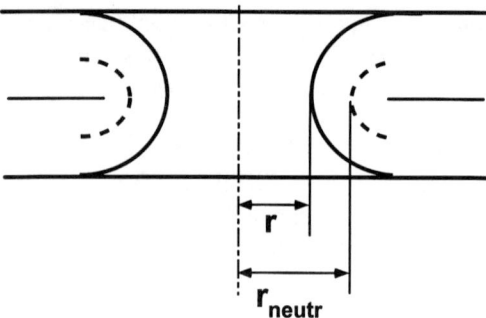

Figure 11.12. A pore edge with two different definitions of pore radius: r designates the radius of the pore opening, while r_{neutr} designates the distance from the pore center to the neutral surface of the bent monolayer.

Let us assume that the neutral surface of a pore is generated by the revolution of the semicircle of radius $h/2$ around the pore axis (Fig. 11.12) where h is the monolayer thickness. Then the principle curvatures c_1 and c_2 are defined according to Fig. 11.6 and integration gives the total elastic energy of the pore:

$$W_{edge} = 2\pi\kappa\left\{\frac{2(\rho + 2)^2}{\sqrt{(\rho + 2)^2 - 1}}\tan^{-1}\sqrt{\frac{\rho + 3}{\rho + 1}} - 4 - c_0 h[0.5\pi(\rho + 2) - 2]\right\}. \quad (11.104)$$

We can now introduce the dimensionless radius of the pore $\rho = 2r/h$.

The dependence of elastic energy of the pore W_{edge} on pore radius is presented in Fig. 11.13. One can see that at zero radius the curve starts at a finite energy and starts growing with a gradually increasing slope. This slope characterizes the linear tension of the pore:

$$f = \frac{1}{2\pi}\frac{dW_{edge}}{dr}, \quad (11.105)$$

and hence, linear tension increases with pore radius. It is very instructive to consider a series expansion of function $W_{edge}(\rho)$. At small radii it can be presented as

$$W_{edge} = 2\pi\kappa[1.14(0.73 - c_0 h) + 1.57(0.60 - c_0 h)\rho + \ldots], \quad (11.106)$$

and at large radii the expansion is

$$W_{edge} = 2\pi\kappa[1.57(1 - c_0 h)\rho + 1.14(0.12 - c_0 h) + \ldots]. \quad (11.107)$$

Using this expression one can find the linear tension of small and large pores:

$$f_{small} = 1.57\kappa(0.60 - c_0 h) \quad \text{and} \quad f_{large} = 1.57\kappa(1 - c_0 h). \quad (11.108)$$

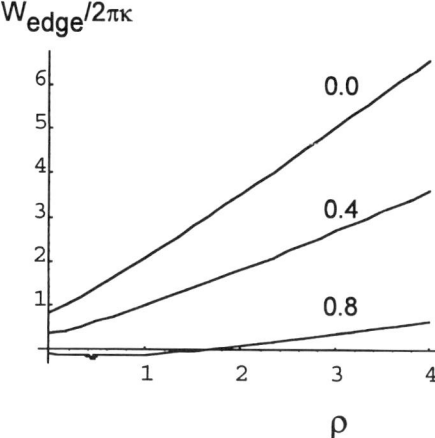

$W_{edge}/2\pi\kappa$

Figure 11.13. Dependence of the pore energy on the pore radius. The values of $c_0 h$ are indicated at the curves.

Therefore, linear tension is always confined to the range defined by equations (11.108) and the bending stiffness κ is a very good estimate of it. In the absence of spontaneous curvature, linear tension is always positive and hence the pore always has a tendency to close. However, if spontaneous curvature c_0 exceeds the value $0.60/h$, the tension can be negative in the beginning and positive afterwards. The elastic energy of the pore in this case displays a minimum (Fig. 11.13). This means that such a pore can become stable and can exist in the membrane indefinitely.

In the model considered above linear tension increased with the radius of the pore, and this resulted in an intermediate stable state depending on the value of the spontaneous curvature. There is also the possibility of an intermediate stable state existing without relying on spontaneous curvature. This can be obtained if the linear tension *decreases* with pore radius. Recall that the elastic energy of a pore with zero radius is not zero but rather equal to $W_{edge} = 2.28\pi\kappa(0.73 - c_0 h)$, which reduces to $1.66\pi\kappa$ in the absence of spontaneous curvature. This means that a pore carries excess energy from the very beginning. Therefore, at small radii it might be advantageous to have a hydrophobic pore instead of a hydrophilic inverted pore. It would have higher linear tension and hence the total energy would increase steeply, but it would start from zero! After a while the pore can change from a hydrophobic to a hydrophilic, inverted pore and at this transition its energy can drop (Fig. 11.14). This can result in the intermediate metastable state, and can manifest itself in the so-called reversible electrical breakdown of lipid membranes (Glaser *et al.*, 1988).

There are other reasons why the linear tension of the pore can vary. Earlier we had assumed that membrane potential does not influence the linear tension. However, further analysis demonstrates it can. This is evident from the definition of f as an energy excess originating from the division of the membrane into two regions ζ and ν. These two-dimensional phases in the reference system are assumed to be homogeneous up to the dividing line L. Since this is not the case in the real system,

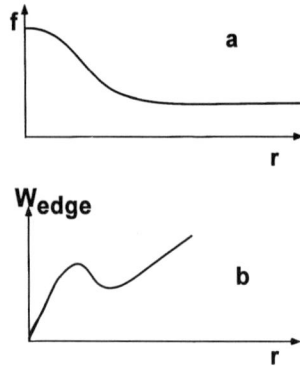

Figure 11.14. Metastable state of a pore. (*a*) Surface tension decreases with radius. (*b*) The edge energy displays a minimum when a hydrophobic pore changes to a hydrophilic one.

the extra energy is ascribed to the dividing line and is accounted for by the term $f\,dL$. Because the membrane energy depends on electrical potential, its distribution and hence the energy excess should also depend on membrane potential. That is why the linear tension f can also depend on the membrane potential. This factor introduces a new aspect into the analysis of the pore formation.

In conclusion, it should be noted that when deriving formula (11.101) we used a traditional assumption for the theory of nucleus formation: that is, that the nucleus and the pore in this case were considered as a system in equilibrium. In the case of a charged membrane this is a rather strong assumption, since it implies the absence of current flow through the pore. This is certainly acceptable only for very narrow pores which prevent ions from permeating through the pore. However, for larger pores one should consider the stationary state of a system instead of the equilibrium state and change from thermodynamic to kinetic analysis (Chizmadzhev and Pastushenko, 1988).

11.4. MEMBRANE FUSION

The cellular membrane evolved as a wall separating the cell's contents from the external medium. This barrier function was conveniently assigned to a lipid bilayer, an ingenious biological invention combining the remarkable property of self-organization with unhindered lateral movement of membrane components. This boundary is surprisingly robust, but occasionally can undergo rupture, a catastrophic event leading to the loss of important cellular components and usually to cell death.

Despite the importance of membrane integrity, a number of crucial physiological events require the breakdown of two apposing membranes and their subsequent reconnection to one another (Poste and Nicolson, 1978). Examples include fertilization, cell division, endocytosis, exocytosis, and the entry of enveloped viruses into cells. Although membrane fusion has long been a focus of attention, even recent reviews inevitably come to the conclusion that "the physical and molecular mechanisms of membrane fusion remain obscure."

Fusion of cellular membranes is preceded by the removal of membrane proteins from the contact region and by the formation of a pure lipid bilayer (Gingell and Ginsberg, 1978; Lucy, 1978; Papahajopoulos, 1978). This preliminary fusion stage and the forces operating on the membrane were investigated by Markin and Glaser (1980) and Kozlov et al. (1989). We will discuss the direct fusion of lipid bilayers and establish the conditions necessary for fusion, i.e., properties of the membranes and external triggering effects produced by the environment (Markin et al., 1984).

Three particularly important properties of membranes affecting fusion can be identified: membrane charge, the phase state of the lipid bilayer, and the geometric characteristics of membranes, namely, their curvature. If the fusion process is initiated by fusogens, it is necessary that at least a portion of the lipid molecules in the membrane be charged. For example, in the case of calcium-initiated fusion of liposomes consisting of a mixture of a neutral lipid phosphatidylethanolamine (PE) and charged phosphatidylserine (PS), the PS content has to constitute at least 50% (Papahadjopoulos et al., 1974). However, it is also possible to bring about the fusion of neutral membranes with the aid of alkylbromides and lysolecithin (Lucy, 1970; Mason et al., 1979) or by hydrostatic pressure (Neher, 1974). As for the phase state of the bilayer, most authors assume that it should be liquid; thus, the fusion of spherical membranes begins at temperatures significantly higher than that of phase transition (Breisblatt and Ohki, 1975). A membrane in the liquid state has a high fluidity which tends to disturb the bilayer stability (Ahkong et al., 1975). Accordingly, the effect of substances such as glycerolmonooleate or oleic acid is explained by their capacity to change membrane fluidity. In contrast, Papahadjopoulos (1978) has presumed it is not the liquid state that provides the condition for fusion. In his opinion, the decisive circumstance is the simultaneous presence in the membrane of liquid and gel-like (crystalline) phases, the boundaries of which prove to be unstable. The third condition relates to the geometrical characteristics of a membrane. Portis et al. (1979) have discussed the concept that large membrane curvatures promote their fusion. However, a comparative study of small and large liposomes (Wilshut et al., 1980) has revealed that a large curvature is by no means absolutely necessary. Nevertheless, the rate of fusion and the number of liposomes involved in this process were found to be higher for smaller rather than larger liposomes.

Classification of Fusion Mechanisms. The absence of a single viewpoint concerning the conditions necessary for membrane fusion suggests that different mechanisms underlie this process. Several hypotheses have been put forward. Some of them are mutually exclusive while others partially overlap. They deal with the reorganization of membranes during fusion and with various kinds of intermediate structures. One of these structures, known as a trilaminary structure (Gingell and Ginsberg, 1978), deserves special attention. It involves formation of a single bilayer from two apposing bilayers (Fig. 11.15). The appearance of the trilaminary structure

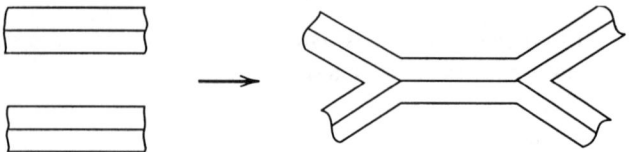

Figure 11.15. A trilaminary structure arising from two approaching bilayers.

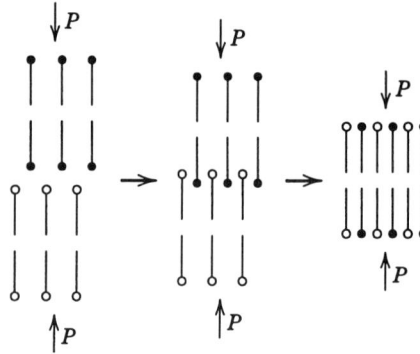

Figure 11.16. A hypothetic squeezing of two bilayers into one another: under normal conditions this process is improbable.

was observed during fusion of two BLMs (Melikyan *et al.*, 1982, 1984; Neher, 1974) and during fusion of chromaffin granules (Edwards *et al.*, 1974). It is an intermediate structure, but can also be the final result of fusion (Grishin *et al.*, 1979). This semi-completed type of fusion is called a monolayer fusion or a hemi-fusion.

How can a trilaminary structure arise? Let us imagine the simplest case where there is mutual penetration of two bilayers (Fig. 11.16). For the sake of simplicity we shall consider a bilayer consisting of neutral lipid molecules with polar heads having a dipole moment μ. As seen in Fig. 11.16, during the process of reciprocal penetration, the head groups of lipids pass over from the medium with a high dielectric permeability (water, $\varepsilon_w = 80$) into another medium with a low dielectric permeability (lipid membrane, $\varepsilon_m = 2$). It is easy to evaluate the energy of polar heads in a medium with dielectric permeability ε:

$$W = \frac{\mu^2}{\varepsilon_0 \varepsilon \delta^3},$$ (11.109)

where δ is the size of the polar head. It is well known that the energy of a polar head is inversely proportional to the dielectric permeability of the surrounding medium. Consequently, for reciprocal penetration of bilayers with surface molecule density n

and thickess $2h$ the work performed by the external pressure P per unit membrane area has to be

$$Ph = \left(\frac{1}{\varepsilon_m} - \frac{1}{\varepsilon_w} \right) \frac{\mu^2 n}{\varepsilon_0 \delta^3}. \tag{11.110}$$

Assuming $\delta \approx 0.5$ nm, $n \approx 0.02$ nm^{-2}, and the dipole moment μ set up by two elementary charges e_0 and $-e_0$ separated by the distance of 0.3 nm, we obtain $P \approx 10^7$ Pa. To obtain such a pressure, a water column 1 km high would be needed! In the case of charged lipids, even higher energy would be needed, since the energy of a charge in a medium is higher than that of a dipole. Consequently, the above mechanism of reciprocal penetration of bilayers without disturbing their structure appears impossible. It must be that the role of hydrostatic pressure in fusion, as reported by Neher (1974) and others, is to establish tight contact between two membranes over a relatively wide area, such that eventually monolayer fusion occurs.

The first type of fusion mechanism associated with the formation of a lipid bridge between two membranes is illustrated in Fig. 11.17. The lipid bridge formed by a curved monolayer (Fig. 11.17(a)) or a bilayer (Fig. 11.17(b)) is called a stalk (Gingell and Ginsberg, 1978; Hui *et al.*, 1982) and this type of a fusion mechanism is called a stalk mechanism. A stalk can be formed either by a single monolayer or by an entire bilayer. The evolution of a monolayer stalk consisting of the expansion of the curved region leads to the formation of a trilaminary structure and to a hemifusion. If the bilayer stalk was formed in the first place, its expansion leads to a complete fusion (Fig. 11.17(b)). If a stalk is to be formed between membranes, the local bulging defects have to appear at both membranes and grow to each other like stalactites and stalagmites. These protrusions can close upon one another directly or, as presumed in the model of Lucy (1970) as modified by Gingell and Ginsberg (1978), through another lipid formation, e.g. micelles.

A mechanism based on inverted micellar intermediate structures was proposed by Verkleij and co-workers (Verkelij, 1984; Verkleij *et al.*, 1980), and Siegel (1984, 1986a,b,c) developed a kinetic theory of this mechanism.

Mechanisms based on an extended tight contact between membranes are called adhesional mechanisms. Initially a tetralayer structure is formed, where two cis-monolayers can, apparently, partially penetrate one another (Fig. 11.18). These two perturbed cis-monolayers can be removed by different forces and as a result the structure can be transformed into a bilayer giving rise to a trilaminary structure. These mechanisms were analysed by Kozlov and Markin (1984).

At the present time the most probable mechanism of lipid bilayer fusion is thought to be a stalk mechanism and we will consider it in detail.

Stalk Mechanism. The theoretical basis for this mechanism was formulated by Markin *et al.* (1984) and was later further developed and experimentally verified by Chernomordik *et al.* (1985, 1987), Kozlov *et al.* (1989b), Leikin *et al.* (1987), and Nanavati *et al.* (1992). The main ideas of this mechanism are as follows.

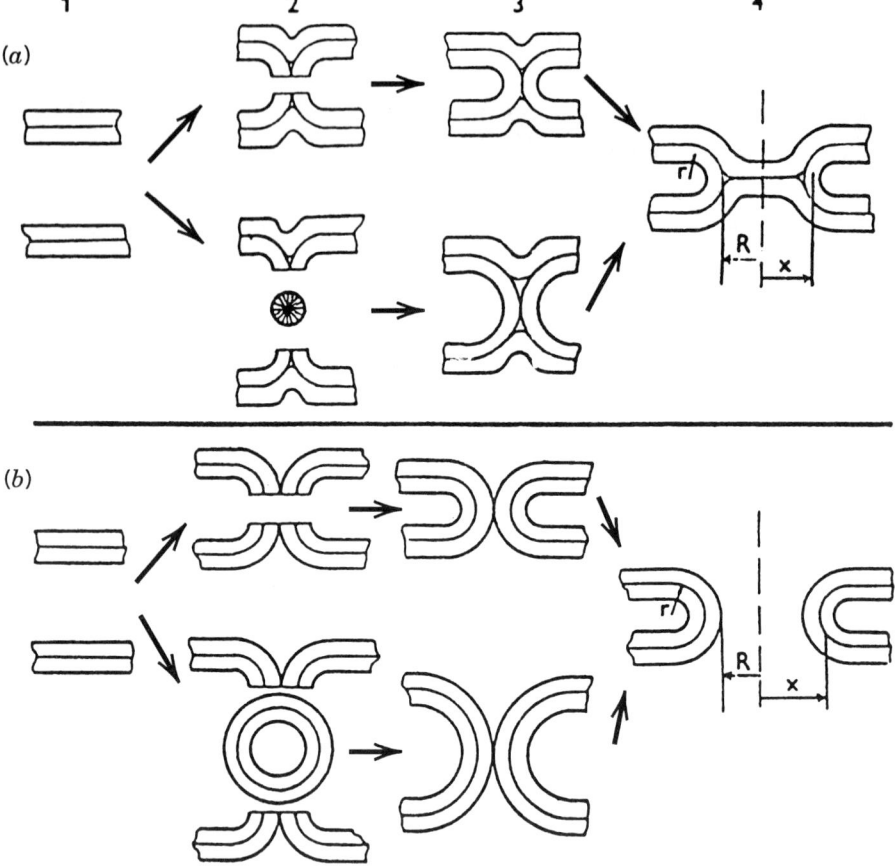

Figure 11.17. The stalk mechanism of membrane fusion; the figure shows cross-sections of the bodies of revolution arising during the process of fusion. (a) The monolayer variant: 1) approaching original membranes; 2) simultaneous formation of bulging defects capable of closing upon one another directly or of interacting through a micelle; 3) the appearance of a stalk with a zero radius; 4) expansion of the stalk. R is the radius of the stalk, $H = 2r$ is its height, x is the distance from an arbitrary point on the neutral surface to the axis of revolution. (b) The bilayer variant. The stages of fusion are the same as in the case of a monolayer stalk. The closure of bulging defects may proceed through liposomes. The neutral surface of the stalk coincides with the surface separating the monolayers.

Let us consider a stalk formed as a result of a direct closure of two bulging defects in opposing membranes having initial curvature, c_{init}, and a spontaneous curvature, c_0. The following analysis is equally applicable to both monolayer and bilayer stalks but for the sake of simplicity we will refer here to a monolayer stalk [Fig 11.17(a)]. We assume that the stalk and surrounding membranes form an axisymmetrical body of revolution. In the middle of the monolayer transformed into a stalk we draw a

(a)

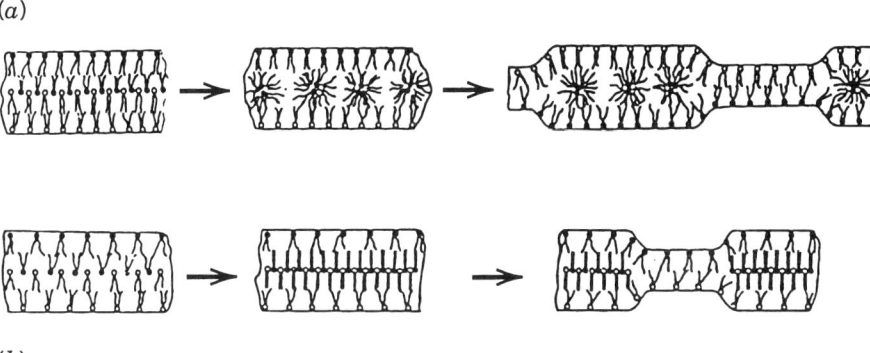

(b)

Figure 11.18. The adhesion mechanism of fusion: (a) adhesion-micellar; (b) adhesion-condensational.

neutral surface. Suppose that a cross-section of this neutral surface by a meridional plane is a part of a circle with radius r as presented in Fig. 11.17(a). The quantity $H = 2r$ is called the length of the stalk. The shortest distance from the axis to the hydrophobic surface of the monolayer R is called the radius of the stalk; this is the minimal radius of the channel formed by the inner surface of the stalk (inner radius). Initially, it is zero, but it can increase when the stalk transforms into a trilaminar structure. Parameter h is the thickness of the stalk wall. In the case of a monolayer stalk its thickness is one monolayer, while in a bilayer stalk it has a thickness of two monolayers.

Because membrane monolayers resist bending, the stalk can accumulate a certain elastic energy very similar to what we have found in the previous section for the pore edge. However, there are some differences related to additional parameters involved in this structure. As before, elastic energy can be presented as $w = (\kappa/2)(c_m + c_p - c_0)^2$, where c_m and c_p are principle curvatures along the meridian and parallel to the body of revolution that represents the stalk.

The meridional curvature of the stalk membrane is $c_m = 1/r$, and the other, "parallel," curvature is related to it by the equation (Pogorelov, 1965)

$$c_p + x\frac{dc_p}{dx} = c_m, \tag{11.111}$$

where x is the distance from a given point at the neutral surface to the axis of revolution, in other words this is the coordinate in the system. The boundary condition for equation (11.111) takes into consideration that at the equatorial plane with coordinate $x = R + h/2$ the "parallel" curvature must be $c_p = 1/(R + h/2)$. Solving equation (11.111) one obtains

$$c_p = \frac{r + R + h/2}{rx} - \frac{1}{r}. \tag{11.112}$$

The energy of the stalk is defined as its elastic energy minus the initial elastic energy of two membranes without stalks, and we also have to take into consideration the contribution of membrane tension γ:

$$W_S = \pi\kappa\left[\int_{\text{stalk}} dA\left(\frac{r+R+h/2}{rx} - \frac{2}{r} - c_0\right)^2 - \int_{\text{stalk}} dA(c_{\text{init}} - c_0)^2\right] - 2\pi\gamma(R+h/2)^2.$$

$$(11.113)$$

The integrals are taken over the surface of the stalk. The first integral represents the bending energy of the stalk membrane and the second integral is equal to the bending energy of the initial membrane. The role of membrane tension was considered in the previous section. Here we separately consider the role of the stalk bending energy W_{SB}:

$$W_{SB}(R, r) = 2\pi\kappa\left\{\left[\left(\frac{2}{r} + c_0\right)^2 - (2c_{\text{init}} - c_0)^2\right]\left[\frac{\pi}{2}r(r+R+h/2) - r^2\right]\right.$$

$$- \pi\left(\frac{2}{r} + c_0\right)(r+R+h/2)$$

$$\left. + \frac{2}{r}\frac{(r+R+h/2)^2}{\sqrt{(R+h/2)(2r+R+h/2)}}\tan^{-1}\sqrt{\frac{(2r+R+h/2)}{(R+h/2)}}\right\}. \qquad (11.114)$$

Evolution of a Stalk. The energy of the stalk $W_{SB}(R, r)$ depends on its width, radius, R, and length $H = 2r$ with the spontaneous curvature c_0 being a parameter. With increasing R at constant H the surface area of the stalk grows, and the surface energy density declines due to the decreasing curvature c_p while c_m remains constant. The interplay of these two tendencies can bring about an interesting behavior of the stalk energy as presented in Fig. 11.19. The energy may monotonically increase with the radius (curve 1); then, the formation and widening of the stalk is impossible. Under other conditions the function $W_s(R)$ may reach a minimum (curves 2–4). In such a situation the stalk tends to assume a radius at which its energy is minimal. Finally, there may be cases in which the energy decreases with increasing radius (curve 5); this corresponds to the tendency of the stalk to expand unlimitedly.

To get more specific data we consider a particular case where the length of stalk $H = 2r$ is constant and equal to h. This corresponds to the case when membranes approach very close to each other. Presuming for the sake of simplicity that the initial curvature of a membrane is negligible, and introducing dimensionless radius of the stalk $\rho = 2R/h$ one can rewrite expression (11.114) in the following form:

$$W_{SB}(\rho) = 2\pi\kappa\left\{\frac{2(\rho+2)^2}{\sqrt{(\rho+2)^2 - 1}}\tan^{-1}\sqrt{\frac{\rho+3}{\rho+1}} - 4 + \frac{c_0 h}{2}[\pi(\rho+2) - 4]\right\}. \qquad (11.115)$$

This equation reminds us of equation (11.103) for the elastic energy of the pore in

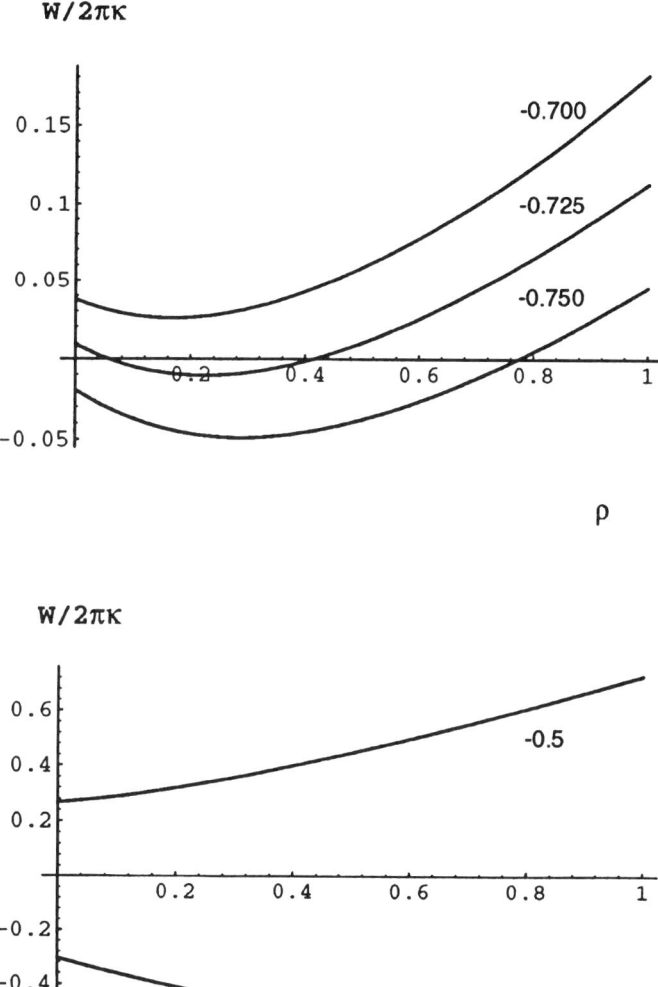

Figure 11.19. The energy of the stalk versus its dimensionless radius $\rho = 2R/h$. The value of parameter c_0h is indicated at the curves.

the bilayer with the only difference being that the last term with spontaneous curvature has an opposite sign. This is quite understandable because the stalk and the pore edge have similar geometry but the direction of bending of the monolayer in the stalk is opposite to the bending of the monolayer in the pore. Therefore, the spontaneous curvature that favors stalk formation inhibits the formation of a pore and vice versa. This fact was experimentally demonstrated by Chernomordik *et al.* (1985) by changing the spontaneous curvature and following the rate of stalk formation and mechanical breakdown.

Analysis of expression (11.115) (illustrated in Fig. 11.19) shows that the stalk energy $W_s(\rho)$ can have a minimum (curves 2, 3, 4), if the spontaneous curvature satisfies the inequality

$$-1 < c_0 h < -0.6. \tag{11.116}$$

In this case, the stalk formed will be expanding until its energy becomes minimal.

It is important to know the stalk energy at the very beginning, when its radius is zero. If $-0.74 < c_0 h < -0.6$ the energy $W_{SB}(\rho)$ for $\rho = 0$ is positive (curves 2, 3). This means that for a stalk to be formed the membranes have to overcome an energy barrier. If the minimal energy is also positive (curve 2), the stalk is unstable, because its energy is higher than the energy of two separate membranes. However, when the minimum of the stalk energy is negative (curve 3), the stalk is stable. If $-1 < c_0 h < -0.74$ even the initial stalk has energy lower than two separate membranes. In the case of high negative curvature

$$c_0 h < -1, \tag{11.117}$$

the stalk will expand without any limit (curve 5). One result of such an expansion may be monolayer fusion, in the case of a monolayer stalk, or a complete fusion in the case of a bilayer stalk. It is worth noting that the stalk radius at a minimum energy has the order of h.

These numerical examples address the dependence of stalk energy on one parameter only—its width while ignoring length. This dependence was analyzed by Chizmadzhev *et al.* (1995). If both variables are free to change, then once formed the stalk should quickly reach semi-stable dimensions. In the absence of spontaneous curvature these dimensions correspond to the ratio $(R + h/2)/r$ equal to 0.6. Thereafter, the length and width should gradually increase until the barrier vanishes and the stalk completely opens.

The stalk model has been extensively discussed in the literature, and a number of important modifications made. Most recently, Siegel (1993) provided a more detailed geometrical description of the stalk, accurately matching the shapes of cis- and trans-monolayers. He explicitly introduced void spaces at the extremes of the stalk, bringing to light a possible pressure dependence of membrane fusion.

The mechanical approach to the fusion problem has shed light on a number of phenomena now under intensive study. One of these involves fluctuating fusion pores, which can exist for milliseconds but eventually close without membrane fusion (Markin and Hudspeth, 1993; Nanavati *et al.*, 1992). The existence of *exocytosis interruptus* implies that a system of two membranes can find itself in a minimum of elastic energy as a function of a pore radius, and that this minimum is broad enough to ensure rather large fluctuations in pore size. Studying mechanical effects should be especially important for exocytosis, a process in which the controlled swelling of the secretory granule matrix probably plays a key role.

The problem of membrane fusion is by no means closed; considerable work remains to be done. For example, in the existing models the shape of a stalk was not calculated but rather postulated. Although the configuration used seems quite reasonable, it is unlikely to represent the absolute minimum of elastic energy. It would therefore be valuable to determine explicitly the stalk shape which requires minimum energy. Another important question is the applicability of Helfrich's simple equation (11.61) to cases of very large membrane curvature such as seen in a fusion stalk where the radius of curvature is only a few nanometers.

A final important question is how to incorporate the mechanical analysis of the fusion intermediates into an understanding of the total fusion process. Each preceding and following step in the fusion process may affect another, making fusion a cooperative process rather than a chain of independent events. This problem is still far from being solved, and is so extensive that we refer the reader to recent reviews on the subject (Chernomordik et al., 1995; Chernomordik and Zimmerberg, 1995; Rand and Parsegian, 1986).

BIBLIOGRAPHY

Abidor, I. G., V. B. Arakelyan, L. V. Chernomordik, Yu. A. Chizmadzhev, V. F. Pastushenko and M. R. Tarasevich (1979). Electric breakdown of bilayer membranes: 1. The main experimental facts and their qualitative discussion. *Bioelectrochem. Bioenerg.*, **6**, 37–59.

Abraham, M. H. and A. F. Danil de Namor (1976). Solubility of electrolytes in 1,2-dichloroethane and 1,1-dichloroethane and derived free energies of transfer. *J. Chem. Soc. Faraday Trans.*, **72**, 955–62.

Abraham, M. H. and J. Liszi (1978). Calculations on ionic solvation. Part 2. Entropies of solvation of gaseous ions using a one-layer continuum model. *J. Chem. Soc., Faraday Trans. 1*, **74**, 2858–67.

Abraham, M.H. and J. Liszi (1980). Calculation of ionic solvation. *J. Chem. Soc. Faraday Trans. 1*, **76**, 1219–31.

Abraham, M.H. and J. Liszi (1981). Calculation of ionic solvation. V. The calculation of partition coefficients of ions. *J. Inorg. Nucl. Chem.*, **43**, 143–52.

Abraham, M.H., J. Liszi and E. Kristof (1982). Calculation of ionic solvation. VII. The free energy of solvation of ions calculated from various local solvent dielectric constant–distance functions. *Austral. J. Chem.*, **35**, 1273–9.

Abraham, M. H., J. Liszi and S. Meszaros (1979). Calculations on ionic solvation. III. The electrostatic free energy of solvation of ions, using a multilayered continuum model. *J. Chem. Phys.*, **70**, 2491–6.

Abraham, M. H., G. S. Whiting, R. Fuchs and E. J. Chambers (1990). Thermodynamics of solute transfer from water to hexadecane. *J. Chem. Soc.*, 291–300.

Abramson, A. A. (1981). *Surface Active Compounds. Properties and Applications*. Khimiya, Leningrad.

Adamson, A. W. (1990). *Physical Chemistry of Surfaces*. Wiley, New York.

Addison, C. C. (1943). The properties of freshly formed surfaces. I. Application of the vibrating-jet technique to surface tension measurements on mobile liquids. *J. Chem. Soc.*, 535–40.

Addison, C. C. (1945). Measurement of surface and interfacial tensions at fresh surfaces by the vibrating-jet method. *Phil Mag.*, **36**, 7–13.

Ahkong, O. F., W. Tampion and J. A. Lucy (1975). Promotion of cell fusion by divalent cation ionophores. *Nature*, **256**, 208–9.

Akhadov, A. Yu. (1972). *The Dielectric Properties of Pure Liquids*. Izdatelstvo Standartov, Moscow.

492

Albery, W. J., A. M. Couper, J. Hadgraft and C. Ryan (1974). Transport and kinetics in two phase systems. *J. Chem. Soc. Faraday Trans.*, I, **70**, 1124–31.

Albery, W. J., R. A. Choudhery and P. R. Fisk (1984). Kinetics and mechanism of interfacial reactions in the solvent extraction of copper. *Faraday Discuss. Chem. Soc.*, **77**, 53–65.

Alent'ev, A. Yu. and E. S. Filatov (1991). Radiometric method for studying interfaces between two immiscible liquids. *Sov. Radiochem.*, **33**, 657–61.

Alexander, A. E. (1940). A study of films at the liquid/liquid interfaces. *Trans. Faraday Soc.*, **35**, 727–37.

Alexander, R., A. J. Parker, J. H. Sharpe and W. E. Waghorne (1972). Solvation of ions. XVI. Solvent activity coefficients of single ions. A recommended extrathermodynamic assumption. *J. Am. Chem. Soc.*, **94**, 1148–58.

Aliev, M. K., L. I. Boguslavsky, A. G. Volkov, I. A. Kozlov, D. O. Levicky and S. T. Metelsky (1976). Study of electrogenic function of Ca^{2+-}ATPase at the octane/water interface. *Bioorg. Khimiya*, **2**, 1132–7.

Allen, M. P. and D. J. Tildesley (1987). *Computer Simulation of Liquids.* Clarendon, Oxford.

Amblard, G., R. Sandeaux, J. Sandeaux and C. Gavach (1985). Transport of potassium ions across planar lipid membranes by the antibiotic grisorixin: 1. The equilibrium state and self-diffusion K^+ fluxes. *J. Membr. Biol.*, **88**, 15–23.

Andersen, O. S. and M. Fuchs (1975). Potential energy barriers to ion transport within lipid bilayers. Studies with tetraphenilborate. *Biophys. J.*, **15**, 795–830.

Anderson, A. F. H. and M. Calvin (1964). The aggregation of chlorophylls. *Arch. Biochem. Biophys.*, **107**, 251–9.

Andrietti, F., A. Peres and R. Pezzotta (1976). Exact solution of the unidimensional Poisson–Boltzmann equation for a 1:2 electrolyte. *Biophys. J.*, **16**, 1121–4.

Antonini, E. M., C. Brunory, C. Greenwood and B. G. Malmstrom (1970). Catalytic mechanism of cytochrome oxidase. *Nature*, **228**, 936–7.

Antonow, G. (1907). Sur la tension superficielle a la limite de deux couches. *J. Chim. Phys.*, **5**, 372–85.

Antonow, G. (1931). Das Gesetz des Gleichgewichts zwischen Zwei Phasen. *Kolloid-Z.*, **59**, 7–12.

Aoki, K. (1995). Linear dependence of the standard ion-transfer potentials of polyanions at the oil/water interface on the surface interaction energy and the charge. *J. Electroanal. Chem.*, **386**, 17–23.

Arai, K. (1994). Electroanalytical study of the electrical potential oscillation across a liquid membrane and drug transfer at an oil–water interface. *Bunseki Kagaku*, **43**, 729–30.

Arai, K., F. Kusu and K. Takamura (1996). Electrochemical behavior of drugs at the oil/water interface. In: *Liquid–Liquid Interfaces. Theory and Methods* (A. G. Volkov and D. W. Deamer, eds.) pp. 375–400. CRC Press, Boca Raton, New York, London, Tokyo.

Arakelyan, V. B. and S. B. Arakelyan (1983a). Energetic profile of dipole molecules at the border between two phases. *Biolog. Zh. Armenii*, **36**, 553–9.

Arakelyan, V. B. and S. B. Arakelyan (1983b). Energetic profile of a dipole molecule in the thin membrane. *Biolog. Zh. Armenii*, **36**, 775–9.

Arakelyan, V. B., S. B. Arakelyan, T. M. Avakyan and V. M. Aslanyan (1985). Electrostatic effect on transport of water across bilayer lipid membrane. *Biophysics*, **30**, 186–7.

Arakelyan, V. B., S. B. Arakelyan, T. M. Avakyan and V. M. Aslanyan (1986). Electrostatic energy of a dipole molecule in an ionic channel of the membrane. *Biophysics*, **31**, 394–5.

Aratono, M., M. Yamanaka, N. Matubayasi, K. Motomura and R. Matuura (1980). Thermodynamic study on the adsorption on dodecylammonium chloride of water/hexane interface. *J. Colloid Interface Sci.*, **74**, 489–94.

Attard, P., D. Wei and G. N. Patey (1992). On the existence of exact conditions in the theory of electrical double layers. *J. Chem. Phys.*, **96**, 3767–74.

Baker, G. F. and W. F. Widdas (1973). The asymmetry of the facilitated transfer system for hexoses in human red cells and the simple kinetics of the two component model. *J. Physiol. (London)*, **231**, 143–65.

Bakker, G. (1928). *Handbuch der Experimental Physik*, Vol. 6, p. 90, Academie Verlag, Leipzig.

Barber, J., J. Mills and A. Love (1977). Electrical diffuse layers and their influence on photosynthetic processes. *FEBS Lett.*, **74**, 174–81.

Barker, N., J. Hadgraft and N. Wotton (1984). Facilitated transport across liquid/liquid interfaces and its relevance to drug diffusion across biological membranes. *Faraday Discuss. Chem. Soc.*, **77**, 97–104.

Baur, E. and S. Korman (1917). Über die Ionenadsorptionspotentiale. *Z. Phys. Chem.*, **92**, 81–97.

Bayliss, W. M. (1923). *Interfacial Forces and Phenomena in Physiology*. Methuen & Co. Ltd., London.

Bazarov, I. P. (1964). *Thermodynamics*. Pergamon Press. Oxford.

Bebhardt, C., H. Gruler and E. Sakmann (1977). On domain structure and local curvature in lipid bilayers and biological membranes. *Z. Naturforsch.*, **32c**, 581–96.

Bell, A. J., J. G. Frey and T. J. VanderNoot (1992). 2nd harmonic generation by paranitrophenol at water/air and water/heptane interfaces. *J. Chem. Soc. Faraday Trans.*, **88**, 2027–30.

Bell, R. P. (1928). Reaction velocity at a liquid–liquid interface. *J. Phys. Chem.*, **32**, 882–93.

Bell, R. P. (1931). The electrical energy of dipole molecules in solution and solubilities of ammonia, hydrogen chloride, and hydrogen sulfite in various solvents. *J. Chem. Soc.*, **32**, 1371–82.

Bender, M. (ed.) (1991). *Interfacial Phenomena in Biological Systems*. Marcel Dekker, Inc., New York.

Benjamin, I. (1991). Molecular dynamics study of the free energy functions for electron-transfer reactions at the liquid–liquid interface. *J. Phys. Chem.*, **95**, 6675–83.

Benjamin, I. (1992a). Dynamics of ion transfer across a liquid–liquid interface: a comparison between molecular dynamics and a diffusion model. *J. Chem. Phys.*, **96**, 577–85.

Benjamin, I. (1992b). Theoretical study of the water/1,2-dichloroethane interface: structure, dynamics, and conformational equilibria at the liquid–liquid interface. *J. Chem. Phys.*, **97**, 1432–45.

Benjamin, I. (1993). Mechanism and dynamics of ion transfer across a liquid–liquid interface. *Science*, **261**, 1558–60.

Benjamin, I. (1994). Solvent dynamics following charge transfer at the liquid–liquid interface. *J. Chem. Phys.*, **180**, 287–96.

Benjamin, I. (1996a). Molecular dynamics of charge transfer at the liquid/liquid interlace. In: *Liquid–Liquid Interfaces. Theory and Methods* (A. G. Volkov and D. W. Deamer, eds.) pp. 179–211. CRC Press, Boca Raton, New York, London, Tokyo.

Benjamin, I. (1996b). Chemical reactions and solvation at liquid interfaces: a microscopic perspective. *Chem. Rev.*, **96**, 1449–75.

Ben-Naim, A. (1980). *Hydrophobic Interactions*. Plenum Press, New York.

Ben-Naim, A. and Y. Marcus (1984). Solvation thermodynamics of nonionic solutes. *J. Chem. Phys.*, **81**, 2016–27.

Benz, R., F. Beckers and U. Zimmermann (1979). Reversible electrical breakdown of lipid bilayer membranes: a charge-pulse relaxation study. *J. Membrane Biol.*, **48**, 181–204.

Berndl, K., J. Käs, R. Lipowsky, E. Sackmann and U. Seifert (1990). Shape transformations of giant vesicles: extreme sensitivity to bilayer asymmetry. *Europhys. Lett.*, **13**, 659–64.

Bessis, M. (1973). *Living Blood Cells and their Ultrastructure.* Springer Verlag, Berlin, Heidelberg, New York.

Beutner, R. (1913a). Neue Erscheinungen der Elektrizitätserregung, welche einige bioelektrische Phänomene erklaren. *Z. Elektrochem.*, **19**, 319–78.

Beutner, R. (1913b). Neue Erscheinungen der Elektrizitatserregung, welche einige bioelektrische Phänomene erklaren. Zweiter Teil. *Z. Elektrochem.*, **19**, 467–506.

Beutner, R. (1918). Kann die Elektrizitätserregung, durch organische Substanzen und lebende gewebe auf Grund bekannter thermodynamischer Gesetze erklärt werden? *Z. Elektrochem.*, **24**, 94–100.

Beveridge, D. L. and G. W. Schnuelle (1975). Free energy of a charge distribution in concentric dielectric continue. *J. Phys. Chem.*, **79**, 2562–6.

Bewig, K. W. (1964). Ionization method of measuring contact potential differences. *Rev. Sci. Instrum.*, **35**, 1160–2.

Bhuiyan, L. B., D. Bratko and C. W. Outhwaite (1991). Electrolyte surface tension in the modified Poisson–Boltzmann approximation. *J. Phys. Chem.*, **95**, 336–40.

Bhuiyan, L. B. and C. W. Outhwaite (1994). The cylindrical electric double layer in the modified Poisson–Boltzman theory. *Phil. Mag. B-Phys.*, **69**, 1051–8.

Bhuiyan, L. B., C. W. Outhwaite and D. Bratko (1992). Structure and thermodynamics of micellar solutions in the modified Poisson–Boltzmann theory. *Chem. Phys. Lett.*, **193**, 203–10.

Bhuiyan, L. B., C. W. Outhwaite, M. Molero and E. Gonzaleztovar (1994). The primitive model of ionic fluids near its critical point in the Poisson–Boltzmann and modified Poisson–Boltzmann theories. *J. Chem. Phys.*, **100**, 8301–6.

Bikerman, J. J. (1970). *Physical Surfaces.* Academic Press, New York.

Birdi, K. S. (1988). *Lipid and Biopolymer Monolayer at Liquid Interfaces.* Plenum Press, New York.

Blandamer, M. J. and M. C. R. Symons (1963). Significance of new values for ionic radii to solvation phenomena in aqueous solution. *J. Phys. Chem.*, **67**, 1304–6.

Blank, M. (1966). Some effects due to the flow of current across a water–nitrobenzene interface. *J. Colloid Interface Sci.*, **22**, 51–7.

Blank, M. (1991). Membrane transport: insights from surface science. In: *Interfacial Phenomena in Biological Systems* (M. Bender, ed.). M. Dekker, Inc., New York.

Blank, M. and S. Feig (1963). Electric field across water–nitrobenzene interfaces. *Science*, **141**, 1173–4.

Blum, L. (1974). Solution of a model for the solvent–electrolyte interactions in the mean spherical approximation. *J. Chem. Phys.*, **61**, 2129–33.

Blum, L. (1977). Theory of electrified interfaces. *J. Phys. Chem.*, **81**, 136–47.

Blum, L. (1990). Structure of the electric double layer. *Adv. Chem. Phys.*, **78**, 171–222.

Blum, L., D. Henderson and R. Parsons (1984). The mean spherical approximation capacitance of the double layer for an electrolyte at the high concentration. *J. Electroanal. Chem.*, **161**, 369–92.

Boguslavsky, L. I., Yu. B. Americ and M. I. Gugeshashvili (1973). Observation of a phase transition in a liquid crystal by means of the Volta-potential measurements. *Sov.*

Electrochem., **9**, 637–637.

Boguslavsky, L. I., V. G. Boitsov and M. I. Gugeshashvili (1976a). Potential changes during the cooling of water/heptane/water concentration cells containing valinomycin. *Sov. Electrochem.*, **12**, 428–9.

Boguslavsky, L. I., A. N. Frumkin and M. I. Gugeshashvili (1974a). Contact phenomena at the heptane water interface in the presence of valinomycin. *Bioelectrochem. Bioenerg.*, **1**, 506–14.

Boguslavsky, L. I., A. N. Frumkin, M. I. Gugeshashvili (1976b). Investigation of the adsorption of tetraalkylammonium salts at water/nitrobenzene interface. *Electrokhimiya*, **12**, 856–60.

Boguslavsky, L. I. and M. I. Gugeshashvili (1972). The EMF of a transfer chain and the Volta-potential in the water–isoamyl alcohol system. *Sov. Electrochem.*, **8**, 1433–5.

Boguslavsky, L. I. and M. I. Gugeshashvili (1988). Spontaneous formation of the space charge in hydrocarbon upon contact with water containing ionic surface-active substances (electrete state of a liquid insulator). *Sov. Electrochem.*, **24**, 1077–84.

Boguslavsky, L. I. and M. I. Gugeshashvili (1989). Spontaneous space charge formation in a hydrocarbon brought into contact with water, containing ionogenic surfactants. The electretic state of a liquid insulator. *J. Electroanal. Chem.*, **263**, 373–82.

Boguslavsky, L. I., A. A. Kondrashin, I. A. Kozlov, S. T. Metelsky, V. P. Skulachev and A. G. Volkov (1975a). Charge transfer between water and octane phases by soluble mitochondrial H^+-ATPase, bacteriorhodopsin and respiratory chain enzymes. *FEBS Lett*, **50**, 223–6.

Boguslavsky, L. I., B. T. Lozhkin and B. A. Kiselev (1975b). Generation of dark potential on bilayer lipid membranes containing chlorophyll. *Doklady Akad. Nauk SSSR*, **222**, 228–31.

Boguslavsky, L. I. and A. G. Volkov (1975). Photoinduced proton transfer across decane/water interface in the presence of chlorophyll. *Doklady Akad. Nauk SSSR*, **224**, 1201–4.

Boguslavsky, L. I. and A. G. Volkov (1977). Electron and proton transfer across water/lipid interfaces in functioning enzyme membrane systems and photosynthetic pigments. In: *IFIAS Workshop on Physico-Chemical Aspects of Electron Transfer Processes*, pp. 20–7. IFIAS, Stockholm.

Boguslavsky, L. I. and A. G. Volkov (1987). Redox and photochemical reactions at the interface between immiscible liquids. In: *The Interface Structure and Electrochemical Processes at the Boundary Between Two Immiscible Liquids* (V. E. Kazarinov, ed.), pp. 143–78. Springer Verlag, Berlin.

Boguslavsky, L. I., A. G. Volkov, V. G. Boytsov, I. A. Kozlov and S. T. Metelsky (1976c). Light-dependent translocation of H^+ from water to octane by bacteriorhodopsin. *Bioorg. Khimiya.*, **2**, 1125–31.

Boguslavsky, L. I., A. G. Volkov and M. D. Kandelaki (1976d). Injection of electrons from water into octane catalyzed by chlorophyll. *Biophysics*, **21**, 808–11.

Boguslavsky, L. I., A. G. Volkov and M. D. Kandelaki (1976e). Transfer of electrons and protons at the decane/water interface in the presence of chlorophyll. *FEBS Lett.*, **65**, 155–8.

Boguslavsky, L. I., A. G. Volkov and M. D. Kandelaki (1977a). Electron transfer by chlorophyll through the interface between two immiscible liquids. *Bioelectrochem. Bioenerg.*, **4**, 68–72.

Boguslavsky, L. I., A. G. Volkov, M. D. Kandelaki and E. A. Nizhnikovsky (1976f). Photooxidation of water and proton transport through the interface between two immiscible liquids in the presence of chlorophyll. *Doklady AN SSSR*, **227**, 727–30.

Boguslavsky, L. I., A. G. Volkov, M. D. Kandelaki, E. A. Nizhnikovsky and M. A. Bibikova (1977b). Photooxidation of water in the presence of chlorophyll and the iron complex of coproporphyrin-III adsorbed at the octane/water interface. *Biofizika*, **22**, 223–7.

Boguslavsky, L. I., A. G. Volkov, A. A. Kondrashin, S. T. Metelsky and A. A. Yasaitis (1976g). Electrogenic function of submitochondrial particles at the octane/water interfaces. *Biokhimiya*, **41**, 1047–51.

Boguslavsky, L. I., A. G. Volkov, A. A. Kondrashin, V. P. Skulachev and A. A. Yasaitis (1975c). Charge transfer through the octane/water interface by the enzyme complexes of the respiratory chain. *Bioorg. Khim.*, **1**, 1783–91.

Boguslavsky, L. I., A. G. Volkov, I. A. Kozlov, A. V. Kargopolov and E. I. Mileykovskaya (1976h). H$^+$-ATPase from *Micrococcus lysodeikticus* in aqueous solution and at the octane/water interface. *Bioorg. Khim.*, **2**, 846–54.

Boguslavsky, L. I., A. G. Volkov, I. A. Kozlov and A. N. Mal'yan (1976i). ATP-hydrolysis coupled proton transfer from water to octane by soluble ATPase from chloroplasts. *Biophysics*, **21**, 286–9.

Boguslavsky, L. I., A. G. Volkov, I. A. Kozlov, S. T. Metelsky and V. P. Skulachev (1974b). Injection of positive charges from water into octane, coupled with hydrolysis of ATP by soluble ATPase. *Doklady Akad. Nauk SSSR*, **218**, 963–6.

Boguslavsky, L. I., A. G. Volkov, I. A. Kozlov, S. T. Metelsky and V. P. Skulachev (1975d). Mitochondrial H$^+$-ATPase at the octane–water interface. *Bioorg. Khim.*, **1**, 1369–78.

Boguslavsky, L. I., A. G. Volkov, I. A. Kozlov and E. I. Mileykovskaya (1975e). Transfer of protons from water into octane by bacterial ATPase. *Doklady Akad. Nauk SSSR*, **222**, 726–9.

Boguslavsky, L. I., Zhuravlev, M. D. Kandelaki and K. Shengeliya (1978). Study of water photooxidation at the octane/water interface in the presence of chlorophyll by a mass-spectrometric method. *Doklady Akad. Nauk SSSR*, **240**, 1453–6.

Bonfillon, A. and D. Langevin (1993). Viscoelasticity of monolayers at oil/water interfaces. *Langmuir*, **9**, 2172–7.

Bonhoeffer, K. F., M. Kahlweit and H. Strehlov (1953). Über electrochemische Analogien zwischen nichwassrigen Elektrolytlosungen und Ionenaustauschern. *Z. Electrochem.*, **57**, 614–5.

Booth, F. (1951a). The dielectric constant of water and the solvation effect. *J. Chem. Phys.*, **19**, 391–4.

Booth, F. (1951b). Erratum: The dielectric constant of water and the solvation effect. *J. Chem. Phys.*, **19**, 1327–8.

Booth, F. (1951c). Erratum: The dielectric constant of water and the solvation effect. *J. Chem. Phys.*, **19**, 1615.

Born, M. (1920). Volumen und Hydrationswarme der Ionen. *Z. Phys.*, **1**, 45–8.

Boruvka, L. and A. W. Neumann (1977). Generalization of the classical theory of capillary. *J. Chem. Phys.*, **66**, 5464–76.

Boruvka, L., Y. Rotenberg and A. W. Neumann (1985). Free energy formulation of the theory of capillarity. *Langmuir*, **1**, 40–4.

Bozic, B., S. Svetina, B. Zeks and R. E. Waugh (1992). Role of lamellar membrane structure in tether formation from bilayer vesicles. *Biophys. J.*, **61**, 963–73.

Bradley, P. A. and G. Parry-Jones (1974). A system for the investigation of nonlinear dielectric effects using digital techniques. *J. Phys. E: Sci. Instrum.*, **7**, 449–52.

Breisblatt, W. and S. Ohki (1975). Fusion in phospholipid spherical membranes. Effect of temperature and lysolecithin. *J. Membrane Biol.*, **23**, 385–401.

Brevet, P. F. and H. H. Girault (1996). Second harmonic generation at liquid/liquid interfaces. In: *Liquid–Liquid Interfaces: Theory and Methods* (A. G. Volkov and D. W. Deamer, eds.), pp. 103–37. CRC Press, Boca Raton, New York, London, Tokyo.

Bridgman, P. W. (1952). *The Physics of High Pressure*. Bell Publ., London.

Brown, A. R., L. J. Yellowlees and H. H. Girault (1993). Photoinduced electron-transfer across the interface between two immiscible electrolyte solutions. *J. Chem. Soc. Faraday Trans.*, **89**, 207–12.

Brust, M., M. Walker, D. Bethell, D. J. Schiffrin and R. Whyman (1994). Synthesis of thiol-derivatized gold nanoparticles in a 2-phase liquid-liquid system. *J. Chem. Soc. Chem. Commun.*, 801–2.

Buckingham, A. D. (1957). A theory of ion-solvent interaction. *Discuss. Faraday Soc.*, **24**, 151–7.

Buff, F. P. (1956). Curved fluid interfaces. I. The generalized Gibbs–Kelvin equation. *J. Chem. Phys.*, **25**, 146–53.

Burbage, J. D. and M. H. Wirth (1992a). Effect of wetting on the reorientation of acridine orange at the interface of water on a hydrophobic surface. *J. Phys. Chem.*, **96**, 5943–8.

Burbage, J. D. and M. H. Wirth (1992b). Reorientation of acridine orange at liquid alkane–water interfaces. *J. Phys. Chem.*, **96**, 9022–5.

Butt, H. J. (1991). Measuring electrostatic Van-der-Waals and hydration forces in electrolyte solutions with an atomic force microscope. *Biophys. J.*, **60**, 1438–44.

Cameron, A. and R. F. Crouch (1963). Interaction of hydrocarbon and surface active agents. *Nature*, **198**, 475–6.

Canham, P. B. (1970). The minimum energy of bending as a possible explanation of the biconcave shape of the human red blood cell. *J. Theor. Biol.*, **26**, 61–81.

Carnie, S. and D. Y. C. Chan (1980). The structure of electrolytes at charged surfaces: Ion–dipole mixtures. *J. Chem. Phys.*, **73**, 2949–57.

Carnie, S. L. and D. Y. C. Chan (1981). The Stillinger–Lovett condition for non-uniform electrolytes. *Chem. Phys. Lett.*, **77**, 437–40.

Carnie, S. L., D. Y. C. Chan, D. J. Mitchell and B. W. Ninham (1981a). The structure of electrolytes at charged surfaces: The primitive model. *J. Chem. Phys.*, **74**, 1472–8.

Carnie, S. L., D. Y. C. Chan and G. R. Walker (1981b). The statistical mechanics of ion-dipole-tetrahedral quadrupole mixtures. *Mol. Phys.*, **43**, 1115–38.

Carnie, S. L. and G. M. Torrie (1984). The statistical mechanics of the electrical double layer. In: *Advances in Chemical Physics* (I. Prigogine and S. A. Rice, eds.) Vol. 56, pp. 141–253. J. Wiley, New York.

Carpenter, I. L. and W. J. Hehre (1990). A molecular dynamics study of the hexane/water interface. *J. Phys. Chem.*, **94**, 531–6.

Carrasquer, G., S. Ahn, M. Schwartz and W. S. Rehm (1985). Electropenicity of the Na–K–ATPase pump in bullfrog cornea epithelium. *Am. J. Physiol.*, **249**, F185–F191.

Carrasquer, G., W. S. Rehm and M. Schwartz (1986). Amphotericin B enhanced anomalous potential difference response to changes in aqueous K^+ in frog cornea. *Biochim. Biophys. Acta*, **862**, 178–84.

Case, B. and R. Parsons (1974). The medium effect for single ionic species. *Trans. Faraday Soc. 1*, **70**, 1636–48.

Cass, A. and A. Finkelstein (1967). Water permeability of thin lipid membranes. *J. Gen. Physiol.*, **50**, 1765–84.

Cevc, G. (1990a). Membrane electrostatics. *Biochim. Biophys. Acta*, **1031**, 311–82.

Cevc, G. (1990b). The molecular mechanism of interaction between monovalent ions and polar surfaces. *Chem. Phys. Lett.*, **170**, 283–8.

Cevc, G., M. Hauser and A. A. Kornyshev (1995). Effects of the interfacial structure on the hydration forces between laterally uniform surfaces. *Langmuir*, **11**, 3103–10.

Chan, D. J. C., D. Mitchell and B. W. Ninham (1979). A model of solvent structure around ions. *J. Chem. Phys.*, **70**, 2946–57.

Chandler, D. (1987). *Introduction to Modern Statistical Mechanics.* Oxford University Press, Oxford.

Chang, C. A., E. K. Wang and Z. C. Pang (1989). Interfacial potential difference for the liquid/liquid ion partition process. *J. Electroanal. Chem.*, **266**, 143–55.

Chapman, D. L. (1913). A contribution to the theory of electrocapillarity. *Phil. Mag.*, **25**, 475–81.

Chasovnikova, L. V., V. Ye. Formazyuk, V. I. Sergiyenko and Yu. A. Valdimirov (1990). Modeling the process of diffusion of anticataractal drugs deep into the eye lens. *Biofizika*, **35**, 464–8.

Chattoray, D. K. and K. S. Birdi (1984). *Adsorption and Gibbs Surface Excess.* Plenum Press, New York.

Cheesman, D. F. (1952). The behavior of proteins at the liquid–liquid interface. *Biochem. J.*, **50**, 667–71.

Cheesman, D. F. and J. T. Davies (1954). Physicochemical and biological aspects of protein at interfaces. In: *Advances in Protein Chemistry* (M. L. Anson, K. Bailey and J. T. Edsall, eds.) Vol. 9, pp. 439–501. Academic Press Inc., New York.

Chen, Q. Z., K. Iwamoto and M. Seno (1991). Kinetic analysis of electron transfer between hexacyanoferrate(III) in water and ferrocene in nitrobenzene by ac impedance measurements. *Electrochim. Acta*, **36**, 291–6.

Chen, Y. L., Z. H. Xu and J. Israelachvili (1992). Structure and interactions of surfactant-covered surfaces on nonaqueous (oil surfactant water) media. *Langmuir*, **8**, 2966–75.

Cheng, Y. and D. J. Schiffrin (1991). Electron transfer between bis(pyridine)meso-tetraphenylporphyrinato iron(II) and ruthenium(III) and the hexacyanofeffate couple at the 1,2-dichloroethane water interface. *J. Electroanal. Chem.*, **314**, 153–63.

Cheng, Y. and D. J. Schiffrin (1993). A.C. impedance study of rate constants for two-phase electron-transfer reactions. *J. Chem. Soc. Faraday Trans.*, **89**, 199–205.

Cheng, Y. and D. J. Schiffrin (1994). Redox electrocatalysis by tetracyanoquinodimethane in phospholipid monolayers adsorbed at a liquid/liquid interface. *J. Chem. Soc. Faraday Trans.*, **90**, 2517–23.

Chernenko, A. A. (1981a). Structure of the diffuse double layer part in the model of hydrated ions. *Elektrokhimiya*, **17**, 1227–31.

Chernenko, A. A. (1981b). The diffuse layer in the model of hydrated ions. *Elektrokhimiya*, **17**, 623–8.

Chernomordik, L. V., M. M. Kozlov, G. B. Melikyan, I. G. Abidor, V. S. Markin and Yu. A. Chizmadzhev (1985). The shape of lipid molecules and monolayer fusion. *Biochim. Biophys. Acta*, **812**, 643–55.

Chernomordik, L., M. M. Kozlov and J. Zimmerberg (1995). Lipids in biological membrane fusion. *J. Membrane Biol.*, **146**, 1–14.

Chernomordik, L. V., G. B. Melikyan and Yu. A. Chizmadzhev (1987). Biomembrane fusion: a new concept derived from model studies using two interacting planar lipid bilayers. *Biochim. Biophys. Acta*, **906**, 309–52.

Chernomordik, L. V. and J. Zimmerberg (1995). Bending membranes to the task: structural intermediates in bilayer fusion. *Current Opinion in Structural Biology*, **5**, 541–7.

Cherny, V. V., V. S. Sokolov and I. G. Abidor (1980). Determination of surface charge of bilayer lipid membranes. *Bioelectrochem. Bioenerg.*, **7**, 413–20.

Chizmadzhev, Y. A., F. S. Cohen, A. Shcherbakov and J. Zimmerberg (1995). Membrane mechanics can account for fusion pore dilation in stages. *Biophys. J.*, **69**, 2489–500.

Chizmadzhev, Yu. A. and V. F. Pastushenko (1988). Theory of electrical breakdown of planar lipid bilayers. In: *Thin Liquid Films*, I. Ivanov. (ed.) pp. 1059–120. Marcel Dekker, New York and Basel.

Churaev, N. V. and B. V. Deryaguin (1985). Inclusion of structural forces in the theory of stability of colloids and films. *J. Colloid Interface Sci.*, **103**, 542–53

Clark, W. M. (1960). *Oxidation–Reduction Potentials of Organic Systems*. Williams & Wilkins, Baltimore.

Clarke, R. J. (1992). An adsorption isotherm for the interaction of membrane-permeable hydrophobic ions with lipid vesicles. *Biophys. Chem.*, **42**, 63–72.

Clunie, J. S., J. F. Goodman and P. C. Symons (1967). Solvation forces in soap films. *Nature*, **216**, 1203–4.

Coetzee, J. F. and W. R. Sharpe (1971). Solute–solvent interactions. VI. Specific interactions of tetraphenylarsonium, tetraphenylphosphonium, and tetraphenilborate ions with water and other solvents. *J. Phys. Chem.*, **75**, 3141–6.

Cohen, L. A. (1970). Chemical modification as a probe of structure and function. In: *The Enzymes* (P. D. Boyer, ed.) Vol. 1, Chap. 3, pp. 148–211. 3rd edition. Academic Press, New York.

Colson-Guastalla, H., M. Dupeyrat and J. Guastalla (1970). Quelques propriétés électrochimiques de syst'emes constitues par des solutions ioniques non miscibles en contact. *J. Chim. Phys.*, **67**, 1508–10.

Conboy, J. (1996). Application of second harmonic generation to the electrochemical interface between two immiscible electrolyte solutions. *Interface*, **5**, 49–50.

Conway, B. E. (1981). *Ionic Hydration in Chemistry and Biophysics*. Elsevier, New York.

Conway, B. E., J. O'M. Bockris and I. A. Ammar (1951). The dielectric constant of the solution in the diffuse and Helmhotz double layer at a charged interface in aqueous solution. *Trans. Faraday Soc.*, **47**, 756–66.

Costa, M., L. Priami and S. Bordi (1976). Apparatus for the detection of surface potential. *J. Electroanal. Chem.*, **70**, 229–32.

Craxford, O. Gatty and Rothschild (1938). Adsorption potentials. *Nature*, **141**, 1098–9.

Cremer, Z. (1906). Über die Ursache der elektromotorischen Eigenschaftender Gewebe, zugleicht ein Beitrag zur Lehre von den polyphasischen Elektrolytketten. *Z. Biol.*, **47**, 562–608.

Croxton, T. L. and D. A. McQuarrie (1979a). A theory of the electrical double layer through the Born-Green-Yvon equation. *Chem. Phys. Lett.*, **68**, 489–94.

Croxton, T. L. and D. A. McQuarrie (1979b). Numerical solution of the Born–Green–Yvon equation for the restricted primitive model of ionic solutions. *J. Phys. Chem.*, **83**, 1840–1.

Croxton, T. L. and D. A. McQuarrie (1981). The electrical double layer in the Born-Green-Yvon equation. *Mol. Phys.*, **42**, 141–51.

Cui, Q. Z., G. Y. Zhu and E. K. Wang (1994). The application of the MPB4 theory to the interface between two immiscible electrolyte solutions. 1. The differential capacitance of the water/nitrobenzene interface. *J. Electroanal. Chem.*, **372**, 15–9.

Cui, Q., G. Y. Zhu and E. K. Wang (1995). The application of the MPB4 theory to the interface between two immiscible electrolyte solutions. 2. The differential capacitance of the water/1,2-dichloroethane interface. *J. Electroanal. Chem.*, **383**, 7–12.

Cumper, C. W. N. and A. E. Alexander (1951). Proteins at interfaces. *Rev. Pure Appl. Chem.*, **1**, 121–51.

Cunnane, V. and L. Murtomaki (1996). Electrocatalysis and electrolysis. In: *Liquid–Liquid Interfaces. Theory and Methods* (A. G. Volkov and D. W. Deamer, eds.) pp. 401–16. CRC Press, Boca Raton, New York, London, Tokyo.

Cunnane, V. J., D. J. Schiffrin, C. Beltran, G. Geblewicz and T. Solomon (1988a). The role of phase transfer catalysts in two phase redox reactions. *J. Electroanal. Chem.*, **247**, 203–14.

Cunnane, V. J., D. J. Schiffrin, M. Fleischmann, G. Geblewicz and D. Williams (1988b). The kinetics of ionic transfer across adsorbed phospholipid layers. *J. Electroanal. Chem.*, **243**, 455–64.

Cunnane, V. J., G. Geblewicz and D. J. Schiffrin (1995). Electron and ion transfer potentials of ferrocene and derivatives at a liquid–liquid interface. *Electrochim. Acta*, **40**, 3005–14.

Czapkiewicz, J. and B. Czapkiewicz-Tutaj (1980). Relative scale of free energy of transfer of anions from water to 1,2-dichloroethane. *J. Chem. Soc. Faraday Trans.*, **76**, 1663–8.

Da, Y. Z., K. Ito and H. Fujiwara (1992). Energy aspects of oil/water partition leading to the novel hydrophobic parameters for the analysis of quantitative structure-activity relationships. *J. Med. Chem.*, **35**, 3382–7.

Damaskin, B. B., O. A. Petrii and V. V. Batrakov (1971). *Adsorption of Organic Compounds on Electrodes*. Plenum Press, New York.

Danielli, J. F. (1945). Reactions at interfaces and their significance in biology. *Nature*, **156**, 468–70.

Danielli, J. F. (1954). The present position in the field of facilitated diffusion and selective active transport. *Colston Papers*, **7**, 1.

Danielli, J. F. and J. T. Davies (1951). Reactions at interfaces in relation to biological problems. *Advanc. Enzymol. Relat. Subj. Biochem.*, **11**, 35–89.

Danielli, J. F. and H. Davson (1935). A contribution to the theory of permeability of thin films. *J. Cell. Comp. Physiol.*, **5**, 495–508.

Danielli, J. F., K. G. A. Pankhurst and A. C. Riddiford (1958). *Surface Phenomena in Chemistry and Biology*. Pergamon Press, New York.

Danil de Namor, A. F., T. Hill and E. Sigstad (1983). Free energies of transfer of 1:1 electrolytes from water to nitrobenzene. *J. Chem. Soc. Faraday Trans. I*, **79**, 2713–22.

Das, T., D. Bratko, L. B. Bhuiyan, C. W. Outhwaite (1995). Modified Poisson–Boltzmann theory applied to linear polyelectrolyte solutions. *J. Phys. Chem.*, **99**, 410–8.

Davies, J. T. (1951a). Measurement of contact potentials at the oil–water interface. *Nature*, **167**, 193–4.

Davies, J. T. (1951b). Stable contact potentials at the oil-water interface. *Z. Electrochem.*, **55**, 559–60.

Davies, J. T. (1953a). Interfacial potentials. I. Dependence on the character of the non-aqueous phase. *Trans. Faraday Soc.*, **49**, 683–6.

Davies, J. T. (1953b). On the shapes of molecules of polyamino acids and proteins at interfaces. *Biochim. Biophys. Acta*, **11**, 165–77.

Davies, J. T. and E. K. Rideal (1955). Interfacial potentials. *Can. J. Chem.*, **33**, 947–60.

Davies, J. T. and E. K. Rideal (1963). *Interfacial phenomena*. 2nd edition. Academic Press, New York.

Davies, M. (1971). Electric field effects in biomolecular systems. *Acta Phys. Pol.*, **A40**, 561–5.

Davies, M. (1976). Some aspects of recent high electric field studies in molecular systems. *Acta Phys. Pol.*, **A50**, 241–54.

De Armond, M. K. and A. H. De Armond (1996). Excited state electron transfer at the interface of two immiscible electrolyte solutions. In: *Liquid–Liquid Interfaces. Theory and Methods* (A. G. Volkov and D. W. Deamer, eds.) pp. 255–76. CRC Press, Boca Raton, New York, London, Tokyo.

Deamer, D. W. (1987). Proton permeation of lipid bilayers. *J. Bioenerg. Biomembrane*, **19**, 457–78.

Deamer, D. W. (1992). Role of water in proton flux mechanism. In: *Biomembrane Structure & Function —State of Art* (B. P. Gaber and K. R. K. Easwaran, eds.) pp. 209–25. Adenine Press, New York.

Deamer, D. W. and J. W. Nichols (1989). Proton flux mechanisms in model and biological membranes. *J. Membrane Biol.*, **107**, 91–103.

Deamer, D. W. and A. G. Volkov (1995). Proton permeation of lipid bilayers. In: *Permeability and Stability of Lipid Bilayers* (E. A. Disalvo and S. A. Simon, eds.) pp. 161–78, CRC Press, Boca Raton, Ann Arbor, London, Tokyo.

Deamer, D. W. and A. G. Volkov (1996). Oil/water interfaces and the origin of life. In: *Liquid–Liquid Interfaces. Theory and Methods* (A. G. Volkov and D. W. Deamer, eds.) pp. 375–400. CRC Press, Boca Raton, New York, London, Tokyo.

Dean, R. B. (1939). Potentials at oil–water interfaces. *Nature*, **144**, 32.

Dean, R. B. (1940). Adsorption potentials. II. Oil–water potentials. *Trans. Faraday Soc.*, **36**, 166–73.

Dean, R. B., O. Gatty and E. K. Rideal (1940). Adsorption potentials. I. General theory. *Trans. Faraday Soc.*, **36**, 161–6.

Defay, R., I. Prigogine, A. Bellemans and D. H. Everett (1966). *Surface Tension and Adsorption.* Wiley, New York.

Dehmlow, E. V. and S. S. Dehmlow (1993). *Phase Transfer Catalysis.* VCH, New York.

Delahay, P. (1965). *Double Layer and Electrode Kinetics.* Interscience Publishers, New York.

d'Epenoux, B., Seta, P., Amblad, G. and C. Gavach (1979). The transfer mechanism of tetraalkylammonium ions across a water–nitrobenzene interface and the structure of the double layer. *J. Electroanal. Chem.*, **99**, 77–84.

Deryaguin, B. V. and A. V. Prokhorov (1981). On the theory of the rupture of black films. *J. Coll. Interface Sci.*, **81**, 108.

Deuling, H. J. and W. Helfrich (1976). The curvature elasticity of fluid membranes: a catalogue of vesicle shapes. *J. Physique*, **37**, 1335–45.

Deuling, H. J. and W. Helfrich (1977). A theoretical explanation for the myelin shapes of red blood cells. *J. Physique*, **37**, 1335–45.

Dilger, J. P., S. McLaughlin, T. J. McIntosh and S. A. Simon (1979). Dielectric constant of phospholipid-bilayers and the permeability of membranes to ions. *Science*, **206**, 1196–8.

Dogonadze, R. R. and A. A. Kornyshev (1974). Polar-solvent structure in theory of ion solvation. *J. Chem. Soc. Faraday Trans., II*, **70**, 1121–32.

Dogonadze, R. R., E. Kalman, A. A. Kornyshev and J. Ulstrup, eds. (1980). *The Chemical Physics of Solvation.* Elsevier, Amsterdam.

Donnan, F. G. (1911). Theorie der Membrangleichgewichte und Membranpotentiale bei Vorhandensein von nicht dialysierenden Electrolyten. Ein Beitrag zur physikalisch-chemischen Physiologie. *Z. Electrochem.*, **17**, 572–81.

Du Bois-Reymond, E. (1848). *Untersuchungen über Thierishe Elektrizität.* Vol. 1. G. Reiner, Berlin.

Du Noüy, L. (1919). A new apparatus for measuring surface tension. *J. Gen. Physiol.*, **1**, 521–4.

Dupeyrat, M. and J. Michel (1969). Study of the mechanism of electroadsorption at the water–nitrobenzene interface. I. Electroadsorption without an inorganic salt. *J. Colloid Interface Sci.*, **29**, 605–12.

Dupeyrat, M. and E. Nakache (1980). Electrocapillarity and electroadsorption. *J. Colloid Interface Sci.*, **73**, 332–44.

Earnshaw, J. C., W. G. Johnson, B. J. Carrol and P. J. Doyle (1996). The drop volume method for interfacial tension determination: an error analysis. *J. Colloid Interface Sci.*, **177**, 150–5.

Edwards, W., J. Phillips and S. Morris (1974). Structural changes in chromaffine granules induced by divalent cations. *Biochim. Biophys. Acta*, **356**, 164–73.

Egberts, E., S. J. Marrink and H. J. Berendsen (1994). Molecular dynamics simulation of a phospholipid membrane. *Eur. Biophys. J.*, **22**, 423–36.

Ehrensvärd, G. and L. G. Sillén (1938). Adsorption potentials at polar liquid–liquid interfaces. *Nature*, **141**, 788–9.

Einarsdottir, O. (1995). Fast reactions of cytochrome oxidase. *Biochim. Biophys. Acta*, **1229**, 129–47.

Elamrani, K. and A. Blume (1983). Effect of the lipid phase transition on the kinetics of H^+/OH^--diffusion across phosphatidic acid bilayer. *Biochim. Biophys. Acta*, **727**, 22–30.

Elkes, J. J., A. C. Frazer, J. H. Schulman and H. C. Stewart (1945). Reversible adsorption of proteins at the oil–water interface. I. Preferential adsorption of proteins at charged oil–water interfaces. *Proc. R. Soc.*, **A184**, 102–15.

Elkin, V. V., V. Alexeev, E. A. Solomatin, V. Ya. Mishuk, D. Leikis and L. I. Knotz (1975). Application of nonlinear A.C. methods in the investigation of the electrical double layer properties. *J. Electroanal. Chem.*, **65**, 11–20.

Engelhardt, H. A., P. Feulner, H. Pfnür and D. Menzel (1977). An accurate and versatile vibrating capacitor for surface and adsorption studies. *J. Phys. E: Sci. Instrum.*, **10**, 1133–41.

Eriksson, J. C. (1966). Thermodynamics of surface phase systems. I. Considerations of the classical thermodynamic theory of plane surface phase systems. *Arkiv Kemi*, **25**, 331–41.

Eriksson, J. C. (1969). Thermodynamics of surface phase systems. V. Contribution to the thermodynamics of the solid–gas interface. *Surf. Sci.*, **14**, 221–46.

Eriksson, J. C. (1983). Some comments on the application of Hansen's thermodynamic formalism for plane interfaces to surfactant solution interfaces. *J. Colloid Interface Sci.*, **93**, 582–6.

Escabi-Perez, J. P., A. Romero, S. Lukak and J. H. Fendler (1979). Aspects of artificial photosynthesis. Photoionization and electron transfer in dihexadecyl phosphate vesicles. *J. Am. Chem. Soc.*, **101**, 2231–3.

Esin, O. and V. Shikhov (1943). Interion interactions in the electric double layer. *Zh. Fizicheskoi Khim.*, **17**, 236–46.

Evans, E. A. (1974). Bending resistance and chemically induced moments in membrane bilayers. *Biophys. J.*, **14**, 923–31.

Evans, E. (1992). Composite membranes and structured interfaces: From simple to complex design in biology. In: *Biomembrane Structure and Function: The State of the Art* (B. P. Gaber and K. R. K. Easwaran, eds.) pp. 81–101. Adenine Press, New York.

Evans, E. and D. Needham (1987). Physical properties of surfactant bilayer membranes:

thermal transitions, elasticity, rigidity, cohesion, and colloidal interactions. *J. Phys. Chem.*, **91**, 4219–28.

Everett, D. H. (1987). Application of thermodynamics to interfacial phenomena. *Pure Appl. Chem.*, **59**, 45–52.

Faraday, M. (1857). X. The Bakerian Lecture. Experimental relations of gold and other metals to light. *Phil. Trans. R. Soc. London*, **147**, 145–81.

Fawcett, W. R. and L. Blum (1992). Application of the mean spherical approximation to the estimation of single ion thermodynamic quantities of solvation for monoatomic monovalent ions in aqueous solutions. *J. Electroanal. Chem.*, **328**, 333–40.

Feldman, V. J., A. A. Kornyshev and M. B. Partenskii (1985). Density functional simulation of interfacial relaxation and capacity of a model metal/electrolyte interface. *Solid State Commun.*, **53**, 157–64.

Feldman, V. J., M. B. Partenskii and A. A. Kornyshev (1987). On the non-local response to charging of a relaxing capacitor. *J. Electroanal. Chem.*, **237**, 1–11.

Fendler, J. H. (1982). *Membrane Mimetic Chemistry*. Wiley, New York.

Figaszewski, Z. (1982). System for measuring separate impedance characteristics with a three- or four-electrode potentiostat. *J. Electroanal. Chem.*, **139**, 309–15.

Figaszewski, Z., Z. Koczorowski and G. Geblewicz (1982). System for electrochemical studies with a four-electrode potentiostat. *J. Electroanal. Chem.*, **139**, 317–22.

Figaszewski, Z. and I. Paleska (1989). Studies of the water/1,2-dichloroethane interface impedance—effect of transfer of the tetramethylammonium ion. *J. Electroanal. Chem.*, **266**, 253–64.

Finkelstein, A. (1987). *Water Movement through Lipid Bilayers, Pores, and Plasma Membranes: Theory and Reality*. Wiley-Interscience, New York.

Fisk, S. and B. Widom (1969). Structure and free energy of the interface between fluid phases in equilibrium near the critical point. *J. Chem. Phys.*, **50**, 3219–27.

Flewelling, R. F. and W. L. Hubbel (1986). The membrane dipole potential in total membrane potential model. Applications to hydrophobic ion interactions with membranes. *Biophys. J.*, **49**, 541–52.

Fong, F. K., ed. (1982). *Light Reaction Path of Photosynthesis*. Springer Verlag, Berlin, New York.

Ford, G. R. and J. D. Scribner (1983). A simple method for predicting hydration energies of organic cations derived from protonation or alkylation of neutral oxygen and nitrogen bases. *J. Org. Chem.*, **48**, 2226–33.

Fragata, M. (1978). A far red absorbing form of chlorophyll a detected in phosphatidylcholine vesicles. *J. Colloid Interface. Sci.*, **66**, 470–4.

Franklin, J. C. and D. S. Cafiso (1993). Internal electrostatic potentials in bilayers: measuring and controlling dipole potentials in lipid vesicles. *Biophys. J.*, **65**, 289–99.

Frankowiak, D. J. and E. Rabinowitch (1966). The methylene blue-ferrous iron reaction in a two phase system. *J. Phys. Chem.*, **70**, 3012–4.

Fraser, M. J. (1957). Behavior of proteins at interfaces. *J. Pharmacy Pharmacol.*, **9**, 497–521.

Fraser, M. J. and J. H. Schulman (1956). Activity of trypsin-lipide complexes at oil–water interfaces. *J. Colloid Sci.*, **11**, 451–70.

Freeman, S. A., M. A. Wang and J. C. Weaver (1994). Theory of electroporation of planar bilayer membranes: predictions of the aqueous area, change in capacitance, and pore–pore separation. *Biophys. J.*, **67**, 42–56.

Freundlich, H. (1926). *Colloid and Capillary Chemistry*, Methuen, London.

Fruge, D. R., G. D. Fong and F. K. Fong (1979). Photosynthesis of polyatomic organic molecules from carbon dioxide and water by photocatalytic action of visible light illuminated platinized chlorophyll *a* dihydrate polycrystals. *J. Am. Chem. Soc.*, **101**, 3694–7.

Frumkin, A. N. (1919). *Electrocapillary Studies and Electrode Potentials.* Saposhnikov Publ. House, Odessa.

Frumkin, A. N. (1979). *Zero Charge Potentials.* Nauka Publ., Moscow.

Frumkin, A. N., V. S. Bagotsky, Z. A. Iofa and B. N. Kabanov (1952). *Kinetics of Electrode Processes*, Moscow State University Press, Moscow.

Frumkin, A. N., M. I. Gugeshashvili and L. I. Boguslavsky (1971). Adsorption potentials at the heptane–water interface in the presence of valinomycin. *Doklady Acad. Nauk SSSR*, **198**, 1452–4.

Frumkin, A. N., O. A. Petrii and B. B. Damaskin (1980). In: *Comprehensive Treatise of Electrochemistry* (J. O'M. Bockris, B. Conway and E. Yeager E., eds.) pp. 221–89, Vol. 1. Plenum Press, New York.

Gaevskii, A. Yu. (1984). Interaction of an ion with the polar liquid–dielectric interface. Effect of the dielectric non-uniformity of the ionic surroundings. *Zh. Fiz. Khim.*, **58**, 157–61.

Gavach, C. (1973). Overpotentials at the interface between an aqueous solution and a water-immiscible solution. *J. Chim. Phys. Physicochim. Biol.*, **70**, 1478–82.

Gavach, C., P. Seta and B. d'Epenoux (1977). The double layer and ion adsorption at the interface between two nonmiscible solutions. Part I. Interfacial tension measurements for the water–nitrobenzene tetraalkylammonium bromide systems. *J. Electroanal. Chem.*, **83**, 225–35.

Gavach, C., P. Seta and F. Henry (1974). A study of the ionic transfer across an aqueous solution/liquid membrane interface by chronopotentiometric and impedance measurements. *Bioelectrochem. Bioenerg.*, **1**, 329–42.

Gawrish, K., D. Ruston, J. Zimmerberg, V. A. Parsegian, R. P. Rand and N. Fuller (1992). Membrane dipole potentials, hydration forces, and the ordering of water at membrane surfaces. *Biophys. J.*, **61**, 1213–23.

Geblewicz, G., Z. Figaszewski and Z. Koczorowski (1984). Study of the impedance of the water-1,2-dichloroethane interface. Influence of the picrate ion transfer. *J. Electroanal. Chem.*, **177**, 1–12.

Geblewicz, G. and D. J. Schiffrin (1988). Electron transfer between immiscible solutions. The hexacyanofeffate–lutetium biphthalocyanine system. *J. Electroanal. Chem.*, **244**, 27–37.

Gemant, A. (1962). *Ions in Hydrocarbons.* Wiley Interscience Publishers, New York.

Gennis, R. B. (1989). *Biomembranes. Molecular Structure and Function.* Springer Verlag, New York.

Georgallas, A., J. D. MacArthur, X. P. Ma, C. V. Nguyen and G. R. Palmer (1987). The diffusion of small ions through phospholipid bilayers. *J. Chem. Phys.*, **86**, 7218–26.

Gesser, H. D., T. A. Wildman and Y. B. Tewari (1977). Photooxidation of *n*-hexadecane by xanthone. *Env. Sci. Technol.*, **11**, 605–8.

Gibbs, J. W. (1928). *Collected Works.* Vols. 1 and 2. Langmans Green and Co., New York.

Gingell, D. and J. A. Fornes (1975). Demonstration of intermolecular forces in cell adhesion using a new electrochemical technique. *Nature*, **256**, 210–1.

Gingell, D. and L. Ginsberg (1978). Problems in the physical interpretation of membrane interaction and fusion. In: *Membrane Fusion* (G. Poste and G. L. Nicolson, eds.) pp. 369–85, Elsevier-North-Holland Biomedical Press, New York.

Gingell, D., I. Todd and V. A. Parsegian (1977). Long-range attraction between red cells and a hydrocarbon surface. *Nature*, **268**, 767–9.

Girault, H. H. J. (1995). Solvent reorganization energy for heterogeneous electron-transfer reactions at liquid/liquid interfaces. *J. Electroanal. Chem.*, **388**, 93–100.

Girault, H. H. J. and D. J. Schiffrin (1983). Thermodynamic surface excess of water and ionic solvation at the interface between immiscible liquids. *J. Electroanal. Chem.*, **150**, 41–9.

Girault, H. H. J. and D. J. Schiffrin (1984a). Adsorption of phosphatidylcholine and phosphatidylethanolamine at the polarized water/1,2-dichloroethane interface. *J. Electroanal. Chem.*, **179**, 277–84.

Girault, H. H. J. and D. J. Schiffrin (1984b). The measurement of the potential of zero charge at the interface between immiscible electrolyte solutions. *J. Electroanal. Chem.*, **161**, 415–7.

Girault, H. H. J. and D. J. Schiffrin (1984c). Thermodynamics of a polarized interface between two immiscible electrolyte solutions. *J. Electroanal. Chem.*, **170**, 127–41.

Girault, H. H. J. and D. J. Schiffrin (1988). Electron transfer reactions at the interface between two immiscible electrolyte solutions. *J. Electroanal. Chem.*, **244**, 15–26.

Girault, H. H. J. and D. J. Schiffrin (1989). Electrochemistry of liquid–liquid interfaces. In: *Electroanalytical Chemistry* (A. J. Bard, ed.) pp. 1–141. M. Dekker, Inc., New York.

Glaser, R. W., S. L. Leikin, L. V. Chernomordik, V. F. Pastushenko and A. I. Sokirko (1988). Reversible electrical breakdown of lipid bilayers: formation and evolution. *Biochim. Biophys. Acta*, **940**, 275–87.

Glueckauf, E. (1964a). Heats and entropies of ions in aqueous solutions. *Trans. Faraday Soc.*, **60**, 572–7.

Glueckauf, E. (1964b). Effect of the dielectric constant on the activity coefficients of electrolytes in aqueous solutions. *Trans. Faraday Soc.*, **60**, 776–82.

Glueckauf, E. (1964c). Bulk dielectric constant of aqueous electrolyte solutions. *Trans. Faraday Soc.*, **60**, 1637–45.

Goedheer, J. C. (1966). In: *The Chlorophylls* (L. P. Vernon and G. R. Seely, eds.) Ch. 6. Academic Press, New York.

Goldschmidt, V. M. (1926). *Geochem. Vert. Ges. der Elemente*, Oslo.

Good, R. J. (1976). Thermodynamics of adsorption and Gibbsian distance parameters in two- and three-phase systems. *Pure Appl. Chem.*, **48**, 427–33.

Good, R. J. (1982). Thermodynamics of adsorption and Gibbsian distance parameters. *J. Colloid Interface Sci.*, **85**, 128–46.

Good, R. J. (1986). Thermodynamics of adsorption and Gibbsian distance parameters. IV: Reply to Motomura. Regarding surface excess of volume and Gibbs dividing surfaces. *J. Colloid Interface Sci.*, **110**, 298–300.

Goodall, M. C. and G. Sachs (1977). A new method of membrane reconstitution, *Biophys. J.*, **17**, 182a.

Goodrich, F. C. (1969). The thermodynamics of fluid interfaces. In: *Surface and Colloid Science* (E. Matijevic and F. R. Eirich, eds.) pp. 1–37. Wiley-Interscience, New York.

Gordillo, G. J., F. V. Molina and D. Posadas (1990). A modified Poisson–Boltzmann surface excess calculation with a field dependent dielectric constant. *Anales Asoc. Quimica Argentina*, **78**, 237–43.

Gourary, B. S. and F. S. Adrian (1960). Wave functions for electron-excess color centers in alkali halide crystals. *Solid State Phys.*, **10**, 127–247.

Gouy, G. (1910). Constitution of the electric charge at the surface of an electrolyte. *Compt. Rend. Acad. Sci.*, **149**, 654–7.

Govington, A. K. and K. E. Newman (1977). NMR studies of the structure of electrolyte solution, in: *Modern Aspects of Electrochemistry*, Eds. J. O'M. Bockris and B. E. Conway, Vol. 12, pp. 41–129. Plenum Press, New York.

Grabe, A. and R. G. Horn (1993). Double-layer and hydration forces measured between silica sheets subjected to various surface treatments. *J. Colloid Interface Sci.*, **157**, 375–83.

Grafov, B. M. and B. B. Damaskin (1994). Consistency of theoretical models of the electrical double layer with the Gibbs adsorption equation. *Electrokhimiya*, **30**, 1413–8.

Grahame, D. C. (1947). The electrical double layer and the theory of capillarity. *Chem. Rev.*, **41**, 441–501.

Grahame, D. C. (1950). Effect of dielectric saturation upon the diffuse double layer and the free energy of hydration of ions. *J. Chem. Phys.*, **18**, 903–9.

Grahame, D. C. and R. W. Whitney (1942). The thermodynamic theory of electrocapillarity. *J. Am. Chem. Soc.*, **64**, 1548–52.

Granfeldt, M. K. and B. Jonsson (1992). The interaction of dipolar layer-hydration forces and image charges. *Phys. Lett.*, **195**, 174–8.

Gratzel, M. (1989). *Heterogeneous Photochemical Electron Transfer.* CRC Press, Boca Raton.

Greberg, H. and R. Kjellander (1994). Electric double layer properties calculated in the anisotropic reference hypernetted chain approximation. *Mol. Phys.*, **83**, 789–801.

Greibrokk, K. (1972). N-Carbomethoxy aziridines. *Acta Chem. Scand.*, **26**, 3305–8.

Greyson, J. (1967). Solvent transport and the electromotive force of anion exchange membrane cells containing heavy and normal water solutions. *J. Phys. Chem.*, **71**, 4549–51.

Grishin, A. F., V. A. Nenashev and G. N. Berestovsky (1979). Interaction between liposomes and bimolecular membranes. *Biofizika*, **24**, 467–71.

Gros, M., Gromb, S. and C. Gavach (1978). The double layer and ion adsorption at the interface between two non-miscible solutions. Part II. Electocapillary behaviour of some water–nitrobenzene systems. *J. Electroanal. Chem.*, **89**, 29–36.

Grubb, S. G., M. W. Kim, Th. Raising and Y. R. Shen (1988). Orientation of molecular monolayers at the liquid–liquid interface as studied by optical second harmonic generation. *Langmuir*, **4**, 452–4.

Gruen, D. W. R. and S. Marcelja (1983a). Spatially varying polarization in ice. *J. Chem. Soc. Faraday Trans.*, **2**, 79, 211–24.

Gruen, D. W. R. and S. Marcelja (1983b). Spatially varying polarization in water. A model for the electric double layer and the hydration force. *J. Chem. Soc. Faraday Trans.*, **2**, 79, 225–42.

Gruen, D. W. R., S. Marcelja and V. A. Parsegian (1984). Water structure near the membrane surface. In: *Cell Surface Dynamics* (A. S. Perelson, C. DeLisi and F. W. Weigel, eds.) pp. 59–91. M. Dekker, New York.

Grundwald, E., G. Baugman and G. Kohnstam (1960). The solvation of electrolytes in dixane–water mixtures as deduced from effect of solvent change on the standard partial molar free energy. *J. Am. Chem. Soc.*, **82**, 5801–11.

Guainazzi, M., G. Silvestri and G. Serravalle (1975). Electrochemical metallization at the liquid–liquid interfaces of non-miscible electrolyte solutions. *J. Chem. Soc. Chem. Commun.*, 200–1.

Guastalla, J. (1969). Irreversibilites dans les propriétés electriques de l'interface entre des solutions ioniques non miscibles. *Compt. Rend. Acad. Sci. Paris*, **269**, 1360–3.

Guastalla, J. (1970). Interface hysteresis and negative differential conductance at liquid–liquid junction between non-miscible ionic solutions. *Nature*, **227**, 485–6.

Gugeshashvili, M. I. (1974). Contact phenomena at the interface between two immiscible liquids. *Ph.D. Thesis* in Chemistry, IELAN, Moscow.

Gugeshashvili, M. I. and L. I. Boguslavsky (1973). Contact phenomena at the heptane/water interface in the presence of sodium dodecylsulfate. *Elektrokhimiya*, **9**, 687–9.

Gugeshashvili, M. I. and L. I. Boguslavsky (1987). Kinetics of development of Volta potentials controlled by space-charge formation in valinomycin-containing water–octane systems. *Sov. Electrochem.*, **24**, 1056–8.

Gugeshashvili, M. I., L. I. Boguslavsky and A. N. Frumkin (1972). High adsorption potentials at the heptane–water interface in the presence of valinomycin. *Doklady Akad. Nauk SSSR*, **206**, 985–7.

Gugeshashvili, M. I., A. V. Indenbom and L. I. Boguslavsky (1988). Adsorption and electric double layer at the interface between two immiscible liquids. In: *Progress in Science and Technology. Electrochemistry.* (V. E. Kazarinov, ed.) Vol. 28, pp. 172–215. VINITI, Moscow.

Gugeshashvili, M. I., Lozhkin, B. T. and L. I. Boguslavsky (1974a). Elimination of diffusion potentials in the water/nitrobenzene/water system. *Elektrokhimiya*, **10**, 1272–4.

Gugeshashvili, M. I., M. A. Manvelyan and L. I. Boguslavsky (1974b). Adsorption and distribution potentials in the water/nitrobenzene system in the presence of tetraalkylammonium salts. *Sov. Electrochem.*, **10**, 782–5.

Gugeshashvili, M. I., V. I. Portnov, A. G. Volkov, L. N. Chekulaeva and V. S. Markin (1991). Emulsion photobioelectrochemistry: bacteriorhodopsin phototransfer of protons through the water/lipid interface. *Bioelectrochem. Bioenerg.*, **26**, 139–58.

Gugeshashvili, M. I., A. G. Volkov, A. Tessier, P. F. Blanchet, D. Cote, G. Munger and R. M. Leblanc (1992). Light energy conversion with chlorophyll *a* monolayers. *Biol. Membrane*, **9**, 862–73.

Gugeshashvili, M. I., A. G. Volkov, B. Zelent, Ju. Galant, G. Munger and R. M. Leblanc (1995). Self organized molecular amphiphilic assemblies of wet chlorophyll *a* in monolayers and thin films. *Membrane Cell Biol.*, **9**, 1–12.

Gugeshashvili, M. I., A. G. Volkov, L. S. Yaguzhinsky, A. F. Mironov and L. I. Boguslavsky (1983a). Coupling of two redox reaction at the octane/water interface with participation of NADH, the iron complex of ethioporphyrin and oxygen. *Elektrokhimiya*, **19**, 1629–32.

Gugeshashvili, M. I., A. G. Volkov, L. S. Yaguzhinsky, A. F. Mironov and L. I. Boguslavsky (1983b). Coupling of two redox reactions at the octane/water interface with the participation of NADH and a ferri-complex ethioporphyrin and oxygen. *Bioelectrochem. Bioenerg.*, **10**, 493–8.

Guggenheim, E. A. (1957). *Thermodynamics.* 3rd edition. North-Holland, Amsterdam.

Gurevich, Yu. Ya. and Yu. I. Kharkats (1986a). Ion transfer through a phase boundary: a stochastic approach. *J. Electroanal. Chem.*, **200**, 3–16.

Gurevich, Yu. Ya. and Yu. I. Kharkats (1986b). Theory of ion transfer across interfaces between two media. *Sov. Electrochem.*, **22**, 463–70.

Gurevich, Yu. Ya. and Yu. I. Kharkats (1990). Theory of ion injection into dielectric ion-conducting materials. *Doklady Akad. Nauk SSSR*, **315**, 128–32.

Gutknecht, J. (1984). Proton/hydroxide conductance through lipid bilayer membranes. *J. Membrane Biol.*, **82**, 105–12.

Guyat, M. J. (1924). Effect Volta metal-electrolyte et couches monomoleculaires. *Ann. Phys. (Paris)*, **2**, 506–638.

Haber, F. (1908). Über feste Electrolyte, ihre zersetsung durch den Strom und ihr electromotorisches Verhalten in galvanischen Ketten. *Ann. Phys.*, **26**, 927–73.

Haber, F. and Z. Klemensiewicz (1909). Über elektrische Phasengrenzkrafte. *Z. Phys. Chem.*, **67**, 385–470.

Hachisu, S. (1984). In A. Kitahara and A. Watanabe (eds.). *Electrical Phenomena at Interfaces. Fundamentals, Measurements, and Applications*, Marcel Dekker, New York, p. 3.

Hajkova, P., D. Homolka, V. Mareček and Z. Samec (1983). The double layer at the interface between two immiscible electrolyte solutions. Capacity of the water/1,2-dichloroethane interface. *J. Electroanal. Chem.*, **151**, 277–82.

Hajkova, P., D. Homolka, V. Mareček, A. G. Volkov and Z. Samec (1985). Measurements of electric double-layer capacity at the water/1,2-dichloroethane interface in the presence of metal–porphyrin complexes. *Sov. Electrochem.*, **21**, 209–15.

Halley, J. W. and D. Price (1987). Quantum theory of the double layer: model including solvent structure. *Phys. Rev.*, **B35**, 9095–102.

Hamilton, R. T. and E. W. Kaler (1990a). Alkali metal ion transport through thin bilayers. *J. Phys. Chem.*, **94**, 2560–6.

Hamilton, R. T. and E. W. Kaler (1990b). Facilitated ion transport through thin bilayers. *J. Membrane Sci.*, **54**, 259–69.

Hanna, G. J. and R. D. Noble (1985). Measurement of liquid–liquid interfacial kinetics. *Chem. Rev.*, **85**, 583–98.

Hansen, R. S. (1962). Thermodynamics of interfaces between condensed phases. *J. Phys. Chem.*, **66**, 410–5.

Harkins, W. D. (1952). *The Physical Chemistry of Surface Films*. Reinhold Publishing Corporation, New York.

Harkins, W. D. and F. E. Brown (1919). The determination of surface tension (free surface energy), and the weight of falling drops: the surface tension of water and benzene by the capillary height method. *J. Am. Chem. Soc.*, **41**, 499–524.

Harris, L. B. and J. Fiason (1984). Vibrating capacitor measurement of surface charge. *J. Phys. E: Sci. Instrum.*, **17**, 788–92.

Hauser, H., D. Oldani and M. C. Phillips (1973). Mechanism of ion escape from phosphatidylcholine and phosphatidylserine single bilayer vesicles. *Biochemistry*, **12**, 4507–17.

Haydon, D. A. (1964). The electrical double layer and electrokinetic phenomena. In: *Recent Progress in Surface Science* (J. F. Danielli, K. G. A. Pankhurst and A. C. Riddiferd, eds.) Chap. 3. Academic Press, New York.

Haydon, D. A. and S. B. Hladky (1980). Ion transport across thin lipid membranes: a critical discussion of mechanism in selected systems. *Quart. Rev. Biophys.* **5**, 187–282.

Haydon, D. A. and J. L. Taylor (1963). The stability and properties of biomolecular lipid leaflets in aqueous solutions. *J. Theor. Biol.*, **4**, 281–96.

Haydon, D. A. and J. L. Taylor (1968). Contact angles for thin lipid films and determination of London-Van der Waals Forces. *Nature*, **217**, 739–40.

Hayoun, M., M. Meyer, M. Mareschal, C. Ciccotti and P. Turq (1987). Molecular dynamics simulation of a liquid–liquid interface. In: *Chemical Reactivity in Liquids* (M. Moreau and P. Turq, eds.) pp. 279–95. Plenum Press, New York.

Hayoun, M., M. Meyer and P. Turq (1994). Molecular dynamics study of a solute-transfer reaction across liquid–liquid interface. *J. Phys. Chem.*, **98**, 6626–32.

Heinrich, V., S. Svetina and B. Zeks (1993). Nonaxisymmetric vesicle shapes in a generalized bilayer-couple model and the transition between oblate and prolate axisymmetric shapes. *Phys. Rev. E: Sci. Instrum.*, **48**, 3112–23.

Heinz, E. (1981). Electrical potentials in biological membrane transport. In: *Molecular Biology, Biochemistry and Biophysics*, Vol. 33. Springer Verlag, Berlin, Heidelberg, New York.

Helenius, V. M., P. H. Hynninen and J. E. I. Korppi-Tommola (1993). Chlorophyll *a* aggregates in hydrocarbon solution, a picosecond spectroscopy and molecular modeling study. *Photochem. Photobiol.*, **58**, 867–73.

Helenius, V. M., J. O. Sikki, P. H. Hynninen and J. E. I. Korppitommola (1994). Femtosecond study of relaxation of hydrated chlorophyll a aggregate in hydrocarbon solution. *Chem. Phys. Lett.*, **226**, 137–43.

Helfrich, W. (1973). Elastic properties of lipid bilayers: theory and possible experiments. *Z. Naturforsch.*, **28c**, 693–703.

Helfrich, W. (1974). Blocked lipid exchanges in bilayers and its possible influence on the shape of vesicles. *Z. Naturforsch.*, **29c**, 510–5.

Helmholtz, H. (1879). Studien über electrische Grenz Schichten. *Ann. Phys.*, **7**, 337–82.

Henderson, D. (1983). Recent progress in the theory of the electric double layer. *Prog. Surf. Sci.*, **13**, 197–224.

Henderson, D. (1990). Explicit formulae for the ion and solvent profiles in the electric double layer using the mean spherical approximation. *Chem. Phys.*, **141**, 79–86.

Henderson, D. and L. Blum (1978). Some exact results and the application of the mean spherical approximation to charged hard spheres near a charged hard wall. *J. Chem. Phys.*, **69**, 5441–9.

Henderson, D. and L. Blum (1981). The application of the generalized mean spherical approximation to the theory of the diffuse double layer. *Can. J. Chem.*, **59**, 1906–17.

Hermann, R. B. (1972). Theory of hydrophobic bonding. II. The correlation of hydrocarbon solubility in water with solvent cavity surface area. *J. Phys. Chem.*, **76**, 2754–9.

Hideshima, T. (1990). Oscillatory reaction of alcohol dehydrogenase in an oil/water system. *Biophys. Chem.*, **38**, 265–8.

Higgins, D. A. and R. M. Corn (1993). Second harmonic generation studies of adsorption at a liquid/liquid electrochemical interface. *J. Phys. Chem.*, **97**, 489–93.

Hill, R. and P. Bendall (1960). Function of the two cytochrome components in chloroplasts: a working hypothesis. *Nature*, **186**, 136–7.

Hill, T. L. (1964). *Thermodynamics of Small Systems*. Benjamin, New York.

Hills, C. J. and P. J. Ovenden (1966). Electrochemistry at high pressures. In: *Advances in Electrochemistry and Electrochemical Engineering*. P. Delahay (ed.). Vol. 4, pp. 185–248. Wiley, New York.

Hoffmanova, A., M. I. Gugeshashvili and L. I. Boguslavsky (1985). Investigation of the water–octane system in the presence of valinomycin. *Electrokhimiya*, **21**, 303–7.

Homola, A. and A. A. Robertson (1976). A compression method for measuring forces between colloid particles. *J. Colloid Interface Sci.*, **54**, 286–408.

Homolka, D., P. Hajkova, V. Mareček and Z. Samec (1983). The double layer at the interface between two immiscible electrolyte solution. Structure of the water/nitrobenzene interface. *J. Electroanal. Chem.*, **159**, 233–8.

Horn, R. G. and J. N. Israelashvili (1981a). Direct measurement of structural forces between two interfaces in a nonpolar liquid. *J. Chem. Phys.*, **75**, 1400–11.

Horn, R. G. and J. N. Israelashvili (1981b). Forces due to structure in a thin liquid crystal film. *J. Physique*, **42**, 39–52.

Horn, R. G. and J. N. Israelashvili (1981c). Molecular organization and viscosity of a thin film of molten polymer between two surfaces by force measurements. *Macromolecules*, **21**, 2836–41.

Hui, S. W., T. P. Stewart and L. T. Boni (1982). Membrane fusion through point defects in bilayers. *Science*, **212**, 921–3.

Hundhammer, B., S. K. Dhawan, A. Bekele and H. J. Seidlitz (1987). Investigation of ion transfer across the membrane-stabilized interface of two immiscible electrolyte solutions. *J. Electroanal. Chem.*, **217**, 253–9.

Hundhammer, B., H. J. Seidlita, S. Becker, S. K. Dhawan and T. Solomon (1984). The dependence of the Galvani potential difference between water and nitrobenzene on salt partition. A model for liquid state ion-selective electrodes. *J. Electroanal. Chem.*, **180**, 355–62.

Hundhammer, B. and T. Solomon (1983). Determination of standard Gibbs energies of ion partition between water and organic solvent by cyclic voltammetry. Part I. *J. Electroanal. Chem.*, **157**, 19–26.

Hundhammer, B., T. Solomon and B. Alemayehu (1982). Voltametric studies of ion transfer across the water-nitrobenzene interface using crystal violet tetraphenylborate as supporting electrolyte in the organic phase. *J. Electroanal. Chem.*, **135**, 301–4.

Hundhammer, B., T. Solomon and H. Alemu (1983). Investigation of the ion transfer across the water–nitrobenzene interface by AC cyclic voltametry. *J. Electroanal. Chem.*, **149**, 179–83.

Hung, L. Q. (1980). Electrochemical properties of the interface between two immiscible solutions. Part 1. Equilibrium situation and Galvani potential difference. *J. Electroanal. Chem.*, **115**, 159–74.

Hung, L. Q. (1983). Electrochemical properties of the interface between two immiscible electrolyte solutions. Part III. The general case of the Galvani potential difference at the interface and of the distribution of an arbitrary number of components interacting in both phases. *J. Electroanal.*, **149**, 1–14.

Hurwitz, H. D. and A. Steinchen-Sanfeld (1965). In: *Fundamental Problems of Contemporary Theoretical Electrochemistry* (A. N. Frumkin, ed.) p.174, Mir, Moscow.

Hwang, S. B., J. I. Korenbrot and W. Stoeckenius (1977). Proton translocation by bacteriorhodopsin through an interface film. *J. Membrane Biol.*, **36**, 137–58.

Indenbom, A. V. (1985). Topography of the interface of liquid electrolytes and the structure of the electric double layer. *Doklady AN SSSR*, **282**, 133–6.

Indenbom, A. V. (1995). Interpretation of ion transfer reactions across the interface between two immiscible electrolyte solutions by the model of "elastic" inner layer. *Electrochim. Acta*, **40**, 2985–91.

Indenbom, A. V. and A. G. Volkov (1986). Topography of the interface between two immiscible electrolyte solutions and the electrocapillarity. *Extended Abstracts. 37th ISE Meeting*, Vilnuis, August 24–31, Vol. 4, pp. 428–30.

Israelashvili, J. N. (1992). *Intermolecular and Surface Forces*. Academic Press, San Diego.

Jaycock, M. J. and G. D. Parfitt (1981). *Chemistry of Interfaces*. Ellis Harwood, New York.

Johnsson, B. (1981). Monte Carlo simulations of liquid water between two rigid walls. *Chem. Phys. Lett.*, **82**, 520–5.

Jones, M. N. (1975). *Biological Interfaces. An Introduction to the Surface and Colloid Science of Biochemical and Biological Systems*. Elsevier, Amsterdam, Oxford, New York.

Joos, P. (1976). Theory of electrical rectification at a liquid–liquid interface. *J. Electroanal. Chem.*, **86**, 75–80.

Joos, P. and R. Vanden Bogaert (1976). Alternating electric current across a nitrobenzene–water interface. I. Adsorption kinetics. *J. Colloid Interface Sci.*, **56**, 206–12.

Joos, P. and M. Vanuffelen (1995). Theory on the growing drop technique for measuring dynamic interfacial tensions. *J. Colloid Interface Sci.*, **171**, 297–305.

Joos, P. and Y. Verburgh (1978). Distribution of electrolytes over two liquid phases. *Bull. Soc. Chim. Belg.*, **87**, 737–45.

Kahlweit, M. and H. Strehlow (1954). Ueber elektrische Potentialdifferenzen an der Phasengrenze nichtmischbarter Flüssigkeiten. *Z. Elektrochem.*, **58**, 658–65.

Kakiuchi, T. (1992). Current–potential characteristic of ion transfer across the interface between two immiscible electrolyte solutions based on the Nernst–Planck equation. *J. Electroanal. Chem.*, **322**, 55–61.

Kakiuchi, T. (1993a). DC and AC responses of ion transfer across an oil water interface with a Goldman-type current potential characteristic. *J. Electroanal. Chem.*, **344**, 1–12.

Kakiuchi, T. (1993b). Mechanism of the transfer of alkali-earth-metal and alkaline-earth-metal ions across the nitrobenzene/water interface facilitated by hexaethylene and octaethylene glycol dodecyl ethers. *J. Colloid Interface Sci.*, **156**, 406–14.

Kakiuchi, T. (1993c). Successive complex formation of multivalent ions with octaethylene glycol dodecyl ether at the nitrobenzene/water interface. *J. Electroanal. Chem.*, **345**, 191–203.

Kakiuchi, T. (1996a). Partition equilibrium of ionic components in two immiscible electrolyte solutions. In: *Liquid–Liquid Interfaces. Theory and Methods* (A. G. Volkov and D. W. Deamer, eds.) pp. 1–18. CRC Press, Boca Raton, New York, London, Tokyo.

Kakiuchi, T. (1996b). Phospholipid monolayers and phospholipases. In: *Liquid–Liquid Interfaces. Theory and Methods* (A. G. Volkov and D. W. Deamer, eds.) pp. 317–31. CRC Press, Boca Raton, New York, London, Tokyo.

Kakiuchi, T., M. Kobayashi and M. Senda (1987). The effect of the electrical potential difference on the adsorption of the hexadecyltrimethylammonium ion at the polarized nitrobenzene–water interface. *Bull Chem. Soc. Jpn.*, **60**, 3109–15.

Kakiuchi, T., M. Kobayashi and M. Senda (1988a). Effect of counterionic species on the adsorption of hexadecyltrimethylammonium ions at the polarized nitrobenzene–water interface. *Bull Chem. Soc. Jpn.*, **61**, 1545–50.

Kakiuchi, T., T. Kondo and M. Senda (1990a). Divalent cation induced phase transition of phosphatidylserine monolayer at the polarized oil–water interface and its influence on the ion transfer processes. *Bull. Chem. Soc. Jpn.*, **63**, 3270–6.

Kakiuchi, T., M. Nakanishi and M. Senda (1988b). The electrocapillary curve of the phosphatidylcholine monolayer at the polarized oil–water interface. 1. Measurement of interfacial tension using a computer-aided pendant-drop method. *Bull. Chem. Soc. Jpn.*, **61**, 1845–51.

Kakiuchi, T., M. Nakanishi and M. Senda (1989). The electrocapillary curve of the phosphatidylcholine monolayer at the polarized oil–water interface. II. Double layer structure of dilauroylphosphatidylcholine monolayer at the nitronbenzene–water interface. *Bull. Chem. Soc. Jpn.*, **62**, 403–9.

Kakiuchi, T., J. Noguchi, M. Kotani and M. Senda (1990b). AC polarographic determination of the rate of ion transfer for a series of alkylammonium ions at the nitrobenzene/water interface. *J. Electroanal. Chem.*, **296**, 517–35.

Kakiuchi, T., J. Noguchi and M. Senda (1992a). Kinetics of the transfer of monovalent anions across the nitrobenzene–water interface. *J. Electroanal. Chem.*, **327**, 63–71.

Kakiuchi, T., J. Noguchi and M. Senda (1992b). Double-layer effect on the transfer of some monovalent ions across the polarized oil–water interface. *J. Electroanal. Chem.*, **336**, 137–52.

Kakiuchi, T. and M. Senda (1983a). Polarizability and electrocapillary measurements of the nitrobenzene–water interface. *Bull. Chem. Soc. Jpn.*, **56**, 1322–6.

Kakiuchi, T. and M. Senda (1983b). Thermodynamics of the electrocapillarity of oil–water interfaces. *Bull. Chem. Soc. Jpn.*, **56**, 2912–8.

Kakiuchi, T. and M. Senda (1983c). Structure of double layer at the interface between nitrobenzene solution of tetrabutylammonium tetraphenylborate and aqueous solution of lithium chloride. *Bull. Chem. Soc. Jpn.*, **56**, 1753–60.

Kakiuchi, T. and M. Senda (1991). Polarizability and nonpolarizability of oil–water interfaces with relevance to AC impedance measurements. *Coll. Czech. Comm.*, **56**, 112–29.

Kakiuchi, T., Y. Teranishi and K. Niki (1995). Adsorption of sorbitan fatty acid esters and a sucrose monoalkanoate at the nitrobenzene–water interface and its effect on the rate of ion transfer across the interface. *Electrochim. Acta*, **40**, 2869–74.

Kakiuchi, T., T. Usui and M. Senda (1990c). Electrodeposition of tetraethylene glycol monodecyl ether at the polarized nitrobenzene–water interface. *Bull. Chem. Soc. Jpn.*, **63**, 2044–50.

Kakiuchi, T., T. Usui and M. Senda (1990d). Effect of poly(oxyethylene) chain length on electrosorption of hexa and octa(ethylene glycol) mono normal-dodecyl ethers at the polarized nitrobenzene–water interface. *Bull. Chem. Soc. Jpn.*, **63**, 3264–9.

Kakiuchi, T., M. Yamane, T. Osakai and M. Senda (1987). Monolayer formation of dilauroylphosphatidylcholine at the polarized nitrobenzene-water interface. *Bull Chem. Soc. Jpn.*, **60**, 4223–8.

Kakutani, T., T. Osakai and M. Senda (1983). A potential-step chronoamperometric study of ion transfer at the water/nitrobenzene interface. *Bull. Chem. Soc. Jpn.*, **56**, 991–6.

Kalweit, and H. Strelow (1954). Über elektrische Potentialdifferenzen an der Phasengrenze nichtmischbarer Flüssigkeiten. *Z. Electrochem.*, **58**, 658–65.

Kamienski, B., I. Kulawik, J. Kulawik, J. Mikulski and J. Pawelek (1967a). Application of a [239]Pu electrode for the determination of interfacial electric potentials. *Bull. Acad. Pol. Sci. Ser. Sci. Chim.*, **15**, 249–52.

Kamienski, B., I. Kulawik, J. Kulawik, J. Mikulski and J. Pawelek (1967b). Some remarks on the method of applying a radioactive source for measuring the interfacial potential. *Bull. Acad. Pol. Sci. Ser. Sci. Chim.*, **15**, 253–6.

Kandelaki, M. D. and A. G. Volkov (1991). The influence of dielectric permittivity of the nonaqueous phase on the photooxidation of water at the interface of two immiscible liquids in the presence of a hydrated oligomer of chlorophyll. *Can. J. Chem.*, **69**, 151–6.

Kandelaki, M. D. and A. G. Volkov (1993). Influence of permittivity of the nonaqueous phase on water photooxidation at interfaces between immiscible liquids in the presence of a hydrated chlorophyll oligomers. *Rus. J. Electrochem.*, **29**, 1158–64.

Kandelaki, M. D., A. G. Volkov, A. L. Levin and L. I. Boguslavsky (1983a). Photooxidation of water by chlorophyll adsorbed on an octane water interface. *Doklady AN SSSR*, **271**, 462–5.

Kandelaki, M. D., A. G. Volkov, A. L. Levin and L. I. Boguslavsky (1983b). Oxygen evolution in the presence of chlorophyll adsorbed at the octane/water interface. *Bioelectrochem. Bioenerg.*, **11**, 167–72.

Kandelaki, M. D., A. G. Volkov, A. L. Levin and L. I. Boguslavsky (1984). Oxygen evolution in the presence of chlorophyll adsorbed at the octane/water interface. Bioconversion of Solar Energy (I. V. Berezin, ed.) pp. 18–21. Publisher: Akad. Nauk SSSR, Pushchino.

Kandelaki, M. D., A. G. Volkov, V. V. Shubin and L. I. Boguslavsky (1987). Chlorophyll–water interaction during oxygen photoevolution at the octane/water interface. *Biochim. Biophys. Acta*, **893**, 170–6.

Kandelaki, M. D., A. G. Volkov, V. V. Shubin, A. L. Levin and L. I. Boguslavsky (1988). Interaction of chlorophyll with water during oxygen photoevolution reaction at the octane/water interface. *Electrokhimiya*, **24**, 288–94.

Kandidow, P. (1911). Ober elektrokapillare Erscheinungen an der Grenze nicht mischbarer Flüssigkeiten. *Chem. Zentr. (II)*, 1095–6.

Kandidow, P. (1913). Ober elektrokapillare Erscheinungen an der Grenze nichtmischbarer Flüssigkeiten. *Z. Physik. Chem.*, **83**, 587–91.

Karpfen, F. M. and J. E. B. Randles (1953). Ionic equilibria and phase-boundary potentials in oil–water systems. *Trans. Faraday Soc.*, **49**, 823–31.

Käs, J. and E. Sackmann (1991). Shape transitions and shape stability of giant phospholipid vesicles in pure water induced by area-to-volume changes. *Biophys. J.*, **60**, 825–44.

Katano, H., K. Maeda and M. Senda (1995). Effect of diffuse layer on the rate of electron transfer across an electrolyte solution interface. *J. Electroanal. Chem.*, **396**, 391–6.

Kelvin Lord (1898). Contact electricity of metals. *Phil. Mag.*, **46**, 82–121.

Kenrick, F. B. (1886). Die potentialsprüge zwischen Gasen und Flüssigkeiten. *Z. Phys. Chem.*, **19**, 625–56.

Kettere, B., B. Neumke and P. Lauger (1971). Transport mechanism of hydrophobic ions through lipid bilayer membranes. *J. Membrane Biol.*, **5**, 225–45.

Keynes, R. D. and E. Rojas (1974). Kinetics and steady-state properties of the charged system controlling sodium conductance in the squid giant axon. *J. Physiol. (London)*, **239**, 393–434.

Kharkats, Yu. I. (1976). On the calculation of the rate constant of the charge transfer across the boundary of two dielectric media. *Sov. Electrochem.*, **12**, 1370–7.

Kharkats, Yu. I. (1978). The calculation of the solvent energy reorganization of reactions with complex distribution of charges in reactants. *Sov. Electrochem.*, **14**, 1721–4.

Kharkats, Y. I. (1990). Mechanism of the acceleration of electron-transfer processes occurring near an interface between immiscible liquids. *Sov. Electrochem.*, **26**, 1032–9.

Kharkats, Yu. I. and A. M. Kuznetsov (1996). Quantum theory of charge transfer. In: *Liquid–Liquid Interfaces. Theory and Methods* (A. G. Volkov and D. W. Deamer, eds.) pp. 139–54. CRC Press, Boca Raton, New York, London, Tokyo.

Kharkats, Yu. and J. Ulstrup (1991). The electrostatic Gibbs energy of finite-size ions near a planar boundary between two dielectric media. *J. Electroanal. Chem.*, **308**, 17–26.

Kharkats, Yu. I. and J. Ulstrup (1993). The electrostatic free energy of finite-size ions near a planar boundary between two dielectric media. *Elektrokhimiya*, **29**, 299–303.

Kharkats, Yu. I. and A. G. Volkov (1985). Interfacial catalysis: multielectron reactions at liquid/liquid interface. *J. Electroanal. Chem.*, **184**, 435–9.

Kharkats, Yu. I. and A. G. Volkov (1987). Membrane catalysis: synchronous multielectron reactions at the liquid–liquid interface. Bioenergetical mechanism. *Biochim. Biophys. Acta*, **891**, 56–67.

Kharkats, Yu. I. and A. G. Volkov (1989). Cytochrome oxidase: the molecular mechanism of functioning. *Bioelectrochem. Bioenerg.*, **22**, 91–103.

Kharkats, Yu. I. and A. G. Volkov (1994). 2:1:1 molecular mechanism of cytochrome oxidase functioning. In: *Charge and Field Effects in Biosystems-4* (M. J. Allen, S. F. Cleary and A. E. Sowers, eds.) pp. 70–7. World Scientific, Singapore, New Jersey, London.

Kharkats, Yu. I., A. G. Volkov and L. I. Boguslavsky (1975). Use of vibrating capacitor to investigate kinetics of catalytic reactions involving charge transfer across a water–oil interface. *Doklady AN SSSR*, **220**, 1441–4.

Kharkats, Yu. I., A. G. Volkov and L. I. Boguslavsky (1976). Enzyme charge transfer through the interface between two immiscible liquids. *Biophysics*, **21**, 634–8.

Kharkats, Yu. I., A. G. Volkov and L. I. Boguslavsky (1977). Transfer of ions and electrons across the interface between two immiscible liquids in functioning enzyme membrane systems. *J. Theor. Biol.*, **65**, 379–91.

Kihara, S., M. Suzuki, K. Maeda, K. Ogura, M. Matsui and Z. Yoshida (1989). The electron transfer at a liquid/liquid interface studied by current-scan polarography at the electrolyte dropping electrode. *J. Electroanal. Chem.*, **271**, 107–25.

Kimmich, G. A. (1975). Preparation and characterization of isolated intestinal epithelial cells and their use in studying intestinal transport. *Methods Membrane Biol.*, **5**, 51–115.

Kingery, W. D. and M. Humerick (1953). Surface tension at elevated temperatures: 1. Furnace and method for use of the sessille drop method: surface tension of silicon, iron and nickel. *J. Phys. Chem.*, **57**, 359–63.

Kinosita, K., Jr. and T. Y. Tsong (1977). Formation and resealing of pores of controlled sizes in human erythrocyte membrane. *Nature*, **268**, 438–41.

Kirkwood, J. G. (1934). On the theory of strong electrolyte solutions. *J. Chem. Phys.*, **2**, 767–81.

Kirkwood, J. G. (1965). *Dielectrics, Intermolecular Forces, Optical Rotation.* Gordon and Breach, New York.

Knots, L. L. and G. G. Dubovik (1985). A method of exciting self-oscillations in a cell for measuring contact potential difference by the capacitor method. *Sov. Electrochem.*, **1**, 701–5.

Koczorowski, Z. (1985). On the surface and zero charge potentials at the water–nitrobenzene interfaces. *J. Electroanal. Chem.*, **190**, 257–60.

Koczorowski, Z. (1987). Galvani and Volta potentials at the interface separating immiscible electrolyte solutions. In: *The Interface Structure and Electrochemical Processes at the Boundary Between Two Immiscible Liquids* (V. E. Kazarinov, ed.) pp. 77–106. Springer Verlag, Berlin, Heidelberg.

Koczorowski, Z. (1996). Volta and surface potentials at liquid/liquid interfaces. In: *Liquid–Liquid Interfaces. Theory and Methods* (A. G. Volkov and D. W. Deamer, eds.) pp. 375–400. CRC Press, Boca Raton, New York, London, Tokyo.

Koczorowski, Z. and G. Geblewicz (1980). Chronopotentiometric studies of the tetrabutylammonium ion transfer from water to 1,2-dichloroethane. *J. Electroanal. Chem.*, **108**, 117–20.

Koczorowski, Z. and G. Geblewicz (1982). Electrochemical studies of the tetrabutyl- and tetramethyl-ammonium ion transfer across the water–1,2-dichloroethane interface. A comparison with the water–nitrobenzene interface. *J. Electroanal. Chem.*, **139**, 177–91.

Koczorowski, Z. and G. Geblewicz (1983). Studies of Galvani potentials of the water–nitrobenzene and water–1,2-dichloroethane interfaces. *J. Electroanal. Chem.*, **152**, 55–66.

Koczorowski, Z., G. Geblewicz and I. Paleska (1984a). Electrochemical study of water–isobutylmethyl ketone interface. *J. Electroanal. Chem.*, **172**, 327–37.

Koczorowski, Z., A. Kalinska and Z. Figaszewski (1982). Effect of constant polarization on the alternating voltage generated by a vibrating interface separating immiscible electrolyte solutions in water and nitrobenzene. *J. Electroanal. Chem.*, **139**, 303–7.

Koczorowski, Z. and J. Kotowski (1978). Mechanoelectrical energy conversion by the water–nitrobenzene interface. *J. Colloid Interface Sci.*, **66**, 584–5.

Koczorowski, Z. and S. Minc (1963). Investigation of polar-oil/water electrochemical interphase properties. II. Solvation energies of 1:1 electrolytes in water saturated nitrobenzene. *Electrochim. Acta*, **8**, 645–9.

Koczorowski, Z., I. Paleska and G. Geblewicz (1984b). Electrochemical study of the immiscible electrolyte solution interface between water and mixed organic solvent. *J. Electroanal. Chem.*, **164**, 201–4.

Koczorowski, Z., I. Paleska and J. Kotowski (1987). Streaming method study of interfaces between two immiscible electrolyte solutions. *J. Electroanal. Chem.*, **235**, 287–98.

Koczorowski, Z. and I. Zagorska (1983). Investigations on Volta potentials in water–nitrobenzene systems and on surface potentials of these solvents. *J. Electroanal. Chem.*, **159**, 183–93.

Koczorowski, Z., I. Zagorska and A. Kalinska (1989). Differences between surface potentials of water and some organic solvents. *Electrochim. Acta*, **34**, 1857–62.

Koenig, F. O. (1934). The thermodynamics of the electrocapillary curve. *J. Phys. Chem.*, **38**, 111–28.

Kolodziej, M. A., G. Parry-Jones and M. Davies (1975). High field dielectric measurements in water. *J. Chem. Soc. Faraday Trans. II*, **71**, 269–74.

Kolthoff, I. M. (1971). A review of electrochemistry in non-aqueous solvents. *Pure Appl. Chem.*, **25**, 305–25.

Kolthoff, I. M. and F. G. Thomas (1965). Electrode potentials in acetonitrile. Estimation of the liquid junction potential between acetonitrile solutions and the aqueous saturated calomel electrode. *J. Phys. Chem.*, **69**, 3049–58.

Kondo, S. (1956). Thermodynamical fundamental equation for spherical interface. *J. Chem. Phys.*, **66**, 5464–76.

Kondo, T. and T. Kakiuchi (1995). Effect of electrostatic interaction on hydrolytic activity of phospholipase A_2 at the polarized oil–water interface. *Bioelectrochem. Bioenerg.*, **36**, 53–6.

Kondo, T., T. Kakiuchi and M. Senda (1994). Hydrolytic activity of phospholipase D from plants and *Streptomyces* spp. against phosphatidylcholine monolayers at the polarized oil/water interface. *Bioelectrochem. Bioenerg.*, **34**, 93–100.

Kontturi, A.-K., K. Konturri, L. Murtomaki and D. J. Schiffrin (1995). Reaction of H^+ with tetraphenylborates: an example of an EC mechanism at liquid/liquid interfaces, *J. Chem. Soc. Faraday Trans.*, **91**, 3433–9.

Kontturi, A.-K., K. Konturri and D. J. Schiffrin (1988). Kinetics of K ion transfer at the water/1,2-dichloroethane interface. *J. Electroanal. Chem.*, **255**, 331–6.

Kontturi, K. and L. Murtomaki (1992). Electrochemical determination of partition coefficients of drugs. *J. Pharm. Sci.*, **81**, 970–5.

Kooyman, R. P. H., T. J. Schaafsma and J. F. Kleibeuker (1977). Fluorescence spectra and zero-field magnetic resonance of chlorophyll *a*-water complexes. *Photochem. Photobiol.*, **26**, 235–40.

Kornyshev, A. A. (1981). Nonlocal screening of ions in a structurized polar liquid. New aspects of solvent description in electrolyte theory. *Electrochim. Acta*, **26**, 1–20.

Kornyshev, A. A. (1988). Non-local enhancement of the dipole–dipole interaction at the interface of two dielectrics. *J. Electroanal. Chem.*, **255**, 297–302.

Kornyshev, A. A. (1989). Metal electrons in the double layer theory. *Electrochim. Acta*, **34**, 1829–47.

Kornyshev, A. A., W. Shmicler and M. A. Vorotyntsev (1982). Nonlocal electrostatic approach to the problem of a double layer at the metal–electrolyte interface. *Phys. Rev.*, **B25**, 5244–56.

Kornyshev, A. A. and J. Ulstrup (1985a). Solvent structural effects on the diffuse double-layer capacitance of metal/electrolyte interfaces. *Chem. Scripta*, **25**, 58–62.

Kornyshev, A. A. and J. Ulstrup (1985b). Solvent structural effect on the deviation from linear Parsons–Zobel plots with increasing electrolyte concentration. *J. Electroanal. Chem.*, **183**, 387–9.

Kornyshev, A. A. and A. G. Volkov (1984). On the evaluation of standard Gibbs energies of ion transfer between two solvents. *J. Electroanal. Chem.*, **180**, 363–81.

Koryta, J. (1979). Electrochemical polarization phenomena at the interface of two immiscible electrolyte solutions. *Electrochim. Acta*, **24**, 293–300.

Koryta, J. (1987). Electrochemistry of liquid membranes: interfacial aspects. *Electrochim. Acta*, **32**, 419–24.

Kotov, N. A. and M. G. Kuzmin (1990a). The photocurrent kinetics across the polarizable interface of immiscible electrolyte solutions in the protoporphyrine-quinone system. *Sov. Electrochem.*, **26**, 1484–8.

Kotov, N. A. and M. G. Kuzmin (1990b). A Photoelectrochemical effect at the interface of immiscible electrolyte solutions. *J. Electroanal. Chem.*, **285**, 223–40.

Kotov, N. A. and M. G. Kuzmin (1991). Analysis of photocurrent kinetics in photoelectrochemical effects at polarizable interface between electrolyte solutions by mathematical modelling method. *Sov. Electrochem.*, **27**, 76–81.

Kotov, N. A. and M. G. Kuzmin (1992). Nature of the processes of charge-carrier generation at ITIES by the photoexitation of porphyrins. *J. Electroanal. Chem.*, **338**, 99–124.

Kotov, N. A. and M. G. Kuzmin (1996). Photoelectrochemical effect at interface between two immiscible electrolyte solutions. In: *Liquid–Liquid Interfaces. Theory and Methods* (A. G. Volkov and D. W. Deamer, eds.) pp. 213–53. CRC Press, Boca Raton, New York, London, Tokyo.

Kott, K. L., D. A. Higgins, R. J. McMahon and R. M. Corn (1993). Observation of photoinduced electron transfer at a liquid–liquid interface by optical second harmonic generation. *J. Am. Chem. Soc.*, **115**, 5342–5.

Kozlov, M. M., L. V. Chernomordik and V. S. Markin (1989a). Mechanism of formation of protein-free spots at erythrocyte membrane: membrane skeleton rupture. *Biol. Membranes*, **6**, 597–611.

Kozlov, M. M., S. L. Leikin, L. V. Chernomordik, V. S. Markin and Yu. A. Chizmadzhev (1989b). Stalk mechanism of membrane fusion. Intermixing of aqueous contents. *Eur. Biophys. J.*, **129**, 411–25.

Kozlov, M. M., S. L. Leikin and V. S. Markin (1989c). Elastic properties of interfaces. Elasticity moduli and spontaneous geometric characteristics. *J. Chem. Soc. Faraday Trans. 2*, **85**, 277–92.

Kozlov, M. M. and V. S. Markin (1984). On the theory of membrane fusion. The adhesion-condensation mechanism. *Gen. Physiol. Biophys.*, **5**, 379–402.

Kozlov, M. M. and V. S. Markin (1989). Definition of the force factors for an interface with non-uniform curvature. *J. Chem. Soc. Faraday Trans. 2*, **85**, 261–76.

Kozlov, M. M. and V. S. Markin (1990a). Elastic properties of membranes—monolayers, bilayers, vesicles. *J. Colloid Interface Sci.*, **138**, 332–45.

Kozlov, M. M. and V. S. Markin (1990b). Mechanics of interfaces—the stability of a spherical shape. *J. Phys.*, **51**, 559–74.

Kozlov, M. M. and V. S. Markin (1991). Interface mechanics—the shape of membrane cylinder. *J. Physique II*, **1**, 805–20.

Kozlov, I. A. and V. P. Skulachev (1977). H^+-adenosine triphosphatase and membrane energy coupling. *Biochim. Biophys. Acta*, **463**, 29–89.

Kozlov, M. M., V. S. Markin and S. L. Leikin (1989d). Elastic properties of interfaces. *J. Chem. Soc., Faraday Trans. 2.*, **85**, 277–92.

Kozlov, M. M., M. Winterhalter and D. Lerche (1992). Elastic properties of strongly curved monolayers. Effect of electric surface charges. *J. Physique II*, **2**, 175–85.

Kozlov, V., V. Vilinskaya and G. Tedoradze (1982). Differential capacity of the mercury electrode in dilute 1:1 valent electrolytes. *Coll. Czech. Chem. Commun.*, **47**, 190–5.

Krishtalik, L. I. (1986). Energetics of multielectron reactions. Photosynthetic oxygen evolution. *Biochim. Biophys. Acta*, **849**, 162–71.

Krishtalik, L. I. and A. M. Kuznetsov (1986). Energetics of the elementary reaction act and the "configurational" electrode potential. *Sov. Electrochem.*, **22**, 218–21.

Krotov, V. V., V. A. Prokhorov, S. Yu. Pavlov and A. I. Rusanov (1995). Method of touching drops: a new method for measuring the surface and interfacial tension of liquids. *Colloids Surf. A.*, **104**, 165–8.

Kruyt, H. R. (1952). *Colloid Science.* Elsevier, Amsterdam.

Krylov, V. S. (1964). Theory of the electric double layer with a discrete structure of the specifically adsorbed charge. *Electrochim. Acta*, **9**, 1247–58.

Krylov, V. S., V. A. Myamlin, L. I. Boguslavsky and M. A. Manvelyan (1977). Some patterns in adsorption of ions at the interface between two immiscible electrolyte solutions. *Elektrokhimiya*, **13**, 834–40.

Krysinski, P. and H. Ti Tien (1986). Voltammetric studies of electron-conducting modified bilayer lipid membranes. *Bioelectrochem. Bioenerg.*, **16**, 185–91.

Kushnev, V. V., A. V. Indenbom and L. I. Boguslavskii (1984). Apparatus for potentiodynamic measurements of the capacitance of the electric double layer and investigations of the transfer of charges through an interface between immiscible electrolytes. *Elektrokhimiya*, **20**, 553–7.

Kutyurin, V. M., I. Yu. Artamkina, A. D. Korsun, I. N. Anisimova and I. V. Matveeva (1973). On the interaction between oxidized form of chlorophyll and water. *Doklady Akad. Nauk SSSR*, **212**, 243–5.

Kuznetsov, A. M. (1983). Variation of the charge of the adsorbed hydrogen atom in the process of the activation in the elementary act of hydrogen ion discharge. *J. Electroanal. Chem.*, **159**, 241–55.

Kuznetsov, A. M. and Yu. I. Kharkats (1987). Problems of a quantum theory of charge transfer reactions at the interface between two immiscible liquids. In: *The Interface Structure and Electrochemical Processes at the Boundary Between Two Immiscible Liquids* (V. E. Kazarinov, ed.) pp. 11–46. Springer-Verlag, Berlin.

Lakshminarayanaiah, N. (1969). *Transport Phenomena in Membranes.* Academic Press, New York.

Lakshminarayanaiah, N. (1976). *Membrane Electrodes.* Academic Press, New York.

Lakshminarayanaiah, N. (1984). *Equations of Membrane Biophysics.* Academic Press, Orlando.

Landau, L. D. and E. M. Lifshitz (1958). *Statistical Physics.* Pergamon, New York.

Landau, L. D. and E. M. Lifshitz (1984). *Electrodynamics of Continuous Media*, 2nd edition. Pergamon, New York.

Lando, J. L. and H. T. Oakley (1967). Tabulated correction factors for the drop-weight-volume determination of surface and interfacial tension. *J. Colloid Interface Sci.*, **25**, 526–30.

Lange, E. and K. P. Miščenko (1930). Zur Thermodynamik der Ionensolvatation. *Z. Phys. Chem.*, **149**, 1–41.

Langmuir, I. (1917). The constitution and fundamental properties of solids and liquids. II. Liquids. *J. Am. Chem. Soc.*, **39**, 1848–906.

Langmuir, J., Waugh, D. F. (1938). The adsorption of proteins at oil–water interfaces and artificial protein-lipid membranes. *J. Gen. Physiol.*, **21**, 745–55.

Lee, C. Y. and H. L. Scott (1980). The surface tension of water: a Monte Carlo calculation using an umbrella sampling algorithm. *J. Chem. Phys.*, **73**, 4591–6.

Lee, L. T., D. Langevin and B. Farnoux (1991). Neutron reflectivity of an oil–water interface. *Phys. Rev. Lett.*, **67**, 2678–81.

Lee, L. T., D. Langevin, E. K. Mann and B. Farnoux (1994). Neutron reflectivity at liquid interfaces. *Physica B*, **198**, 83–8.

Leikin, S. L., M. M. Kozlov, L. V. Chernomordik, V. S. Markin and Yu. A. Chizmadzhev (1987). Membrane fusion: overcoming of the hydration barrier and local restructuring. *J. Theor. Biol.*, **129**, 411–25.

LeNeveu, D. M. and P. R. Parsegian (1976). Measurement of forces between lecithin bilayers. *Nature*, **259**, 601–3.

Leo, A., C. Hansch and D. Elkins (1971). Partition coefficients and their uses. *Chem. Rev.*, **71**, 525–612.

Leontovich, M. A. (1983). *Introduction to Thermodynamics*, Nauka Publ. House, Moscow.

Lev, A. A. (1975). *Ionic Selectivity of Cellular Membranes*. Nauka, Leningrad.

Levesque, D., J. J. Weis and G. N. Patey (1980). Charged hard spheres in dipolar hard sphere solvents. A model for electrolyte solutions. *J. Chem. Phys.*, **72**, 1887–99.

Levine, S. and C. W. Outwaite (1978). Comparison of theories of the aqueous electric double layer at a charged plane interface. *J. Chem. Soc. Faraday Trans. II*, **74**, 1670–89.

Levine, S., C. W. Outhwaite and L. B. Bhuiyan (1981). Statistical mechanical theories of the electric double layer. *J. Electroanal. Chem.*, **123**, 105–19.

Lewis, G. N. and M. Rendall (1961). *Thermodynamics*. 2nd ed., McGraw-Hill, New York.

Lhotsky, A., K. Holub, P. Neuzil and V. Marecek (1996). Ac impedance analysis of tetraethylammonium ion transfer at liquid/liquid microinterfaces. *J. Chem. Soc., Faraday Trans.*, **92**, 3851–7.

Likhtenshtein, G. I. (1988). *Chemical Physics of Redox Metalloenzyme Catalysis*. Springer Verlag, Berlin, New York.

Lilly, M. D. (1982). Two-liquid-phase biocatalytic reactions. *J. Chem. Tech. Biotechnol.*, **32**, 162–9.

Lin, M., J. L. Filpo, P. Mansoura and J. F. Baret (1979). A phase transition of the adsorbed layer: high pressure effect of fatty alcohol adsorption at an oil–water interface. *J. Chem. Phys.*, **71**, 2202–9.

Linse, P. (1987). Monte Carlo simulation of liquid–liquid benzene–water interface. *J. Chem. Phys.*, **86**, 4177–87.

Lipowsky, R. (1991). The conformation of membranes. *Nature*, **349**, 475–81.

Livshits, V. A. and B. G. Dzikovskii (1994). Molecular ordering and dynamics on interphase boundaries oil–water emulsions. *Zh. Fiz. Khim.*, **68**, 1650–7.

Livshits, V. A., B. G. Dzikovskii and V. P. Tsybyshev. (1994). Molecular order and dynamics at liquid–liquid interfaces: dependence on the surfactant concentration on the surface and a spin probe charge. *Zh. Fiz. Khim.*, **68**, 1644–9.

Llopis, J. (1971). Surface potential at liquid interfaces. In: *Modern Aspects of Electrochemistry*. (J. O'M. Bockris and B. E. Conway, eds.) Vol. 6, p. 71. Plenum Press, New York.

Lohnstein, T. (1906). Zur Theorie des Abtrophens. Nachtrag und weitere Belege. *Ann. Phys.*, **20**, 606–18.

Lohnstein, T. (1908). Kritisches über das sogenannte Gesetz von Tate. *Z. Phys. Chem.*, **64**, 686–92.

Lohnstein, T. (1913). Nochmals das sogenannte Gesetz von Tate. *Z. Phys. Chem.*, **84**, 410–8.

Lorenz, W. (1961). Zur allgemeinen Theorie des Impedanzspektrums potentialabhängiger Phasengrenzreaktionen. *Z. Phys. Chem.*, **218**, 272–6.

Lorenz, W. (1962). Zur Anwendung der Gibbsschen Adsorptionsthermodynamik auf Adsorptionsreaktionen mit partiellem Ladungsübergang. *Z. Phys. Chem.*, **219**, 421–3.

Lorenz, W. (1963). Adsorptionskinetik und Ladungsübergangskoeffizienten von Jodid- und Bromidionen an Quecksilberelektroden. *Z. Phys. Chem.*, **224**, 145–76.

Lorenz, W. (1973). Zur Formulierung von Elementarreaktionen in flüssigen und elektrochemischen Systemen. *Z. Phys. Chem.*, **253**, 243–56.

Lucretius, T. C. (1953). *The Nature of Things (De rerlun natura)*. Harvard University Press, Cambridge.

Lucy, J. A. (1970). T'he fusion of biological membranes. *Nature*, **227**, 814–7.

Lucy, J. A. (1978). Mechanisms of chemically induced cell fusion. In: *Membrane Fusion* (G. Poste and G. L. Nicolson, eds.) pp. 267–304, Elsevier-North-Holland Biomedical Press, New York.

Luther, R. (1896). Elektromotorische Kraft und Verteilungsgleichgewicht. *Z. Phys. Chem.*, **19**, 529–71.

Mackay, R. A., S. A. Myers, L. Bodalbhai and A. Brajter-Toth (1990). Microemulsion structure and its effect on electrochemical reactions. *Anal. Chem.*, **62**, 1084–90.

MacRitchie, F. (1990). *Chemistry at Interfaces*. Academic Press, San Diego, New York.

Maeda, H., T. Kakiuchi and M. Senda (1987). The structure of electrical double layer and the effect of counterion on the adsorption of cetyltrimethylammonium ion at the polarized nitrobenzene–water interface. *Rev. Polarogr. (Kyoto)*, **33**, 38.

Maeda, K., S. Kihara, M. Suzuki and M. Matsui (1991). Voltametric interpretation of ion transfer coupled with electron-transfer at a liquid/liquid interface. *J. Electroanal. Chem.*, **303**, 171–84.

Maher, K. A. and J. D. Stevenson (1988). Impact frustration of the origin of life. *Nature*, **331**, 612–4.

Makosza, M. (1977). Reactions of carbanions and halogenocarbenes in two-phase systems. *Russ. Chem. Rev.*, **46**, 2174–202.

Malev, V. V., V. P. Kapranchik and V. V. Proyaev (1991). Rotating disk filter as a method of studying the kinetics of redistribution of a substance between immiscible liquids. *Sov. Electrochem.*, **27**, 27–33.

Malsch, J. (1928). Über die Messung der Dielektrizitätskonstanten von Flüssigkeiten bei hohen electrischen Feldstärken. II. *Phys. Z.*, **29**, 770–7.

Malsch, J. (1929). Dielektrizitätskonstante und Assoziation. *Phys. Z.*, **30**, 837–9.

Marcelja, S. and N. Radic (1976). Repulsion of interfaces due to boundary water. *Chem. Phys. Lett.*, **42**, 129–30.

Marcus, R. A. (1956). On the theory of oxidation–reduction reactions. *J. Chem. Phys.*, **24**, 966–78.

Marcus, R. A. (1990a). Reorganization free energy for electron transfers at liquid–liquid and dielectric semiconductor–liquid interfaces. *J. Phys. Chem.*, **94**, 1050–5.

Marcus, R. A. (1990b). Theory of electron-transfer rates across liquid–liquid interfaces. *J. Phys. Chem.*, **94**, 4152–5.

Marcus, R. A. (1991). Theory of electron-transfer rates across liquid–liquid interfaces. 2. Relationships and application. *J. Phys. Chem.*, **95**, 2010–3.

Marcus, R. A. (1995). Additions and corrections. *J. Phys. Chem.*, **99**, 5742.

Marcus, Y. (1983). Thermodynamic functions of transfer of single ions from water to non-aqueous and mixed solvents. Part I. Gibbs free energies of transfer to nonaqueous solvent. *Pure Appl. Chem.*, **55**, 977–1021.

Marcus, Y. (1985). *Ion Solvation*. Wiley, Chichester.

Marcus, Y. (1996). Ion solvation. In: *Liquid–Liquid Interfaces. Theory and Methods* (A. G. Volkov and D. W. Deamer, eds.) pp. 39–61. CRC Press, Boca Raton, New York, London, Tokyo.

Mareček, V., A. Lhotsky and S. Racinsky (1995a). Fluctuation analysis and faradaic impedance at micro liquid/liquid interface-I. *Electrochim. Acta*, **40**, 2905–8.

Mareček, V., A. Lhotsky and S. Racinsky (1995b). Fluctuation analysis and faradaic impedance at micro liquid/liquid interface-II, *Electrochim. Acta*, **40**, 2909–14.

Mareček, V. and Z. Samec (1983). Evaluation of ohmic potential drop and capacity of interface between two immiscible electrolyte solutions by the galvanostatic pulse method. *J. Electroanal. Chem.*, **149**, 185–92.

Mareček, V. and Z. Samec (1985). Fast performance galvanostatic pulse technique for evaluation of the ohmic potential drop and capacitance of the interface between two immiscible electrolyte solutions. *J. Electroanal. Chem.*, **185**, 263–71.

Markin, V. S. (1981). Lateral organization of membranes and cell shapes. *Biophys. J.*, **36**, 1–19.

Markin, V. S. and Yu. A. Chizmadzhev (1974). *Induced Ion Transport*. Nauka, Moscow.

Markin, V. S., Y. A. Chizmadzhev and E. Neumann (1990). Thermodynamics of pore formation in a charged membrane. *Biol. Membranes*, **7**, 543–8.

Markin, V. S. and R. Glaser (1980). Forces and membrane particle displacement in the fluid mosaic model of cell membranes. *Stud. Biophys.*, **80**, 201–11.

Markin, V. S., M. I. Gugeshashvili, A. G. Volkov, G. Munger and R. M. Leblanc (1992a). The standard Gibbs free energy of adsorption equilibrium and isotherms of adsorption of amphiphilic molecules and clusters at the oil/water and gas/water interfaces. Adsorption of dry and hydrated chlorophyll. *J. Colloid Interface Sci.*, **154**, 264–75.

Markin, V. S., M. I. Gugeshashvili, A. G. Volkov, G. Munger and R. M. Leblanc (1992b). Isotherm of adsorption of amphiphilic molecules and clusters at the oil/water and gas/water interfaces. Adsorption of pheophytin. *J. Electrochem. Soc.*, **139**, 455.

Markin, V. S., M. I. Gugeshashvili, A. G. Volkov, G. Munger and R. M. Leblanc (1992c). Isotherm of adsorption of amphiphilic molecules and clusters at the oil/water and gas/water interfaces. Adsorption of pheophytin. *Extended Abstracts, Fall Meeting, Toronto, Canada, October 1992*, Vol. 92-2, p. 978. The Electrochemical Society, Inc.

Markin, V. S. and A. J. Hudspeth (1993). To fuse or not to fuse. *Biophys. J.*, **65**, 1752–4.

Markin, V. S. and M. M. Kozlov (1985). Pore statistics in bilayer lipid membranes. *Biol. Membranes*, **2**, 205–23.

Markin, V. S. and M. M. Kozlov (1989). Is it really impermissible to shift the Gibbs dividing surface in the classical theory of capillarity? *Langmuir*, **5**, 1130–2.

Markin, V. S., M. M. Kozlov and V. L. Borovjagin (1984). On the theory of membrane fusion. The stalk mechanism. *Gen. Physiol. Biophys.*, **5**, 361–77.

Markin, V. S., M. M. Kozlov and S. L. Leikin (1988). Definition of surface tension at a non-spherical interface. *J. Chem. Soc. Faraday Trans. 2*, **84**, 1149–62.

Markin, V. S., M. M. Kozlov and S. L. Leikin (1989). Determination of surface tension at a nonspherical interface. *Colloid J.*, **51**, 768–77.

Markin, V. S. and B. Martinac (1991). Mechanosensitive channels as reporters of bilayer expansion: a theoretical model. *Biophys. J.*, **60**, 1120–7.

Markin, V. S., V. M. Mirsky and Y. A. Chizmadzhev (1987a). Photoelectric activity of bacteriorhodopsin in planar lipid bilayer. In: *Receptors and Ion Channels* (Yu. A. Ovchinnikov and F. Hucho, eds) pp. 247–254. Walter de Gruyter and Co., Berlin, New York.

Markin, V. S., V. F. Pastushenko and Yu. A. Chizmadzhev (1977a). Physics of the nerve impulse. *Sov. Phys. Usp.*, **20**, 836–60.

Markin, V. S. and V. S. Sokolov (1990). A new concept of electrochemical membrane equilibrium-coupled transport and membrane potentials. *Bioelectrochem. Bioenerg.*, **23**, 1–16.

Markin, V. S., V. S. Sokolov, L. I. Boguslavsky and L. S. Yaguzhinsky (1975). Dimer mechanism of the work of nigericin on the bilayer lipid membranes. *J. Membrane Biol.*, **25**, 23–7.

Markin, V. S., V. S. Sokolov, L. I. Boguslavsky and L. S. Yaguzhinsky (1977b). Dimer mechanism of the work of nigericin on the bilayer lipid membranes. *Biofizika*, **22**, 46–52.

Markin, V. S., S. Svetina and B. Zeks (1987b). Osmotic shrinkage of giant lipid vesicles and the bilayer couple hypothesis. *Biol. Membranes*, **4**, 280–9.

Markin, V. S. and A. G. Volkov (1987a). The standard free energy of ion resolvation and non-linear dielectric effects. *Electrokhimiya*, **23**, 1105–12.

Markin, V. S. and A. G. Volkov (1987b). Theoretical description of Gibbs energy of ion transfer. *Russ. Chem. Rev.*, **56**, 1953–72.

Markin, V. S. and A. G. Volkov (1987c). Interfacial potentials at the interface between two immiscible electrolyte solutions. *Electrokhimiya*, **23**, 1405–13.

Markin, V. S. and A. G. Volkov (1987d). The standard Gibbs energy of ion resolvation and non-linear dielectric effects. *J. Electroanal. Chem.*, **235**, 23–40.

Markin, V. S. and A. G. Volkov (1988a). Electrocapillarity at ITIES. In: *The Interface Structure and Electrochemical Processes at the Boundary between Two Immiscible Liquids* (V. E. Kazarinov, ed.) pp. 94–130, Vol. 28. VINITI, Moscow.

Markin, V. S. and A. G. Volkov (1988b). Electrocapillary properties of the interface between two immiscible electrolyte solutions. The electrocapillary theory and the Gibbs rule. *Electrokhimiya*, **24**, 318–24.

Markin, V. S. and A. G. Volkov (1988c). Electrocapillary properties of the ITIES. The electrocapillary equation in Hansen's representation. *Electrokhimiya*, **24**, 325–31.

Markin, V. S. and A. G. Volkov (1988d). Electrocapillary properties of the ITIES. Surface excesses and the interface charge. *Electrokhimiya*, **24**, 478–84.

Markin, V. S. and A. G. Volkov (1988e). Hansen's method in the electrocapillarity theory as generalization of the Gibbs method (review). *Electrokhimiya*, **24**, 579–617.

Markin, V. S. and A. G. Volkov (1988f). Interfacial potentials. *Russ. Chem. Rev.*, **57**, 1963–89.

Markin, V. S. and A. G. Volkov (1989a). Electrocapillary phenomena at the interface between two immiscible liquids. *Prog. Surf. Sci.*, **30**, 233–356.

Markin, V. S. and A. G. Volkov (1989b). The Gibbs free energy of ion transfer between two immiscible liquids (review). *Electrochim. Acta*, **34**, 93–107.

Markin, V. S. and A. G. Volkov (1989c). The specific volume and entropy of formation of the interface between two immiscible electrolyte solution. *J. Colloid Interface Sci.*, **135**, 305–12.

Markin, V. S. and A. G. Volkov (1989d). The Hansen method for the description of adsorption at a curved interface between immiscible liquids: an extension of the Gibbs method. *J. Colloid. Interface Sci.*, **135**, 553–61.

Markin, V. S. and A. G. Volkov (1989e). Interfacial potentials at the interface between two immiscible electrolyte solutions. Some problems in definitions and interpretation. *J. Colloid Interface Sci.*, **131**, 382–92.

Markin, V. S. and A. G. Volkov (1990a). Specific volume and entropy of the formation of the interface between two immiscible liquids. *J. Colloid Interface Sci.*, **135**, 305–12.

Markin, V. S. and A. G. Volkov (1990b). The Hansen method for the description of adsorption at a curved interface between immiscible liquids: an extension of the Gibbs method. *J. Colloid Interface Sci.*, **135**, 553–61.

Markin, V. S. and A. G. Volkov (1990c). Electrocapillary phenomena at polarizable and reversible ITIES. The generalized electrocapillarity equation in Hansen representation. *Electrochim. Acta*, **35**, 715–24.

Markin, V. S. and A. G. Volkov (1990d). Specific volume and entropy of formation of the interface of two immiscible liquids. *Colloidny J.*, **52**, 51–8.

Markin, V. S. and A. G. Volkov (1990e). Hansen method for description of adorption on a curved interface between immiscible liquids, as a generalization of the Gibbs method. *Colloidny J.*, **52**, 237–44.

Markin, V. S. and A. G. Volkov (1990f). Potentials at the interface between two immiscible electrolyte solutions. *Adv. Colloid Interface Sci.*, **31**, 111–52.

Markin, V. S. and A. G. Volkov (1996). Adsorption isotherms and the structure of oil/water interface. In: *Liquid–Liquid Interfaces. Theory and Methods* (A. G. Volkov and D. W. Deamer, eds.) pp. 63–75. CRC Press, Boca Raton, New York, London, Tokyo.

Marra, J. and J. N. Israelachvili (1985). Direct measurements of forces between phosphatidyl-choline and phosphatidylethanolamine bilayers in aqueous electrolyte. *Biochemistry*, **24**, 4608–18.

Marrink, S. J. and H. J. C. Berendsen (1994). Simulation of water transport through a lipid membrane. *J. Phys. Chem.*, **98**, 4155–68.

Marsh, D. (1989). Water adsorption isotherms and hydration forces for lysolipids and diacyl phospholipids. *Biophys. J.*, **55**, 1093–100.

Martinez, M. M., L. B. Bhuiyan and C. W. Outhwaite (1990). Thermodynamic consistency in the symmetric Poisson–Boltzmann equation for primitive model electrolytes. *J. Chem. Soc. Faraday Trans.*, **86**, 3383–90.

Martynov, G. A. and R. R. Salem (1983). *Electrical double layer at a metal dilute electrolyte solution interface. Lecture Notes in Chemistry*, Vol. 33. Springer Verlag, Berlin.

Mason, W. T., S. B. Hladky and D. A. Haydon (1979). Fusion of photoreceptor membrane vesicles. *J. Mol. Biol.*, **46**, 171–81.

Masterdon, W. L., D. Bolocofsky and T. P. Lee (1971). Ionic radii from scaled particle theory of the salt effect. *J. Phys. Chem.*, **75**, 2809–15.

Mathai, K. G. and E. R. Rabinowitch (1962). Studies of the thionine-ferrous iron reaction in heterogeneous system. *J. Phys. Chem.*, **66**, 663–4.

Matubayasi, N., K. Motomura, M. Aratono and R. Matuura (1978). Thermodynamic study of the adsorption of 1-octadecanol at hexane/water interface. *Bull. Chem. Soc. Jpn.*, **51**, 2800–3.

Matubayasi, N., K. Motomura, M. Aratono and R. Matuura (1979). Thermodynamic study on the adsorption of dioctadecyl ether at hexane/water interface. *Bull. Chem. Soc. Jpn.*, **52**, 1597–600.

Matubayasi, N., K. Motomura, S. Kaneshina, M. Nakamura and R. Matuura R. (1977). Effect of pressure on interfacial tension between oil and water. *Bull. Chem. Soc. Jpn.*, **50**, 523–4.

Maximychev, A. V., S. K. Chamorovsky, A. S. Kholmanskii, V. V. Erokhin, E. V. Levin, L. N. Chekulaeva, A. A. Kononenko and N. G. Rambidi (1991). Oriented multilayer purple membrane films from Halobacteria obtained by Langmuir–Blodgett technique and electrosedimentation. *Biol. Membranes*, **5**, 18–41.

McClain, B. R., D. D. Lee, B. L. Carvalho, S. G. Mochrie, S. H. Chen and J. D. Litster (1994). X-ray reflectivity study of an oil–water interface in equilibrium with a middle-phase microemulsion. *Phys. Rev. Lett.*, **72**, 246–9.

McCormack, D., S. L. Carnie and D. Y. C. Chan (1995). Calculations of electric double layer force and interaction free energy between dissimilar surfaces. *J. Colloid Interface Sci.*, **169**, 177–96.

McLaughlin, S. (1989). The electrostatic properties of membranes. *Ann. Rev. Biophys. Biophys. Chem.*, **13**, 113–36.

Melikyan, G. B., I. G. Abidor, L. V. Chernomordik and L. M. Chailakhyan (1982). Fusion and fission of bilayer membranes. *Doklady Akad. Nauk SSSR*, **263**, 1009–13.

Melikyan, G. B., M. M. Kozlov, L. V. Chernomordik and V. S. Markin (1984). Fussion of the bilayer lipid tube. *Biochim. Biophys. Acta*, **776**, 169–75.

Melrose, J. C. (1968). *Ind. Eng. Chem.*, **60**, 53.

Menger, F. M. (1970). Interfacial physical organic chemistry. Imidazole-catalyzed ester hydrolysis at a water–heptane boundary. *J. Am. Chem. Soc.*, **92**, 5965–71.

Menger, F. M. (1972). Reactivity of organic molecules at phase boundaries. *Chem. Soc. Rev.*, **1**, 229–53.

Meyer, M., M. Mareschal and M. Hayoun (1988). Computer modeling of a liquid–liquid interface. *J. Chem. Phys.*, **89**, 1067–73.

Miao, L., U. Seifert, M. Wortis and H. G. Dobereiner (1994). Budding transitions of fluid bilayer vesicles: the effect of Aret-difference elasticity. *Phys. Rev. E: Sci. Instrum.*, **49**, 5389–407.

Michael, D. and I. Benjamin (1995). Solute orientational dynamics and surface roughness of water/hydrocarbon interfaces. *J. Phys. Chem.*, **99**, 1530–6.

Miklavic, S. J. and P. Attard (1994). A practical approach to solving the double-layer problem that includes affects of ion size and correlation. *J. Phys. Chem.*, **98**, 4320–6.

Miller, S. L., H. C. Urey and J. Oro (1976). Origin of organic compounds on the primitive Earth and in meteorites. *J. Mol. Evol.*, **9**, 59–72.

Miller, S. T. and P. Neogi (1985). *Interfacial Phenomena*. Marcel Dekker, New York.

Millich, F. and C. E. Carraher (1977). *Interfacial Synthesis*. Vols. 1 and 2. Marcel Dekker, New York.

Minc, S. and Z. Koczorowski (1960). Measurements of the interphase potential differences water/polar oil by the dynamic condenser method. *Rocz. Chem.*, **34**, 349–51.

Minc, S. and Z. Koczorowski (1963). Investigation of polar oil–water electrochemical interphase properties—I. Application of the dynamic condenser method for nitrobenzene–water distribution potential difference measurements. *Electrochim. Acta*, **8**, 575–82.

Minc, S. and Z. Koczorowski (1965). Surface potentials of ionized monolayers. I. The interface of anions on the electric double layer structure at the water–decane interface in the presence of CTAB monolayer. *Rocz. Chem.*, **39**, 469–78.

Mirsalikhova, N. M., L. I. Boguslavsky, A. G. Volkov, B. A. Tashmuchamedov and F. A. Umarova (1976). Generation of a potential jump by Na^+, K^+-ATPase on an octane/water interface. *Doklady Akad. Nauk SSSR*, **229**, 1473–6.

Mishuk, V. Ya., V. V. Elkin and D. Leikis (1980). Measurement of the potential of the minimum of the C versus ϕ curves by the method of amplitude demodulation. *Elektrokhimiya*, **16**, 1243–6.

Mitchell, P. (1961). Coupling of phosphorylation to electron and hydrogen transfer by a chemi-osmotic type of mechanism. *Nature*, **191**, 144–8.

Mitchell, P. (1976). Vectorial chemistry and the molecular mechanics of chemiosmotic coupling: power transmission by proticity. *Biochem. Soc. Trans.*, **4**, 399–430.

Mohilner, D. M. (1962). Thermodynamic treatment of the electrocapillary curve for reversible electrodes and properties of the double layer. *J. Phys. Chem.*, **66**, 724–6.

Mohilner, D. M. (1966). The electrical double layer. Part I. Elements of double layer theory. In: *Electroanalytical Chemistry* (A. J. Bard, ed.) Vol. 1, pp. 241–409. Marcel Dekker, New York.

Mohilner, D. M., H. Nakadonari and P. R. Mohilner (1977). Electrosorption of 2-butanol at the mercury–solution interface. 2. Theory of noncongruent electrosorption. *J. Phys. Chem.*, **81**, 244–52.

Molero, M. C. and W. Outhwaite (1990). A mean field analysis of an ion dipole mixture against a charged hard wall with specific adsorption. 2. Non-linear results. *J. Chem. Soc. Faraday Trans.*, **86**, 35–42.

Molero, M. C., W. Outhwaite and L. B. Bhuiyan (1992). Individual ionic activity coefficients from a symmetric Poisson–Boltzmann theory. *J. Chem. Soc. Faraday Trans.*, **88**, 1541–7.

Morf, W. E. (1981). *The Principles of Ion-Selective Electrodes and of Membrane Transport.* Akademiai Kiado, Budapest.

Motomura, K. (1977). Thermodynamic studies on adsorption at interfaces. *J. Colloid Interface Sci.*, **64**, 348–55.

Motomura, K. (1980). Thermodynamics of interfacial monolayers. *Adv. Colloid Interface Sci.*, **12**, 1–42.

Motomura, K. (1986). Comments on "Thermodynamics of adsorption and Gibbsian distance parameters" by R. J. Good. *J. Colloid Interface Sci.*, **110**, 294–7.

Motomura, K. and M. Aratono (1987). Geometric formalism of the thermodynamics of adsorption at interfaces between two fluid phases. *Langmuir*, **3**, 304–6.

Motomura, K., M. Aratono, N. Matubayasi and R. Matuura (1978a). Thermodynamics studies on adsorption at interfaces. *J. Colloid Interface Sci.*, **67**, 247–54.

Motomura, K., H. Iyota, M. Aratono, M. Yamanaka and R. Matuura (1983). Thermodynamic consideration of the pressure dependence of interfacial tension. *J. Colloid Interface Sci.*, **93**, 264–9.

Motomura, K., I. Kajiwara, N. Ikeda and M. Aratono (1989). Thermodynamic study of the adsorption of sodium perfluorooctanoate at the water/hexane interface. *Colloid Surf.*, **38**, 61–9.

Motomura, K., N. Matubayasi, R. Matuura and M. Aratono (1978b). Thermodynamic studies on adsorption at interfaces. *J. Colloid Interface Sci.*, **64**, 356–61.

Mueller, P. and D. P. Rudin (1968). Action potentials induced in bimolecular lipid membranes, *Nature*, **271**, 713–9.

Mueller, P., D. O. Rudin, H. T. Tien and W. C. Wescott (1962). Reconstitution of cell membrane structure in vitro and its transformation into excitable system. *Nature*, **194**, 979–80.

Mui, B. L.-S., H.-G. Dobereiner, T. D. Madden and P. R. Cullis (1995). Influence of transbilayer area asymmetry on the morphology of large unilamellar vesicles. *Biophys. J.*, **69**, 930–41.

Mullins, L. J. (1979). The generation of electric currents in cardiac fibers by Na/Ca exchange. *Am. J. Physiol.*, **23**, C103–C110.

Mullins, L. J. and K. Noda (1963). The influence of sodium-free solutions on the membrane potential of frog muscle. *J. Gen. Physiol.*, **47**, 117–32.

Munger, G., R. M. Leblanc, B. Zelent, J. Gallant, H. A. Tajmirriahi, J. Aghion, M. I. Gugeshashvili and A. G. Volkov (1992). The hydrated chlorophyll *a* oligomer in monolayers and Langmuir–Blodgett films. *Biol. Membranes*, **9**, 874–80.

Munger, G., R. M. Leblanc, B. Zelent, A. G. Volkov, M. I. Gugeshashvili, J. Gallant, H. A. Tajmirriahi and J. Aghion (1992). Characterization of monolayers and Langmuir–Blodgett films of dry and wet chlorophyll *a*. *Thin Solid Films*, **210**, 739–42.

Murokami, N. and T. Konishi (1988). DCCD-sensitive Na$^+$-transport in the membrane vesicles of Halobacterium halobium. *J. Biochem.*, **103**, 231–6.

Nagle, J. F. (1987). Theory of passive proton conductance in lipid bilayer. *J. Bioenerg. Biomembranes*, **19**, 413–26.

Nagle, J. F. and H. J. Morowitz (1978). Molecular mechanisms for proton transport in membrane. *Proc. Natl. Acad. Sci. USA*, **75**, 298–302.

Nagle, J. F. and S. Tristram-Nagle (1983). Hydrogen bonded chain mechanisms for proton conduction and proton pumping. *J. Membrane Biol.*, **74**, 1–14.

Nakanaga, T. and T. Takenaka (1977). Resonance Raman spectra of monolayers of a surface-active dye adsorbed at the oil–water interface. *J. Phys. Chem.*, **81**, 645–9.

Nakanishi, K. and R. Matsuno (1986). Kinetics and equilibrium of enzymatic synthesis of peptides in aqueous/organic biphasic systems—thermolysin catalyzed synthesis of N-(benzyloxycarbonyl)-L-aspartyl-L-phenylalanine methyl ester. *Eur. J. Biochem.*, **161**, 541–9.

Nakanishi, K. and R. Matsuno (1990). Continuous peptide synthesis in a water–immiscible organic solvent with an immobilized enzyme. *Ann. N.Y. Acad. Sci.*, **613**, 652–5.

Nakatani, K., T. Uchida, H. Misawa, N. Kitamura and H. Masuhara (1994). Laser trapping and electrochemistry of a single oil droplet in water—electron transfer across the oil-droplet electrode interface. *J. Electroanalyt. Chem.*, **367**, 109–14.

Nanavati, C., V. S. Markin, A. F. Oberhauser and J. M. Fernandez (1992). The exocytotic fusion pore modeled as a lipidic pore. *Biophys. J.*, **63**, 1118–32.

Naumann, R., A., R. Jonczyk, J. Kopp, H. van Esch, W. Ringsdorf, W. Knoll and P. Graber (1995). Incorporation of membrane proteins in solid-supported lipid layers. *Angew. Chem. Int. Ed. Engl.*, **34**, 2056–8.

Neher, E. (1974). Asymmetric membranes resulting from the fusion of two black lipid membranes. *Biochim. Biophys. Acta*, **373**, 327–36.

Nernst, W. (1888). Zur Kinetik der in Losung befindlichen Korper. I. Theorie der Diffusion. *Z. Physik. Chem.*, **2**, 613–37.

Nernst, W. (1889). Die elektromotorische Wirksamkeit der Ionen. *Z. Physik. Chem.*, **4**, 129–81.

Nernst, W. and E. H. Riesenfeld (1902). Über elektrolytische Erscheinungen an der Grenzflache zweiter Losungmittel. *Ann. Phys.*, **8**, 600–8.

Neumann, E. and K. Rosenheck (1972). Permeability changes induced by electric impulses in vesicular membranes. *J. Membrane Biol.*, **10**, 279–90.

Neumke, B. and P. Laüger (1969). Nonlinear electrical effects in lipid bilayer membranes. II. Integration of the generalized Nernst–Planck equations. *Biophys. J.*, **9**, 1160–70.

Nichols, J. W. and D. W. Deamer (1978). Proton and hydroxyde permeability coefficient measured for unilamellar lyposomes. In: *Frontiers of Biological Energetics* (P. L. Dutton, J. S. Leigh and A. Scarpa, eds.) Vol. 2, pp. 1273–83. Academic Press, New York.

Nichols, J. W. and D. W. Deamer (1980). Net proton–hydroxyl permeability of large unimolecular lyposomes measured by an acid–base titration technique. *Proc. Natl. Acad. Sci. USA*, **77**, 2038–42.

Nicholson, R. S. (1965). Theory and application of cyclic voltammetry for measurement of electrode reaction kinetics. *Anal. Chem.*, **37**, 1351–5.

Nicholson, R. S. and I. Shain (1964). Theory of stationary electrode polarography. Single scan and cyclic methods applied to reversible, irreversible and kinetic systems. *Anal. Chem.*, **56**, 706–23.

Ninham, B. W. (1989). Hydration forces—real and imagined. *Chem. Scripta*, **29A**, 15–21.

Ono, S. and S. Kondo (1960). *Handbuch der Physik*. Vol. X. Springer Verlag, Berlin.

Osakai, T. and K. Ebina (1996). Quantum chemical approach to the Gibbs energy of ion transfer between two immiscible liquids. *J. Electroanal. Chem.*, **412**, 1–9.

Osakai, T., S. Himeno and A. Saito (1993a). Linear dependence of the standard ion transfer potentials of heteropoly and isopoly anions at the 1,2-dichloroethane/water interface on their surface charge densities. *J. Electroanal. Chem.*, **360**, 299–307.

Osakai, T., T. Kakutani, Y. Nishiwaki and M. Senda (1983). Determination of the standard free energies of transfer of alkylammonium ions from nitrobenzene to water using polarographic methods with immiscible electrolyte solution interface. *Bunseki Kagaku*, **32**, E81–E84.

Osakai, T., T. Kakutani and M. Senda (1984). A.C. polarographic study of ion transfer at the water/nitrobenzene interface. *Bull. Chem. Soc. Jpn.*, **57**, 370–6.

Osakai, T., H. Katano, K. Maeda, S. Himeno and A. Saito (1993b). A hydrophobicity scale of heteropoly and isopolyanions based on voltammetric studies of their transfer at the nitrobenzene/water interface. *Bull. Chem. Soc. Jpn.*, **66**, 1111–5.

Ostwald, W. (1890). Elektrische Eigenschaften halbdurchlässiger Scheidewände. *Z. Phys. Chem.*, **6**, 71–82.

Ottova-Leitmannova, A. and H. Ti Tien (1993). Bilayer lipid membranes: an experimental system for biomolecular electronic devices development. *Prog. Surf. Sci.*, **41**, 337–445.

Outhwaite, C. W. (1970). A modified Poisson–Boltzmann equation in the double layer. *Chem. Phys. Lett.*, **7**, 636–8.

Outhwaite, C. W. (1981). A preliminary treatment of solute and solvent interactions in the diffuse part of electric double layer. *Can. J. Chem.*, **59**, 1854–9.

Outhwaite, C. W. (1983). An improved modified Poisson–Boltzman equation in electric-double-layer theory. *J. Chem. Soc., Faraday Trans. II*, **79**, 707–18.

Outhwaite, C. W. and L. B. Bhuiyan (1982). A further treatment of the exclusion-volume term in the modified Poisson–Boltzmann theory of the electric double layer. *J. Chem. Soc. Faraday Trans. II*, **78**, 775–85.

Outhwaite, C. W. and L. B. Bhuiyan (1983). An improved modified Poisson–Boltzmann equation in electric-double-layer theory. *J. Chem. Soc. Faraday Trans. II*, **79**, 707–18.

Outhwaite, C. W. and L. B. Bhuiyan (1991a). The electric double layer around an isolated spherical macroion. *Electrochim. Acta*, **36**, 1747–9.

Outhwaite, C. W. and L. B. Bhuiyan (1991b). A modified Poisson–Boltzmann analysis of the electric double layer around an isolated spherical macroion. *Mol. Phys.*, **74**, 367–81.

Outhwaite, C. W., L. B. Bhuiyan and S. Levine (1980). Theory of the electric double layer using a modified Poisson–Boltzmann equation. *J. Chem. Soc., Faraday Trans. II*, **76**, 1388–408.

Outhwaite, C. W. and M. Molero (1989). An ion-dipole mixture against a charged hard wall with specific adsorption. *J. Chem. Soc. Faraday Trans.*, **85**, 1585–99.

Outhwaite, C. W. and M. Molero (1990). Another comment on the anomalous behaviour of the inner-layer capacity at low electrolyte concentrations. *J. Electroanal. Chem.*, **286**, 239–43.

Outhwaite, C. W. and M. Molero (1991a). Application of a symmetric Poisson–Boltzmann equation to electrolyte mixtures. *Chem. Phys. Lett.*, **184**, 566–70.

Outhwaite, C. W. and M. Molero (1991b). A mean field analysis of an ion dipole mixture against a charged hard wall with specific dipole adsorption-classification of differential capacity plots. *Electrochim. Acta*, **36**, 1685–7.

Outhwaite, C. W., M. Molero and L. B. Bhuiyan (1991). Symmetrical Poisson–Boltzmann and modified Poisson–Boltzmann theories. *J. Chem. Soc. Faraday Trans.*, **87**, 3227–30.

Outhwaite, C. W., M. Molero and L. B. Bhuiyan (1993). Primitive model electrolytes in the modified Poisson–Boltzmann theory. *J. Chem. Soc. Faraday Trans.*, **89**, 1315–20.

Outhwaite, C. W., M. Molero and L. B. Bhuiyan (1994). Primitive model electrolytes in the modified Poisson–Boltzmann theory. *J. Chem. Soc. Faraday Trans.*, **90**, 2002.

Overbeek, J. Th. G. (1952). In: *Colloid Science* (H. R. Kruyt, ed.) Vol. 1, p. 116. Elsevier, Amsterdam.

Padday, J. F. (1969). Theory of surface tension. In: *Surface and Colloid Science* (E. Matievic and F. R. Eirich, eds.), Vol. 1, pp. 39–251. Wiley-Interscience, New York.

Paleska, I., J. Kotowski, Z. Koczorowski, E. Nakache and M. Dupeyrat (1990a). Electrochemical study of the water/nitroethane interface. *J. Electroanal. Chem.*, **287**, 129–35.

Paleska, I., Z. Koczorowski and W. Wawrzynczak (1990b). Note on zero charge potentials at interfaces of immiscible electrolyte solutions measured by various methods. *J. Electroanal. Chem.*, **280**, 439–42.

Papahadjopoulos, D. (1978). Calcium-induced phase changes and fusion in natural and model membranes. In: *Membrane Fusion* (G. Poste and G. L. Nickolson, eds.), p. 765, Elsevier-North-Holland Biomedical Press, New York.

Papahajopoulos, D., G. Poste, B. E. Schaeffer and W. Vail (1974). Membrane fusion and molecular segregation in phospholipid vesicles. *Biochim. Biopphys. Acta*, **352**, 10–21.

Parker, A. J. (1969). Protic-dipolar aprotic solvent effects on traces of bimolecular reactions. *Chem. Rev.*, **69**, 1–32.

Parker, A. J. (1976). Solvation of ions—enthalpies, entropies and free energies of transfer. *Electrochim. Acta*, **21**, 671–9.

Parsegian, A. (1969). Energy of ion crossing a low dielectric membrane: solutions to four relevant electrostatic problems. *Nature*, **221**, 844–6.

Parsegian, V. A., R. P. Rand and N. L. Fuller (1991). Direct osmotic stress measurements of hydration and electrostatic double layer forces between bilayers of double-chained ammonium acetate surfactant. *J. Phys. Chem.*, **95**, 4777–82.

Parsons, R. (1954). Equilibrium properties of electrified interfaces. In: *Modern Aspects of Electrochemistry* (J. O'M. Bockris and B. E. Conway, eds.) pp. 103–79. Butterworths Scientific Publications, London.

Parsons, R. (1961). The structure of the electrical double layer and its influence on the rates of electrode reactions. In: *Advances in Electrochemistry and Electrochemical Engineering* (P. Delahay, ed.) Vol. 1, pp. 1–64. Wiley, New York.

Parsons, R. (1980). Thermodynamic methods for the study of interfacial regions in electrochemical systems. In: *Comprehensive Treatise of Electrochemistry* (J. O'M. Bockris, B. E. Conway and E. Eager, eds.) Vol. 1, pp. 1–44. Plenum Press, New York.

Parsons, R. (1990). Electrical double layer—recent experimental and theoretical developments. *Chem. Rev.*, **90**, 813–26.

Parsons, R. and B. T. Rubin (1974). The medium effect for single ionic species. *J. Chem. Soc. Faraday Trans. 1*, **70**, 1636–48.

Parsons, R. and F. G. R. Zobel (1965). The interphase between mercury and aqueous sodium dihydrogen phosphate. *J. Electroanal. Chem.*, **9**, 323–48.

Partenskii, M. B., V. Dorman and P. C. Jordan (1996). The question of negative capacitance and its relation to instabilities and phase transitions at electrified interfaces. *Int. Rev. Phys. Chem.*, **15**, 153–82.

Partenskii, M. B. and V. I. Feldman (1983). An application of the sum rules and the statistical electron theory in the calculations of the capacity of the double layer at the metal/electrolyte solution interface. *Sov. Electrochem.*, **24**, 340–3.

Partenskii, M. B. and V. I. Feldman (1988). Use of the sum rules and of the statistical theory of electronic response in calculations of double-layer capacitance at interfaces between metals and surface-inactive electrolyte solutions. *Elektrokhimiya*, **24**, 369–73.

Partenskii, M. B. and P. C. Jordan (1993). The admissible sign of the differential capacity, instabilities, and phase transitions at electrified interfaces. *J. Chem. Phys.*, **99**, 2992–3002.

Patey, G. N. and G. M. Torrie (1989). Water and salt near charged surfaces—a discussion of some recent theoretical results. Chem. Scripta, **29A**, 39–47.

Patra, C. N. and S. K. Ghosh (1994a). A nonlocal density-functional theory of electric double layer-charge-asymmetric electrolytes. J. Chem. Phys, **101**, 4143–9.

Patra, C. N. and S. K. Ghosh (1994b). A nonlocal density functional theory of electric double layer-symmetric electrolytes. J. Chem. Phys., **100**, 5219–29.

Patra, C. N. and S. K. Ghosh (1995). Electric double layer at a metal/electrolyte interface—a density functional approach. J. Chem. Phys., **102**, 2556.

Paula, S., A. G. Volkov, A. N. Van Hoek, T. H. Haines and D. W. Deamer (1996). Permeation of protons, potassium ions, and small polar molecules through phospholipid bilayers as a function of membrane thickness. Biophys. J., **70**, 339–48.

Pauling, L. (1927). The sizes of ions and the structure of ionic crystals. J. Amer. Chem. Soc., **49**, 765–90.

Pereira, C. M., A. Martins, M. Rocha, B. J. Silva and F. Silva (1994). Differential capacitance of liquid/liquid interfaces effect of electrolytes present in each phase. J. Chem. Soc. Faraday Trans., **90**, 143–8.

Perkins, W. R. and D. Cafiso (1986). An electrical and structural characterization of H^+/OH^- currents in phospholipid vesicles. Biochemistry, **25**, 2270–6.

Petrov, A. G., M. D. Mitov and A. Derzhanski (1980). Edge energy and pore stability in bilayer membranes. In: Advances in Liquid Crystal Research and Applications (Bata, L., ed.), pp. 695–702. Pergamon, Oxford.

Phipps, J. S., R. M. Richardson, T. Cosgrove and A. Eaglesham (1993). Neutron reflection studies of copolymers at the hexane/water interface. Langmuir, **9**, 3530–7.

Piasecki, D. A. and M. Wirth (1993). Reorientation of acridine orange in a sodium dodecyl sulfate monolayer at the water–hexadecane interface. J. Phys. Chem., **97**, 7700–5.

Pimentel, G. C. and A. L. McClellan (1960). The Hydrogen Bond. Freeman, San Francisco.

Pinchuk, V. M. (1993). Quantum chemical modeling of the photodecomposition of the water dimer in chlorophyll complexes of magnesium and manganese. J. Struct. Chem., **34**, 203–7.

Planck, M. (1890). Ueber die Potencialdifferenz zwischen zwei verdunnten Losungen binarer Electrolyte. Ann. Phys. Chem. N.F., **40**, 561–76.

Planck, M. (1891). Ueber das Princip der Vermehrung der Entropie. Ann. Phys., **44**, 385–428.

Pliny the Elder (1962). Natural History. Harvard University Press, Cambridge.

Plischke, M. and D. Henderson (1989). Pair correlation functions and the structure of the electric double layer. Electrochim. Acta, **34**, 1863–7.

Pogorelov, A. V. (1965). The Theory of Shells under Supercritical Deformations. Nauka Publishers, Moscow.

Popov, A. N. (1977). Electrocapillary phenomena at the interfacial boundary between aqueous and organic phases. Elektrokhimiya, **13**, 1393–5.

Popov, A. N. (1987). Counterions and adsorption of ion-exchange extractans at the water/oil interface. The Interface Structure and Electrochemical Processes at the Boundary Between Two Immiscible Liquids (V. E. Kazarinov, ed.), pp. 143–205. Springer Verlag, Berlin-Heidelberg.

Popov, A. N. (1996). Electrodialysis through liquid ion-exchange membranes and the oil/water interface. In: Liquid–Liquid Interfaces. Theory and Methods (A. G. Volkov and D. W. Deamer, eds.) pp. 333–61. CRC Press, Boca Raton, New York, London, Tokyo.

Popov, A. N., G. Yu. Drachev and Yu. A. Shchipunov (1983). Electrocapillary phenomena at the *n*-heptane/water interface in the presence of phospholipids. *Zh. Fiz. Khimii*, **57**, 2288–91.

Popovych, O. (1970). Estimation of medium effects for single ions in non-aqueous solvents. *CRC Crit. Rev. Anal. Chem.*, **1**, 73–117.

Portis, A., C. Newton, W. Pangbom and D. Papahadjopoulos (1979). Studies of mechanism of membrane fusion. Evidence for an intermediate Ca^{2+}-PS complex. Synergism with Mg^{2+} and inhibition by spectrin. *Biochemistry*, **18**, 780–90.

Post, A., S. E. Young and R. N. Robertson (1984). Light-induced proton translocation by bacteriorhodopsin at the interface of an octane-in-water emulsion and inhibition by a retinotoxin. *Photobiochem. Photobiophys.*, **8**, 153–62.

Poste, G. and G. L. Nicolson (1978). *Membrane Fusion*. Elsevier-North-Holland Biomedical Press, New York.

Pratt, L. R. (1985). Theory of hydrophobic effects. *Annu. Rev. Phys. Chem.*, **36**, 433–49.

Princen, H. M., I. Y. Z. Zia and S. G. Mason (1967). Measurement of interfacial tension from the shape of a rotating drop. *J. Colloid Interface Sci.*, **23**, 99–107.

Rabinowitch, E. and I. Weiss (1937). Reversible oxidation of chlorophyll. *Proc. R. Soc. (London) A*, **162**, 251–67.

Radic, N. and S. Marcelja (1978). Solvent contribution to the Debye screening length. *Chem. Phys. Lett.*, **55**, 377–9.

Rais, J. (1971). Individual extraction constants of univalent ions in the system water–nitrobenzene. *Coll. Czech. Chem. Commun.*, **36**, 3253–62.

Ramanathan, P. S. and H. L. Friedman (1971). Study of a refined model for aqueous 1-1 electrolytes. *J. Chem. Phys.*, **54**, 1086–99.

Rand, R. P. and V. A. Parsegian (1986). Mimicry and mechanism in phospholipid models of membrane fusion. *Ann. Rev. Physiol.*, **48**, 201–12.

Rand, R. P. and V. A. Parsegian (1989). Hydration forces between phospholipid bilayers. *Biochim. Biophys. Acta*, **988**, 351–76.

Randles, J. E. B. (1956). The real hydration energies of ions. *Trans. Faraday Soc.*, **52**, 1573–81.

Reeves, J. P. and C. C. Hale (1984). The stoichiometry of the cardiac sodium–calcium exchange system. *J. Biol. Chem.*, **259**, 7733–9.

Rehbinder, P. (1927a). Grenzfläshenaktivität (Adsorbierbarkeit) und Dielektrizitätskonstante. I. Abhängigkeit der Grenzflächenaktivität und der Adsorption an verschiedenen Trennungsflächen von der Polarität bzw. Dielektrizitätskonstante der beiden die Grenzfläche bildenden Phasen und des adsorbierten Stoffes. *Z. Physik. Chem.*, **129**, 161–75.

Rehbinder, P. (1927b). Über Grenzfläshenaktivität bzw. -energie an verschiedenen Grenzflächen und deren spezifisches Adsorptionvermögen. *Biochem. Z.*, **187**, 19–31.

Reid, J. D., O. R. Melroy and R. P. Buck (1983). Double layer charge and potential profiles at immiscible liquid/liquid electrolyte interfaces. *J. Electroanal. Chem.*, **147**, 71–82.

Reid, J. D., P. Vanysek and R. P. Buck (1984a). Potential dependence of capacitance at a polarizable (blocked) liquid/liquid interface. *J. Electroanal. Chem.*, **161**, 1–15.

Reid, J. D., P. Vanysek and R. P. Buck (1984b). Potential dependence of capacitance at a liquid/liquid interface. Part II. Unblocked interface. *J. Electroanal. Chem.*, **170**, 109–25.

Riesenfeld, E. H. (1902a). Bestimmung der Ueber Uhrungszahl einiger Salze in Phenol. *Ann. Phys.*, **8**, 609–15.

Riesenfeld, E. H. (1902b). Konzentrationsketten mit nichtmischbiren Losungsmitteln. *Ann. Phys.*, **8**, 616–24.

Riesenfeld, E. H. and B. Reinhold (1909). Bestimmung von Ober Uhrungszahlen aus elektromotorischen Kraften in Losungsmitteln, welche mit Wasser nur beschränkt mischbar sind. *Z. Phys. Chem.*, **68**, 459–70.

Rivera, S. R. and T. S. Sorensen (1994). Grand canonical ensemble Monte-Carlo simulations of ion distribution functions and of the electric double layer in spherical charged or uncharged pores with restricted primitive model electrolytes. *Mol. Simulat.*, **14**, 35–66.

Riveros, O. J. (1991). Dynamics of electron transfer in the oxidation of water by chlorophyll *a* dimer. *Biophys. Chem.*, **40**, 109–15.

Robinson, R. A. and R. H. Stokes (1959). *Electrolyte Solutions*, 2nd edition. Butterworths, London.

Rosano, H. L. (1967). Mechanisms of water transport through nonaqueous liquid membranes. *J. Colloid Interface Sci.*, **23**, 73–9.

Rosenberg, M. (1991). Basic and applied aspects of microbial adhesion at the hydrocarbon/water interface. *Crit. Rev. Microbiol.*, **18**, 159–73.

Rosenberg, T. and W. Wilbrandt (1957). Uphill transport induced by counterflow. *J. Gen. Physiol.*, **41**, 289–96.

Rossignol, M., P. Thomas and C. Grignon (1982). Proton permeability of lyposomes from natural phospholipid mixture. *Biochim. Biophys. Acta*, **684**, 195–9.

Rotenberg, Y., L. Boruvka and A. W. Neumann (1985). Invariance of the free energy of calillarity. *Langmuir*, **2**, 533–5.

Rowlinson, J. S. (1983). The thermodynamics of liquid lens. *J. Chem. Soc. Faraday Trans.* ii, **79**, 77–90.

Rowlinson, J. S. and B. Widom (1982). *Molecular Theory of Capillarity*. Clarendon Press, Oxford.

Rumer, Iu. B. and M. Sh. Ryvkin (1980). *Thermodynamics, Statistical Physics and Kinetics*. Mir, Moscow.

Rusanov, A. I. (1960). *Thermodynamics of Surface Phenomena*. Leningrad State University Press, Leningrad.

Rusanov, A. I. (1962). Equilibria in ionic systems with discontinuities. I. Planar discontinuity surfaces. *Russ. J. Phys. Chem.*, **36**, 285–8.

Rusanov, A. I. (1967). *Phase Transitions and Surface Phenomena*. Khimiya, Leningrad.

Rusanov, A. I. and F. C. Goodrich, eds. (1981). *The Modern Theory of Capillarity. To the Centennial of a Gibbs theory of Capillarity*. Academic Verlag, Berlin.

Rusanov, A. I., S. S. Dukhin and A. E. Yaroshchuk (1984). Problem of the surface layer in liquid mixtures and the electric double layer. *Kolloidnyi Zh.*, **46**, 490–4.

Rusanov, A. I. and F. M. Kuni (1982). Theory of nucleation on charged nuclei. 1. General thermodynamic relationships. *Kolloidnyi Zh.*, **44**, 934–41.

Rusanov, A. I. and V. L. Kuz'min (1976). Effect of an electric field on the surface tension of a polar liquid. *Kolloidnyi Zh.*, **39**, 388–90.

Rusanov, A. I. and V. A. Prokhorov (1994). *Interfacial Tensiometry*. Khimiya, St. Petersburgh.

Rusanov, A. N. (1983). In: *Surface Forces and Boundary Layers of Liquids* (B. V. Deryaguin, ed.) p. 152. Nauka Publishers, Moscow.

Rutkowsky, C. A., L. M. Williams, T. H. Haynes and H. Z. Cummins (1991). The elasticity of synthetic phospholipid vesicles obtained by photon correlation spectroscopy. *Biochemistry*, **30**, 5688–96.

Rutland, M. W. and H. K. Christenson (1990). Effect of nonionic surfactant of ion adsorption and hydration forces. *Langmuir*, **6**, 1083–7.

Sabela, A., V. Mareček, Z. Samec and R. Fuoco (1992). Standard Gibbs energies of transfer of univalent ions from water to 1,2-dichloroethane. *Electrochim. Acta*, **37**, 231–5.

Samec, Z. (1979a). Charge transfer between two immiscible electrolyte solutions. Part I. Basic equation for the rate of the charge transfer across the interface. *J. Electroanal. Chem.*, **99**, 197–205.

Samec, Z. (1979b). Charge transfer between two immiscible electrolyte solutions. Part III. Stationary curve of current vs. potential of electron transfer across interface. *J. Electroanal. Chem.*, **103**, 1–9.

Samec, Z. (1980). Charge transfer between two immiscible electrolyte solutions. Part V. Convolution potential sweep voltammetry of ion or electron transfer. *J. Electroanal. Chem.*, **111**, 211–6.

Samec, Z. (1988). Electrical double layer at the interface between two immiscible electrolyte solutions. *Chem. Rev.*, **88**, 617–32.

Samec, Z. (1996). Kinetics of charge transfer. In: *Liquid–Liquid Interfaces. Theory and Methods* (A. G. Volkov and D. W. Deamer, eds.) pp. 155–78. CRC Press, Boca Raton, New York, London, Tokyo.

Samec, Z., Yu. Ya. Gurevich and Yu. I. Kharkats (1986). Stochastic approach to the ion transfer kinetics across the interface between two immiscible electrolyte solutions. Comparison with the experimental data. *J. Electroanal. Chem.*, **204**, 257–66.

Samec, Z., D. Homolka and V. Mareček (1982). Charge transfer between two immiscible electrolyte solutions. Part VIII. Transfer of alkali and alkaline earth-metal cations across the water/nitrobenzene interface facilitated by synthetic neutral ion carriers. *J. Electroanal. Chem.*, **135**, 265–83.

Samec, Z. and T. Kakiuchi (1995). Charge transfer kinetics at water–organic solvent phase boundaries. In: *Advances in Electrochemical Science and Electrochemical Engineering* (H. Gerischer and C. W. Tobias, eds.) Vol. 4, Chap. 5, pp. 297–361. VCH, Weinheim.

Samec, Z. and V. Mareček (1987). A study of the electrical double layer at the interface between two immiscible electrolyte solutions by impedance measurements. In: *The Interface Structure and Electrochemical Processes at the Boundary Between Two Immiscible Liquids* (V. E. Kazarinov, ed.) pp. 123–41. Springer Verlag, Berlin.

Samec, Z., V. Mareček, K. Holub, S. Racinsky and P. Hajkova (1987). The double layer at the interface between two immiscible electrolyte solutions Part III Capacitance of the water/1,2-dichloroethane interface. *J. Electroanal Chem.*, **225**, 65–78.

Samec, Z., V. Mareček and D. Homolka (1981). The double layer at the interface between two immiscible electrolyte solutions. Part I. Capacity of the water/nitrobenzene interface. *J. Electroanal. Chem.*, **126**, 121–9.

Samec, Z., V. Mareček and D. Homolka (1983). Charge transfer between two immiscible electrolyte solutions. Part IX. Kinetics of the transfer choline and acetylcholine cations across the water/nitrobenzene interface. *J. Electroanal. Chem.*, **158**, 25–36.

Samec, Z., V. Mareček and D. Homolka (1984a). Double layers at a liquid/liquid interfaces. *Faraday Discuss. Chem. Soc.*, **77**, 197–208.

Samec, Z., V. Mareček and D. Homolka (1984b). The use of the mean spherical approximation in calculation of the double-layer capacitance for the interface between two immiscible electrolyte solutions. *J. Electroanal. Chem.*, **170**, 383–6.

Samec, Z., V. Mareček and D. Homolka (1985). The double layer at the interface between two immiscible electrolyte solutions. Part II. Structure of the water/nitrobenzene interface in the presence of 1:1 and 2:2 electrolytes. *J. Electroanal. Chem.*, **187**, 31–51.

Samec, Z., V. Mareček and J. Weber (1979a). Detection of an electron transfer across the interface between two immiscible electrolyte solutions by cyclic voltammetry with four-electrode system. *J. Electroanal. Chem.*, **96**, 245–7.

Samec, Z., V. Mareček and J. Weber (1979b). Charge transfer between two immiscible electrolyte solutions. Part IV. Electron transfer between hexacyanoferrate(III) in water and ferrocene in nitrobenzene investigated by cyclic voltammetry with four-electrode system. *J. Electroanal. Chem.*, **103**, 11–8.

Sandeaux, R., P. Seta, C. Jeminet, M. Alleaume and C. Gavach (1978). The influence of pH on the conductance of lipid bimolecular membranes in location to the alkaline ion transport induced by carboxylic carriers grisorixin, alborixin and monensin. *Biochim. Biophys. Acta*, **511**, 499–508.

Sanfeld, A. (1968). *Introduction to the Thermodynamics of Charged and Polarized Layers.* Wiley, New York.

Sassaman, J. L. and M. J. Wirth (1994). Reorientation of acridine orange at an alcohol-modified liquid/liquid interface. *Colloids Surf. A*, **93**, 49–58.

Schnuelle, G. W. and D. L. Beveridge (1975). A statistical thermodynamic supermolecule–continuum study of ion hydration. 1. Site method. *J. Phys. Chem.*, **79**, 2566–73.

Schuhmann, D. and P. Seta (1979). On the impedance of the interface between two immiscible electrolytic solutions in the presence of an adsorbed ion soluble in both media. *J. Colloid Interface Sci.*, **69**, 448–59.

Schuhmann, D. and P. Seta (1980). On the impedance of the interface between two immiscible electrolytic solutions in the presence of an adsorbed ion soluble in both media. *Physico. Chem. Hydrodyn.*, **1**, 57–68.

Schulman, J. H. and E. C. Cockbain (1940). Molecular interactions at oil/water interfaces. *Trans. Faraday Soc.*, **36**, 651–61.

Schulman, J. H. and E. C. Cockbain (1940). Molecular interactions at oil/water interfaces. *Trans. Faraday Soc.*, **36**, 661–8.

Schweighofer, K. J. and I. Benjamin (1995). Electric field effects on the structure and dynamics at a liquid/liquid interface. *J. Electroanal. Chem.*, **391**, 1–10.

Seelich, F. (1948). Die Adsorption an der Grenzfläche zweiter Flüssigkeiten I. Über die Berechnung der Adsorption aus konzentrierten Lösungen grenzflächenaktiver Stoffe nach der Gibbsschen Gleichung. *Monats. Chem.*, **79**, 338–47.

Seely, G. R. (1977). Chlorophyll in model systems: clues to the role of chlorophyll in photosynthesis. In: *Primary Processes of Photosynthesis* (J. Barber, ed.), pp. 1–53. Elsevier, Amsterdam.

Seifert, U. (1994). Fluid membranes—theory of vesicle conformations. *Ph.D. Thesis.* Institut für Festkörperforschung, Jülich.

Seifert, U., K. Berndl and R. Lipowsky (1991a). Shape transformations of vesicles: Phase diagrams for spontaneous-curvature and bilayer-coupling models. *Phys. Rev. A*, **44**, 1182–202.

Seifert, U., L. Miao, H.-G. Dobereiner and M. Wortis (1991b). Budding transition for bilayer fluid vesicles with area-difference elasticity. In: (R. Lipowsky, D. Richter and K. Kremer, eds), Vol. 66, pp. 93–6. *The Structure and Conformation of Amphiphilic Membranes. Springer Proceedings in Physics* Berlin: Springer.

Senda, M., T. Kakiuchi and T. Osakai (1991). Electrochemistry at the interface between two immiscible electrolyte solutions. *Electrochim. Acta*, **36**, 253–62.

Senda, M. and Y. Yamamoto (1996). Amperometric ion-selective electrode sensors. In: *Liquid–Liquid Interfaces. Theory and Methods* (A. G. Volkov and D. W. Deamer, eds.) pp. 375–400. CRC Press, Boca Raton, New York, London, Tokyo.

Seta, P., B. d'Epenoux and C. Gavach (1979). The double layer and ionic adsorption at the interface between two immiscible solutions. Part II. Long chain alkyl-trimethylammonium halides at partition equilibrium between water and nitrobenzene. *J. Electroanal. Chem.*, **95**, 191–9.

Seta, P. and C. Gavach (1972). Measure de l'impedance des interfaces entre deux solutions electrolytiques non miscibles. *Compt. Rend. Acad. Sci. Ser. C*, **275**, 1231–4.

Sharp, K., A. Jean-Charles and B. Honig (1992). A local dielectric constant model for solvation free energies which accounts for solute polarizability. *J. Phys. Chem.*, **96**, 3822–8.

Shchipunov, Y. A. and A. F. Kolpakov (1991). Phospholipids at the oil–water interface. Adsorption and interfacial phenomena in electric field. *Adv. Colloid Interface Sci.*, **35**, 31–138.

Shchipunov, Yu. A. (1996). Immiscible liquid interface and self-organized assemblies of lecithin. In: *Liquid–Liquid Interfaces. Theory and Methods* (A. G. Volkov and D. W. Deamer, eds.) pp. 295–315. CRC Press, Boca Raton, New York, London, Tokyo.

Sheetz, M. P. and S. J. Singer (1974). Biological Membranes as Bilayer Couples. A Molecular Mechanism of Drug-Erythrocyte Interactions. *Proc. Nati. Acad. Sci. USA*, **71**, 4457–61.

Shirai, O., S. Kihara, Y. Yoshida and M. Matsui (1995). Ion transfer through a liquid membrane or a bilayer lipid membrane in the presence of sufficient electrolytes. *J. Electroanal. Chem.*, **389**, 61–70.

Shirai, O., Y. Yoshida, M. Matsui, K. Maeda and S. Kihara (1996). Voltammetric study on the transport of ions of various hydrophobicity types through bilayer lipid membranes composed of various lipids. *Bull. Chem. Soc. Jpn.*, **69**, 3151–62.

Siegel, D. P. (1984). Inverted micellar structures in bilayer membranes: Formation rates and half-lives. *Biophys. J.*, **45**, 399–420.

Siegel, D. P. (1986a). Inverted micellar intermediates and the transitions between lamellar, inverted hexagonal, and cubic lipid phases. I. Mechanism of the L-to-H_{II} phase transition. *Biophys. J.*, **49**, 1155–70.

Siegel, D. P. (1986b). Inverted micellar intermediates and the transitions between lamellar, inverted hexagonal, and cubic lipid phases. II. Implications for membrane–membrane interactions and membrane fusion. *Biophys. J.*, **49**, 1171–83.

Siegel, D. P. (1986c). Inverted micellar intermediates and the transitions between lamellar, inverted hexagonal, and cubic lipid phases. III. Formation of isotropic and inverted cubic phases and membrane fusion via intermediates in transitions between L_α and H_{II} phases. *Chem. Phys. Lipids*, **42**, 279–301.

Siegel, D. P. (1993). Energetics of intermediates in membrane fusion: comparison of stalk and inverted micellar intermediate mechanisms. *Biophys. J.*, **65**, 2124–40.

Silva, F. (1984). Thermodynamics of the ideally polarized interface between two immiscible solutions of electrolytes. The electrocapillary equation and surface excesses. *Rev. Port. Quim.*, **26**, 25–9.

Silva, F. and C. Moura (1984). On the measurement of the impedance of ITIES. The nitrobenzene–water and 1,2-dichloroethane–water interfaces. *J. Electroanal. Chem.*, **177**, 317–23.

Sinfelt, J. H. and H. G. Drickamer (1955). Resistance in a liquid–liquid interface. *J. Chem. Phys.*, **23**, 1095–9.

Sisskind, B. and J. Kasarnowsky (1933). Studying of gases solubilities. 2. The solubility of argon. *Zh. Fiz. Khim.*, **4**, 683–90.

Sivukhin, D. V. (1990). *General Physics. Thermodynamics and Molecular Physics.* Vol. 2. Nauka, Moscow.

Sjolin, S. (1942). The oil/water interface, with and without monomolecular films, as a model of the living cell membrane. *Acta Physiol. Scand.*, **4**, 365–72.

Skulachev, V. P. and I. A. Kozlov (1977). H^+-*ATPases*. Nauka, Moscow.

Smith, G. W. (1944). The measurement of boundary tension by the pendent-drop method. 2. Hydrocarbons. *J. Phys. Chem.*, **48**, 168–72.

Sokolov, V. S., V. V. Cherny and I. G. Abidor (1980). Measurement of bilayer lipid membrane surface charge. *Doklady Acad. Nauk SSSR*, **251**, 236–9.

Sokolov, V. S., V. V. Cherny, M. V. Simonova and V. S. Markin (1990). Elcctrical potential distribution over the bilayer lipid membrane due to amphiphilic ion adsorption. *Bioelectrochem. Bioenerg.*, **23**, 27–44.

Sokolov, V. S. and V. S. Markin (1984). Electrogenic membrane transport of K and H ions by antibiotics nigericine and grisorixin. *Biol. Membranes*, **1**, 1071–87.

Sokolov, V. S. and V. S. Markin (1986). Membrane potential in the presence of coupled electrogenic transport. Thermodynamic equilibrium. *Biol. Membranes*, **3**, 638–49.

Solomon, T., H. Alemu and B. Hundhammer (1984a). Standard Gibbs energies of transfer of ions across the water–acetophenone interface. *J. Electroanal. Chem.*, **169**, 303–9.

Solomon, T., H. Alemu and B. Hundhammer (1984b). Standard Gibbs energies of transfer of ions across the water/chlorobenzene + nitrobenzene interface. *J. Electroanal. Chem.*, **169**, 311–4.

Solomon, T. and A. J. Bard (1995). Reverse (uphill) electron transfer at the liquid/liquid interface. *J. Phys. Chem.*, **99**, 17487–9.

Sparnay, M. J. (1958). Corrections of the theory of the flat diffuse double layer. *Rec. Trav. Chem.*, **77**, 872–88.

Spitzer, J. J. (1992). Theory of dissociative electrical double layers—the limit of close separations and hydration forces. *Langmuir*, **8**, 1659–62.

Starks, C. M., C. L. Liotta and N. M. Halper (1994). *Phase Transfer Catalysis*. Chapman & Hall, New York.

Stein, W. D. (1967). *The Movement of Molecules Across Cell Membranes*. Academic Press, New York.

Stein, W. D. (1986). *Transport and Diffusion Across Cell Membranes*. Academic Press, New York.

Stern, O. (1924). Zur Theorie der elektrolytischen Doppelschicht. *Z. Elektrochem.*, **30**, 508–16.

Stiles, P. J. (1980). Contributions from dielectric inhomogenity to the free energy of ionic solution. *Austral. J. Chem.*, **33**, 1389–91.

Stilinger, F. H. and J. G. Kirkwood (1960). Theory of the diffuse double layer. *J. Chem. Phys.*, **33**, 1282–90.

Strehlow, H. (1952). Comparison of electromotive series in different solvents. *Z. Electrochem.*, **56**, 827–33.

Sugden, S. (1921). The determination of surface tension from the rise in capillary tubes. *J. Chem. Soc.*, **119**, 1483–92.

Svensson, B., B. Jonsson and C. E. Woodward (1990). Monte-Carlo simulations of an electrical double layer. *J. Phys. Chem.*, **94**, 2105–13.

Svetina, S., A. Ottova-Leitmanova and R. Glaser (1982). Membrane bending energy in relation to bilayer couples concept of red blood cell shape transformations. *J. Theor. Biol.*, **94**, 13–23.

Svetina, S. and B. Zeks (1983). Bilayer couple hypothesis of red cell shape transformations and osmotic hemolysis. *Biochim. Biophys. Acta*, **42**, 86–90.

Svetina, S. and B. Zeks (1989). Membrane bending energy and shape determination of phospholipid vesicles and red blood cells. *Eur. Biophys. J.*, **17**, 101–11.

Tanford, C. (1980). *The Hydrophobic Effect: Formation of Micelles and Biological Membranes*. Wiley, New York.

Tang, C. W. and A. C. Albrecht (1975). Chlorophyll *a* photovoltaic cell. *Nature*, **254**, 507–9.

Tate, T. (1864). On the magnitude of a drop of liquid formed under different circumstances. *Phil. Mag.*, **27**, 176–80.

Taupin, C., M. Dvolaitzky and C. Sauterey (1981). Osmotic pressure induced pores in phospholipid vesicles. *Biochemistry*, **14**, 4771.

Tien, H. Ti. (1974). *Bilayer Lipid Membranes (BLM). Theory and Practice*. Marcel Dekker, Inc., New York.

Tien, H. Ti. (1980). Energy conversion via pigmented bilayer lipid membranes. *Separation, Sci. Technol.*, **15**, 1035–58.

Tien, H. Ti. (1986). Redox reactions in lipid bilayers and membrane bioenergetics. *Bioelectrochem. Bioenerg.*, **15**, 19–38.

Tien, H. Ti. (1989). Membrane photobiophysics and photochemistry. *Prog. Surf. Sci.*, **30**, 1–199.

Tien, H. Ti. and S. P. Verma (1970). Electronic processes in bilayer lipid membranes. *Nature*, **227**, 1232–4.

Timiriazeff, C. (1885). Colourless chlorophyll. *Nature*, **32**, 342.

Timiriazeff, C. (1886). Chlorophyll. *Nature*, **34**, 52.

Tkachenko, A. G., O. V. Syrkova, V. K. Tsvetkov and V. G. Korsakov (1990). Formation of CoO and Fe_3O_4 at the interface between immiscible solutions. *J. Appl. Chem.*, **63**, 239–42.

Tolman, R. (1949). The effect of droplet size on surface tension. *J. Chem. Phys.*, **17**, 333–7.

Toro, M., E. Erzt, J. Cerbon, G. Alegria, R. Alva, Y. Meas and O. S. Estrada (1987). Formation of ion-translocating oligomers in nigericin. *J. Membrane Biol.*, **95**, 3–8.

Torrie, G. M. (1992). Negative differential capacities in electrical double layers. *J. Chem. Phys.*, **96**, 3772–4.

Torrie, G. M., P. G. Kusalik and G. N. Patey (1989). Theory of the electric double layer: ion size effects in a molecular solvent. *J. Chem. Phys.*, **91**, 6367–75.

Torrie, G. M. and G. N. Patey (1991). Molecular solvent models of electrical double layers. *Electrochim. Acta*, **36**, 1677–84.

Torrie, G. M. and G. N. Patey (1993). Molecular solvent model for an electrical double layer —asymmetric solvent effects. *J. Phys. Chem.*, **97**, 12909–18.

Torrie, G. M., A. Perera and G. N. Patey (1989). Reference hypernetted-chain theory for dipolar hard spheres at charged surfaces. *Mol. Phys.*, **67**, 1337–53.

Torrie, G. M. and J. P. Valleau (1979). A Monte Carlo study of an electrical double layer. *Chem. Phys. Lett.*, **65**, 343–6.

Torrie, G. M. and J. P. Valleau (1980). Electrical double layers. 1. Monte Carlo study of a uniformly charged surface. *J. Chem. Phys.*, **73**, 5807–16.

Torrie, G. M. and J. P. Valleau (1982). Electrical double layers. 4. Limitations of the Gouy–Chapman theory. *J. Phys. Chem.*, **86**, 3251–7.

Torrie, G. M. and J. P. Valleau (1986). Double layer structure at the interface between two immiscible electrolyte solutions. *J. Electroanal. Chem.*, **206**, 69–79.

Toyoshima, Y., M. Marino, H. Motoki and M. Sukigara (1977). Photo-oxidation of water in phospholipid bilayer membranes containing chlorophyll *a*. *Nature*, **265**, 187–9.

Trasatti, S. and R. Parsons (1988). Interphases in systems of conducting phases. *Pure Appl. Chem.*, **58**, 437–54.

Trissl, H. W. and P. Graber (1980). Properties of chloroplasts spread at the heptane/water interface. Measurements of the photosynthetic charge separation in the nanosecond range. *Bioelectrochem. Bioenerg.*, **7**, 167–86.

Trissl, H. W. and P. Lauger (1970). Photoelectric effects in thin chlorophyll films. *Z. Naturf.*, **25b**, 1059–60.

Trosper, T. L. (1972). Some properties of chlorophyll *a* at hydrocarbon–water interfaces and in black lipid membranes. *J. Membrane Biol.*, **8**, 133–48.

Tsong, T. Y. (1991). Electroporation of cell membranes. *Biophys. J.*, **60**, 297–306.

Uhlig, H. H. (1937). The solubilities of gases and surface tension. *J. Phys. Chem.*, **41**, 1215–25.

Valleau, J. P., L. K. Cohen and D. N. Card (1980). Primitive model electrolytes. II. The symmetrical electrolyte. *J. Chem. Phys.*, **72**, 5942–84.

Vanbuuren, A. R., S. J. Marrink and H. J. C. Berendsen (1993). A molecular dynamics study of the decane/water interface. *J. Phys. Chem.*, **97**, 9206–12.

VanderNoot, T. J. and D. J. Schiffrin (1990). Non-linear regression of impedance data for ion transfer across liquid–liquid interfaces. *Electrochim. Acta*, **35**, 1359–67.

VanderNoot, T. J., D. J. Schiffrin and R. S. Whiteside (1990). Design and evaluation of a four-electrode potentiostat/voltage clamp suitable for ac impedance measurements at the interface of immiscible electrolytes. *J. Electroanal. Chem.*, **278**, 137–50.

Vanysek, P. (1985). *Electrochemistry on Liquid/Liquid Interfaces*. Springer Verlag, Berlin.

Verburgh, Y. and P. Joos (1980). Effect of a supporting electrolyte on the interfacial transference numbers during electroadsorption at the nitrobenzene/water interface. *J. Colloid Interface Sci.*, **74**, 384–95.

Verkleij, A. J. (1984). Lipidic intramembranous particles. *Biochim. Biophys. Acta*, **779**, 43–63.

Verkleij, A. J., C. J. A. VanEchteld, W. J. Gerritsen, P. R. Cullis and B. DeKruijff (1980). The lipidic particle as an intermediate structure in membrane fusion processes and bilayer to hexagonal H_{II} transitions. *Biochim. Biophys. Acta*, **600**, 620–4.

Verschaffeit, J. E. (1936a). La thermomécanique de la couche superficielle. I. Généralités; corps purs. *Acad. R. Belg. Bull, Classe Sci.*, **22**, 373–89.

Verschaffeit, J. E. (1936b). La thermomécanique de la couche superficielle. II. La formule d'adsorption. *Acad. R. Belg. Bull, Classe Sci.*, **22**, 390–401.

Verschaffeit, J. E. (1936c). La thermomécanique de la couche superficielle. III. Phases mixtes. *Acad. R. Belg. Bull, Classe Sci.*, **22**, 402–11.

Verwey, E. J. W. (1950). Theory of the electric double layer of stabilized emulsions. *Proc. Konikl. Nederl. Acad. Wet.*, **53**, 376–85.

Verwey, E. J. W. and K. F. Niessen (1939). Electrical double layer at the interface of two immiscible liquids. *Phil. Mag.*, **28**, 435–46.

Verwey, E. J. W. and J. Th. G. Overbeek (1948). *Theory of the Stability of Lipophobic Colloids*. Amsterdam, Elsevier.

Vetter, K. J. (1961). *Electrochemische Kinetik*. Springer Verlag, Berlin, Heidelberg.

Volkov, A. G. (1984a). A possible mechanism of the photooxidation of water sensitized by chlorophyll adsorbed at the interface. *J. Electroanal. Chem.*, **173**, 15–24.

Volkov, A. G. (1984b). Water photooxidation by hydrated chlorophyll oligomer. In: *Solar Energy Bioconversion* (I. V. Berezin, ed.) pp. 149–52. NCBI AN SSSR, Pushchino.

Volkov, A. G. (1985a). Possible mechanism of the photooxidation of water sensitized by chlorophyll adsorbed on an interface. *Sov. Electrochem.*, **21**, 91–8.

Volkov, A. G. (1985b). The possible mechanism of functioning of photosystem II in higher plants. *Biophysics*, **30**, 491.

Volkov, A. G. (1986a). Molecular mechanism of the photooxidation of water during photosynthesis: cluster catalysis of synchronous multielectron reactions. *Mol. Biol.*, **20**, 728–36.

Volkov, A. G. (1986b). The electrochemical mechanism of functioning of photosystem II in higher plants. *J. Electroanal. Chem.*, **205**, 245–57.

Volkov, A. G. (1986c). The molecular mechanism of functioning of photosystem II in higher plants. *Photobiochem. Photobiophys.*, **11**, 1–7.

Volkov, A. G. (1987a). Calculation of the interfacial potentials from measurements of the free Gibbs adsorption energy at ITIES. *Electrokhimiya*, **23**, 271–5.

Volkov, A. G. (1987b). Electrocapillary phenomena at the interface between two immiscible electrolyte solutions. *Electrokhimiya*, **23**, 90–7.

Volkov, A. G. (1987c). Thylakoid membrane: electrochemical mechanisms of photosynthesis. The mechanism of oxygen evolution in the reaction center of photosystem II of green plants. *Biol. Membranes*, **4**, 984–93.

Volkov, A. G. (1988). The molecular mechanism of oxygen evolution (Review). *Uspekhi Sov. Biol. (Progress in Modern Biology)*, **105**, 467–87.

Volkov, A. G. (1989). Oxygen evolution in the course of photosynthesis. *Bioelectrochem. Bioenerg.*, **21**, 3–24.

Volkov, A. G. (1996). Potentials of thermodynamic and free zero charge at the interface between two immiscible electrolytes. *Langmuir*, **12**, 3315–9.

Volkov, A. G., M. A. Bibikova, A. F. Mironov and L. I. Boguslavsky (1983a). Adsorption and catalytic properties of an iron complex of coproporfyrin III at the octane/water interface. *Elektrokhimiya*, **19**, 1398–401.

Volkov, A. G., M. A. Bibikova, A. F. Mironov and L. I. Boguslavsky (1983b). Adsorption and catalytic properties of iron coproporphyrin II complex at the octane/water interface. *Bioelectrochem. Bioenerg.*, **10**, 477–83.

Volkov, A. G. and L. I. Boguslavsky (1976). Porphyrins as a model of enzymes of the respiratory chain of mitochondria. In: *Biochemistry of Mitochondria* (S. E. Severin, ed.) p. 14. Nauka Publ., Moscow.

Volkov, A. G. and D. W. Deamer (1994). Mechanisms of the passive ion permeation of lipid bilayers: partition or transient aqueous pores. In: *Charge and Field Effects in Biosystems-4* (M. J. Allen, S. F. Cleary and A. E. Sowers, eds.) pp. 145–55. World Scientific, Singapore, New Jersey, London.

Volkov, A. G. and D. W. Deamer (eds.) (1996). *Liquid–Liquid Interfaces: Theory and Methods.* CRC Press, Boca Raton, London, Tokyo.

Volkov, A. G. and D. W. Deamer (1997). Redox chemistry at liquid/liquid interfaces. *Prog. Colloid Polym. Sci.*, **103**, 21–7.

Volkov, A. G., D. W. Deamer, D. L. Tanelian and V. S. Markin (1996). The electrical double layer at the oil/water interface. *Prog. Surf. Sci.*, **53**, 1–134.

Volkov, A. G., M. I. Gugeshashvili, M. D. Kandelaki, V. S. Markin, B. Zelent, G. Munger and R. M. Leblanc (1991). Artificial photosynthesis at octane/water interface in the presence of hydrated chlorophyll *a* oligomer thin film. *Proc. Soc. Photo-Opt. Instrum. Eng.*, **1436**, 68–79.

Volkov, A. G., M. I. Gugeshashvili, A. F. Mironov and L. I. Boguslavsky (1982a). Reduction of hydrophobic porphyrin at the octane/water interface controlled by the specific adsorption of halogen anions. *Bioelectrochem. Bioenerg.*, **9**, 551–8.

Volkov, A. G., M. I. Gugeshashvili, A. F. Mironov, A. N. Nizhnik and L. I. Boguslavsky (1982b). Reduction of hydrophobic porphyrin at the octane–water interface controlled by specific adsorption of halide anions. *Elektrokhimiya*, **18**, 1628–34.

Volkov, A. G., M. I. Gugeshashvili, A. F. Mironov and L. I. Boguslavsky (1983c). Oxidation of NADH and reduction of vitamin K_3 at the octane/water interface catalyzed by metal complexes of ethioporphyrin III. *Elektrokhimiya*, **19**, 1194–8.

Volkov, A. G., M. I. Gugeshashvili, A. F. Mironov and L. I. Boguslavsky (1983d). Oxidation of NADH and reduction of vitamin K_3 at the octane/water interface with ethioporphyrin complexes as catalysts. *Bioelectrochem. Bioenerg.*, **10**, 485–91.

Volkov, A. G., M. I. Gugeshashvili, V. I. Portnov, V. S. Markin and L. N. Chekulaeva (1992a). Emulsion bioelectrochemistry: Bacteriorhodopsin phototransfer of protons through the interface water/lipid in octane. In: *Charge and Field Effects in Biosystems-3* (M. J. Allen, S. F. Cleary, A. E. Sowers and D. D. Shillady, eds.) pp. 365–72. Birkhauser, Boston, Basel, Berlin.

Volkov, A. G., M. I. Gugeshashvili, G. Munger, R. M. Leblanc, B. Zelent, Ju. Galant, H.-A. Tajmir-Riahi and J. Aghion (1992b). The hydrated chlorophyll *a* oligomer in monolayers and Langmuir–Blodgett films. *Biol. Membranes*, **9**, 874–80.

Volkov, A. G., M. I. Gugeshashvili, G. Munger, R. M. Leblanc (1992c). Light-dependent oxygen uptake by hydrated oligomer of chlorophyll. *Biol. Membranes*, **9**, 576–80.

Volkov, A. G., M. I. Gugeshashvili, G. Munger and R. M. Leblanc (1993). The light-dependent oxygen reduction by monolayers of hydrated chlorophyll *a* oligomer. *Bioelectrochem. Bioenerg.*, **29**, 305–14.

Volkov, A. G., M. I. Gugeshashvili and D. W. Deamer (1995a). Energy conversion at liquid–liquid interfaces: artificial photosynthetic systems. *Electrochim. Acta*, **40**, 2849–68.

Volkov, A. G., M. I. Gugeshashvili, B. Zelent, D. Cote, G. Munger, A. Tessier, P. F. Blanchet and R. M. Leblanc (1995b). Light energy conversion with chlorophyll *a* and pheophytin *a* monolayers at the optically transparent SnO_2 electrode: artificial photosynthesis. *Bioelectrochem. Bioenerg.*, **38**, 333–42.

Volkov, A. G. and R. A. Haack (1995a). Insect induces bioelectrochemical signals in potato plants. *Bioelectrochem. Bioenerg.*, **35**, 55–60.

Volkov, A. G. and R. A. Haack (1995b). Bioelectrochemical signals in potato plants. *Plant Physiol.*, **42**, pp. 17–23.

Volkov, A. G. and Yu. I. Kharkats (1985). Catalytic properties of the interface between two immiscible liquids during redox reaction. *Kinetica Kataliz*, **26**, 1322–6.

Volkov, A. G. and Yu. I. Kharkats (1986). Synchronous multielectron reactions at the liquid/liquid interface. *Chem. Phys.*, **5**, 964–71.

Volkov, A. G. and Yu. I. Kharkats (1987). Membrane catalysis: synchronous multielectron reactions at the interface between two liquid phases. Bioenergetic mechanisms. In: *Chemical Physics of Enzyme Catalysis* (A. Aaviksaar, ed.) pp. 87–90. Publ. House Estonian Academy of Sciences, Tallinn.

Volkov, A. G. and Yu. I. Kharkats (1988). Cytochrome oxidase: Mechanism of functioning. *Biol. Membranes*, **5**, 920–31.

Volkov, A. G., V. D. Kolev, A. L. Levin and L. I. Boguslavsky (1985c). Oxygen photoevolution at the octane/water interface in the presence of β-carotene and chlorophyll *a*. *Photobiochem. Photobiophys.*, **10**, 105–11.

Volkov, A. G., V. D. Kolev, A. L. Levin and L. I. Boguslavsky (1986). Oxygen photoevolution at the octane/water interface in the presence of β-carotene and chlorophyll *a*. *Electrokhimiya*, **22**, 1303–7.

Volkov, A. G., A. A. Kondrashin, E. M. Glagoleva and L. I. Boguslavsky (1976a). Electrogenic function of the solubilized mitochondrial trangidrogenase. In: *Biochemistry of Mitochondria* (S. E. Severin, ed.) p. 86. Nauka Publ., Moscow.

Volkov, A. G. and A. A. Kornyshev (1985). Dependence of the free Gibbs energy of resolvation during ion transfer from one solvent to another on the ion size. *Electrokhimiya*, **21**, 814–7.

Volkov, A. G., B. T. Lozhkin and L. I. Boguslavsky (1975a). Proton and electron transfer through bilayer membranes and decane/water interface in the presence of chlorophyll. *Doklady Akad. Nauk SSSR*, **220**, 1207–10.

Volkov, A. G., V. S. Markin, R. M. Leblanc, M. I. Gugeshashvili, B. Zelent and G. Munger (1994). Monolayers of pheophytin *a* at the oil/water, gas/water and SnO₂/water interfaces: adsorption and photoelectrochemistry. The general isotherm of adsorption of amphiphilic molecules. *J. Solut. Chem.*, **23**, 223–48.

Volkov, A. G., E. I. Mileykovskaya, L. I. Boguslavsky and I. A. Kozlov (1975b). Reversible H⁺-ATPases on lipid/water interface. In: *Bioenergetics and Mitochondria* (E. Rusanov, ed.) pp. 47–52. BAN, Varna.

Volkov, A. G., A. F. Mironov and L. I. Boguslavsky (1976b). Electron and proton transfer across liquid/liquid interface in the presence of porphyrins. *Elektrokhimiya*, **12**, 1326–9.

Volkov, A. G., S. Paula and D. W. Deamer (1997). Two mechanisms of permeation of small neutral molecules and hydrated ions across phospholipid bilayers. *Bioelectrochem. Bioenerg.* **42**, 153–160.

Vorotyntsev, M. A., Yu. A. Ermakov, V. S. Markin and A. A. Rubashkin (1993). Distribution of the interfacial potential drop in a situation when ionic solution components enter a surface layer of finite thickness with fixed space charge. *Elektrokhimiya*, **29**, 596–610.

Vorotyntsev, M. A. and S. N. Ivanov (1987). Interaction between charges near dielectric/electrolyte solution interfaces. Ions in the dielectric. *Elektrokhimiya*, **23**, 215–21.

Vorotyntsev, M. A. and S. N. Ivanov (1989a). Energy of the image forces and the interaction of an ion with a charged group at the insulator/electrolyte solution interface. *Elektrokhimiya*, **25**, 550–4.

Vorotyntsev, M. A. and S. N. Ivanov (1989b). Ionic adsorption isotherms at homogeneous insulator/electrolyte solution interfaces. *Elektrokhimiya*, **25**, 554–7.

Vorotyntsev, M. A. and A. A. Kornyshev (1979). Physical significance of an effective dielectric constant that depends on the distance from the electrode. *Electrokhimiya*, **15**, 660–4.

Vorotyntsev, M. A. and A. A. Kornyshev (1984). Models to describe the collective properties of the metal/solvent interface in electric double-layer theory. *Electrokhimiya*, **20**, 3–47.

Vorotyntsev, M. A. and A. A. Kornyshev (1993). *Electrostatics of a Medium with the Spatial Dispersion*. Nauka, Moscow.

Vorotyntsev, M. A. and P. V. Mityushev (1991). Electric double layer structure in a surface-inactive electrolyte solution—effect of the Stern layer and spatial correlations of solvent polarization. *Electrochim. Acta*, **36**, 401–9.

Waddington, T. C. (1966). Ionic radii and the method of the undetermined parameter. *Trans. Faraday Soc.*, **62**, 1482–92.

Walter, A. and J. Gutknecht (1986). Permeability of small nonelectrolytes through lipid bilayer membranes. *J. Membrane Biol.*, **90**, 207–17.

Wandlowski, T., V. Mareček and Z. Samec (1988). Adsorption of phospholipids at the interface between two immiscible electrolyte solutions. Part 1. Equilibrium adsorption of phosphatidylcholines at the water/nitrobenzene interface. *J. Electroanal. Chem.*, **242**, 277–90.

Wandlowski, T., V. Mareček, K. Holub and Z. Samec (1989). Ion transfer across liquid–liquid phase boundaries: electrochemical kinetics by faradaic impedance. *J. Phys. Chem.*, **93**, 8204–12.

Wandlowski, T., V. Mareček and Z. Samec (1990). Galvani potential scales for water–nitrobenzene and water–1,2-dichloroethane interfaces. *Electrochim. Acta*, **35**, 1173–5.

Wandlowski, T., V. Mareček, Z. Samec and R. Fuoco (1992). Effect of temperature on the ion transfer across an interface between two immiscible electrolyte solutions: ion transfer dynamics. *J. Electroanal. Chem.*, **331**, 765–82.

Wandlowski, T., S. Racinsky, V. Mareček and Z. Samec (1987). Adsorption of phospholipids at the interface between two immiscible electrolyte solutions. *J. Electroanal. Chem.*, **227**, 281–5.

Wang, C. C. and L. J. Bruner (1978). Dielectric saturation of aqueous boundary layer adjacent to charged bilayer membranes. *J. Membranes Biol.*, **38**, 311–31.

Wang, E. K. and Y. Q. Liu (1990). Ion transfer with the coupled chemical reactions at the water–nitrobenzene interface. *Electrochim. Acta*, **35**, 1965–9.

Warburg, E. (1899). Ueber das Verhalten sogenannter unpolarisirbarer Electroden gegen Wechselstrom. *Ann. Phys. Chem. N.F.*, **67**, 493–9.

Wardsmith, R. S., M. J. Hey and J. Mitchell (1994). Protein–polysaccharide interactions at the oil–water interface. *Food Hydroc.*, **8**, 309–15.

Watanabe, A. (1984). Electrochemistry of oil–water interfaces. *Surf. Colloid Sci.*, **13**, 1–70.

Watanabe, A., M. Matsumoto, H. Tamai and R. Goto (1967a). Electrocapillary phenomena at oil–water interfaces. Part I. Electrocapillary curves at oil–water systems containing surface active agent. *Kolloid. Z. Z. Polym.*, **220**, 152–9.

Watanabe, A., M. Matsumoto, H. Tamai and R. Goto (1967b). Electrocapillary phenomena at oil–water interfaces. Part II. The counterion binding at oil–water interfaces. *Kolloid. Z. Z. Polym.*, **221**, 47–52.

Watanabe, A., M. Matsumoto, H. Tamai and R. Goto (1968). Electrocapillary phenomena at oil–water interfaces. Part III. The behaviour of lecithin at oil-water interfaces. *Kolloid. Z. Z. Polym.*, **228**, 58–63.

Watarai, H. and Y. Chida (1994). Simple devices for the measurements of absorption spectra at liquid–liquid interfaces. *Anal. Sci.*, **10**, 105–7.

Watson, W. F. (1953). Reversible bleaching of chlorophyll by metallic salts. *J. Am. Chem. Soc.*, **75**, 2522–3.

Watts, A. and T. J. VanderNoot (1996). In: *Liquid–Liquid Interfaces. Theory and Methods* (A. G. Volkov and D. W. Deamer, eds.) pp. 77–102. CRC Press, Boca Raton, New York, London, Tokyo.

Weaver, J. C. (1993). Electroporation: a general phenomenon for manipulating cells and tissue of cell membranes. *J. Cell. Biochem.*, **51**, 426–35.

Webb, T. J. (1926). The free energy of hydration of ions and the electrostriction of the solvent. *J. Am. Chem. Soc.*, **48**, 2589–603.

Wei, C., A. J. Bard and M. V. Mirkin (1995). Scanning electrochemical microscopy. Application of SECM to the study of charge transfer processes at the liquid/liquid interface. *J. Phys. Chem.*, **99**, 16033–42.

Wei, D. Q., G. N. Patey and G. M. Torrie (1990). Double-layer structure at an ion-adsorbing surface. *J. Phys. Chem.*, **94**, 4260–8.

Wei, D. Q., G. M. Torrie and G. N. Patey (1993). Molecular solvent model for an electrical double layer—effects of ionic polarizability. *J. Chem. Phys.*, **99**, 3990–7.

Wertheim, M. S. (1979). Equilibrium statistical mechanics of polar fluids. *Ann. Rev. Phys. Chem.*, **30**, 417–501.

Wesson, L. G. (1948). *Tables of Electric Dipoles Moments*. The Technology Press, MIT, Cambridge, MA.

West, J. C. (1980). Energy coupling in secondary active transport. *Biochim. Biophys. Acta*, **604**, 91–126.

Widdas, W. F. (1954). Facilitated transport of hexoses across the human erythrocyte membrane. *J. Physiol. (London)*, **125**, 163–80.

Wiener, M. C., G. I. King and S. White (1991). Structure of a fluid dioleoylphosphatidylcholine bilayer determined by joint refinement of x-ray neutron diffraction data. 1. Scaling of neutron data and the distribution of double bonds and water. *Biophys. J.*, **60**, 568–76.

Wiese, W., W. Harbich and W. Helfrich (1992). Budding of lipid bilayer vesicles and flat membranes. *J. Phys.: Cond. Matter*, **4**, 1647–57.

Wilhelmy, L. (1863). *Ann. Physik*, **119**, 177.

Wiles, M. C., D. J. Schiffrin, T. J. VanderNoot and A. F. Silva (1990). Experimental artifacts associated with impedance measurements at liquid–liquid interfaces. *J. Electroanal. Chem.*, **278**, 151–9.

Wilner, I., W. E. Ford, J. W. Otvos and M. Calvin (1979). Photoinduced electron transfer across a water–oil boundary as a model for redox reaction separation. *Nature*, **280**, 823–4.

Wilshut J., N. Düzgünes, R. Fraley, D. Papahadjopoulos (1980). Studies of mechanism of membrane fusion. Kinetics of calcium ion induced fusion of phosphatidyiserine vesicles followed by a new assay for mixing of aqueous vesicle contents. *Biochemistry*, **19**, 6011–21.

Wirth, M. J. and J. D. Burbage (1992). Reorientation of acridine orange at liquid alkane water interfaces. *J. Phys. Chem.*, **96**, 9022–5.

Yaguzhinsky, L. S., L. I. Boguslavsky, A. G. Volkov and A. B. Rakhmaninova (1975a). Synthesis of ATP connected with the functioning of membrane proton pumps at the octane–water interface. *Doklady. Akad. Nauk SSSR*, **221**, 1465–8.

Yaguzhinsky, L. S., L. I. Boguslavsky, A. G. Volkov and A. B. Rakhmaninova (1975b). Synthesis of ATP coupled with action of membrane protonic pumps at the octane–water interface. *Nature*, **259**, 494–6.

Yaguzhinsky, L. S., A. G. Volkov and L. I. Boguslavsky (1976). On the mechanism of action of p-(N.N-di-2chloroethyl)amino-phenyl-acetic acid, a specific inhibitor of respiration and phosphorylation in mitochondria. *Biochemistry*, **41**, 1203–7.

Yaguzhinsky, L. S., A. G. Volkov and L. I. Boguslavsky (1977). Chlorphenacyl-inhibitor of the proton transfer step by membrane cationic pumps. *Bioelectrochem. Bioenerg.*, **4**, 225–30.

Yamaguchi, T. and T. Shinbo (1989). Novel interfacial engine between two immiscible liquid. *Chem. Lett.*, 935–8.

Yamins, H. G. and W. A. Zisman (1933). A new method of studying the electrical properties of monomolecular films on liquids. *J. Chem. Phys.*, **1**, 656–61.

Yan, N. X. and J. H. Masliyah (1994). Adsorption and desorption of clay particles at the oil–water interface. *J. Colloid Interface Sci.*, **168**, 386–92.

Ye, R. and A. S. Verkman (1989). Simultaneous optical measurement of osmotic and diffusional water permeability in cells and liposomes. *Biochemistry*, **28**, 824–9.

Yoshida, Z. and H. Freiser (1984). Ascending water electrode studies of metal extractants. Faradaic ion transfer of protonated 1,10-phenanthroline and its derivatives across an aqueous 1,2-dichloroethane interface. *J. Electroanal. Chem.*, **162**, 307–19.

Yoshida, Z. and S. Kihara (1987). The role of non-ionic polyoxyethylene ether surfactants on ion transfer across aqueous/organic solution interfaces studied by polarography with electrolyte dropping electrode. *J. Electroanal. Chem.*, **227**, 171–81.

Yoshikawa, K. and Makino, M. (1989). Self-pulsing at an oil/water interface in the presence of phospholipid. *Chem. Phys. Lett.*, **160**, 623–6.

Young, T. (1805). An essay on the cohesion of films. *Phil. Trans. R. Soc. (London)*, **95**, 65–82.

Young, T. F. and W. D. Harkins (1928). Interfacial tension for solid–liquid and liquid–liquid interfaces. *Int. Crit. Tables*. Vol. 4, pp. 436–9. McGraw-Hill, New York.

Yufei, C., V. J. Cunnane, D. J. Schiffrin, L. Murtomäki and K. Kontturi (1991). Interfacial capacitance and ionic association at electrified liquid/liquid interfaces. *J. Chem. Soc. Faraday Trans.*, **87**, 107–14.

Yufit, S. S. (1984). *Mechanism of Phase Transfer Catalysis*. Nauka, Moscow.

Zagorska, I. and Z. Koczorowski (1986). Experimental study of Volta potentials in water/nitrobenzene systems with anion adsorption. *J. Electroanal. Chem.*, **204**, 273–9.

Zagorska, I., Z. Koczorowski and I. Paleska (1990). Volta potentials in water/1,2-dichloroethane and water/nitroethane systems. *J. Electroanal. Chem.*, **282**, 51–8.

Zaitsev, N. K., N. F. Gorelik, N. A. Kotov and M. G. Kuzmin (1988). A photoelectrochemical effect at the polarizable interface between liquid electrolyte solutions in protoporphyrin-quinone systems. *Sov. Electrochem.*, **24**, 1243–7.

Zaitsev, N. K., I. I. Kulakov and M. G. Kuzmin (1985). The photoelectrochemical effect at the interface between immiscible liquid electrolyte solutions. *Elektrokhimiya*, **21**, 1293–7.

Zamaraev, K. I. and V. N. Parmon (1983). Molecular photocatalytic systems for solar energy conversion: catalysts for the evolution of hydrogen and oxygen from water. *Usp. Khimii*, **52**, 1433–67.

Zelent, B. Ju. Gallant, A. G. Volkov, M. I. Gugeshashvili, G. Munger, H.-A. Tajmir-Riahi and R. M. Leblanc (1993). Hydrated chlorophyll alpha oligomers in solutions, monolayers, and thin films. *J. Mol. Struct.*, **297**, 1–11.

Zimmermann, U. and G. A. Neil (1996). *Electromanipulation of Cells*. CRC Press, Boca Raton, New York, London, Tokyo.

Zisman, W. A. (1932). A new method of measuring contact potential differences in metals. *Rev. Sci. Instrum.*, **3**, 367–70.

Zlochower, I. A. and J. H. Schulman (1967). A study of molecular interactions and mobility at liquid–liquid interfaces by NMR spectroscopy. *J. Colloid Interface Sci.*, **24**, 115–24.

APPENDIX

UNITS AND SYMBOLS IN THE SYSTEME INTERNATIONAL (SI)

TABLE A1. Base units of the International System of units (SI).

Meter	m	Length
Kilogram	kg	Mass
Second	s	Time
Ampere	A	Electrical current
Kelvin	K	Thermodynamic temperature
Mole	mol	Amount of substance

TABLE A2. Derived SI units.

Quantity	SI Unit	Symbol	Definition of Unit
Energy	Joule	J	$\text{kg m}^2 \text{ s}^{-2}$
Force	Newton	N	$\text{kg m s}^{-2} = \text{J m}^{-1}$
Power	Watt	W	$\text{kg m}^2 \text{ s}^{-3} = \text{J s}^{-1}$
Pressure	Pascal	Pa	N m^{-2}
Electric charge	Coulomb	C	A s
Electric potential	Volt	V	$\text{J A}^{-1} \text{ s}^{-1} = \text{J C}^{-1}$
Electric field	Volt/meter		V m^{-1}
Frequency	Hertz	Hz	s^{-1}
Electrical resistance	Ohm	Ω	$\text{kg m}^2 \text{ s}^{-3} \text{ A}^{-2} = \text{V A}^{-1}$
Electrical conductance	Siemans	S	$\text{kg}^{-1} \text{ m}^{-2} \text{ s}^3 \text{ A}^2 = \text{A V}^{-1} = \Omega^{-1}$
Electrical capacitance	Farad	F	$\text{kg}^{-1} \text{ m}^{-2} \text{ s}^4 \text{ A}^2 = \text{C V}^{-1}$

TABLE A3. Multiplicative prefixes.

a	10^{-18}	atto-
f	10^{-15}	femto-
p	10^{-12}	pico-
n	10^{-9}	nano-
μ	10^{-6}	micro-
m	10^{-3}	milli-
k	10^{3}	kilo-
M	10^{6}	mega-
G	10^{9}	giga-
T	10^{12}	tera-
P	10^{15}	peta-
E	10^{18}	exa-

TABLE A4. Fundamental constants.

Constant	Symbol	SI	CGS
Avogadro's constant	N_A	$6.022 \times 10^{23}\,\text{mol}^{-1}$	$6.022 \times 10^{23}\,\text{mol}^{-1}$
Boltzmann's constant	k	$1.381 \times 10^{-23}\,\text{J K}^{-1}$	$1.381 \times 10^{-16}\,\text{erg deg}^{-1}$
Elementary charge	e	$1.602 \times 10^{-19}\,\text{C}$	$4.803 \times 10^{-10}\,\text{esu}$
Faraday constant	$F = N_A e$	$9.649 \times 10^{4}\,\text{C mol}^{-1}$	$9.649 \times 10^{4}\,\text{C mol}^{-1}$
Molar gas constant	$R = N_A k$	$8.314\,\text{J K}^{-1}\,\text{mol}^{-1}$	$8.314 \times 10^{7}\,\text{erg mol}^{-1}\,\text{deg}^{-1}$
Permittivity of free space	ε_0	$8.854 \times 10^{-12}\,\text{C}^2\,\text{J}^{-1}\,\text{m}^{-1}$	1
Planck's constant	h	$6.626 \times 10^{-34}\,\text{J s}$	$6.626 \times 10^{-27}\,\text{erg s}$

TABLE A.5. Conversion from SI to CGS.

$1\text{ nm} = 10^{-9}\text{ m} = 10\text{ Å} = 10^{-7}\text{ cm}$
$1\text{ J} = 10^{7}\text{ erg} = 0.239\text{ cal} = 6.242 \times 10^{18}\text{ eV} = 5.034 \times 10^{22}\text{ cm}^{-1} = 7.243 \times 10^{22}\text{ K}$
$1\text{ kJ mol}^{-1} = 0.239\text{ kcal mol}^{-1}$
$1\text{ N} = 10^{5}\text{ dyne}$
$1\text{ Pa} = 1\text{ N m}^{-2} = 9.872 \times 10^{-6}\text{ atm} = 7.50 \times 10^{-3}\text{ Torr} = 10\text{ dyne cm}^{-2}$
$1\text{ bar} = 10^{5}\text{ Nm}^{-2} = 10^{-5}\text{ Pa} = 0.9868\text{ atm} = 750.06\text{ mm Hg}$
$1\text{ cm H}_2\text{O} = 98.07\text{ Pa}$

TABLE A6. Conversion from CGS to SI.

1 Å (angstrom) $= 10^{-10}$ m $= 10^{-8}$ cm $= 10^{-4}$ μm $= 10^{-1}$ nm

1 litre $= 10^{-3}$ m^3 $= 1$ dm^3

1 erg $= 10^{-7}$ J

1 cal $= 4.184$ J

1 kcal mol^{-1} $= 4.184$ kJ mol^{-1}

1 dyne $= 10^{-5}$ N

1 dyne cm^{-1} $= 1$ erg cm^{-2} $= 1$ mN m^{-1} $= 1$ mJ m^{-2} (unit of surface tension)

1 dyne cm^{-2} $= 10^{-1}$ Pa (N m^{-2})

1 torr $= 1$ mm Hg $= 1.316 \times 10^{-3}$ atm $= 133.3$ Pa (N m^{-2})

0 °C $= 273.15$ K (triple-point of water)

1 poise (P) $= 1$ g cm^{-1} s^{-1} $= 10^{-1}$ kg m^{-1} s^{-1} $= 10^{-1}$ N s m^{-2} (unit of viscosity)

TABLE A7. Useful quantities and relations.

Standard volume of ideal gas $= 22.414$ dm^3 mol^{-1} (1 mol^{-1})

$kT = RT/N_A = 4.048 \times 10^{-21}$ J $= 4.114 \times 10^{-14}$ erg at 293.15 K (20°C)

1 kT per molecule $= 2.438$ kJ mol^{-1} at 293.15 K (20°C)

$RT = 2.438$ kJ·mol^{-1} at 293.15 K (20°C)

$RT/F = kT/e = 25.27$ mV at 293.15 K (20°C)

$(kT/e) \ln (10) = (RT/F) \ln(10) = 58.18$ mV at 293.15 K (20°C)

1 C m^{-2} $= 1$ unit charge per 0.16 nm^2 (16 Å^2)

κ^{-1} (Debye length) $= 0.304/\sqrt{M}$ nm for 1:1 electrolyte at 298 K, where

   1 M $= 1$ mol dm^{-3} $\equiv 6.022 \times 10^{26}$ molecules per m^3

Debye (D) $= 10^{-18}$ esu $= 3.336 \times 10^{-30}$ C m (unit of electric dipole moment)

INDEX